# Arthropod Diversity and Conservation

TOPICS IN BIODIVERSITY AND CONSERVATION

Volume 1

*The titles published in this series are listed at the end of this volume.*

# Arthropod Diversity and Conservation

*Edited by*

David L. Hawksworth

and

Alan T. Bull

Reprinted from *Biodiversity and Conservation,* volume 15:1 (2006), excluding the paper by R. O'Malley et al., The diversity and distribution of the fruit bat fauna (Mammalia, Chiroptera, Megachiroptera) of Danjugan Island, Cauayan, Negros Occidental, Philippines (with notes on the Microchiroptera). pp. 43–56.

Springer

A C.I.P. Catalogue record for this book is available from the library of Congress.

ISBN-10 1-4020-5203-0 (HB)
ISBN-13 978-1-4020-5203-3 (HB)
ISBN-10 1-4020-5204-9 (e-book)
ISBN-13 978-1-4020-5204-0 (e-book)

Published by Springer,
P.O. Box 17, 3300 AA Dordrecht, The Netherlands.

*www.springer.com*

*Printed on acid-free paper*



Printed in the Netherlands.

# Contents
# Arthropod Diversity and Conservation

**Aims and Scope**

*Biodiversity and Conservation* is an international journal devoted to the publication of articles on all aspects of biological diversity, its description, analysis and conservation, and its controlled rational use by humankind. The scope of *Biodiversity and Conservation* is wide and multidisciplinary, and embraces all life-forms. Research papers, as well as Editorials, Comments and Research notes, on biodiversity and conservation and contributions which deal with the practicalities of conservation management, economic, social and political issues and with case studies are welcome. The journal provides a forum for examining the conflict between sustainable development and human dependence on biodiversity, in fields such as agriculture, environmental management and biotechnology. The Editors encourage contributions from developing countries in order to realize proper global perspectives on matters of biodiversity and conservation.



---

*Biodiversity and Conservation* is published monthly (bimonthly in June and December).
Subscriptions should be sent to **Springer, P.O. Box 322, 3300 AH Dordrecht, The Netherlands**, or at **P.O. Box 358, Accord Station, Hingham, MA 02018-0358, USA**, or to any subscription agent. Private subscriptions should be sent directly to the publishers.
Changes of mailing address should be notified together with our latest label.
For advertisement rates, prices of back volumes, and other information, please apply to **Springer, P.O. Box 17, 3300 AA Dordrecht, The Netherlands**.
Periodicals postage is paid at Rahway, NJ. USPS No. 008067.
US Mailing Agent: Mercury Airfreight International Ltd, 365 Blair Road, Avenel, NJ 07001.
*Postmaster*: Please send all address corrections to: *Biodiversity and Conservation*, c/o Mercury Airfreight International Ltd, 365 Blair Road, Avenel, NJ 07001, USA.

Published by Springer, Van Godewijckstraat 30, P.O. Box 17, 3300 AA Dordrecht, The Netherlands, and 101 Philip Drive, Norwell, MA 02061, USA.

*Printed on acid-free paper.*

Biodiversity and Conservation (2006) 15:1–2
DOI 10.1007/s10531-006-5741-5


# Introduction

## Topics in Biodiversity and Conservation: Arthropod diversity and conservation

'Biodiversity' is another way of saying 'the amazing variety of life on this planet'; what it is made up of (compositional biodiversity), the form it takes (structural biodiversity) and how it works (functional biodiversity). Of the 8 million or so species on Earth, only about a quarter of a million are plants. Nevertheless, it is the plants which largely underlie and generate the rest of biodiversity through their enormous bulk and complexity of architecture. The remaining components of biodiversity are the smaller, largely unseen, silent majority, most of which are arthropods. They make up a jointed-legged exuberance of life forms with all their morphological glory. Vertebrates, in contrast, are a mere blip on the biotic horizon, elevated in importance in the bigger scheme of things only by our psyche.

These biases in nature are brought home when one considers some ratios. The total biomass of plants to animals is about 99.999 to 0.001, while the total number of vascular plant species to animal species is around 0.026 to 99.974. This means that the basis of at least terrestrial life hinges on the massive city of plants. In contrast, the compositional variety of life is mostly arthropods, which inhabit virtually every recess and plane on the planet. Indeed, the arthropods are the functional woof and weft of life, featuring somewhere at one level or more in virtually every food web. Many even eat each other. They chew, suck, mine or filter virtually anything organic from rhino dung to the toughest plants, and even sit on the deep sea floor capitalizing on the fine rain of organic debris.

With such a resourceful and adaptive tangle of life and so many life forms, how can they possibly be in need of conservation? The point is that probably many are not. Yet in the face of the human demographic meteorite, still as many as a quarter may be imperiled, principally though loss of tropical forest. Our overall task is thus one of triage, choosing areas and themes that will enable us to conserve the most with the minimum of resources. Of course, 'the most' is shrouded in mist. We have to peer carefully into this haze and carefully define our conservation goals. Then, with a little, we can conserve a lot, both the irreplaceable present (an ecological goal) while also providing for the future, so enabling evolution. Much of this conservation will be for our own good (e.g. preservation of pollinators, maintenance of soil quality), but much should also be driven out of the goodness of our own hearts, by recognizing nature's intrinsic value.

This compilation of peer-reviewed papers, drawn from researchers around the world, provides insight into a huge range of different aspects of arthropod diversity and its conservation; as much a celebration of the diversity of arthropod biologies as arthropod diversity itself. These papers are not on any single theme, but reflect some of the exciting new research taking place, and also in some of the most biodiverse corners of the planet. We hope these studies will stimulate further research in this wonderfully complex and fascinating field, with its important mission.

MICHAEL J. SAMWAYS
*Department of Entomology and Centre for Agricultural Biodiversity,*
*University of Stellenbosch,*
*Private Bag XI,*
*7602 Matieland,*
*South Africa.*
*E-mail: samways@sun.ac.za*

Biodiversity and Conservation (2006) 15:3–23
DOI 10.1007/s10531-004-1234-6


# Arthropod diversity in Lama forest reserve (South Benin), a mosaic of natural, degraded and plantation forests

THIBAULT LACHAT[1,*], SERGE ATTIGNON[1,2], JULIEN DJEGO[2], GEORG GOERGEN[3], PETER NAGEL[1], BRICE SINSIN[2] and RALF PEVELING[1]

*[1]Institut für Natur-, Landschafts- und Umweltschutz (NLU) – Biogeographie, Universität Basel, St. Johanns-Vorstadt 10, 4056 Basel, Switzerland; [2]Laboratoire d'Ecologie Appliquée, Faculté des Sciences Agronomiques (FSA), Université d'Abomey-Calavi, Cotonou, Benin; [3]International Institute of Tropical Agriculture (IITA), Biodiversity Centre, Cotonou, Benin; *Author for correspondence (e-mail: thibault.lachat@unibas.ch; fax: ++41-61-267-0801; phone: ++41-61-267-0800)*

Received 18 November 2003; accepted in revised form 25 June 2004

**Key words:** Arthropod assemblages, Biodiversity conservation, Dahomey gap, Degraded forest, Forest plantations, Indicator species, Natural forest

**Abstract.** Arthropod assemblages were examined in Lama forest reserve, a protected area situated in the Dahomey gap, southern Benin, composed of plantations, degraded forest and remnants of natural forest. The objectives were to compare assemblages in relation to forest type and use, to elucidate the value of forest plantations for biodiversity conservation and to identify indicator species for specific forest habitats. Arthropods were collected over an 11-month period, using standardized sets of traps (pitfall, emergence, Malaise and flight intercept traps). Nine different habitats were studied, including natural and degraded forest, forest plantations (*Tectona grandis* and *Senna siamea*) of different age, and isolated forest fragments. Our analysis focused on detritivorous and xylophagous arthropods but also included ground beetles and heteropterans, totalling 393 species. We found no differences in species richness among natural and degraded forest habitats in the centre of the reserve (*Noyau central*). Outside of the *Noyau central*, species richness was highest in old teak plantations and isolated forest fragments and lowest in young teak and fuelwood plantations. Detrended correspondence analysis (DCA) separated three main groups: (1) natural forest, (2) degraded forest and young plantations, and (3) old plantations and isolated forest fragments. Multiple regression of DCA scores of the first two axes on environmental variables identified one natural and three disturbance-related predictors of arthropod assemblages in Lama forest: soil type (texture), canopy height, naturalness (proportion of Guineo-Congolian plant species) and understorey vegetation cover. We identified 15 indicator species for six different forest habitats. The highest numbers were found in abandoned settlements and old teak plantations. $\beta$-diversity was similar among the three DCA ordination groups (degraded forest excluded). Values for $\beta$-diversity were relatively high, suggesting that all major forest habitats contribute significantly to regional species pools and should therefore be protected. To enhance arthropod diversity, we propose that management practices in Lama forest should aim to encourage the development of species-rich understorey vegetation of the Guineo-Congolian phytogeographical region.

## Introduction

West African forests are listed among the 25 hotspots considered as priority areas for biodiversity conservation (Myers et al. 2000). At the same time, there is consensus that secondary forests are becoming increasingly important for biodiversity conservation (Zapfack et al. 2002; Gemerden et al. 2003), and that the contribution of forest plantations to the conservation of tropical forests must be evaluated.

Forest plantations are extending world-wide. Since 1990, the area planted has quadrupled. In Africa, plantations account for only 1.2% of the total forest cover (FAO 2000), but the proportion in Benin is relatively high (4.2%). Forest plantations provide a range of forest products on a relatively limited land surface, and can therefore contribute to reducing deforestation and degradation of natural forests (FAO 2001). Being a man-made type of forest, plantations are considered to support low biodiversity and hence be of little interest for biodiversity research and conservation. Thus, only few studies on the biodiversity in tropical forest plantations have been completed to date (e.g. Watt et al. 1997; Lawton et al. 1998; Davis et al. 2001). These studies showed that forest plantations are not necessarily 'biodiversity deserts' (Speight and Wylie 2001) but that they can support a rich and varied fauna. The importance of tropical forest plantations for the conservation of wildlife and as nuclei for natural forest regeneration has been demonstrated in Madagascar (Goodman et al. 1996), Sri Lanka (Ashton et al. 1993), Thailand (Elliott et al. 1998) and Australia (Tucker and Murphy 1997).

Several authors studied the response of insects (ants, termites, moths, dung and carrion beetles) to the degradation of tropical forests (e.g. Nummelin and Hanski 1989; Holloway et al. 1992; Vasconcelos et al. 2000; Eggleton et al. 2002). These studies showed that the composition and species richness of arthropod assemblages vary depending on disturbance levels, regional species pools and the spatial and temporal scale of the study. However, little information is available on the effects of different forest management regimes on the composition of arthropod assemblages in West Africa.

As one of the last remnants of natural forest within the Dahomey gap (Ballouche et al. 2000), and an important refuge for several endangered plants and animals, Lama forest reserve is of key concern for biodiversity conservation in Benin. Only few studies have been conducted so far, the majority focusing on natural forest. Despite their larger size, degraded forests and forest plantations have received little attention. With the exception of litter-dwelling arthropods (Attignon et al. 2004) and butterflies (Fermon et al. 2001), arthropod assemblages have not been well studied, let alone surveyed in different habitat types and successional stages of forest in the reserve.

The goal of the present study was to enhance the scientific understanding of the Lama forest arthropod fauna as a basis for improved, conservation-oriented forest management. The specific objectives were (1) to compare arthropod diversity and assemblages in the principal forest types of Lama forest, (2) to assess the value of forest plantations for biodiversity conservation and (3) to identify indicator taxa for specific forest habitats.

## Materials and methods

*Study site*

The Lama forest reserve is situated about 80 km north of Cotonou (6°55.8′ to 6°58.8′N and 2°4.2′ to 2°10.8′E), covering an area of 16,250 ha in a shallow depression between the Allada and Abomey plateaus. The forest is located in the Dahomey gap, a low rainfall zone separating the western and eastern part of the humid Guineo-Congolian evergreen and semi-evergreen forests of West and West-central Africa (White 1983). The climate is relatively dry (ca. 1200 mm rainfall), with a pronounced dry season from November to March (Sayer et al. 1992). The prevailing soils are hydromorphic vertisols rich in nutrients and clay, as well as sandy ferralsols. The natural vegetation is a semi-deciduous forest, belonging to the drier peripheral semi-evergreen Guineo-Congolian rain forest system (White 1983).

Despite having been legally protected since 1946, deforestation for agriculture in Lama forest continued until 1988, leading to the reduction of natural forest cover from 11000 to 1900 ha. Since then, a central part covering 4800 ha, the so-called *Noyau central*, lies under strict protection. Peasants living in the *Noyau central* were resettled in nearby agroforestry schemes. Reflecting the traditional farming system, the *Noyau central* is composed of a small-scale mosaic of natural and degraded forest patches of variable size and successional stages (Specht 2002). The *Noyau central* is surrounded by young and old teak plantations (7000 ha, *Tectona grandis*) and fuelwood plantations (2400 ha, mainly *Senna siamea*, interspersed with *T. grandis*, and a few stands of *Acacia auriculiformis*) (Figure 1). The present study focused on nine different forest types representing all major habitats within the reserve boundaries plus a few forest remnants outside of the reserve. Five forest types were situated within the *Noyau central* and four outside:

1. Semi-deciduous forest (1937 ha) is dominated by tree species such as *Afzelia africana, Ceiba pentandra, Dialium guineense, Diospyros mespiliformis, Drypetes floribunda, Celtis brownii, Mimusops andongensis*. The understorey vegetation can be dense and canopy height reaches 16–21 m (Table 1).
2. *Cynometra megalophylla* lowland forest (< 100 ha) grows in seasonally flooded areas. This forest is characterized by *C. megalophylla* and other plants adapted to seasonal flooding. The understorey vegetation is usually less dense than in semi-deciduous forest.
3. *Anogeissus leiocarpa* dry forest (1222 ha) is a secondary forest developing on former slash-and-burn patches. *A. leiocarpus* can reach 20 m in height and promotes the establishment of shade-tolerant plants.
4. Abandoned settlements (166 ha) of the resettled population present a characteristic vegetation type composed of cultivated trees such as oil palm (*Elaeis guineensis*) and guava (*Psidium guajava*), and secondary regrowth.
5. Perennial *Chromolaena odorata* thicket (1452 ha) grows on former farmland. *C. odorata* is an alien, invasive species of neotropical origin

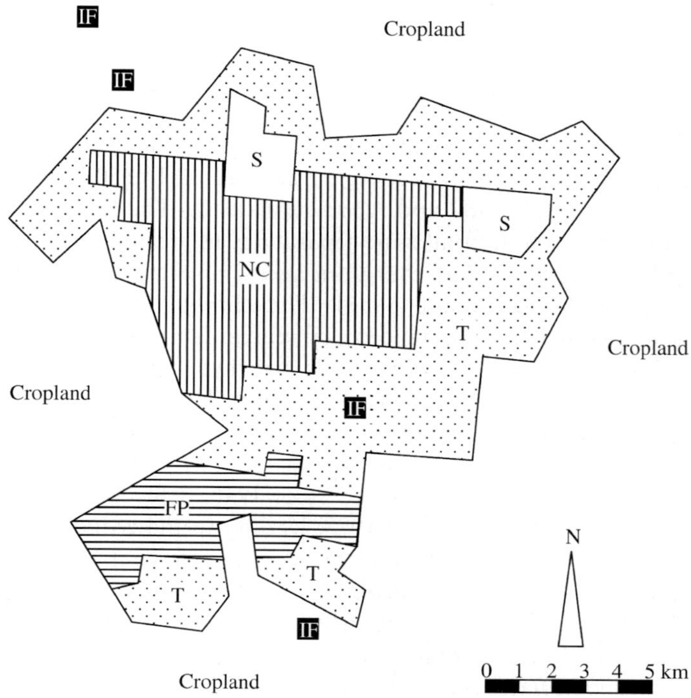


*Figure 1.* Schematic view of Lama forest reserve. NC, *Noyau central*; T, Teak plantations; FP, Fuelwood plantations; S, Settlements; IF, Isolated forest fragments (the latter not to scale).

encroaching open-canopy forest patches, clearings, as well as fallow land. It is rapidly shaded out if the forest canopy cover exceeds about 40%.

6. Young teak plantations (7200 ha) were planted between 1985 and 1995 on vertisol around the *Noyau central.*
7. Old teak plantations (2200 ha) were planted between 1955 and 1965 on sandy ferralsol (transition between vertisols in the valley and ferralitic soils on the surrounding plateaus).
8. Fuelwood plantations (2400 ha) were planted between 1988 and 1996. They are composed of *S. siamea*, *T. grandis* and *A. auriculiformis*.
9. Most isolated forest fragments are located outside of the reserve. These widely scattered sacred groves are embedded in a matrix of farmland and degraded savannah. They are used as ceremonial places and remain in a relatively natural state. The remnants are usually very small (1–2 ha).

Sampling sites were selected according to three criteria: (1) spatial representativeness, (2) patch size and (3) accessibility. Each forest type was represented by four replicates. Thus, the total number of sites was 36 (Table 1). Distances between sites of the same forest type ranged from 0.3 to 19.0 km. A minimum distance of 20 m (small patches) or 50 m (large patches) was maintained between sampling sites and patch borders.

*Table 1.* Site characteristics of the forest types studied in Lama forest.

| | Classified forest | | | | | | | | Sacred grove |
|---|---|---|---|---|---|---|---|---|---|
| | Inside of the *Noyau central* | | | | | Outside of the *Noyau Central* | | | |
| Forest type: | Semi-deciduous forest | Lowland forest | *A. leiocarpa* dry forest | Abandoned settlements | *C. odorata* thicket | Young teak plantations | Old teak plantations | Fuelwood plantations | Isolated forest fragments |
| Forest code: | SF1–4 | LF1–4 | DF1–4 | AS1–4 | CT1–4 | YT1–4 | OT1–4 | FP1–4 | IF1–4 |
| Soil type | Vertisol | Vertisol | Vertisol | Vertisol | Vertisol | Vertisol | S. ferralsol | Vertisol[b] | S. ferralsol[c] |
| % clay[m] | 75 (45–75) | 75 | 75 | 75 | 75 | 60 (45–75) | 11 (8–15) | 35 (8–75) | 25 (15–45) |
| Naturalness (%)[a] | 67 | 67 | 23 | 23 | 23 | 13 | 36 | 20 | 48 |
| Nearest distance to natural forest (km)[m] | 0.0 | 0.0 | 0.75 (0.05–0.10) | 0.10 (0.05–0.20) | 0.08 (0.05–0.20) | 0.98 (0.20–2.20) | 3.80 (2.50–6.40) | 4.90 (4.30–5.70) | 3.28 (1.90–7.00) |
| Temperature (°C)[d] | 25.9 | 25.0[d] | 26.9 | 27.0[e] | 26.5 | 26.8 | 27.4 | 27.2 | 26.6 |
| Min–Max | 24.1–28.9 | 24.0–26.5 | 24.5–30.4 | 25.2–29.3 | 24.7–29.3 | 24.8–30.3 | 25.4–30.6 | 25.2–28.6 | 24.5–29.7 |
| Relative humidity (%)[d] | 92.2 | 94.3[e] | 85.2 | 86.2[f] | 86.2 | 86.8 | 84.5 | 83.5 | 85.6 |
| Min–Max | 81.0–99.0 | 90.3–98.1 | 72.9–94.1 | 78.6–95.4 | 76.1–93.4 | 73.8–95.3 | 73.6–91.9 | 75.6–91.5 | 75.2–93.4 |
| Tree cover (%)[m] | 55 (50–67) | 69 (65–80) | 60 (48–65) | 55 (55–77) | 12 (3–45) | 78 (65–85) | 67 (47–72) | 63 (55–65) | 55 (40–72) |
| Canopy height (m)[m] | 17.5 (16 21) | 21.0 (21–23) | 17.5 (16–20) | 17.0 (15–19) | 20.0 (8–25) | 18.5 (13–21) | 24.0 (22–26) | 14.0 (12–20) | 24.5 (18–28) |
| Tree species richness[m] | 7 (6–8) | 6 (4–11) | 7 (6–10) | 7 (6–10) | 8 (5–14) | 2 (2–3) | 1 (1–2) | 2 (2–5) | 9 (8–11) |
| UV cover (%)[m] | 83 (37–92) | 68 (55–75) | 75 (29–90) | 72 (45–80) | 81 (17–97) | 20 (8–30) | 35 (27–75) | 66 (45–70) | 85 (65–97) |
| UV height (m)[m] | 1.3 (1.0–1.5) | 1.5 (1.4–1.5) | 1.1 (1.0–1.3) | 1.0 (0.9–1.4) | 1.2 (0.7–1.3) | 1.0 (0.5–1.3) | 1.4 (1.3–1.4) | 1.3 (1.2–1.4) | 1.5 (1.2–1.5) |
| UV species richness[m] | 35 (28–52) | 45 (30–47) | 42 (29–52) | 38 (28–43) | 23 (14–40) | 42 (21–45) | 35 (34–37) | 36 (17–40) | 49 (41–63) |

[a]Dominance of plant species of the Guineo-Congolian phytogeographic type.
[b]Except FP4 (sandy vertisol).
[c]Except IF4 (vertisol).
[d]Measured on only one site per forest type (monthly average).
[e]November–March not available.
[f]August–November not available.
[m]Median (range); S. ferralsol = Sandy ferralsol; UV = Understorey vegetation.

A botanical survey was also conducted within the scope of our study, using the Braun–Blanquet system. Two hundred ninety plant species in 77 families were sampled (unpublished data).

*Sampling methods*

On each site, we installed an equal number of sampling devices, comprising (1) three funnel pitfall traps, each consisting of a collecting jar inside a plastic sleeve, an 11 cm diameter funnel and a transparent plastic roof 20 cm in diameter (Southwood 1978), (2) one Malaise trap (after Townes 1972) with a 1.5 $m^2$ black vertical mesh panel, (3) one flight intercept trap intercepting insects flying between 1.0 and 1.5 m above the ground, consisting of two crossed black vertical mesh panels, each measuring 0.25 $m^2$ (0.5 × 0.5 m), and top and bottom funnels 50 cm in diameter (Wilkening et al. 1981), and (4) one pyramid-shaped emergence trap (ground photo-eclector) covering an area of 0.75 $m^2$ (≈ 0.86 × 0.86 m), equipped with a collecting jar on the top and one pitfall trap (Mühlenberg 1993). The traps were spaced out along 30 m north–south transects. The placement design was similar at all sites. We used formalin (0.5%) as a preservative, adding some detergent to lower the surface tension. A preliminary 2-week sampling was conducted in May 2001 to establish the methodology. The sampling period for the present study was 1 week per month from June 2001 to April 2002. Specimens were sorted, counted, labelled and stored in alcohol (75%) for later identification.

*Sorting scope and identification*

Among the wide range of invertebrates sampled, we focused on detritivorous and xylophagous arthropods because of their important role in nutrient cycling in forest ecosystems (Didham et al. 1996). These taxa comprised 14 coleopteran families, as well as representatives of Isoptera, Diplopoda and Isopoda. We also included epigeal predators (Carabidae and Chilopoda), omnivorous beetles (Tenebrionidae) and both herbivorous and predatory bugs (Heteroptera). These additional taxa were retained for a more comprehensive characterization of arthropod assemblages and because some have been used as representative indicators in previous biodiversity assessments (Duelli and Obrist 1998; Giulio et al. 2001; Rainio and Niemelä 2003).

All arthropods were first sorted to 'morphospecies' (*sensu* New 1998) and then taxonomically identified at the International Institute of Tropical Agriculture (IITA) in Benin. Voucher specimens were deposited at the IITA Biodiversity Center and partly at the Museum of Natural History, Basel, Switzerland. The analysis was done at the morphospecies level for taxa with difficult taxonomy (e.g. most Diplopoda).

*Environmental variables*

Soils were classified according to the FAO system (FAO-UNESCO 1974), and the percentage of clay (soil texture) was estimated by touch. Temperature (°C) and relative humidity (RH) were recorded every hour from April 2002 to March 2003, using one data logger (Hobo Pro RH/Temp) per forest type. Loggers were attached to tree trunks about 1 m above ground level. The naturalness (*sensu* Angermeier 1999) of the different forest types was calculated based on the proportion of plant species belonging to the Guineo-Congolian phytogeographical region. We also determined cover, height and species richness of the main vegetation strata (Table 1). The nearest distance between sampling sites and natural forest patches was measured with a geographic information system (ArcView 3.1), using the vegetation map of Specht (2002) (Table 1).

*Measures of diversity*

We used species richness as a measure of $\alpha$-diversity (the number of species within a habitat). $\beta$-diversity (the degree of change in species composition between habitats) was evaluated for selected groups of forest habitats according to Whittaker's formula $\beta_w = \gamma/\bar{\alpha}$, where $\gamma$ is the species pool within a group of habitats ($\gamma$-diversity) and $\bar{\alpha}$ is the average number of species per site (Whittaker 1960).

*Data analysis*

We used the total catch per taxon and per sampling site for statistical analyses. This was done by pooling specimens from all sampling periods and traps within sampling sites.

One-way analysis of variance (ANOVA) was conducted to test differences in arthropod assemblages among forest types (Zar 1999), followed by Bonferroni multiple comparison of means. In view of an unfavourable ratio between factor levels ($n = 9$) and replicates ($n = 4$), we also performed a *post hoc* power analysis (SPSS 12.0).

To determine the similarity of forest types based on their arthropod assemblages, we performed detrended correspondence analysis (DCA) (Hill and Gauch 1980), using PC-ORD 4.27 (McCune and Mefford 1999). Abundances of species rarer than $F_{max}/5$ (where $F_{max}$ is the frequency of the most common species) were down-weighted in proportion to their frequency. Axes were rescaled with a threshold of zero, and the number of segments was set to 26 (default). Reciprocal averaging (RA), also known as correspondence analysis, revealed the same grouping, but DCA was preferred because it squashed the arch effect associated with RA and corrected the compression of the axis ends. The proportion of variance represented by the ordination axes was calculated according to an after-the-fact method, using the relative Euclidean distance (McCune and Mefford 1999). Stepwise multiple regression with forward selection (SPSS 12.0) was conducted to relate DCA ordination scores of the first two axes to the environmental variables listed in Table 1.

A hierarchical cluster analysis based on presence/absence data was employed to distinguish groups of sites in the DCA ordination plot (SPSS, 12.0, settings: Ward's method, squared Euclidean distance). Clusters were grouped in probability ellipses whose axes are proportional in length to a specified percentage of the $x$ and $y$ coordinates (Jennrich and Turner 1969). The inclusion probability was set to $p = 0.90$. Computation of the ellipses was done with ArcView 3.1.

We used Mantel tests to evaluate the relationship between arthropod assemblages and the distance to the nearest natural forest patch (Mantel 1967). These tests were performed with PC-ORD 4.27 (McCune and Mefford 1999), using binary data (Sørensen distance) and Monte Carlo randomization (1000 runs).

### *Indicator species*

Indicator species for the different forest types were determined according to Dufrêne and Legendre (1997). The method combines data on the frequency of occurrence (faithfulness) and relative abundance (concentration) of species in a particular habitat. The significance of indicator values was tested using Monte Carlo randomization (1000 runs). The threshold level for the indicator value was 25%. This implies that the frequency of occurrence of a species indicative of a particular habitat must be ≥50%, and its relative abundance therein ≥50% of its total abundance at all sites (Dufrêne and Legendre 1997). The significance level was $p \leq 0.01$, as proposed by the authors. The analysis was performed with PC-ORD 4.27 (McCune and Mefford 1999). Note that indicator organisms may include both species restricted to a certain habitat and those more widely distributed yet especially abundant in a particular type of forest.

### *Estimation of true species richness*

True species richness was estimated by computing the abundance-based coverage estimator (ACE) and the incidence-based coverage estimator ICE (1000 runs) for the eleven collecting periods, using EstimateS 6.0 (Colwell 1997). The ACE is based on species with $\leq 10$ individuals in the sample (Chao et al. 1993). The corresponding ICE, likewise, is based on species found in $\leq 10$ sampling sites (Lee and Chao 1994). These estimations were the most appropriate for our data set which was characterized by a large number of singletons (species occurring with one individual only) and uniques (species occurring in one sample only).

## Results

A total of 9431 specimens belonged to the taxonomic groups examined in the present study, representing 393 species (Appendix 1). More than one third of all species were singletons, and only 14 were collected in all forest types. Seventy one percent (ACE) and 66% (ICE) of the estimated true species number were sampled.

### *Coleoptera*

We collected 264 species in 16 families of Coleoptera, representing 67% of the total number of species and 58% (5499 specimens) of the total number of arthropods included in the analysis. Except for carabids, the high number of specimens was due to a few very abundant species. Of 37 scarabaeid species, six represented over 63% of all specimens. One of 17 species of Elateridae represented 97% of the total catch in this family, and one of 12 species of Scolytidae 58% of the total catch. All but one scarabaeid species of the dominant taxa were found in all forest types. Thirty eight percent of all coleopteran species were singletons. Most coleopterans were collected in isolated forest fragments (111 species, 42% of all coleopterans) and old teak plantations (94 species, 36%), and the smallest number in young teak plantations (64 species, 24%) and fuelwood plantations (59 species, 22%). Forest types within the *Noyau central* had about the same number of coleopteran species, with an average of 81 ± 1 species (mean ± SE).

### *Heteroptera*

Within 15 families of Heteroptera, we found 75 species (19% of the total number of arthropod species analyzed) and 558 specimens (6% of the total catch analyzed). Forty-one percent of the Reduviidae (22 species) and 54% of the Lygaeidae (17 species) were represented by three species only. One of five species of Alydidae made up 95% of the total catch for this family. More than 41% of all heteropterans were singletons. Species numbers were highest in isolated forest fragments (32 species) and lowest in semi-deciduous forest (13 species). The other forest types within the *Noyau central* presented an average species number of 23 ± 1 (mean ± SE).

### *Chilopoda, Isopoda, Diplopoda and Isoptera*

For the classes Isopoda, Diplopoda and Chilopoda and for the order Isoptera, the analysis was performed at the morphospecies level. Forty eight percent of all Diplopoda (27 species) were represented by only one species occurring in all forest types. Two of 14 species of Isopoda made up 52% of the total catch of this taxon. Only six species of Isoptera were sampled, with one species representing 96% of all specimens.

*Arthropod diversity*

Despite the unfavourable ratio between factor levels and replicates, the statistical power of the analyses of variance was satisfactory (0.996). Differences in species richness among forest types were significant at $p < 0.001$ ($F_{8,27} = 5.76$). Isolated forest fragments and old teak plantations showed the highest number of species (Figure 2), with an average of 71 ± 5 and 67 ± 2 species, respectively (mean ± SE). Species richness was lowest in young teak (40 ± 6) and fuelwood plantations (43 ± 6). Natural and degraded forest in the *Noyau central* had intermediate levels of species richness, ranging from 50 ± 2 in *C. megalophylla* lowland forest to 57 ± 2 in abandoned settlements. Statistically significant differences were found between young plantations (teak and fuelwood) and old teak plantations or isolated forest fragments (Figure 2).

$\beta$-diversity of arthropod assemblages was computed for the three habitat (site) groups obtained by the DCA ordination (see below). Two of these groups (group one and three, Figure 3) contained two forest types each, including all old-growth forests, and one group (group two) the remaining forest types. Of these, we selected young teak and fuelwood plantations as representatives of young-growth forest. Because some groups comprised only three of the four replicate sites per forest type, we randomly excluded one replicate site of each complete sample to achieve an equal number of sites per group which is a prerequisite for comparing $\beta_w$-diversity. Thus, each group contained three replicate sites of two different forest types ($n = 6$). $\beta$-diversity was very similar among these three groups of forest habitats, ranging from $\beta_w = 2.8$ in group three (old teak plantations and isolated forest fragments) and $\beta_w = 3.1$ in

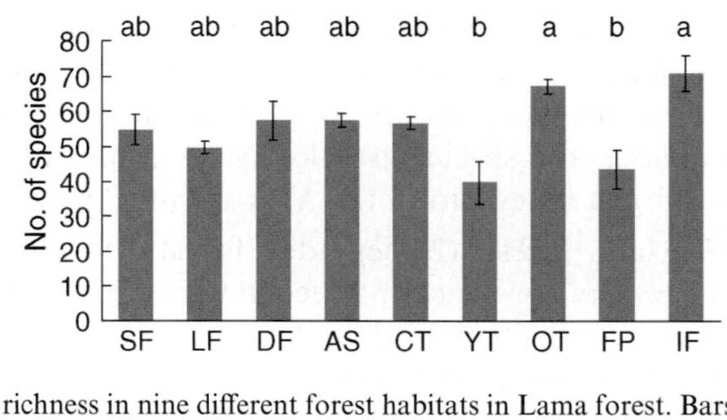


*Figure 2.* Species richness in nine different forest habitats in Lama forest. Bars show means ± SE ($n = 4$). SF, Semi-deciduous forest; LF, *C. megalophylla* lowland forest; DF, *A. leiocarpa* dry forest; AS, Abandoned settlements; CT, *C. odorata* thicket; YT, Young teak plantations; OT, Old teak plantations; FP, Fuelwood plantations; IF, Isolated forest fragments. Means marked with different letters are significantly different at $p < 0.05$.

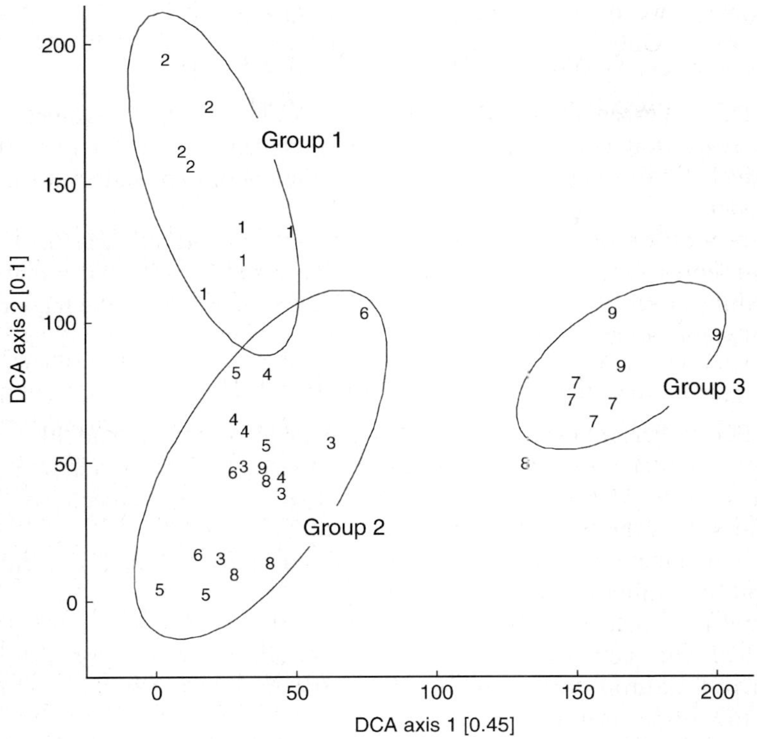


*Figure 3.* Detrended correspondence analysis (DCA) of sampling sites in species space proportion of variance. Groups 1, 2 and 3 are defined by probability ellipses ($p = 90\%$). 1, Semi-deciduous forest; 2, *C. megalophylla* lowland forest; 3, *A. leiocarpa* dry forest; 4, Abandoned settlements; 5, *C. odorata* thicket; 6, Young teak plantations; 7, Old teak plantations; 8, Fuelwood plantations; 9, Isolated forest fragments.

group two (young teak and fuelwood plantations) to $\beta_w = 3.2$ in group one (semi-deciduous forest and *C. megalophylla* lowland forest).

## *Arthropod assemblages*

DCA of sampling sites in species space revealed three distinct groups of forest habitats: (1) natural forest, comprising semi-deciduous forest and *C. megalophylla* lowland forest, (2) degraded forest and young plantations, including *A. leiocarpa* dry forest, abandoned settlements, *C. odorata* thicket, young teak plantations (all but one site) and fuelwood plantations (all but one site), and (3) old teak plantations and isolated forest fragments (all but one site) (Figure 3).

The proportion of variance represented by the first axis of the DCA ordination was 0.45. Multiple regression identified two significant predictors of DCA axis one scores, soil texture and canopy height ($F_{2,33} = 76.6$,

$R^2 = 0.823$, $p < 0.001$). The regression equation is:

$$\text{Axis } 1 = 78.3 - 1.6(\text{PC}) + 3.8(\text{CH}),$$

where PC = Percentage of clay and CH = Canopy height; $t$-values for the partial regression coefficients were −9.0 ($p < 0.001$) and 3.6 ($p = 0.001$), respectively, indicating that soil texture was the major explanatory variable in this model.

The proportion of variance represented by the second axis was 0.10. DCA axis two scores were also best predicted by two variables only, naturalness and understorey cover ($F_{2,33} = 49.7$, $R^2 = 0.751$, $p < 0.001$). For the second axis, the regression equation is:

$$\text{Axis } 2 = 26.8 + 2.3(\text{PG}) - 0.6(\text{PU}),$$

where PG = Percentage of Guineo-Congolian plant species and PU = Percentage of understorey vegetation cover; $t$-values for the partial regression coefficients were 10.0 ($p < 0.001$) and −3.3 ($p = 0.002$), respectively. Thus, naturalness was the more important explanatory variable.

The remaining environmental variables listed in Table 1 had no significant effect on the ordination scores.

Mantel tests revealed that arthropod assemblages of replicate sites within young teak and fuelwood plantations were not correlated with the distance to the nearest natural forest patches ($r_s = 0.436$, $p = 0.260$ and $r_s = 0.533$, $p = 0.169$, respectively).

### *Indicator species*

We identified 15 indicator species for six different forest habitats, ranging from one to five species per forest type (Table 2). No indicators were found for dry forest, *C. odorata* thicket and young teak plantations. The indicators included 11 species of Coleoptera, two species of Isopoda and one species each of Diplopoda and Heteroptera. Most indicator species were recorded for abandoned settlements ($n = 4$) and old teak plantations ($n = 5$).

## Discussion

### *Arthropod diversity*

Against our expectations, α-diversity was similar among the different forest types within the *Noyau central*. We would have expected lower species richness in disturbed habitats such as *C. odorata* thicket. However, not only was species richness similar, but the similarity in species composition was also high, ranging from 38% (*C. odorata* thicket versus lowland forest) to 58% (*C. odorata* thicket versus dry forest) species in common (Lachat, unpublished

*Table 2.* Indicator species *sensu* Dufrêne and Legendre (1997) for Lama forest.

| | Indicator value (%) | *P* |
|---|---|---|
| Semi-deciduous forest | | |
| *Chlaenius* (*Chlaenites* s.l.) sp. (Carabidae, Chlaeniinae) | 80.0 | 0.001 |
| Scolytidae sp. 9 | 58.7 | 0.001 |
| *C. megalophylla* lowland forest | | |
| *Stenocoris southwoodi* Ahmad, Alydidae | 95.8 | 0.001 |
| Abandoned settlement | | |
| *Onthophagus* sp. 1, Scarabaeidae | 66.7 | 0.001 |
| *Onthophagus* sp. 3, Scarabaeidae | 54.8 | 0.003 |
| *Sisyphus* sp. 1, Scarabaeidae | 66.7 | 0.004 |
| Elateridae sp. 3 | 31.1 | 0.007 |
| Old teak plantations | | |
| *Hoplolenus obesus* (Murray, 1858) (Carabidae, Oodini) | 72.2 | 0.001 |
| *Trochalus* sp. 1, Scarabaeidae | 75.0 | 0.008 |
| *Trochalus* sp. 2, Scarabaeidae | 72.1 | 0.002 |
| Tenebrionidae sp. 24 | 60.9 | 0.007 |
| Diplopoda sp. 1 | 31.3 | 0.002 |
| Fuelwood plantations | | |
| Isopoda, Eubelidae sp. 3 | 68.9 | 0.005 |
| Isopoda, Eubelidae sp. 5 | 61.5 | 0.001 |
| Isolated forest fragments | | |
| *Tetragonoderus* (s.str.) *quadrimaculatus* Gory, 1833 (Carabidae, Cyclosomini) | 87.5 | 0.001 |

No indicator species were found for *A. leiocarpa* dry forest, *C. odorata* thickets and young teak plantations.

data). This suggests a high connectivity between natural, secondary and degraded parts of the *Noyau central* whose spatial structure is characterized by contiguous patches of variable size, sometimes covering less than 1 ha (Specht 2002). Moreover, forest cover was relatively high ( $\leq 45\%$) even in degraded patches, which would be expected to facilitate movements between forest habitats. This is corroborated by observations in the Amazon which showed that secondary growth reduces the barrier effect of cleared forest for forest dung and carrion beetles (Klein 1989).

Our results confirm those of other studies that found no major difference in species richness of arthropod assemblages between primary forest and secondary and/or degraded (logged) forest (e.g., Nummelin and Hanski 1989; Holloway et al. 1992; Kalif et al. 2001).

Significant differences in species richness were found only among forest habitats outside of the *Noyau central* (Figure 2). Notably, species richness in old teak plantations was as high as in forest fragments and in the *Noyau central*, which demonstrates the importance of old teak for arthropod diversity conservation. The low species richness in young teak and fuelwood plantations came as no surprise. These forests are more exposed to silvicultural practices.

Moreover, despite a fire exclusion strategy adopted by the forestry authorities, agricultural fires sometimes escape into young teak and fuelwood plantations.

$\beta$-diversity was similar between the three groups obtained by the DCA-ordination (natural forest, young plantations, old plantations and isolated forest fragments). Furthermore, $\beta_w$-values were relatively high (2.8–3.2), compared to a theoretical minimum of $\beta_w = 1$ (each species occurs on all sites) and a maximum of $\beta_w = 6$ (each species occurs on one site only) for $n = 6$ sites per group. $\beta$-diversity – hence species turnover – increases with increasing spatial heterogeneity, resource selectivity and the diversity of refugia available to rare species (Stanton 1979; Deshmukh 1986). From a conservation point of view, high $\beta$-diversity implies that the preservation of diversity is most effective if habitats are protected entirely.

*The importance of plantations for biodiversity conservation*

Several modifications to the design and management of tropical plantations have been proposed that may enhance regional biodiversity without compromising economic benefits (reviewed in Lamb 1998). Among the various approaches, two are pertinent to Lama forest: the establishment of plantations in the vicinity of natural forest – which may act as a reservoir and source of forest species – and the development of understorey vegetation. The second option is only feasible for long-rotation sawlogs such as teak. The growth of understorey vegetation and a concomitant increase in biodiversity is enhanced by the selective harvesting of logs which creates gaps for plant colonization. In Lama forest, this process seems to be supported by the fire exclusion practice. Contrary to timber plantations, understorey development is unlikely in short-rotation forests such as fuelwood plantations (Lamb 1998). Fuelwood in Lama forest is harvested at an age of 20 years or less. These plantations obviously contribute less to biodiversity conservation, as reflected by the low species richness found in our study. Even though, the production of fuelwood itself may reduce the pressure on natural forest resources.

The evidence provided in the present study shows that old teak plantations are important habitats for forest species. This is supported by the presence of typical forest specialists such as *Paussus excavatus*, *P. liber* and *P. bicornis* (Carabidae, Paussinae). Similar observations were made in mature plantations of endemic hardwood in Cameroon where butterfly assemblages were undistinguishable from those found in natural forest (Stork et al. 2003). An elevated arthropod diversity in old plantations may have consequences not only from a biodiversity conservation but also from a pest management perspective. A high degree of naturalness and/or close distance to natural forest may benefit natural enemies of forest pests, thereby reducing the risk of infestations of plantation forests (Speight and Wylie 2001).

The distance between natural forest and sampling sites in young teak plantations varied from 0.2 to 2.2 km, and the distance to sampling sites in fuelwood plantations from 4.3 to 5.7 km. Thus, one might hypothesize that assemblages within plantations differ depending on their distance to the nearest natural forest patch. Yet, a border effect was dismissed on the basis of the Mantel tests performed, suggesting that plantations adjacent to the *Noyau central* may act as dispersal corridors. However, our study design was not conceived to monitor movement pathways between forest patches.

*Influence of environmental variables on arthropod assemblages*

Arthropod assemblages in the different forest types were most strongly related to soil type (DCA axis 1), a natural site character, and to naturalness of the vegetation (DCA axis 2), an indicator of disturbance. The other two statistically significant explanatory variables were canopy height (DCA axis 1) and understorey cover (DCA axis 2), representing disturbance indicators related to land use and management.

Soil may have influenced arthropod assemblages in two ways. First, the prevailing vertisols show distinct seasonal swelling–shrinking cycles. During the dry season, they harden and form deep cracks. In the rainy season, they are saturated with water, leading to flooding in depressions. This in turn might reduce the habitat available to epigeal species not tolerating temporary flooding, forcing them to retreat to mounds of the so-called gilgai micro relief (irregular land surface with alternating mounds and depressions in areas with vertisol). To the contrary, physical properties of the sandy ferralsols in old teak plantations and isolated forest fragments do not change dramatically between seasons, and the soil may offer suitable habitats throughout the year. Second, soil influences arthropods indirectly by affecting the vegetation. However, ordination of our vegetation data did not clearly segregate plant associations of forests stocking on vertisol and ferralsol (Djego, unpublished data), suggesting that soil type had a more pronounced influence on arthropod assemblages than on plant associations.

The importance of soil as a co-determinant of arthropod assemblages was also evidenced by the sites plotted outside of the corresponding probability ellipses in the DCA ordination (Figure 3). For example, the soil of one of the fuelwood plantation sites (label 8) was a sandy vertisol. This site plotted next to group three which also comprised sites on ferralsol. Likewise, the only isolated forest fragment (label 9) located on vertisol – and being embedded in a matrix of young teak plantations – was plotted together with group two sites, all of which shared the same soil. The remaining environmental predictors of arthropod assemblages in Lama forest (naturalness, canopy height, understorey vegetation cover) are related to land use and silvicultural practices. Naturalness is an indicator of human disturbance (clearing for agriculture, conversion to plantation forests and other land uses). Highly disturbed parts of Lama forest have a higher proportion of plant species with wide (sometimes pantropical) distribution, while Guineo-Congolian species dominate in less disturbed parts. Naturalness increases not only from degraded to natural forest, but also from young to old plantations. Unfortunately, little is known about the biogeography of the arthropod species sampled, which makes it difficult to define geographic types and to draw parallels with phytogeographic types.

Canopy height reflects a succession towards old-growth forest. The tallest canopy trees were found in old teak plantations and isolated forest fragments (cf., Table 1). The similarity in vegetation structure may have contributed to the high similarity of arthropod assemblages among these forests.

The development of understorey vegetation is a characteristic of old-growth stands (Lamb 1998). However, understorey cover in Lama forest was highly variable within and among the different forest types (Table 1) and should therefore be interpreted with caution. For example, disturbed, open-canopy forests were often dominated by uniform *C. odorata* thicket, whereas a diversity of native, shade-tolerant plants prevailed in closed-canopy forests (Djego, unpublished data). Thus, understorey cover alone appears to be an insufficient predictor of arthropod assemblages, but together with naturalness it defines environmental conditions relevant to their composition.

### *Edge effects in old teak plantations and isolated forest fragments*

In contrast to other studies (Didham et al. 1998; Barbosa and Marquet 2001), the highest species richness was encountered in isolated forest fragments and old teak. Moreover, the similarity of arthropod assemblages was high, despite long distances among replicate sites (16–19 km).

Another common trait of these forest types – apart from soil type and naturalness – is their adjacency to open country (degraded savannah and/or cropland). Forest edges are likely to attract arthropods from open landscape as well as forest, thereby increasing overall species richness (Laurance et al. 2002). Some forest species may even increase in abundance near edges, in particular those adapted to the microclimate prevailing in open forest or treefall gaps (Kapos 1989; Laurance et al. 2002). Such edge effects – along with the combined effect of the four environmental explanatory variables – may explain the high diversity and distinctiveness of arthropod assemblages in old teak plantations and isolated forest fragments.

### *Indicator species*

Indicator species have been defined as taxa that ‘mirror changes in a wider array of groups as a consequence of environmental change’ or that ‘reflect overall diversity and complexity of an assemblage’ (New 1998). The idea to focus on indicator species is also owed to limitations in processing and identifying the huge numbers of samples typically collected during invertebrate surveys. The four most important criteria for choosing invertebrate indicators are that they have a well-known taxonomy and ecology, are accessible to sampling and respond to environmental changes (New 1998). Strictly speaking, none of the species identified in the present study fulfils all of these criteria. At this initial stage of research, they are therefore simply considered as species characteristic of certain forest habitats. Their role as indicators of successional changes requires validation in future monitoring programmes.

Notably, the highest numbers of indicators were found in abandoned settlements and old teak plantations. This can be interpreted as an indication of the importance of these two forest types to regional biodiversity.

Taxonomically, most species belong to the family Scarabaeidae (chafers, Melolonthinae, and dung beetles, Scarabaeinae). Dung beetles have often been used as indicators because of their reliance on vertebrate dung or carrion and their sensitivity to habitat disturbance (Klein 1989; Nummelin and Hanski 1989; Hill 1995). Carabidae are also well represented. While being common indicators in temperate ecosystems, their suitability for tropical forests is as yet not well established (Rainio and Niemelä 2003).

## Conclusion

This study provides a first overview of the arthropod diversity in Lama forest reserve, one of the last and largest vestiges of natural forest in southern Benin, and highlights its importance for biodiversity conservation. No differences in arthropod species richness were found among habitats within the *Noyau central*, a small-scale mosaic of natural and degraded forest. However, great differences were observed among forest plantations separating the *Noyau central* from the matrix of agricultural land. We identified four environmental variables as significant predictors of arthropod assemblages. Of these, soil type is a natural factor promoting high species richness in old plantations and isolated forest fragments. The remaining variables naturalness, understorey cover and canopy height are related to silvicultural practices and are therefore amenable to an improved, conservation-oriented forest management. To enhance arthropod diversity in Lama forest, we propose that management practices should aim to encourage the development of species-rich understorey vegetation of the Guineo-Congolian phytogeographical region. Animals higher up the food chain, in particular insectivorous reptiles, birds and mammals, may also benefit from increased arthropod diversity. In this respect, arthropod conservation is not an end in itself but a contribution to overall biodiversity conservation.

## Acknowledgements

This study was funded by the Swiss National Science Foundation and the Swiss Agency for Development and Cooperation. It was carried out in the framework of the project *BioLama*. We thank all *BioLama* staff for their support, especially M. Guedegbé, C. Houngbedji, L. Konetche and D. Weibel. We are grateful to P. Houayé, R. Akossou, D. Honfozo, H. Hodonou and A. V. N'velin of the *Office National du Bois* for their backing, to Gwendoline Altherr for statistical help and to C. Schmid for the determination of Isopoda. Further, J. Refisch and two anonymous referees are acknowledged for helpful comments on an earlier version of this article.

## References

Angermeier P.L. 1999. The natural imperative for biological conservation. Conserv. Biol. 14: 373–381.

Ashton P.M.S., Gunatilleke C.V.S. and Gunatilleke I.A.U.N. 1993. A shelterwood method of regeneration for sustained timber production in Mesua-Shorea forest of Southwest Sri Lanka. In: Erdelen W., Preu C., Ishwaran C.M. and Bandara M. (eds), Ecology and Landscape Management in Sri Lanka. Margraf Verlag, Weikersheim, pp. 255–274.

Attignon S.E., Weibel D., Lachat T., Sinsin B., Nagel P., Peveling R. 2004. Leaf litter breakdown in natural and plantation forests of the Lama forest reserve in Benin. Appl. Soil Ecol. in press.

Ballouche A., Akoègninou A., Neumann K., Salzmann U. and Sowunmi M.A. 2000. Le projet 'Dahomey Gap: une contribution à l'histoire de la végétation au Sud-Bénin et Sud-Ouest du Nigeria. Berichte des Sonderforschungsbereichs 268: 237–251.

Barbosa O. and Marquet P.A. 2001. Effects of forest fragmentation on the beetle assemblage at the relict forest of Fray Jorge, Chile. Oecologia 132: 296–306.

Chao A., Ma M.-C. and Yang M.C.K. 1993. Stopping rules and estimation for recapture debugging with unequal failure rates. Biometrica 80: 193–201.

Colwell R.K. 1997. EstimateS: Statistical Estimation of Species Richness and Shared Species from Samples. http://viceroy.eeb.uconn.edu/estimates.

Davis A.J., Holloway J.D., Huijbregts H., Krikken J., Kirk-Spriggs A.H. and Sutton S.L. 2001. Dung beetles as indicators of change in the forests of northern Borneo. J. Appl. Ecol. 38: 593–616.

Deshmukh I. 1986. Ecology and Tropical Biology. Blackwell Scientific Publications, Boston.

Didham R.K., Ghazoul J., Stork N.E. and Davis A.J. 1996. Insects in fragmented forests: a functional approach. Tree 11: 255–260.

Didham R.K., Hammond P.M., Lawton J.H., Eggleton P. and Stork N.E. 1998. Beetle species responses to tropical forest fragmentation. Ecol. Monogr. 68: 295–323.

Duelli P. and Obrist M.K. 1998. In search of the best correlates for local organismal biodiversity in cultivated areas. Biodiv. Conserv. 7: 297–309.

Dufrêne M. and Legendre P. 1997. Species assemblages and indicator species: the need of a flexible asymmetrical approach. Ecol. Monogr. 67: 345–366.

Eggleton P., Bignell D.E., Hauser S., Dibog L., Norgrove L. and Madong B. 2002. Termite diversity across an anthropogenic disturbance gradient in the humid forest zone of West Africa. Agr. Ecosyst. Environ. 90: 189–202.

Elliott S., Blakesley D. and Anusaransunthron V. 1998. Forests for the Future – Growing and Planting Native Trees for Restoring Forest Ecosystems. Forest Restoration Research Unit. Chiang Mai University, Chiang Mai.

FAO. 2000. Global Forest Resources Assessment 2000. Main report, Rome.

FAO. 2001. State of the World's Forests 2001. Rome.

FAO-UNESCO. 1974. Map of the World. Paris.
Fermon H., Schulze C.H., Waltert M. and Mühlenberg M. 2001. The butterfly fauna of the *Noyau central*, Lama Forest (Republic of Benin), with notes on its ecological composition and geographic distribution. Afr. Entomol. 9: 177–185.
Gemerden B.S.V., Shu G.N. and Olff H. 2003. Recovery of conservation values in Central African rain forest after logging and shifting cultivation. Biodiv. Conserv. 12: 1553–1570.
Giulio M.D., Edwards P.J. and Meister E. 2001. Enhancing insects in agricultural grasslands: the roles of management and landscape structure. J. Appl. Ecol. 38: 310–319.
Goodman S.M., Rakotondravony D., Schatz G. and Wilmé L. 1996. Species richness of forest-dwelling birds, rodents and insectivores in a planted forest of native trees: a test case from Ankaratra, Madagascar. Ecotropica 2: 109–120.
Hill C.J. 1995. Linear strips of rain forest vegetation as potential dispersal corridors for rain forest insects. Conserv. Biol. 9: 1559–1566.
Hill M.O. and Gauch H.G.J. 1980. Detrended correspondance analysis: an improved ordination technique. Vegetatio 42: 47–58.
Holloway J.D., Kirk-Spriggs A.H. and Khen C.V. 1992. The response of some rain forest insect groups to logging and conversion to plantation. In: Marshall A.G. and Swaine M.D. (eds), Tropical Rain Forest: Disturbance and Recovery. Philosophical Transactions of the Royal Society of London Series B. 335: 425–436.
Jennrich R.I. and Turner F.B. 1969. Measurement of non-circular home range. J. Theor. Biol. 22: 227–237.
Kalif K.A.B., Azevedo-Ramos C., Moutinho P. and Malcher S.A.O. 2001. The effect of logging on the ground-foraging ant community in eastern Amazonia. Stud. Neotrop. Fauna Environ. 36: 1–5.
Kapos V. 1989. Effects of isolation on the water status of forest patches in the Brazilian Amazon. J. Trop. Ecol. 5: 173–185.
Klein B. 1989. Effects of forest fragmentation on dung and carrion beetle communities in central Amazonia. Ecology 70: 1715–1725.
Lamb D. 1998. Large-scale ecological restoration of degraded tropical forest lands: the potential role of timber plantations. Restor. Ecol. 6: 271–279.
Laurance W.F., Lovejoy T.E., Vasconcelos H.L., Bruna E.M., Didham R.K., Stouffer P.C., Gascon C., Bierregaard R.O., Laurance S.G. and Sampaio E. 2002. Ecosystem decay of Amazonian forest fragments: a 22-year investigation. Conserv. Biol. 16: 605–618.
Lawton J.H., Bignell D.E., Bolton B., Bloemers G.F., Eggleton P., Hammond P.M., Hodda M., Holt R.D., Larsen T.B., Mawdsley N.A., Stork N.E., Srivastava D.S. and Watt A.D. 1998. Biodiversity inventories, indicator taxa and effects of habitat modification in tropical forest. Nature 391: 72–76.
Lee S.-M. and Chao A. 1994. Estimating population size via sample coverage for closed capture-recapture models. Biometrics 50: 88–97.
Mantel N. 1967. The detection of disease clustering and a generalized regression approach. Cancer Res. 27: 209–221.
McCune B. and Mefford M.J. 1999. PC-ORD: Multivariate Analysis of Ecological Data. MjM Software Design, Gleneden Beach, Oregon.
Mühlenberg M. 1993. Freilandökologie. UTB für Wissenschaft. Quelle und Meyer Heidelberg, Wiesbach.
Myers N., Mittermeier R.A., Mittermeier C.G., d. Fonseca G.A.B. and Kent J. 2000. Biodiversity hotspots for conservation priorities. Nature 403: 853–858.
New T.R. 1998. Invertebrate Surveys for Conservation. Oxford, Oxford University Press.
Nummelin M. and Hanski I. 1989. Dung beetles of the Kibale Forest, Uganda; comparison between virgin and managed forests. J. Trop. Ecol. 5: 349–352.
Rainio J. and Niemelä J. 2003. Ground beetles (Coleoptera: Carabidae) as bioindicators. Biodiv. Conserv. 12: 487–506.

Sayer J.A., Harcourt C.S. and Collins N.M. 1992. The Conservation Atlas of Tropical Forests. Africa. IUCN, New York.
Southwood T.R.E. 1978. Ecological Methods: with Particular Reference to the Study of Insect Population. Chapman and Hall, London, New-York.
Specht I. 2002. La forêt de la Lama, Bénin–SIG basé sur Landsat 7. Opuscula Biogeographica Basileensia 2. Basel, Switzerland.
Speight M.R. and Wylie F.R. 2001. Insect Pests in Tropical Forestry. CABI Publishing, Oxford, Brisbane.
Stanton N.L. 1979. Patterns of species diversity in the temperate and tropical litter mites. Ecology 60: 295–304.
Stork N.E., Srivastava D.S., Watt A.D. and Larsen T.B. 2003. Butterfly diversity and silvicultural practice in lowland rainforests of Cameroon. Biodiv. Conserv. 12: 387–410.
Townes H. 1972. A light-weight malaise trap. Entomol. News 83: 239–247.
Tucker N. and Murphy T. 1997. The effect of ecological rehabilitation on vegetation recruitment: some observations from the wet tropics of North Queensland. Forest Ecol. Manage. 99: 133–152.
Vasconcelos H.L., Vilhena J.M.S. and Caliri G.J.A. 2000. Responses of ants to selective logging of a central Amazonian forest. J. Appl. Ecol. 37: 508–514.
Watt A.D., Stork N.E., McBeeath C. and Lawson G.L. 1997. Impact of forest management on insect abundance and damage in a lowland tropical forest in southern Cameroon. J. Appl. Ecol. 34: 985–998.
White F. 1983. The Vegetation of Africa. UNESCO, Paris.
Wilkening A.J., Foltz J.L., Atkinson T.H. and Connor M.D. 1981. An omnidirectional flight trap for ascending and descending insects. Can. Entomol. 113: 453–455.
Whittaker R.H. 1960. Vegetation of the Siskiyou Mountains, Oregon and California. Ecol. Monogr. 30: 279–338.
Zapfack L., Engwald S., Sonke B., Achoundong G. and Madong B.A. 2002. The impact of land conversion on plant biodiversity in the forest zone of Cameroon. Biodiv. Conserv. 11: 2047–2061.
Zar J.H. 1999. Biostatistical Analysis. Pearson Education, New Jersey.

**Appendix 1**. Arthropods included in the analysis, collected in Lama forest from 2001 to 2002.

| Coleoptera | | | | Heteroptera | | | | Other arthropods | | | |
|---|---|---|---|---|---|---|---|---|---|---|---|
| Family | Species | Individuals | Singletons | Family | Species | Individuals | Singletons | Group | Species | Individuals | Singletons |
| Carabidae | 57 | 727 | 18 | Reduviidae | 22 | 114 | 7 | Chilopoda | 7 | 47 | 2 |
| Scarabaeidae | 37 | 1444 | 4 | Lygaeidae | 17 | 132 | 4 | Diplopoda | 27 | 1055 | 2 |
| Cerambycidae | 32 | 70 | 21 | Pentatomidae | 8 | 13 | 7 | Isopoda | 14 | 1885 | 1 |
| Tenebrionidae | 29 | 119 | 15 | Coreidae | 6 | 10 | 3 | Isoptera | 6 | 387 | 1 |
| Erotylidae | 24 | 135 | 8 | Alydidae | 5 | 200 | 1 | | | | |
| Buprestidae | 21 | 137 | 8 | Plataspidae | 5 | 28 | 1 | | | | |
| Anthribidae | 17 | 70 | 7 | Aradidae | 2 | 2 | 2 | | | | |
| Elateridae | 17 | 2211 | 7 | Cydnidae | 2 | 23 | 1 | | | | |
| Scolytidae | 12 | 522 | 1 | Rhopalidae | 2 | 2 | 2 | | | | |
| Scaphidiidae | 8 | 41 | 4 | Pyrrhocoridae | 1 | 9 | 0 | | | | |
| Brentidae | 4 | 4 | 4 | Berytidac | 1 | 1 | 1 | | | | |
| Lycidae | 2 | 13 | 0 | Dinidoridae | 1 | 2 | 0 | | | | |
| Lyctidae | 1 | 1 | 1 | Largidae | 1 | 20 | 0 | | | | |
| Anobiidae | 1 | 1 | 1 | Miridae | 1 | 1 | 1 | | | | |
| Platypodidae | 1 | 3 | 0 | Tingidae | 1 | 1 | 1 | | | | |
| Bostrychidae | 1 | 1 | 1 | | | | | | | | |
| Total | 264 | 5499 | 100 | | 75 | 558 | 31 | | 54 | 3374 | 6 |

Biodiversity and Conservation (2006) 15:25–42
DOI 10.1007/s10531-004-1870-x 

# Ecosystem disturbances and diversity increase: implications for invertebrate conservation

RAFAEL DIAS LOYOLA[1,*], SOFIA-LUIZA BRITO[2] and RODRIGO LOPES FERREIRA[1]
[1]*Laboratório de Ecologia e Comportamento de Insetos, Instituto de Ciências Biológicas, Universidade Federal de Minas Gerais, Caixa Postal 486/CEP 30161-970, Belo Horizonte, MG, Brasil;* [2]*Laboratório de Ecologia do Zooplâncton, Universidade Federal de Minas Gerais, Caixa Postal 486/CEP 30161-970, Belo Horizonte, MG, Brasil;* **Author for correspondence (e-mail: rdloyola@icb.ufmg.br; Phone: + 55-31-34-992579)*

Received 22 December 2003; accepted in revised form 26 July 2004

**Key words:** Conservation, Diversity, Ecosystem disturbances, Management, Neotropics, Pantanal, Soil Arthropods

**Abstract.** The Pantanal is one of the faunistic provinces considered as a priority area for invertebrate conservation. However, it is one of the areas in Brazil where the local fauna is less assessed, thus needing more scientific information that could allow political decisions to be made regarding conservation. The continuous pressure for new pasture areas leads to improper habitat occupation and destruction, like fragmentation of forest areas in the region. Such alterations can cause different impacts on the local fauna, including the soil arthropods. The main objective of this work was to compare the morphospecies composition, diversity and density of the soil arthropod fauna between a secondary single species forest (Cambarazal) and a cultivated pasture with exotic and native grass species, using only pitfall traps as sampling method. We found a great variation on the vegetal cover among environments. A higher humidity in the forest soil was observed, as well as a greater compaction of the soil in the cultivated pasture. A total of 3635 individuals were collected, belonging to 214 different morphospecies. 139 morphospecies were collected in the forest (37% exclusive to this environment), while 134 morphospecies were collected in the cultivated pasture (35% exclusive). The diversity was higher in the forest ($H' = 1.634$) than in the cultivated pasture ($H' = 1.253$). However, considering the area as a whole (forest and pasture) the global diversity was increased. In this paper we discuss about the effects of environmental changes on soil arthropod diversity and propose a hypothetical model for invertebrate management in mosaic ecosystems.

## Introduction

The Pantanal is a vast area dominated by a complex of flora and fauna often called the Pantanal Complex (Rizzini 1997). The whole area is strongly influenced by the rain regime that generates tidal cycles and the resulting floods deposit nutrients through the silt deposits and organic detritus suspended in the water. On top of some higher formations, called 'cordilheiras', there are fields and forests never flooded before. In areas where the water runs relatively faster there are huge fields with the dominance of grass that are mainly used by the many cattle farms in the region. These fields are the native pastures. There is a dominancy of vegetal species in some areas, creating vast conglomerations as the 'Paratudal', a grouping of Paratudo (*Tabebuia aurea*), and the 'Cambarazal', a grouping of Cambará (*Vochysia divergens*) (Rizzini 1997).

Nowadays, the native pastures are very important to the economy of the Brazilian mid-west region, especially in the Cerrado areas. Native pastures occupy around 39% of this environment area, which support a livestock of 66 million individuals or almost 40% of the national livestock (Filgueiras and Wechsler 1992). It is expected that these numbers could increase to 210 million by 2010 (Meirelles 1996). The Pantanal environment is not homogeneous, just as its resources are not continuous. Therefore, native pastures could be understood as a mosaic resource (Filgueiras and Wechsler 1992).

The continuous pressure for new pasture areas leads to improper habitat occupation and destruction, like fragmentation of forest areas in the region, generating direct species extinction (Garay and Dias 2001). In many cases, deforestation occurs only by cutting down trees while in other cases it is preceded by intentional fires. Intentional fires are usual in the Cerrado region where the farmers use fire as a tool to maintain their pasture fields and to anticipate flowering, which usually occurs in the beginning of the rainy season. However, fires can change the floristic cover, decreasing the proportion of occupied amounts of trees and shrubs and increasing the herbaceous layer. Such techniques can lead to soil degradation and the appearance of uncovered areas, which strongly accelerates erosion (Silva 1996).

Most of the time, the objective of deforestation is the creation of cultivated pasture fields (in some cases with exotic species of pasture grass), because animal production in native pastures is much smaller than in cultivated fields (Filgueiras and Wechsler 1992). However, the substitution of native with cultivated pastures includes the intensive use of inputs, with high maintenance and formation costs and also causes major ecosystem alterations. The introduction of exotic species or the seed dispersion of invasive species on the pasture fields, due to the continuous movement of the cattle between the two kinds of pastures, are good examples of such alterations (Filgueiras and Wechsler 1992; Primack and Rodrigues 2001).

Such alterations can cause different impacts on the local fauna, including the soil arthropods. These communities are very special due to their role in nutrient cycling and organic matter decomposition. These organisms are responsible for the fragmentation of the accumulated litter derived from the surrounding vegetation and other available resources in the environment (Seastedt 1984; Moore et al. 1991). The species composition and structure of the arthropod community depend on many factors, such as the vegetation and soil type, local climate and microhabitat diversity (Schowalter and Sabin 1991). Besides, the microarthropod communities respond both in density and diversity to environmental changes. Such results could affect the decomposition process, thus altering the whole functioning of a specific ecosystem (Richards 1974; Primavesi 1982; Silva 1996).

There are a few studies about Pantanal's arthropods communities and these studies focus mainly on arthropod–plant interactions. Butakka and Miyazaki (1998) and Moretti et al. (2003) studied macrophytes-associated communities and Santos et al. (2003) communities that were associated with palms canopies. Matthew et al. (2003) also studied interactions between ants (competition) and between ants and phorids (parasitism). Except for the work of Santos et al. (2003), which presents more detailed description of insect communities associated with *Attalea phalerata* (Arecaceae), the majority of these studies did not aim to describe the studied community and when so, taxonomic identifications were some times restricted to an order level. Therefore, this paper presents new data about soil arthropods communities of Pantanal Matogrossense.

Our main objective was to compare the morphospecies composition, diversity and density of the soil arthropods fauna between a secondary single species forest (Cambarazal) and a cultivated pasture with exotic and native grass species. The existence of differences between humidity and soil compaction on different areas and their possible relationship with the richness and diversity of the arthropods morphospecies were also analyzed. In this paper we discuss about the effects of environmental changes on soil arthropod diversity and propose a hypothetical model for invertebrate management in mosaic ecosystems.

## Methods

### *Study area*

This work was carried out in August 2001 on the Retiro Novo farm, in the Nossa Senhora do Livramento municipality, Mato Grosso, located in an area called Pirizal (16°15′12″ S and 56°22′12″ W). There are two well defined seasons in the region: the dry period, from April to September, and the rainy period, from October to March, with a maximum rainfall corresponding to 1.300 mm. The temperature records show that the hottest month is January (29 °C average temperature) and the coldest July (22 °C average temperature) (Alho et al. 1987).

### *Procedures*

Two study areas were compared. The first is the floodable Cambará forest or Cambarazal, located on transitional areas between 'murundum' fields and clean fields. This area is characterized by the predominance of the species *Vochysia divergens* (Cambará) and cultivated pastures, with native and exotic grass species introduced after fires and deforestation. The exotic species, *Brachiaria humidicola* (Rend.) Schweich, is an perrenial, erect plant, native of East Africa (Lorenzi 2000). It is readily eaten by cattle and is highly digestible, it has high resistance to dry periods, humidity and trampling, and is very commonly used on pasture fields.

In each area a 150 m linear transect was marked, on which 15 sampling stations, 10 m apart from each other, were established. In each sampling station a pitfall trap was placed, which was made up of a 12 cm length and 9.5 cm diameter plastic receptacle, with openings at soil level. Each trap contained saturated saline solution and some detergent drops, and was exposed for a continuous period of 120 h.

To characterize the transect, an area of 4 $m^2$ was marked around each trap, where the species composition of all trees, shrubs or grasses in the area were assessed.

Soil samples were taken from five sampling stations sorted, at random, in each transect. These samples were weighted in the field, put in plastic bags and taken to the laboratory. In the laboratory the samples were dried at 70 °C for 48 h, and weighed again to measure soil humidity. The Student *t*-test was performed to verify any significant differences between the percentages of soil mass loss at each location. One soil sample on each sampling station was also collected to perform a granulometric analysis. The grain classification followed the Wentwhort scale, as proposed by Suguio (1973).

To estimate soil compaction, four measurements around each sampling station were taken using a penetrometer. The significance between the differences found on the soil compaction, at each location, was tested using the Student *t*-test. The average values of the soil penetration measures were correlated with the richness and diversity using simple linear regression.

The organisms were sorted under a stereomicroscope, identified to the possible taxonomic level and stored in 70% alcohol in the Laboratório de Ecologia e Comportamento de Insetos, Departamento de Biologia Geral, Instituto de Ciências Biológicas at Universidade Federal de Minas Gerais.

The arthropods were sorted into morphospecies, counted and diverdity index calculated, using the Shannon-Weaner index ($\log_{10}$) (Wolda 1981). For such calculations the immature forms of insects were considered distinct morphospecies (Ferreira and Marques 1998). The significance between the differences found in the diversity values of the sampled stations was tested using the Student *t*-test, modified by Hutchenson (Zar 1996). To visualize the similarity between the traps in each transect and between the sampled stations, we performed a cluster analysis based on the Jaccard similarity coefficient values and average–medium amalgamation technique (UPGMA). Classification of the families in trophic guilds was based on Borror and White (1970) and White (1983).

The verification of the spatial distribution pattern, random or aggregated, of each sampled morphospecies, was done using the Chi-Squared test ($\chi^2$). All the statistics analyses employed were based on Zar (1996).

## Results

There was a great variation on the vegetal cover along transects, on species composition and also on the structure of the environment. The forest transect showed a greater number of plant species (Table 1), although there is a

*Table 1.* Family, species, common name, character and height of plants found in the forest and cultivated pasture transects.

| Site | Family | Species | Common name | Character | Height (m) |
|---|---|---|---|---|---|
| Forest | Cecropiaceae | *Cecropia pachystachya* | Embaúba | Arboreous | 4.0–7.0 |
| | Euphorbiaceae | *Mabea* sp.1 | Mamona-do-mato | Arboreous | 4.0–8.0 |
| | | *Mabea* sp.2 | Mamona-do-mato | Arboreous | 3.0–6.0 |
| | | *Alchornea discolor* | Uva-brava | Arboreous | 4.0–7.0 |
| | Myrtaceae | *Eugenia* sp. | Pimenteirinha | Shrub | 1.0–4.0 |
| | | *Eugenia florida* | Jamelão do campo | Arboreous | 2.0–6.0 |
| | | *Campomanesia eugenioides* | – | Shrub | 2.0–4.0 |
| | Rubiaceae | *Duroea* sp. | – | Arboreous | 3.0–6.0 |
| | Vochysiaceae | *Vochysia divergens* | Cambará | Arboreous | 5.0–18 |
| Cultivated pasture | Cyperaceae | *Cyperus brevifolius* | Junquinho | Herbaceous | 0.1–0.4 |
| | Euphorbiaceae | *Phyllanthus tenellus* | Quebra-pedra | Herbaceous | 0.2–0.5 |
| | Fabaceae | *Desmodium discolor* | Amoroso | Shrub | 1.5–2.5 |
| | Onagraceae | *Ludwigia inclinata* | Cruz-de-malta | Herbaceous | 0.6–1.2 |
| | Poacea | *Axonopus purpusii* | Grama-tapete | Herbaceous | 0.1–0.3 |
| | Sterculiaceae | *Melochia arenosa* | Guanxuma | Herbaceous | 0.4–1.2 |

dominance of *Vochysia divergens*, and a great abundance of litter and other elements (e.g., falling logs and sticks, manure). However, the transect in the cultivated pasture showed homogeneity, with a vegetal cover composed specially by grass and some pioneer species (Table 1).

The percentages of soil mass loss in the forest ($X = 22.24\%$) were significantly higher than in the cultivated pasture ($X = 12.24\%$) ($t = 4.66$, $p = 0.001$), suggesting a higher humidity in the forest soil.

The granulometric analysis showed that in both areas most of the soil is composed of fine sand (0.125 mm sieve, 47.2%), followed by very fine sand (0.063 mm sieve, 23.1%).

The soil penetration measures were significantly different in the forest ($X = 4.92$ cm) than the cultivated pasture ($X = 3.78$ cm) ($t = 5.41$, $p < 0.001$), which suggests a greater compaction of the soil in the pasture. No correlation was observed between the richness or diversity of morphospecies and the average values of soil penetration.

A total of 3635 individuals were collected, belonging to 214 morphospecies. The cultivated pasture presented a significantly greater abundance of individuals ($n = 2435$, 67%) than the forest ($n = 1200$, 33%) (Table 2). In the forest, 139 morphospecies were collected, from which 80 (37%) were exclusive to this environment. However, in the cultivated pasture, a total of 134 morphospecies were collected, from which 75 (35%) were exclusive to this environment. The most abundant orders in both areas were Hymenoptera ($n = 348$ in the forest and $n = 972$ in the cultivated pastures) and Collembola ($n = 265$ in the forest and $n = 883$ in the cultivated pastures). Only 59 (28% out of 214) morphospecies

*Table 2.* Richness, abundance and trophic guild of soil arthropod fauna sampled in the Forest and Cultivated Pasture. Obs.: unidentified species represented by ' – '.

| Class | Order | Suborder or superfamily | Family | Trophic guild | Site | | | | | |
|---|---|---|---|---|---|---|---|---|---|---|
| | | | | | Forest | | | Cultivated pasture | | |
| | | | | | # Morpho | # Ind. | % Ind. | # Morpho | # Ind. | % Ind. |
| Insecta | Hymenoptera | Apoidea | Apidae | Nectarivore | 1 | 6 | 0.5 | 1 | 1 | 0.0 |
| | | | Halictidae | Nectarivore | 0 | 0 | 0.0 | 2 | 2 | 0.1 |
| | | Chalcidoidea | – | Nectarivore (parasitoid) | 7 | 30 | 2.5 | 0 | 0 | 0.0 |
| | | Chrysidoidea | Chrysididae | Nectarivore (parasitoid) | 0 | 0 | 0.0 | 1 | 1 | 0.0 |
| | | | – | Nectarivore (parasitoid) | 2 | 4 | 0.3 | 3 | 5 | 0.2 |
| | | Ichneumonoidea | Ichneumonidae | Nectarivore (parasitoid) | 1 | 1 | 0.1 | 0 | 0 | 0.0 |
| | | Scolioidea | Formicidae | Omnivore | 6 | 288 | 24.0 | 11 | 957 | 39.3 |
| | | | Mutillidae | Nectarivore | 2 | 4 | 0.3 | 1 | 2 | 0.1 |
| | | Vepoidea | Vespidae | Nectarivore (predator) | 1 | 1 | 0.1 | 0 | 0 | 0.0 |
| | | | Pompilidae | Nectarivore (predator) | 0 | 0 | 0.0 | 1 | 1 | 0.0 |
| | | – | – | – | 3 | 14 | 1.2 | 2 | 3 | 0.1 |
| | Heteroptera | Cimicoidea | Anthocoridae | Predator | 2 | 4 | 0.3 | 0 | 0 | 0.0 |
| | | Reduvioidea | Reduviidae | Predator | 1 | 1 | 0.1 | 0 | 0 | 0.0 |
| | | – | – | Predator | 1 | 3 | 0.3 | 0 | 0 | 0.0 |
| | Lepdoptera | Noctuidoidea | Noctuidae | – | 1 | 1 | 0.1 | 0 | 0 | 0.0 |
| | Homóptera | Cicadoidea | Cicadellidae | Phytophagous | 4 | 17 | 1.4 | 5 | 27 | 1.1 |
| | | | Delphacidae | Phytophagous | 0 | 0 | 0.0 | 1 | 4 | 0.2 |
| | | | – | Phytophagous | 1 | 2 | 0.2 | 1 | 1 | 0.0 |
| | | Fulgoroidea | Fulgoridae | Phytophagous | 1 | 1 | 0.1 | 0 | 0 | 0.0 |
| | | | – | Phytophagous | 1 | 1 | 0.1 | 1 | 8 | 0.3 |
| | | – | – | Phytophagous | 1 | 1 | 0.1 | 3 | 7 | 0.3 |
| | Ensifera | Grylloidea | Gryllidae | Herbivore | 1 | 13 | 1.1 | 4 | 48 | 2.0 |
| | Caelifera | Acridoidea | Acrididae | Herbivore | 1 | 2 | 0.2 | 1 | 2 | 0.1 |
| | Dictyoptera | Blattoidea | Blattidae | omnivore | 2 | 3 | 0.3 | 2 | 3 | 0.1 |
| | Coleoptera | Cucujoidea | Nitidulidae | Frugivore | 3 | 89 | 7.4 | 0 | 0 | 0.0 |

| | | | | | | | | | | |
|---|---|---|---|---|---|---|---|---|---|---|
| | | Curculionoidea | Scolytidae | Herbivore | 1 | 33 | 2.8 | 0 | 0 | 0.0 |
| | | | Curculionidae | Herbivore | 0 | 0 | 0.0 | 1 | 1 | 0.0 |
| | | Staphylinoidea | Staphylinidae | Predator | 6 | 12 | 1.0 | 1 | 8 | 0.3 |
| | | Elateroidea | Elateridae | Scavenger | 3 | 12 | 1.0 | 2 | 4 | 0.2 |
| | | Chrysomeloidea | Chrysomelidae | Herbivore | 1 | 2 | 0.2 | 3 | 3 | 0.1 |
| | | Scarabaeoidea | Scarabaeidae | Herbivore/frugivore | 3 | 3 | 0.3 | 1 | 4 | 0.2 |
| | | Carabaeoidea | Carabidae | Predator | 5 | 8 | 0.7 | 3 | 10 | 0.4 |
| | | Tenebrionoidea | Lagriidae | Herbivore | 0 | 0 | 0.0 | 1 | 1 | 0.0 |
| | | | Tenebrionidae | Herbivore | 2 | 3 | 0.3 | 0 | 0 | 0.0 |
| | | – | – | – | 9 | 14 | 1.2 | 12 | 18 | 0.7 |
| | Díptera | Brachycera | Empididae | Predator | 2 | 30 | 2.5 | 2 | 3 | 0.1 |
| | | | Asilidae | Predator | 2 | 2 | 0.2 | 0 | 0 | 0.0 |
| | | | Bombyliidae | Nectarivore | 0 | 0 | 0.0 | 1 | 11 | 0.5 |
| | | | Dolichopodidae | Predador | 0 | 0 | 0.0 | 2 | 6 | 0.2 |
| | | Nematocera | Dixidae | – | 3 | 17 | 1.4 | 2 | 2 | 0.1 |
| | | | Culicidae | Phytophagous/ hematophagous | 4 | 42 | 3.5 | 2 | 2 | 0.1 |
| | | | Cecidomyiidae | – | 4 | 33 | 2.8 | 5 | 20 | 0.8 |
| | | | Ceratopogonidae | Phytophagous | 0 | 0 | 0.0 | 3 | 5 | 0.2 |
| | | Cyclorrhapha | Phoridae | Omnivore | 3 | 75 | 6.3 | 1 | 9 | 0.4 |
| | | | Tachinidae | Nectarivore | 2 | 2 | 0.2 | 2 | 2 | 0.1 |
| | | – | – | – | 1 | 1 | 0.1 | 4 | 4 | 0.2 |
| | Psocoptera | Trogiomorpha | – | Scavenger | 2 | 34 | 2.8 | 1 | 1 | 0.0 |
| | Collembola | Entomobryomorpha | – | Scavenger | 5 | 263 | 21.9 | 6 | 870 | 35.7 |
| | | Poduriomorpha | Poduridae | Scavenger | 1 | 1 | 0.1 | 0 | 0 | 0.0 |
| | | Sminthuriomorpha | Sminthuridae | Scavenger | 1 | 1 | 0.1 | 2 | 13 | 0.5 |
| | Thysanoptera | – | – | Nectarivore/predator | 1 | 29 | 2.4 | 1 | 179 | 7.4 |
| | Isoptera | – | – | Scavenger/herbivore | 0 | 0 | 0.0 | 2 | 8 | 0.3 |
| Arachnida | Araneida | Labdognatha | Titanoecidae | Predator | 1 | 3 | 0.3 | 0 | 0 | 0.0 |
| | | | Caponidae | Predator | 1 | 1 | 0.1 | 0 | 0 | 0.0 |
| | | | Linyphiidae | Predator | 2 | 2 | 0.2 | 4 | 6 | 0.2 |
| | | | Lycosidae | Predator | 2 | 10 | 0.8 | 4 | 22 | 0.9 |
| | | | Corinnidae | Predator | 6 | 10 | 0.8 | 1 | 1 | 0.0 |

*Table 2.* Continued.

| Class | Order | Suborder or superfamily | Family | Trophic guild | Site | | | | | |
|---|---|---|---|---|---|---|---|---|---|---|
| | | | | | Forest | | | Cultivated pasture | | |
| | | | | | # Morpho | # Ind. | % Ind. | # Morpho | # Ind. | % Ind. |
| | | | Salticidae | Predator | 3 | 6 | 0.5 | 5 | 22 | 0.9 |
| | | | Gnaphosidae | Predator | 1 | 2 | 0.2 | 1 | 1 | 0.0 |
| | | | Araneidae | Predator | 1 | 1 | 0.1 | 1 | 1 | 0.0 |
| | | | Theridiidae | Predator | 1 | 1 | 0.1 | 2 | 2 | 0.1 |
| | | | Miturgidae | Predator | 1 | 1 | 0.1 | 1 | 1 | 0.0 |
| | | | Pisauridae | Predator | 1 | 2 | 0.2 | 0 | 0 | 0.0 |
| | | | Ctenidae | Predator | 1 | 2 | 0.2 | 0 | 0 | 0.0 |
| | | | Tetragnathidae | Predator | 1 | 1 | 0.1 | 0 | 0 | 0.0 |
| | | | Hahniidae | Predator | 0 | 0 | 0.0 | 1 | 1 | 0.0 |
| | | | Oonopidae | Predator | 0 | 0 | 0.0 | 1 | 1 | 0.0 |
| | | | Oxyopidae | Predator | 0 | 0 | 0.0 | 1 | 1 | 0.0 |
| | | | Philodromidae | Predator | 0 | 0 | 0.0 | 1 | 1 | 0.0 |
| | | | Zodariidae | Predator | 0 | 0 | 0.0 | 1 | 1 | 0.0 |
| | Acarina | – | – | Scavenger/predator | 12 | 54 | 4.5 | 9 | 116 | 4.8 |
| | Scorpionida | – | Bothriuridae | Predator | 0 | 0 | 0.0 | 1 | 2 | 0.1 |
| Chilopoda | Scutigeromorpha | – | – | Predator | 1 | 1 | 0.1 | 0 | 0 | 0.0 |
| TOTAL | | | | | 139 | 1200 | 100.0 | 134 | 2435 | 100.0 |

were common to both environments. Therefore, although there is no significant difference between the number of species ($t = -0.088$, $p = 0.930$), the morphospecies composition of both environments is very different.

The diversity of each environment was significantly different ($t = 3.07$, $p = 0.004$) with higher diversity values for the forest ($H' = 1.634$) than for the cultivated pasture ($H' = 1.253$) (Figure 1). The cultivated pasture also presented a lower equitability value ($E = 0.587$) when compared to the forest ($E = 0.759$). The cluster analysis showed a clear separation between the two environments and also a low similarity between them (13%) (Figure 2).

The use of pitfall traps has the disadvantage that catch rate varies with the nature of the surrounding vegetation. This is because, as with all trapping techniques, catch rates are affected by invertebrate density and activity, and vegetation in the vicinity of the trap impedes invertebrate movement. Futhermore, pitfall traps also tend to catch proportionally more large (>3 mm long) invertebrates (Sutherland 1996). This is a limitation of this study. Other

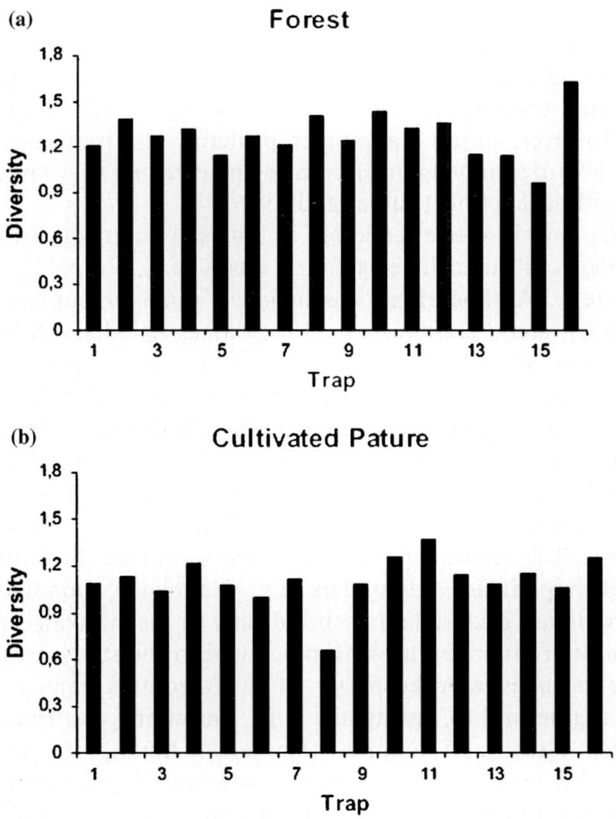


*Figure 1.* Diversity (Shannon-Weaner) of soil arthropod fauna in each pitfall trap. (a) Forest and (b) Cultivated Pasture.

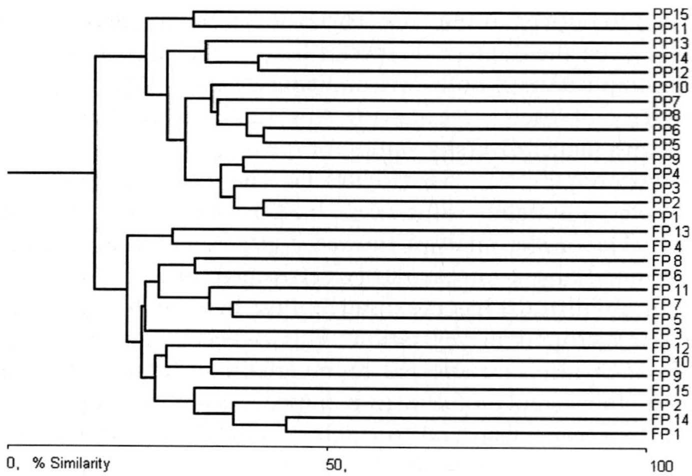


*Figure 2.* Cluster analysis of different pitfall traps placed in the Forest and Cultivated Pasture. Obs.: PP = pasture pitfall and FP = forest pitfall.

limitations are the fact that this is a one year study, with no replication and with one limited trapping period. Moreover, only one sampling method was employed. However, pitfall trapping is probably the most commonly used trapping method for studying invertebrates and catching very large numbers of individuals with little effort (Sutherland 1996).

Most of the morphospecies collected in both environments are predators of small arthropods. That includes solitary wasps (e.g., Vespidae, Pompilidae), Heteroptera (e.g., Anthocoridae, Reduviidae), some Coleoptera (Staphylinidae, Carabidae), some Diptera (e.g., Empididae, Asilidae, Dolichopodidae), some Acari and all spiders (e.g., Titanoecidae, Caponidae, Linyphiidae, Lycosidae, Corinnidae, Salticidae, Gnaphosidae, Araneidae, Theridiidae, Miturgidae, Pisauridae, Ctenidae, Tetragnathidae, Hahniidae, Oonopidae, Oxyopidae, Philodromidae, Zodariidae). Among these groups, the Heteroptera occurred exclusively on the forest, such as the Asilidae, the chilopod (Scutigeromorpha) and some spider families (e.g., Titanoecidae, Canopidae, Pisauridae, Ctenidae, Tetragnathidae). However, the scorpions (Bothriuridae), some Diptera (Dolichopodidae) and spiders (e.g., Hahniidae, Oonopidae, Oxyopidae, Philodromidae, Zodariidae) occurred only in the cultivated pasture.

The second more representative trophic guild in the study area is the detritivores, that includes morphospecies of a Coleoptera family (Elateridae), booklices (Trogiomorpha), springtails (e.g., Entomobryomorpha, Poduridae, Sminthuridae), Acari and termites (Isoptera). Among the morphospecies belonging to this guild, the termites occurred only in the cultivated pasture while the Poduridae were exclusive to the forest.

The third trophic group with the highest number of morphospecies collected is the herbivores, which includes the crickets (Gryllidae), grasshoppers

(Acrididae) and most of the Coleoptera (e.g., Scolytidae, Curculionidae, Chrysomelidae, Scarabaeidae, Lagriidae, Tenebrionidae). Among the herbivore Coleoptera some morphospecies were collected only in the forest (Scolytidae and Tenebrionidae). One morphospecies (Lagriidae) was reported only in the cultivated pasture.

Most of the frugivores occurred in the forest, and that includes only some Coleoptera morphospecies (Nitidulidae). Other frugivore beetles (Sacarabaeidae) occurred in both environments. Omnivore morphospecies occurred in both environments and include roaches (Blattidae), Diptera from the Phoridae family and ants (Formicidae). The phytophagous guild is made of some Diptera (e.g., Culicidae, Ceratopogonidae) and all Homoptera (e.g., Cicadellidae, Delphacidae, Fulgoridae). Some morphospecies occurred only in the cultivated pasture (e.g., Ceratopogonidae, Delphacidae) while others only in the forest (Fulgoridae).

The presence of nectivore arthropods is considered accidental, due to the fact that all of them are not soil arthropods. This guild includes some Hymenoptera (e.g., Apidae, Halictidae, Chalcidoidea, Chrysididae, Ichneumonidae). The Chacidoidea were collected only in the forest while the Halictidae and the Chysididae only in the cultivated pasture.

Most of the collected morphospecies showed a random pattern of spatial distribution ($n$ = 168, 77%) while 51 morphospecies (23%) showed an aggregated pattern. These morphospecies belonged mainly to the following orders: Collembola, Hymenoptera (Formicidae), Isoptera and Thysanoptera. From the morphospecies that showed an aggregated pattern, 34 occurred in the cultivated pasture while 31 occurred in the forest, with 15 morphospecies common to both environments.

## Discussion

The Pantanal Matogrossense is one of the Brazilian areas where the local fauna is less assessed and that suffers a certain anthropogenic influence (up to 32%), thus needing more scientific information that could allow political decisions to be made regarding conservation (CI 1999).

Although Pantanal is one of the faunistic provinces considered as a priority area to invertebrate conservation, only a few insect orders (Hymenoptera, Lepidoptera and Isoptera) are well documented, not considering the ecological processes (e.g., nutrient cycling) in which other invertebrate groups, like the ones sampled in this study, are involved. Data presented here are unknown and add important information about soil arthropod communities. Moreover, this study discusses about effects of disturbances on the species richness and on the composition of studied community. Besides, the studied area is classified as 'anthropogenically influenced cerrado', with no faunistic assessments (CI 1999) or Conservation Units nearby. The spatial heterogeneity of this environment, and consequently, its fragility (e.g., species can depend on one or two landscape elements to complete their life cycles) (Ingham and Samways 1996), magnifies the impacts caused by fishing, agriculture and cattle-raising. Nowadays, these activities have been gradually substituted by the intensive exploration and deforestation of natural areas (CI 1999).

The natural recolonization of the deforested area by ecological succession can be altered with the introduction of exotic species, by means of moving native species through resource competition or by human management (Primack and Rodrigues 2001). Stopping or interrupting natural recolonization can cause considerable alterations on the vegetal cover of the deforested area (Ramos and Rosa 1992), leading to a decrease in species richness and modifications to species composition of the local vegetal community, as observed in this study.

The low similarity of arthropod morphospecies reveals a clear substitution in the faunistic composition between the vegetal formations. The difference in the morphospecies composition of the communities is clearly observed in the cluster analysis where all the pitfall traps placed in the forest were separated from the ones in the pasture. This separation not only indicates a difference between the morphospecies composition of a specific area, but also, indirectly, differences between the two environments tested (Ramos and Rosa 1992).

The burning or/and removal of the vegetal cover destroy the organic matter, diminish soil infiltration capacity, lead to erosion processes and laminar hardening of the landscape (Goudie 1994; Silva 1996) compressing the soil, as observed in this study. Although there were significant differences between soil compaction and humidity in both environments, the fact that there was no correlation with richness and diversity could be explained, mostly, by the selectivity of the pitfall traps.

The high values of arthropod abundance found in the cultivated pasture could be a result of the presence of ants. Although the proximity to a colony can mask the data, similar results were found in an ant community on the central Amazon, suggesting that in cultivated pastures, as well as in disturbed areas, the abundance tends to increase (Vasconcelos 1999). However, some groups like Diptera and Coleoptera presented higher abundance values in the forest environment. These results were also found by Medri and Lopes (2001), and appear to be related to the favorable microclimatic conditions and to a greater resource demand.

The trophic web of the soil arthropods community in the sampled area is composed mainly of herbivores and detritivores. This fact can be related to the continuous contribution of the litter fall in both environments (especially in the forest), leading to very high values of abundance for such organisms, which in turn has a direct influence on predators. Therefore, the higher values of richness of spiders found in this study appear to be correlated to the higher resource availability in the environment. The relationship between the abundance of predators and detritivore organisms was also observed by Ferreira and Souza-Silva (2001). Formicidae is a very abundant group and although it belongs to the omnivore trophic guild, spiders also consume it. It is also important to point out that some groups regarded as accidental (e.g. wasps), play an important role on community structure. Adults of the orders Pompilidae and Vespidae, for instance, are predators (larvae feed upon prey) and Chalcidoidea, Chrysididae and Ichneumonidae are parasites.

The high diversity found in the forest is according to the theoretical expectation that a higher structural diversity of the environment supports higher species diversity (Pianka 1982). Forest environments, in general, provide diverse mirco-habitats which supports higher diversity of invertebrates, due to their complex structure (Elton 1973; Ferreira and Marques 1998). However, the cultivated pasture environment did not provide the same conditions to permit the establishment of a similar fauna. This was also observed in other comparative studies between cultivated *Eucalyptus* sp. forest and secondary forest areas (Ferreira and Marques 1998).

In a smaller scale, the diversity of a specific area, forest or field, decreases after the occurrence of negative environmental impacts (e.g., deforestation) (Haskell 2000; Primack and Rodrigues 2001). However, considering the area as a whole (impacted and non-impacted) the global diversity increases. This fact occurs because the impact can be negative to some species, while to others it creates additional conditions that can be used through out their life cycles (Ingham and Samways 1996). It is known that the borders of a determined habitat present higher number of species than the vegetation inside the habitat due the variety of microhabitats (e.g., increased light levels, temperature, humidity and wind) (Primack and Rodrigues 2001). Therefore, as observed in this study, the diversity of the cultivated pasture area, established after the removal of the forest, decreases. However, the total diversity of the area increases.

This is possible in Pantanal because, as a transitional area, the region shows a mosaic of terrestrial ecosystems, related mainly to the Cerrado and, partially to the Amazon forest and also aquatic and semi-aquatic ecosystems (Rizzini 1991). Once fauna and flora are strongly related, this ecosystem also shows a faunistic mosaic, with species from different environments such as the Cerrado, the Amazon Forest and the Bolivian swamps. The result is the low endemism observed in Pantanal. Therefore, the extinction and recolonization dynamics of a specific area is strongly influenced by these characteristics. In a mosaic environment, most of the species show a distribution pattern different from the vegetation spots. Only a few species of phytophagous insects or the ones directly associated to the litter have their distribution strongly connected to the vegetation (Ingham and Samways 1996). Pantanal species are more adapted to a situation of constant environmental changes and show extreme plasticity of their communities. If the conditions change, new species will appear structuring a new 'sketch' of the community, proper to the new condition. This study presents the readaptation of an important ecologic structural component: the soil community. Ecosystems lacking such characteristics do not show this increase in species numbers on a global scale, because recolonization with species from other environments is reduced and few species proper to the environment adapt to the new situation. This would be the case of the Brazilian Atlantic Forest, Amazon and Caatinga.

Based on such characteristics, the proposition of a management plan that would promote the conservation of invertebrates is of great importance. Here we propose a rotational model of succession areas, simulating three hypothetical situations (Figure 3).

## Recommendations

### *Situation I – Forest and cultivated pasture*

In a property where there is a forest area and a cultivated pasture area we recommend the establishment of a rotational model. The first step is to deforest half of the remaining forested area, where a cultivated pasture will be implemented. The second step is to establish a secondary succession area (A) in a portion of the pasture area that already exists. Such procedure would lead to an increase on the spatial heterogeneity of the area and an increase on the global invertebrate diversity. In addition, there would be a greater possibility of land use, since the trees in the succession area could be cut down, after a certain growing period, and the wood used as fuel or in the production of furniture. After that, this area would be converted back to a cultivated pasture and another area would be left for secondary succession (B). The rotation would be done in a way so that forest, cultivated pasture and successional area would always exist in the ecosystem (C). The landowner would keep the same pasture area, with less soil erosion and another income source with other uses for the land (selling coal or wood). The forest fragment would always be conserved. It would be necessary to establish a small corridor in area A that could allow the passage of cattle. In B and C, the creation of the corridor is not necessary. However, a continuous link between pastures and between the forest and the successional area is generated. This would decrease the landscape mosaic and, consequently, would not provide high diversity values, such as in A. So, the diversity would oscillate in certain periods.

### *Situation II – Forest and natural pasture*

In this situation the deforestation and/or burning of half the forested area is recommended to create a cultivated or natural pasture. As a first option (1), it could be implemented as a secondary succession area and a rotation model (like the one proposed in Situation 1) in the natural pasture. With the implementation of the cultivated pasture, the spatial heterogeneity of invertebrates would increase, since there would be four different environments on the ecosystem (forest, natural pasture, cultivated pasture and successional area). The problem with the implementation of the cultivated pasture is the loss of the

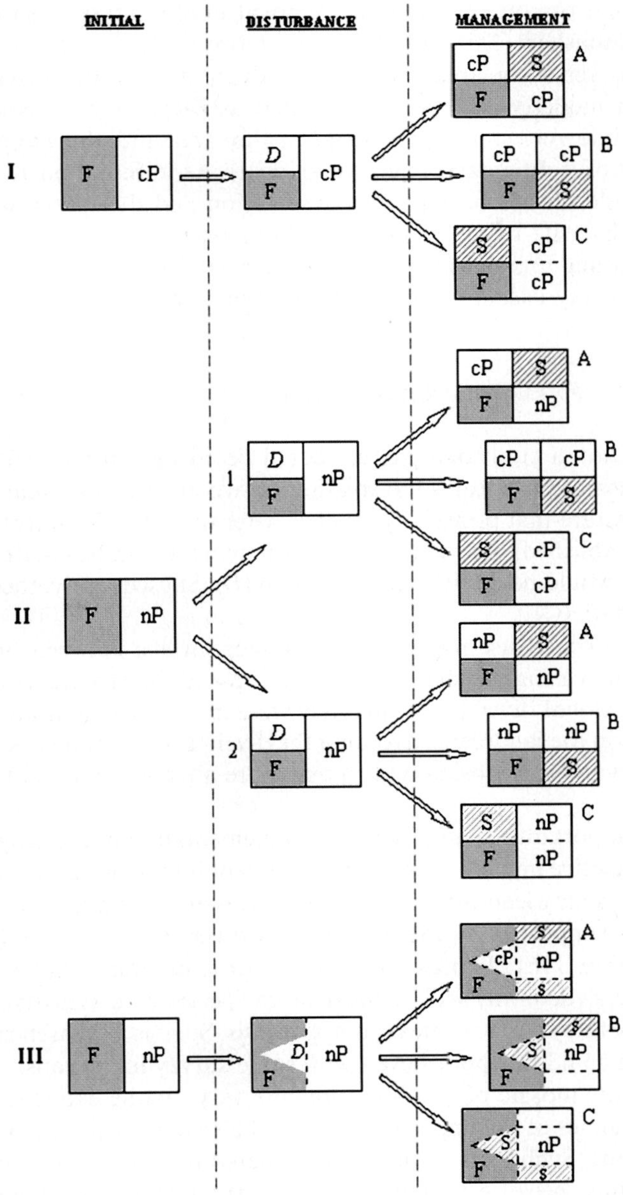


*Figure 3.* Hypothetical rotational model of succession areas for invertebrate conservation. Situations exposed by numbers: (I) Forest and cultivated pasture, (II) Forest and natural pasture, (III) Forest and cultivated pasture. See text for explanations.

natural pasture, since plants from cultivated pasture grow faster and are better competitors, with a faster establishment in the area. Still, the rotation model would favor this loss.

As a second option (2), after the removal of the forest, a natural pasture could be established. This would demand more time, although it would keep the vegetal cover of the area (with no introduction of exotic grass). After that, the rotation model would be implemented as previously proposed. In both cases, the landowner still has the same pasture area and also the possibility to explore part of it differently (e.g., wood, coal). As in Situation 1, a portion of the area would always be kept for conservation and the spatial heterogeneity and global diversity of invertebrate would increase.

It is interesting to point out that since the rotation model functions the same way, the diversity oscillation would also occur in both cases.

*Situation III – Forest and cultivated pasture*

In this situation a cultivated pasture could be created in the deforested area. The management proposal is the creation of two areas of secondary succession (A) in the pasture that previously existed. After that, a different rotation model would be established. One of the succession areas would be cut down and one pasture area would be destined to succession (B). Subsequently, these two areas would alternate (C).

This kind of land use increases the contact surface between the two environments (increasing border effects), the spatial heterogeneity and consequently, the global diversity of invertebrates. In this situation there would be no corridors in the landscape, avoiding the diversity oscillations. Also, it would contribute to less soil erosion and to one more alternative to land exploration in the area.

It is very important to emphasize that implementation of the proposed model will have a positive impact on the conservation of invertebrates, however, its use for vertebrates may lead to negative results. It is also relevant to point out that the model is supported by the microspatial diversity hypothesis (MacArthur and MacArthur 1961; Pianka 1966) which proposes that habitats which are more diverse structurally will support more species, and that the results presented in this study corroborates this hypothesis. So, it is a theoretical supported model which could be applied even with some survey limitations.

The faunistic mosaic of Pantanal presents very unique extinction and colonization dynamics of the impacted areas, and also a great plasticity to adapt to new conditions. Such facts should be considered in conservation and management programs, because the limits between the impacted and non-impacted areas are not clear and show a greater number of species than areas inside forests.

**Acknowledgments**

The authors would like to thank Professor Dr Miguel A. Marini, to Universidade Federal de Minas Gerais and to Universidade Federal do Mato Grosso for the logistical support. We also would like to thank Josué S. Nunes and Luciana Rebelatto, for their help in the field expeditions, to Éder S. S. Álvares and Éwerton O. Machado for their help identifying the Araneida, and to Luciano Cota for language improvements.

## References

Alho C.J.R., Campos Z.M.S. and Gonçalves H.C. 1987. Ecologia de capivara (Hidrochaeris hidrochaeris, Rodentia) do Pantanal: I – Habitats, densidade e tamanho de grupo. Rev. Brasil. Biol. 47(1/2): 87–97.

Butakka C.M.M. and Miyazaki R.D. 1998. Estudos preliminares da entomofauna aquática associada a Eichhornia azurea (SW) Kunth na Baía Mariana no Pantanal de Barão de Melgaço, Mato Grosso, Brasil. In: Anais do IV Simpósio de Ecossistemas Brasileiros. Academia Ciências do Estado de São Paulo, Águas de Lindóia, Brasil.

Borror D.J. and White R.E. 1970. A Field Guide to Insects of North America to Mexico. Peterson Field Guides. Houghton Mifflin Company, New York, USA.

Conservation International do Brasil 1999. Ações Prioritárias para Conservação da Biodiversidade do Cerrado e Pantanal. Brasília, MMA.

Elton C.S. 1973. The structure of invertebrate populations inside neotropical rain forest. J. Anim. Ecol. 42: 55–103.

Ferreira R.L. and Marques M.M.G.S.M. 1998. A fauna de artrópodes de serrapilheira de áreas de monocultura com Eucalyptus sp. e mata secundária heterogênea. Anais da Sociedade Entomológica do Brasil 27(3): 395–403.

Ferreira R.L. and Souza-silva M. 2001. Biodiversity under rocks: the role of microhabitat in structuring invertebrate communities in Brazilian outcrops. Biodiv. Conserv. 7(10): 1171–1183.

Filgueiras T.S. and Wechsler F.S. 1992. Aproveitamento e manejo – Pastagens nativas. In: DIAS B.F.S. (ed.), Alternativas de Desenvolvimento dos Cerrados: Manejo e Conservação dos Recursos Naturais Renováveis. IBAMA, Brasília, 96pp.

Garay I.E.G. and Dias B.F.S. 2001. Conservação da biodiversidade em ecossistemas tropicais: avanços conceituais e revisão de novas metodologias de avaliação e monitoramento. Editoras Vozes, Petrópolis, RJ, 430pp.

Goudie A. 1994. The human impact on the natural environment, 4th ed. The MIT Press, Cambridge, Massachusets.

Haskell D.G. 2000. Effects of forest roads in macroinvertebrate soil fauna in the southern Appalachian Mountains. Conserv. Biol. 14(1): 57–63.

Ingham D.S. and Samways M.J. 1996. Application of fragmentation and variegation models to epigaeic invertebrates in South Africa. Conserv. Biol. 10(5): 1353–1358.

Lorenzi H. 2000. Plantas daninhas do Brasil: terrestres, aquáticas, parasitas e exóticas, 3ª ed. Nova Odessa, SP, 608pp.

MacArthur R.H. and MacArthur J.W. 1961. On birds species diversity. Ecology 42: 594–598.

Matthew R.O., Dahlsten D.L. and Benson W.W. 2003. Ecological interactions among ants in the genus *Linepithema*, their phorid parasitoids, and ant competidores. Ecol. Entomol. 28: 203–210.

Moretti M.S., Goulart M.D.C. and Callisto M. 2003. Avaliação rápida da macrofauna associada a *Echhornia azurea* (Swartz) Kunth, 1843 e *Pontederia lanceolata* Nutt., 1818 (Pontederiaceae) na Baia do Coqueiro, Pantanal de Poconé (MT/Brasil). Rev. Brasil. Zoociências 5(1): 7–21.

Medri I.M. and Lopes J. 2001. Coleopterofauna em floresta e pastagem no norte do Paraná, Brasil, coletada com armadilha de solo. Rev. Brasil. Zool. 18(Suppl 1): 125–133.

Meirelles J.C. 1996. Pecuáriia bovídea – O desafio da produtividade. Folha da EMBRAPA, Brasília, Maio/Junho, p. 2.

Moore J.C., Hunt H.W. and Elliot E.T. 1991. Interactions between soil organisms and herbivores. In: Barbosa P., Kirschik V. and Jones C. (eds.), Multithrophic-level Interactions Among Microorganisms, Plants and Insects. John Wiley, New York, 385pp.

Pianka E.R. 1966. Latitudinal gradients in species diversity: a review of concepts. Am. Nat. 100: 33–46.
Pianka E.R. 1982. Evolutionary Ecology, 3rd ed. Harper & Row, New York, 356pp.
Primack R.B. and Rodrigues E. 2001. Biologia da conservação. Londrina, 328pp. E. Rodrigues.
Primavesi A. 1982. O manejo ecológico do solo. Livraria Nobel, SA, 514pp.
Ramos A.E. and Rosa C.M.M. 1992. Impactos de processos ecológicos – Impacto das queimadas. In: DIAS B.F.S. (ed.), Alternativas de Desenvolvimento dos Cerrados: Manejo e Conservação dos Recursos Naturais Renováveis 96p. IBAMA, Brasília, pp. 34–38.
Richards B.N. 1974. Introduction to the Soil Ecosystem. Longman Group Ltd., New York, 266pp.
Rizzini C.T. 1997. Tratado de fitogeografia do Brasil: aspectos ecológicos, sociológicos e florísticos. Âmbito Cultural Edições Ltda, 747pp.
Santos G.B., Marques M.I., Adis J. and De Musis C.R. 2003. Artrópodos associados à copa de *Attalea phalerata* Mart. (Arecaceae), na região do Pantanal de Poconé, Mato Grosso, Brasil. Rev. Brasil. Entomol. 47(2): 211–224.
Schowalter T.D. and Sabin T.E. 1991. Serrapilheira microarthropod responses to the canopy herbivory, season and decomposition is serrapilheira bags in a regenerating conifer ecosystem in Western Oregon. Biol. Fert. Soils 11: 93–96.
Seastedt T.R. 1984. The role of microarthropods in decomposition and mineralization processes. Ann. Rev. Entomol. 29: 25–46.
Silva L.F. 1996. Solos tropicais: aspectos pedológicos, ecológicos e de manejo. Terra Brasilis, São Paulo, 137pp.
Suguio K. 1973. Introdução à Sedimentologia. ed. Edgard Blucher Ltda. EDUSP, 317pp.
Sutherland W.J. 1996. Ecological Census Techniques: A Handbook. Cambridge University Press, Cambridge, 336pp.
Vasconcelos H.L. 1999. Effects of forest disturbance on the structure of ground-foraging ant communities in central Amazonia. Biodiv. Conserv. 8: 409–420.
Wolda H. 1981. Similarity indices, sample size and diversity. Oecologia 50: 296–302.
White R.E. 1983. A Field Guide to Beetles of North America. Peterson Field Guides. Houghton Mifflin Company, New York, USA.
Zar J.H. 1996. Biostatistical Analysis, 3rd ed.. Prentice-Hall, New Jersey, 718pp.

Biodiversity and Conservation (2006) 15:43–54
DOI 10.1007/s10531-004-2932-9 

# Effects of landscape elements on the distribution of the rare bumblebee species *Bombus muscorum* in an agricultural landscape

TIM DIEKÖTTER[1,]*, KERSTIN WALTHER-HELLWIG[2], MANUEL CONRADI[3], MATTHIAS SUTER[3] and ROBERT FRANKL[3]
*[1]Geobotanical Institute ETH Zurich, Zürichbergstrasse, 38, 8044 Zürich, Switzerland; [2]Department of General and Special Zoology – Animal Ecology, University Giessen, Germany; [3]Department of Biology, Nature Conservation, University Marburg, Germany; *Author for correspondence (e-mail: tim.diekoetter@env.ethz.ch; phone: +41-1-632-4588; fax: +41-1-632-1215)*

Received 24 December 2003; accepted in revised form 10 August 2004

**Key words:** Distribution, Endangered pollinators, Floral resources, Foraging behaviour, Habitat fragmentation, Spatial structure, Landscape analysis

**Abstract.** The regional distribution pattern of *Bombus muscorum* was studied in an agricultural landscape of central Germany, one of two remaining areas with the occurrence of this nationally endangered species in the Land Hesse. To determine the landscape characteristics that facilitate the occurrence of *B. muscorum*, grid-based observation records were analysed in a GIS environment at a regional scale. A significantly negative effect of the number of trees on the occurrence of *B. muscorum* and a significantly positive one of the proportion of arable land, strongly support the species' preference for open landscapes. Yet, apart from open landscapes additional landscape features were shown to be important. A significantly positive effect of ditches in the final model revealed the importance of this landscape element for the occurrence of *B. muscorum*. This finding was additionally supported by recordings of nest-searching queens, nests, and flower visits along dithes. The positive effects of clover and fallow land indicate the species' need for suitable food resources throughout the season. Because *B. muscorum* exhibits small foraging ranges, it is essential that landscape elements that provide nesting sites, foraging habitats and undisturbed hibernation structures are next to each other. The low numbers of individuals of *B. muscorum* recorded indicate that the supply of these habitat elements may have reached a critical threshold in the study region.

## Introduction

Since the intensification of agriculture during the 1950s the hitherto positive correlation between agricultural practices and species diversity became negative (Stoate et al. 2001). Declines in number of species in agricultural landscapes have been shown for plants and numerous animal groups (Sotherton and Self 1999). Pollinators, one of the most important functional groups in the landscape, are also negatively affected by modern agricultural techniques and the concurrent landscape changes (Williams 1989; Osborne and Corbet 1994; Buchmann and Nabhan 1996). In this regard, bumblebees are no exception; several species in Europe show diminishing ranges and declines in numbers (Williams 1986; Rasmont 1988; Williams et al. 1991).

The conversion from hay to silage production as well as the intensification of grassland management (Stoate 1996) has reduced suitable habitats for bumblebees. Furthermore, continuing enlargement of arable fields results in increasing fragmentation of remaining biotopes, such as hedgerows, field boundaries, ditches or path margins. The composition of the landscape is being severely altered (Meeus 1993), yet, the close proximity of certain habitats is often essential for the survival of species in a given landscape. Especially in the case of bumblebees, as central-place foragers, the spatial neighbourhood of nesting sites and foraging habitats, as well as the existence of undisturbed places for hibernation is essential (Svensson et al. 2000; Carvell 2002).

The maintenance of a diverse set of species within the taxon of bumblebees (*Bombus*) is of great value, not only from a conservational point of view, but also from an economical one. Besides many wild flowers, bumblebees pollinate numerous cultivated crops (Free 1993; O'Toole 1993; Watanabe 1994). Flowers with long corollas are especially dependent on the long-tongued species of bumblebees, such as *B. muscorum* (Rasmont et al. 1993). Whereas numerous studies have addressed foraging behaviour and distribution patterns of bumblebees at a small scale, such as within and between patches of flowers (Thomson 1996; Goverde et al. 2002), only recently has movement of bumblebees been studied at a landscape scale (Osborne et al. 1999; Walther-Hellwig and Frankl 2000; Bhattacharya et al. 2003; Kreyer et al. 2004). Analysing the effect of landscape structure on species richness and abundance of all species of bumblebees together, Steffan-Dewenter et al. (2002) did not find any significant result at neither spatial scale considered. However, it is highly probable that bumblebee species display species-specific activity ranges (Walther-Hellwig and Frankl 2000) and therefore react to the landscape structure at species-specific spatial scales.

*B. muscorum*, a species showing small activity ranges (Walther-Hellwig and Frankl 2000), occurs throughout Europe but disappeared from most of its range in the UK (Goulson 2003 and references therein) and is listed as endangered on the red list in Germany (Westrich et al. 1998).

The aim of the present study was to define the landscape characteristics that facilitate the occurrence of the critically endangered species *B. muscorum* in the 'Amöneburger Becken'. A geographic information system (GIS) was used to analyse this intensively used agricultural landscape, accommodating one of the remaining two verified populations of *B. muscorum* in the Land Hesse, Germany (Frommer 2001; Tischendorf 2001).

## Methods

### *Study area and sampling*

The study was conducted in the 'Amöneburger Becken', a basin landscape near Marburg, Hesse (Germany). Bumblebees were surveyed in an area of 60 $km^2$

(Gauss-Krueger coordinates of the centre point: 3492000, 5627500), ranging from 200 to 305 m above sea level. Ditches and brooks drain this formerly wet basin landscape, allowing for intensive agriculture on the predominant loess soils in the area. Except for small villages often surrounded by orchards, the landscape is open showing only few vertical landscape elements (Figure 1).

To create distribution maps of *B. muscorum* Reinig, a 1 km$^2$ grid was projected on the study area. Quadrats to be investigated were chosen systematically. In total, 31 out of 60 quadrats were sampled (Figure 2). Each of the sample quadrats chosen was investigated on five dates.

On each observation date, 70 min were spent in a quadrat to search for *B. muscorum*: spots of aggregated food resources were investigated for 30 min altogether, preferably six different locations were sampled for 5 min each; the remaining 40 min were available for the location of aggregated food resources along linear landscape elements (e.g. road and field margins, hedges, ditches and forest edges). Species name, sex, caste, and visited food resource as well as the quadrant (0.5 km × 0.5 km) in which a bumblebee was encountered were recorded during stationary observations and transect walks. Abundances were noted during transect walks only, to avoid double counts. Walking distances during the search for food resources were kept to around 3 km per quadrat and investigation. Landscape elements checked for bumblebees differed in width but showed a total width of around 2 m most of the time. All accessible linear

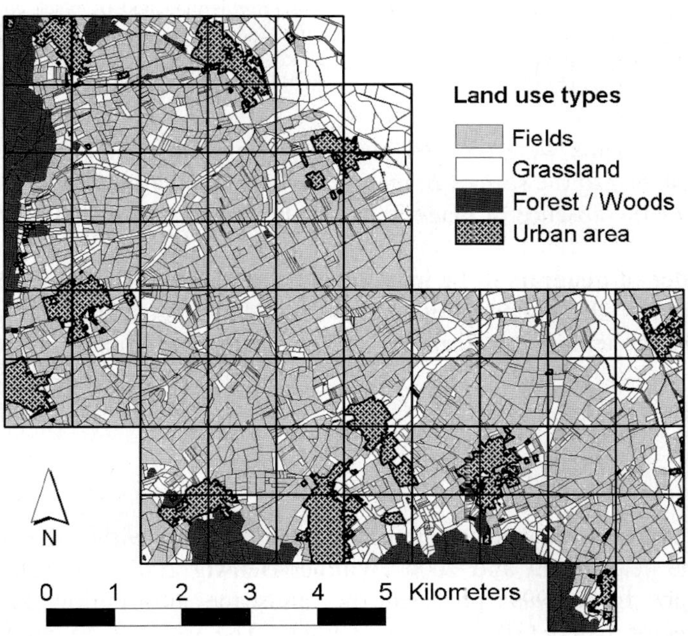


*Figure 1.* General map of the study area 'Amöneburger Becken' and distribution of main land use types.

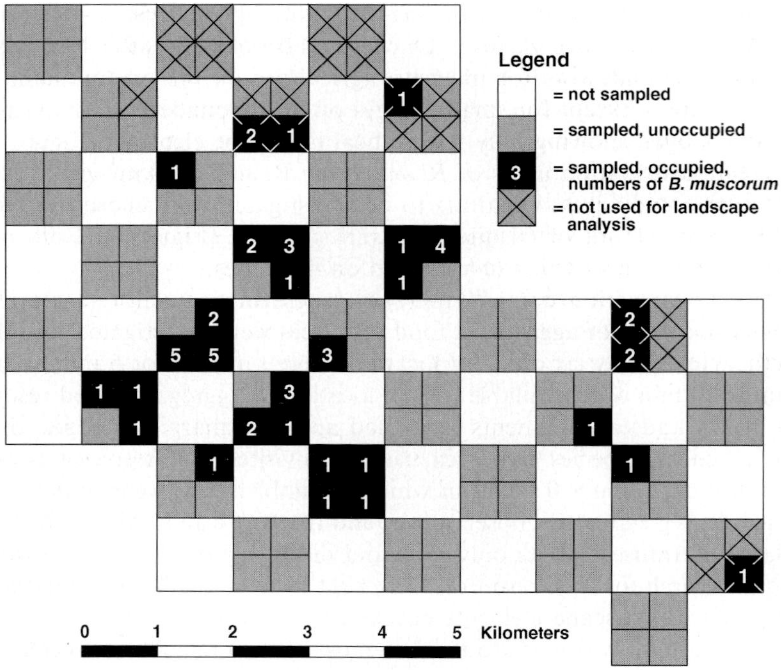


*Figure 2.* Distribution patterns of *Bombus muscorum* within the 'Amöneburger Becken'. Sampling units are underlayed in grey; presence of *Bombus muscorum* is indicated by black colour. Digits show abundances. Due to missing information for the landscape model, crossed sampling units were not used for spatial analyses.

landscape elements were covered over the whole sampling period. All individuals assigned to the species *B. muscorum* were caught with an insect net and checked for the absence of black hairs on the thorax to prevent mistake with *B. humilis*.

The order of quadrats to be investigated was changed randomly. Sampling took place from the beginning of June to the end of August 2001, between 0800 and 2000 h at temperatures above 12 °C on days without rain and stormy wind.

### *Landscape-models and analysis*

Intensive mapping of land use and landscape elements within the study area during the years 1999 and 2000 (Walther-Hellwig 2001), as well as aerial photographs from 1999, provided the landscape information used in the present vector based GIS-landscape-models. The vertical structure, the potential supply of pollen and nectar and the suitability for nesting and hibernation of landscape elements guided the model-building. Models encompassed field crops with a detailed mapping of potential food resources including orchards, semi-natural landscape elements, single trees as well as woodlands and urban structures (Table 1). The topology of woodlands and urban areas as well as of semi-natural landscape elements, such as hedgerows, banks, ditches etc. showed only marginal changes over the mapping period. Superabundant food resources such as rape, clover, beans etc. were updated by additional mapping during fieldwork.

All spatial analyses were performed in ArcView 3.2 (ESRI Geoinformatik, Hannover, Germany) enhanced by several scripts and extensions, using the 0.5 km × 0.5 km grid. Due to missing landscape information on the margins of the study area, only a subset of 100 sampling units (Figure 2) could be considered while analysing the occurrence of *B. muscorum* in relation to landscape

*Table 1.* Landscape elements mapped as polygons, lines or points.

| | LE | V |
|---|---|---|
| Polygon-Layer | [%] | |
| *Camelina sativa* (L.) Crantz | <0.1 | 1 |
| *Helianthus annuus* L. | <0.1 | 1 |
| *Phacelia tanacetifolia* Bentham | <0.1 | 1 |
| *Sinapis arvensis* L. | 0.1 | 1 |
| *Solanum tuberosum* L. | 0.4 | 1 |
| Legumes | 0.1 | 1 |
| *Pisum sativum ssp.* L. | 1.0 | |
| *Vicia faba* L. | 1.0 | |
| *Brassica napus* L. | 6.0 | |
| *Trifolium pratense* L. | 0.9 | 2 |
| *Trifolium repens* L. | 0.2 | 2 |
| Fallow land | 1.1 | |
| Arable land | 51.7 | |
| Grassland | 22.3 | |
| Hedgerows and groves | 0.5 | 3 |
| Woodland | 6.6 | 3 |
| Orchards | 0.5 | 4 |
| Rural settlements | 8.0 | 4 |
| Line-Layer | [km] | |
| Brooks | 47.9 | |
| Ditches | 18.3 | |
| Flower-rich banks | 14.8 | 5 |
| Flower-rich structures | 3.1 | 5 |
| Grassland elements | 6.8 | |
| Flower-rich hedgerows | 1.5 | 6 |
| Hedgerows | 10.3 | 6 |
| Rows of trees | 4.4 | 6 |
| Field paths | 355.9 | |
| Streets | 82.2 | |
| Point-Layer | [*n*] | |
| Trees | 2416 | |

LE = quantities of mapped landscape elements in the Amöneburger Becken; V: '#' variables assigned to new groups.

structure. The reduced area comprised only 24 of the 30 sampling units found to be occupied by *B. muscorum*. A multiple logistic regression was performed on the occurrence of *B. muscorum* in quadrants (0.5 km × 0.5 km) against the numbers of bumblebees encountered besides *B. muscorum* and the landscape variables mapped. Several landscape variables were combined in ecologically meaningful groups prior to analysis to prevent problems in convergence. The following groups were established: flower resources of rare crops (1), clover (2), woody structures (3), adjacent vertical structures (4), linear flower resources (5), and linear vertical structures (6) (Table 1). The new variable (3) was log-transformed to prevent the Hauck–Donner effect (Hauck and Donner 1977). Stepwise backward selection using the stepAIC procedure was applied for model reduction. All statistical analyses were performed using the statistical software R, version 1.7.1. (The Free Software Foundation Inc., Boston, USA).

## Results

### *Presence, absence, abundance and food resources*

During the present study 7 queens, 44 workers and 2 males of *B. muscorum* were recorded, representing 1.7% of the total number of bumblebees observed. *B. muscorum* was present in 30 of 124 quadrants (0.5 km × 0.5 km) (Figure 2). Individuals of *B. muscorum* were found visiting 13 different plant species; eight individuals were observed in flight. *Trifolium pratense* was most frequently used. Four out of 21 flower visits to *T. pratense* were recorded on agricultural cultivations (Table 2).

*Table 2.* Plant species utilized by *Bombus muscorum* and number of individuals observed on each plant species.

| Species | Number of observed individuals of *Bombus muscorum* |
|---|---|
| *Trifolium pratense* L. | 21 |
| *Trifolium repens* L. | 8 |
| *Vicia cracca* L. | 3 |
| *Lotus corniculatus* L. | 2 |
| *Lotus pedunculatus* Cav. | 2 |
| *Phacelia tanacetifolia* Benth. | 2 |
| *Stachys palustris* L. | 1 |
| *Centaurea jacea* L. | 1 |
| *Vicia sepium* L. | 1 |
| *Cirsium vulgare* (Savi) Ten. | 1 |
| *Lythrum salicaria* L. | 1 |
| *Galeopsis pubescens* Bess. | 1 |
| *Lathyrus pratensis* L. | 1 |

*Table 3.* Results of the multiple logistic regression (based on the presence of *Bombus muscorum*, numbers of individuals of other bumblebee species, and landscape composition).

| Variable | Estimate | SE | *z*-value | *p* |
|---|---|---|---|---|
| Intercept | −4.194e + 00 | 1.480e + 00 | −2.834 | 0.005 |
| Clover (2)[a] | 8.674e−05 | 5.813e−05 | 1.492 | 0.136 |
| Arable land | 1.727e−05 | 6.857e−06 | 2.519 | 0.012 |
| Fallow land | 9.305e−05 | 5.800e−05 | 1.604 | 0.109 |
| Trees | −1.002e−01 | 4.326e−02 | −2.318 | 0.020 |
| Ditches | 3.381e−03 | 1.589e−03 | 2.128 | 0.033 |

Originally, all 20 variables were fitted; the model was reduced using a stepwise backward selection with the AIC as criterion to omit terms.
[a]Clover is a combined variable, see Table 1.

### *Landscape analysis and evaluation of landscape elements*

The predominant landuse in the 'Amöneburger Becken' is arable agriculture including several crops (60.7% of study area). Furthermore, intensively used meadows can be found on 22.3% of the area, in large parts along the river Ohm. Rural settlement structures make up an area percentage of 8%. Villages are often surrounded by orchards (0.5%). Forests, mainly found at the southern and western margins of the basin cover 6.6% of the study area. Table 1 contains area percentages (plane), lengths (linear) and numbers (punctual) of all landscape elements investigated.

The stepwise backward selection for the logistic regression retained five of 20 variables that were included in the original model (Table 3). The final model contained the variables arable land and ditches that showed significantly positive effects on the occurrence of *B. muscorum*. Also the presence of clover and fallow land contributed to the occurrence of the species in a positive way. In contrast, the number of trees showed a significantly negative effect on the presence of *B. muscorum* (Table 3).

## Discussion

Most *B. muscorum* were found in the central parts of the study area, matching exactly those areas most intensively used for agricultural purposes. However, this surprising observation mirrors the distinctive habitat requirements of *B. muscorum*.

The significantly negative effect of the number of trees on the occurrence of *B. muscorum* reflects the species' preference for open landscapes (Dylewska 1957; Reinig 1970). This is supported by the observation that out of 24 analysed sampling units occupied by *B. muscorum* two only included rural settlements and none woodland. Accordingly we assume that the significantly positive effect of the landscape variable arable land on the occurrence of *B. muscorum* is not caused by the type of crop cultivated, but the absence of vertical landscape elements on the area under crop within the inner part of the study area.

The avoidance of vertical landscape structures at a regional scale mirrors the species' main distribution along coastal areas at the biogeographical scale (Reinig 1970; Wagner 1971; Peters 1972; Westrich 1990; Pekkarinen and Teras 1993; Plowright et al. 1997). Under natural conditions, *B. muscorum* probably mainly occurs in the open landscapes of coastal areas. It seems that the increase of open landscapes in the interior, due to human activities, caused an expansion of the distribution ranges of *B. muscorum*. Accordingly, *B. muscorum* has been recorded in open regions of the interior by several authors (Dylewska 1957; Reinig 1970; Westrich 1990).

However, despite the omnipresence of open landscapes, such as intensively used agricultural landscapes (Statistisches Bundesamt 2002), *B. muscorum* is regarded as an endangered species in Germany (Jedicke 1997; Westrich et al. 1998) and appears to be very rare in the interior (Wolf 1985; Hagen 1994).

Nowadays, wet lowlands seem to be the only remaining habitat in the interior suitable for *B. muscorum* (Westrich 1990; Treiber 1998). The significantly positive effect of ditches on the occurrence of *B. muscorum*, together with the frequently observed nest-searching behaviour of queens and the two findings of nest along brooks and ditches, leads us to the assumption that these landscape elements are essential to meet the habitat requirements of this bumblebee species (see also Reinig 1970). Comparably high shares of brooks and ditches occur within the central part of the formerly wet floodplain 'Amöneburger Becken' (Rittweger 1997), and are therefore regarded to be the main reason for this landscape harbouring one of the two remaining populations of *B. muscorum* in the Land Hesse (Frommer 2001; Tischendorf 2001).

In addition to their location on the north-facing slope of a small brook the observed nests of *B. muscorum* were close to fields of *T. pratense*. With regard to this observation, the positive effects of clover and fallow land, and the recordings of flower visits, an additional requirement essential for the presence of *B. muscorum* became apparent – the provision of suitable food resources in spatial proximity to nesting-sites. Long-tongued bumblebee species, such as *B. muscorum*, show comparably low competition abilities on super-abundant flower resources like rape (Heinrich 1974; Ranta and Vepsalainen 1981; Plowright et al. 1997). Furthermore, in most years there is only a small overlap of this particular resource with the seasonal occurrence of *B. muscorum*, as this species is emerging relatively late. In accordance, recorded flower visits showed that agricultural cultivations of plants with long corollas such as *T. pratense* and *T. repens* are frequently utilized by this bee. *T. pratense* is grown extensively throughout the 'Amöneburger Becken' and is harvested on demand during a great part of the growing season. Contrary to modern silage production, this technique results in a higher percentage and a greater continuity of flowering plants. However, the periodicity of agricultural resources makes alternative foraging habitats, such as fallow land, essential (Backman and Tiainen 2002; Croxton et al. 2002).

Flower visits to the important forage species *T. pratense* and *T. repens* were also frequently recorded on plants growing on taluses, path- and field-margins. *Vicia cracca*, *V. sepium* and *Lotus corniculatus* occur in these linear grassland elements, too. Furthermore, *B. muscorum* was recorded on *L. pedunculatus*, *Lythrum salicaria*, *Galeopsis pubescens* and *Stachys palustris* all growing along brooks and ditches. This shows that besides the required nesting-sites brooks and ditches also supply valuable food resources. With the exception of *Phacelia tanacetifolia*, all floral resources visited by this bumblebee species underline the importance of non-cultivated flower-rich elements (Kells and Goulson 2003) or semi-natural grasslands (Söderström et al. 2001) within the open landscape.

Continuous enlargement of agricultural fields and the disappearance of extensively used grasslands or non-crop features such as field margins and ditches (Stoate et al. 2001) result in an impoverished landscape not only in terms of nectar and pollen resources but also in terms of suitable sites for nesting and hibernation (Riemann 1987; Jennersten et al. 1993). Increasing levels of competition for the remaining resources between different species of bumblebees or within the guild of pollinators might be a consequence. This seems not yet to be the case in the study area as the present analysis did not reveal any effect of numbers of individuals of other bumblebee species recorded in the area on the occurrence of *B. muscorum*.

Actual sizes of agricultural fields below the regional average (Hessisches Statistisches Landesamt 1999) might indicate a higher proportion of edge structures within the 'Amöneburger Becken' compared to other intensively used agricultural areas in the Land Hesse. This circumstance, besides the numerous brooks and ditches draining this basin landscape, seems to be another landscape characteristic that allows *B. muscorum*, a species supposed to have comparatively small foraging ranges (Walther-Hellwig and Frankl 2000) and presumably low competitive abilities on mass-flowering crops, to still exist in this area.

Although Williams (1986, 1989) argues that the patterns of abundance of bumble species in the UK are best explained by their climatic optima and declining populations as observed in *B. muscorum* might also be affected by climatic shifts (cf. Thomas et al. 2004), we regard the continuing impoverishment of the landscape in terms of semi-natural landscape elements as one of the main factors that negatively affects population sizes and distributions ranges of *B. muscorum*. The survival of *B. muscorum* within the 'Amöneburger Becken' will be crucially dependent on the establishment and the spatial connectivity of landscape elements providing habitats for nesting, foraging and hibernating such as ditches and flower-rich fields, margins or banks.

**Acknowledgements**

This study was supported by grants of the Philipps University Marburg. We would like to thank numerous farmers for their kind cooperation in the field.

T.D. is also grateful for the support by the Geobotanical Institute ETH Zurich and valuable advices by S. Güsewell and R. Billeter. We also would like to thank two anonymous reviewers for valuable comments on an earlier version of this manuscript.

## References

Backman J.P.C. and Tiainen J. 2002. Habitat quality of field margins in a Finnish farmland area for bumblebees (Hymenoptera: *Bombus* and *Psithyrus*). Agric. Ecosyst. Environ. 89: 53–68.

Bhattacharya M., Primack R.B. and Gerwein J. 2003. Are roads and railroads barriers to bumblebee movement in a temperate suburban conservation area? Biol. Conserv. 109: 37–45.

Buchmann S.L. and Nabhan G.P. 1996. The Forgotten Pollinators. Island Press/Shearwater Books, Washington, DC, USA.

Carvell C. 2002. Habitat use and conservation of bumblebees (*Bombus* spp.) under different grassland management regimes. Biol. Conserv. 103: 33–49.

Croxton P.J., Carvell C., Mountford J.O. and Sparks T.H. 2002. A comparison of green lanes and field margins as bumblebee habitat in an arable landscape. Biol. Conserv. 107: 365–374.

Dylewska M. 1957. The distribution of the species of the genus *Bombus* Latr. in Poland. Acta Zool. Cracov. 2: 259–278.

Free J.B. 1993. Insect Pollination of Crops. Academic Press, London, UK.

Frommer U. 2001. Bestandsaufnahme der Bienenfauna im Mittleren Hessen (Hymenoptera, Apidae). Naturwissenschaftliche Veröffentlichungen Darmstadt/Bericht N.F. 24: 129–191.

Goulson D. 2003. Bumblebees. Their Behaviour and Ecology. Oxford University Press, Oxford, UK.

Goverde M., Schweizer K., Baur B. and Erhardt A. 2002. Small-scale habitat fragmentation effects on pollinator behaviour: experimental evidence from the bumblebee *Bombus veteranus* on calcareous grasslands. Biol. Conserv. 104: 293–299.

Hagen E. von 1994. Hummeln: Bestimmen, Ansiedeln, Vermehren, Schützen. Naturbuch Verlag, Augsburg, Germany.

Hauck W.W. and Donner A. 1977. Wald's test as applied to hypotheses in logit analysis. J. Am. Stat. Assoc. 72: 851–853.

Heinrich B. 1979. Bumblebee Economics. Harvard University Press, Cambridge, MA and London.

Hessisches Statistisches Landesamt 2000. Statistisches Handbuch Hessen 1999/2000. Landesamt, Wiesbaden, Germany.

Jedicke E. (ed.), 1997. Die Roten Listen: Gefährdete Pflanzen, Tiere, Pflanzengesellschaften und Biotope in Bund und Ländern. Ulmer, Stuttgart, Germany.

Jennersten O., Loman J., Møller A.P., Robertson J. and Widén B. 1993. Conservation biology in agricultural habitat islands. In: Hansson L. (ed.), Ecological Principles of Nature Conservation: Applications in Temperate and Boreal Environments. Elsevier Applied Science, London, UK.

Kells A.R. and Goulson D. 2003. Preferred nesting sites of bumblebee queens (Hymenoptera : Apidae) in agroecosystems in the UK. Biol. Conserv. 109: 165–174.

Kreyer D., Oed A., Walther-Hellwig K. and Frankl R. 2004. Are forests potential landscape barriers for foraging bumblebees? Landscape scale experiments with *Bombus terrestris agg.* and *Bombus pascuorum* (Hymenoptera, Apidae). Biol. Conserv. 116/1: 111–118.

Meeus J.H.A. 1993. The transformation of agricultural landscapes in Western Europe. Sci. Total Environ. 129: 171–190.

Osborne J.L., Clark S.J., Morris R.J., Williams I.H., Riley J.R., Smith A.D., Reynolds D.R. and Edwards A.S. 1999. A landscape-scale study of bumblebee foraging range and constancy, using harmonic radar. J. Appl. Ecol. 36: 519–533.

Osborne J.L. and Corbet S.A. 1994. Managing habitats for pollinators in farmland. Aspects Appl. Biol. 40: 207–215.

O'Toole C. (ed.), 1993. Diversity of Native Bees and Agroecosystems. CAB International, Wallingford, UK.

Pekkarinen A. and Teras I. 1993. Zoogeography of *Bombus* and *Psithyrus* in Northwestern Europe (Hymenoptera, Apidae). Ann. Zool. Fennici 30: 187–208.

Peters G. 1972. Ursachen für den Rückgang der seltenen heimischen Hummelarten (Hym., *Bombus* et *Psithyrus*). Entomol. Berichte 85–90.

Plowright C.M.S., Plowright R.C. and Williams P.H. 1997. Replacement of *Bombus muscorum* by *Bombus pascuorum* in northern Britain. Can. Entomol. 129: 985–990.

Ranta E. and Vepsalainen K. 1981. Why are there so many species – spatio-temporal heterogeneity and Northern bumblebee communities. Oikos 36: 28–34.

Rasmont P. 1988. Monographie écologique et zoogéographique des bourdons de France et de Belgique (Hymenoptera, Apidae, Bombinae). Faculté des Science Agronomique de l'Ètat, Gembloux, Belgium.

Rasmont P., Leclercq J., Jacob-Remacle A., Pauly A. and Gaspar C. 1993. The faunistic drift of Apoidea in Belgium. In: Bruneau E. (ed.), Bees for Pollination. Commission of the European Communities, Brussels, pp. 65–87.

Reinig W.F. 1970. Ökologische Studien an mittel- und südosteuropäischen Hummeln (*Bombus* Latr., 1802) (Hym., Apidae). Mitteilungen der Münchner Entomologischen Gesellschaft 60: 1–56.

Riemann H. 1987. Bienen, Wespen und Ameisen (Hymenoptera Aculeata) als Besiedler von Böschungen an tiefliegenden Entwässerungsgräben. Abh. Naturwiss. Ver. Bremen 40: 333–346.

Rittweger H. 1997. Spätquartäre Sedimente im Amöneburger Becken – Archive der Umweltgeschichte einer mittelhessischen Altsiedellandschaft. In: Hermann F.-R. (ed.), Materialien zur Vor- und Frühgeschichte von Hessen. Selbstverlag des Landesamt für Denkmalpflege, Wiesbaden, Germany.

Söderström B., Svensson B., Vessby K. and Glimskär A. 2001. Plants, insects and birds in semi-natural pastures in relation to local habitat and landscape factors. Biodiv. Conserv. 10: 1839–1863.

Sotherton N.W. and Self M.J. 1999. Changes in plant and arthropod biodiversity on lowland farmland: an overview. In: Aebischer N.J., Evans A.D., Grice P.V. and Vickery J.A. (eds), Ecology and Conservation of Lowland Farmbirds. British Ornithologists' Union, Tring, pp. 26–35.

Statistisches Bundesamt (ed.), 2002. Statistisches Jahrbuch 2002 für die Bundesrepublik Deutschland. Neuster Verlag, Stuttgart, Germany.

Steffan-Dewenter I., Munzenberg U., Burger C., Thies C. and Tscharntke T. 2002. Scale-dependent effects of landscape context on three pollinator guilds. Ecology 83: 1421–1432.

Stoate C. 1996. The changing face of lowland farming and wildlife, Part 2 1945. Brit. Wildlife 7: 162–172.

Stoate C., Boatman N.D., Borralho R.J., Carvalho C.R., de Snoo G.R. and Eden P. 2001. Ecological impacts of arable intensification in Europe. J. Environ. Manage. 63: 337–365.

Svensson B., Lagerlöf J. and Svensson B.G. 2000. Habitat preferences of nest-seeking bumble bees (Hymenoptera: Apidae) in an agricultural landscape. Agric. Ecosyst. Environ. 77: 247–255.

Thomas C.D., Cameron A., Green R.E., Bakkenes M., Beaumont L.J., Collingham Y.C., Erasmus B.F.N., de Siqueira M.F., Grainger A., Hannah L., Hughes L., Huntley B., van Jaarsveld A.S., Midgley G.F., Miles L., Ortega-Huerta, M.A., Peterson A.T., Phillips O.L. and Williams S.E. 2004. Extinction risk from climate change. Nature 427: 145–148.

Thomson J.D. 1996. Trapline foraging by bumblebees. I. Persistence of flight-path geometry. Behav. Ecol. 7: 158–164.

Tischendorf S. 2001. Wildbienen und Wespen (Hymenoptera: Apidae) im Oberrheinischen Auwaldgebiet "Kühkopf-Knoblauchsaue" (Hessen). Hessische Faunistische Briefe 20: 21–42.

Treiber R. 1998. Verbreitung und Ökologie der Hummeln (Hymenoptera: Apidae) im Landkreis Freudenstadt (Baden-Württemberg). Mit. bad. Landesver. Naturkunde und Naturschute, N.F. 17: 155–180.

Wagner R. 1971. Die Veränderung der Hummelfauna Cuxhavens in diesem Jahrhundert. Der Versuch einer Deutung. Entomologische Mitteilungen des Zoologischen Museums Hamburg 4: 207–232.

Walther-Hellwig K. 2001. Untersuchungen zu Sammelflugweiten und Dichteverteilung von Honigbienen. Unveröffentlichte Auftragsarbeit für den Landesverband Hess. Imker e.V., Kirchhain in Zusammenarbeit mit d. Hess. Dienstleistungszentrum f. Landwirtschaft, Gartenbau u, Gartenbau u. Naturschutz (HDLGN), Kirchhain, Germany.

Walther-Hellwig K. and Frankl R. 2000. Foraging habitats and foraging distances of bumblebees, *Bombus* spp. (Hym., Apidae), in an agricultural landscape. J. Appl. Entomol. 124: 299–306.

Watanabe M.E. 1994. Pollination worries rise as honey bees decline. Science 256: 1170.

Westrich P. 1990. Die Wildbienen Baden-Württembergs. Ulmer, Stuttgart, Germany.

Westrich P., Schwenninger H.R., Dathe H.H., Riemann H., Saure C., Voith J. and Weber K. 1998. Rote Liste der Bienen (Hymenoptera: Apidae). Schriftenreihe für Landschaftspflege und Naturschutz 55: 119–129.

Williams P.H. 1986. Environmental change and the distributions of British bumble bees (*Bombus* Latr). Bee World 67: 50–61.

Williams P.H. 1989. Why are there so many species of bumble bees at Dungeness. Bot. J. Linn. Soc. 101: 31–44.

Williams I.H., Corbet S.A. and Osborne J.L. 1991. Bee-keeping, wild bees and pollination in the European Community. Bee World 72: 170–180.

Wolf H. 1985. Veränderungen der Hummelfauna (Hymenoptera: Apidae) bei Frankfurt (Main) und Marburg (Lahn). Hessische Faunistische Briefe 5: 66–69.

Biodiversity and Conservation (2006) 15:55–68
DOI 10.1007/s10531-004-2177-7


# How ant nests increase soil biota richness and abundance: a field experiment

APRIL M. BOULTON[1,*] and KEITH D. AMBERMAN[2]
[1]*Department of Environmental Science and Policy, University of California, Davis, California 95616;* [2]*Veterinary Hospital, University of Pennsylvania, Philadelphia, Pennsylvania 19103;* **Author for correspondence. Present address: Department of Biology, Villanova University, Villanova, PA 19085 (e-mail: amberman@peoplepc.com)*

Recieved 24 February 2004; accepted in revised form 2 August 2004

**Key words:** Abundance, Ant nests, Field experiment, MANOVA, *Messor andrei*, Richness, Soil foodwebs

**Abstract.** Although many studies have shown that ant nests tend to increase soil nutrient concentrations, only a few have examined ant impact on soil biota. To date, no one has examined the mechanism behind this complex 'ant effect.' In this study, we employed a 2 × 2 complete factorial design (water × food) in the field to mimic the effects of harvester ant nests (*Messor andrei*) on soil. We hypothesized that, in the absence of ants, addition of moisture and food (seeds and insects) would interact to produce conditions found in ant nests. Our results indicated that the addition of food to the soil (regardless of water addition) best mimicked the conditions found inside *M. andrei* nests. Both food-treated and ant-nest soils supported higher numbers of bacteria, nematodes, miscellaneous eukaryotes, and microarthropods compared to the other soil treatments. Microbial richness was also highest in ant and food-treated samples. Moreover, the ant effect in our experiment occurred in just two months. Because ants are a widespread, abundant group with many long-lived species, they could substantially influence soil properties and belowground food webs and may have important restoration/conservation implications for terrestrial communities.

## Introduction

Few studies have examined factors affecting species richness for terrestrial microbiota, and soil biota diversity has largely gone unstudied for multiple subgroups (e.g., Boag and Yeates 1998). Major factors affecting the diversity and/or abundance of belowground biota include soil nutrients, moisture, and temperature (Campbell and Biederbeck 1976), physical soil disturbances (Doran 1987), and interactions among fauna (Beare et al. 1992; Wagner et al. 1997; Laakso and Setälä 1998). Ants (Hymenoptera: Formicidae) are a major structuring force in many terrestrial communities worldwide and have various functional roles, such as scavenging, predation, granivory, and omnivory. Because ants belong to a number of guilds and interact with many different taxa ranging from plants to insects to vertebrates, they play a prominent role in structuring diversity and abundance of other taxa in many communities, such as soil biota (Whitford 1996; Folgarait 1998).

Most ants nest in the soil and may affect soil biota via numerous pathways. For instance, ant activity and respiration increase moisture and temperature in the surrounding soil (Cole 1994; Whitford 1996). Ground-nesting ants increase soil nutrients by carrying aboveground, nutrient-rich material several centimeters belowground (Friese and Allen 1993; Folgarait 1998). Ants also build belowground galleries and tunnels, thereby disturbing and creating new soil structure (Cole 1994). Finally, ants directly interact with soil biota through predation and commensalism (Laakso and Setälä 1997, 1998).

Although belowground and aboveground communities are tightly linked through plants, earthworms, and insect larvae (e.g., Strong et al. 1996; Whitford 1996; Mikola et al. 2001; Preisser 2003; Zak et al. 2003), the two systems are spatially distinct, and ants may be critical in moving aboveground resources to belowground consumers. Unlike other soil-nesters (e.g., termites), most ants are omnivorous and exploit a variety of materials such as seeds, plant tissue, and insect carcasses (Friese and Allen 1993; Whitford 1996). This is especially true for the so-called granivorous ants, which, in addition to seeds, consume an assortment of resources, such as soft- and hard-bodied insects and bird and mammal feces (e.g., MacKay 1981; Hölldobler and Wilson 1990). Ants in the genus *Messor* are major insect granivores and are widespread in arid and semi-arid regions throughout the world (Whitford 1996). Individual colonies of *Messor* are long-lived and can thrive for up to 10 years in a single location (Hölldobler and Wilson 1990, but see also Brown 1999). Harvester ant nests can be one or more meters deep (e.g., MacKay 1981), and an average *Messor andrei* nest is approximately 60 cm wide on the surface (pers. observ.).

We recently reported that *M. andrei* increases abundance and richness of multiple soil taxa and concentrates N, P, and organic matter (OM) in their nests (Boulton et al. 2003). Other researchers have observed similar trends for other ant species (Wagner et al. 1997; Laakso and Setälä 1998; Folgarait 1998). However, we lack experimental evidence for the mechanism behind these 'ant effects,' which could be due to any number of factors, such as ant predation, food storage, excretion/elimination, soil structure, and other ant behaviors or nest qualities. Of all these factors, food and moisture additions to soil are the most reasonable to test experimentally. It would be difficult to mimic ant-nest structure or to add/subtract ant nests (the latter would involve an insecticide, which could damage the soil food web as well). Thus, in this study, we attempted to mimic the effects of *M. andrei* nests on soil and its biota by employing a 2 × 2 complete factorial design (water × food). We hypothesized that, in the absence of ants, moisture, and food additions would interact to produce the 'ant effect' by mimicking many of the qualities found in ant nests.

## Methods

### *Study site*

Our field experiment was carried out from April–June 2001 in northern California at the McLaughlin Natural Reserve (Napa, Lake, and Yolo

Counties). McLaughlin has a high percentage of serpentine soil, which is characteristically high in magnesium and other heavy metals and low in calcium (UC NRS 2000). The Reserve has a Mediterranean climate – hot, dry summers and cool, wet winters. Summer air temperatures can be as high as 40 °C, while winter temperatures can fall below freezing. Mean annual precipitation is 75 cm. The Reserve lies within the California Floristic Province and supports serpentine mixed chaparral, cypress chaparral, and grasslands (UC NRS 2000). Ten *M. andrei* nests, located on serpentine grassland, were selected for the experiment. These nests were at least 5 m from one another.

*Mimicking food and water inputs of ants*

The most important component of our field experiment (described in detail below) was the food (seeds and insects) and water additions to the soil. In order to approximate the amount of food added to the soil by *M. andrei* on a daily basis, we collected foragers returning to these same study nests during spring 2000 in order to describe the variety and types of food returned to their nests. The analysis of food items carried by these workers was used to determine what should be added as experimental food. Because we could not imitate ant-nest structure (which includes compartments for larvae, food storage, etc.), we mixed our food additions with non-ant soil (hereafter referred to as 'implant soil'). Most items (83%) returned to the nest by workers were seed material (mean diameter of seeds 1.8 ± 5.9 mm), 13% were insect carcasses or parts, and 4% were leaf material or unidentifiable. There was roughly a 1:6 insect:-seed ratio in our samples, which we mimicked using commercial poppy seeds and crickets. However, the weight of all insects returned to nest (2.9 ± 4.3 g) often equaled the weight of all seeds returned to the nest (3.7 ± 1.8 g) per day. The average weight of all items returned per day to *M. andrei* nest was 6.9 ± 3.1 g. Based on these weight results, we added 8 g of seeds and 6 g of insects to each implant core. Because we were not able to add food continuously to these cores (like the ants do on a daily basis), we doubled the weight of our seeds and insects. Our implant cores were in place for 60 d, but they had a small volume compared to the volume of an entire *M. andrei* nest. For this reason, we did not multiply the original seed and insect weights by 60, which would have mimicked the daily input of *M. andrei* ants over 60 d.

Poppy seeds were selected for our food additions based on their diameter (~1 mm), which approximated the mean diameter of seeds collected from *M. andrei* foragers returning to the nest. The native California poppy, *Eschsholtzia californica*, also occurs throughout McLaughlin (UC NRS 2000) and is a possible seed source for *M. andrei* in nature. Because many of the seeds and insect carcasses were often damaged or carried back to the nest in parts, experimental food was masticated in a food processor after being microwaved for 3 min – both processes prevented the commercial seeds from germinating in the field. Crickets were used for the insect additions and were frozen for 24 h before being homogenized in a food processor.

The amount of water added to the water and food + water cores was based on saturation of the core (i.e., 50% volumetric water content). In several pilot cores, approximately 8 ml of water was sufficient to saturate a core. Eventhough ant nests are rarely fully saturated, saturation was selected for these treatments for two reasons. First, cores were visited weekly, so our water additions had to last longer than 1 or 2 days. Second, implant soil lacked the structure of non-implant soil due to removal and mixing procedures. Unstructured soils do not hold water as well as normal soil (Brady and Weil 1996), which further necessitated over-watering. However, our homogenized, structureless soil implants resembled ant-nest soils because ant nests also contain unstructured soil due to ant tunneling.

*Experimental design*

We manipulated soil moisture and food additions in soil 3 m away from each nest. Five soil cores were removed at each of the ten nest sites using a bulk density sampler (5 cm × 30 cm). We took one core from the nest center and four cores that were each 3 m from the nest and 1 m from each other. Each core was assigned to one of five treatments: ant nest (no additions), control (no additions), water-addition, food-addition, and food + water-addition cores (see Figure 1). We chose to space our 'satellite' cores 3 m from the nest because ant nests can influence the surrounding soil up to 1–2 m away from the nest center (e.g., Whitford and DiMarco 1995; Dean et al. 1997). Previous work in this system also supports these findings (Boulton et al. 2003), so placing our experimental cores 3 m from the nest center should have constituted non-ant areas.

After the soil core was removed from each site, the soil was mixed and all visible organic matter (e.g., adult ants, pupae, larvae, leaf or seed material) was

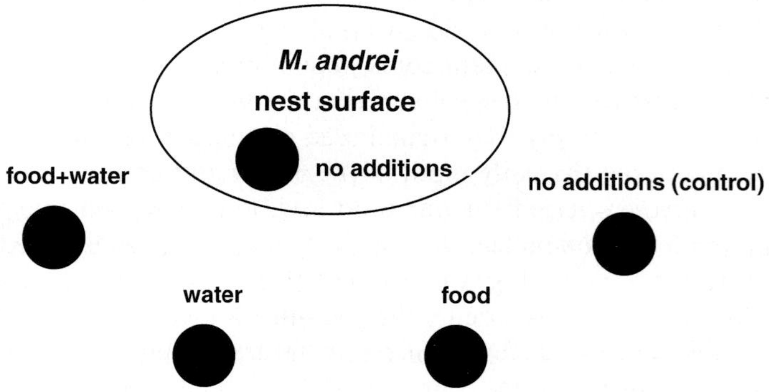


*Figure 1.* Diagram represents experimental design in the field. Black circles represent manipulated soil cores. Each of the four non-ant cores is 1-m from each other and 3 m from the ant-nest center.

removed and placed in a vial for later analysis. We then collected soil 100 m from our study site, removed visible organic matter, mixed, and implanted it into each of the five vacant cores described above (ant-nest core plus four treatment cores) for a total of 50 experimental cores (10 nest sites times 5 cores/nest site). The soil used from this adjacent site was also analyzed ($N = 3$) for chemistry and microbiota (as described below) to ensure that it was typical, non-ant soil. Hereafter, we use the term 'implant soil' to denote soil that was experimentally treated and used to replace original soil in and near each ant nest.

Water-addition cores at each site received 8 ml of tap water after the soil was implanted into the core. As described in the previous section, food-addition cores were mixed with masticated poppy seeds and homogenized cricket prior to implantation. Food + water cores were a combination of these two core treatments (i.e., mixed with seeds and crickets pre-implantation, as well as watered post-implantation). Soil temperature and moisture were monitored weekly for all implants. The water and food + water treatments received 8 ml of water whenever their volumetric soil content fell below the levels found in the adjacent ant nest. After 60 days, we retrieved each core with the bulk density sampler.

### *Soil attributes*

Before sampling, soil temperature and volumetric water content was recorded at 20 cm depth. Water content was measured using the HydroSense soil probe (Campbell Scientific, Inc.). In the laboratory, 0.5 $cm^3$ of soil from the collected core was removed in order to measure its pH using pH indicator paper (LaMotte Soil pH Kit). In the laboratory, a sub-sample (~25 g) of each soil core was passed through a 2-mm mesh sieve, dried, and transported to the Division of Agriculture and Natural Resources (DANR) Analytical Lab at the University of California, Davis for analysis of soil N, P, and OM.

### *Soil biota*

We quantified soil biota richness and abundance following methods described in detail in Boulton et al. (2003). For each soil core, abundance and richness was determined for bacteria, fungi, and other eukaryotes (e.g., ciliates) using phospholipid fatty acid (PLFA) analysis. Nematodes were extracted from soil (~10 g per sample) using Baermann funnels, and the entire suspension was examined at 140× magnification with a dissecting microscope. Each nematode was identified to feeding guild via mouthpart morphology (Yeates et al. 1993; Bongers and Bongers 1998; Jaffee et al. 1998), which is as effective as high-resolution taxonomy in characterizing food web structure (Parmelee et al. 1995). Finally, microarthropods were extracted from soil (~30 g per sample) using Tullgren funnels and were identified as mites, collembolans, or miscellaneous microarthropods (e.g., proturans, larvae, or unidentifiable) under 120× magnification.

*Data analysis*

Data were analyzed using one-way MANOVA with soil treatment as the categorical, independent variable (i.e., ant, control, food, water, and food + water). We used Tukey's test for *a posteriori* comparisons to explore the significant differences across dependent variables due to soil treatment (Sokal and Rohlf 1995). For example, a significant difference between ant soil and another soil category indicated dissimilarity, while no significant difference between ant soils and a given soil category suggested that the two soils were similar to each other for that particular response variable. In order to meet parametric assumptions and to use standard units across dependent variables, all data were transformed into their standard normal deviates, $(Y_i - \mu)/\sigma$. When non-transformed means are reported for a factor, they are followed by their standard deviation. We used the statistical package SPSS (version 10.0.7) for all above analyses. Principal component analysis (PCA) was run using PC-ORD, version 4.0.

## Results

*Pre-treatment soils*

The implant soil was typical of non-ant soil found at this site in that N, P, and OM were significantly reduced and soil taxa were less abundant compared to ant soil. The four non-ant, pre-treatment soil cores ($N = 40$) taken near each nest did not significantly differ from one another in chemical or biological properties. One-way analysis of variance tests on the non-ant, pre-treatment soils yielded insignificant differences for all abiotic and biotic dependent variables (for all tests, df $= 39$, $p > 0.05$). In comparisons between ant and non-ant soil, there were significantly more bacteria, fungi, nematodes, miscellaneous eukaryotes, PLFAs, and microarthropods and higher concentrations of N, P, and OM in ant-nest cores than in non-ant soils (data not shown).

*Post-treatment soils*

In general, ant, food, and food + water soils resembled each other with higher concentrations of N, P, and OM and with more types and abundance of bacteria, miscellaneous eukaryotes, nematodes, and microarthropods (Table 1). Fungal abundance and soil moisture were the only two dependent variables that did not show this trend. The MANOVA analysis also indicated a significant multivariate effect of soil treatment on the dependent variables ($F_{52} = 7.2$, $p = 0.0001$; Table 2).

The *a posteriori* comparisons consistently revealed significant differences between ant-nest samples and control and water-addition cores, while ant,

*Table 1.* Results of one-way MANOVA comparing the five soil categories post-treatment: ant, control, water-addition, food-addition, food + water-addition.

| Effect | Hypothesis df | Error df | *F* | *p* |
|---|---|---|---|---|
| Intercept | 13.0 | 33.0 | 178.5 | 0.0001 |
| Treatment | 52.0 | 129.9 | 7.2 | 0.0001 |

| Variable | Ant-soil core | Control core | Food + water-addition core | Food-addition core | Water-addition core | *F* | *p* |
|---|---|---|---|---|---|---|---|
| Temperature (°C) | 26.1 ± 2.2 | 28.4 ± 1.2 | 28.4 ± 1.8 | 28.1 ± 1.7 | 28.8 ± 1.1 | 5.1 | 0.002 |
| Moisture (%) | 5.7 ± 3.3 | 5.7 ± 9.5 | 4.8 ± 9.2 | 5.3 ± 1.5 | 6.4 ± 1.8 | 1 | 0.444 |
| pH | 6.6 ± 0.4 | 7.0 ± 0.0 | 7.5 ± 0.4 | 7.1 ± 0.2 | 7.0 ± 0.0 | 11.5 | 0.0001 |
| Nitrogen (%) | 0.4 ± 0.1 | 0.1 ± 0.0 | 0.4 ± 0.1 | 0.3 ± 0.1 | 0.1 ± 0.0 | 66.3 | 0.0001 |
| Phosphorous (%) | 26.2 ± 28.4 | 3.5 ± 0.7 | 63.5 ± 47.1 | 42.6 ± 46.7 | 2.8 ± 0.8 | 45 | 0.0001 |
| Organic matter (ppm) | 4.1 ± 0.9 | 2.1 ± 0.2 | 4.6 ± 0.6 | 4.2 ± 0.5 | 2.0 ± 0.2 | 47.9 | 0.0001 |
| Bacteria abundance | 99.8 ± 25.2 | 51.7 ± 9.3 | 131.0 ± 26.3 | 122.0 ± 36.4 | 52.4 ± 12.1 | 24.5 | 0.0001 |
| Fungal abundance | 9.5 ± 4.3 | 5.8 ± 2.3 | 1.6 ± 2.6 | 8.0 ± 21.0 | 6.0 ± 2.4 | 0.9 | 0.454 |
| Misc. eukaryote abundance | 3.7 ± 0.7 | 1.2 ± 0.8 | 3.8 ± 1.4 | 4.0 ± 2.9 | 1.2 ± 0.5 | 8.2 | 0.0001 |
| PLFA richness | 59.9 ± 5.3 | 44.4 ± 3.7 | 59.4 ± 6.0 | 58.2 ± 6.1 | 43.7 ± 3.9 | 21.2 | 0.0001 |
| PLFA abundance | 102.5 ± 24.9 | 46.2 ± 10.5 | 121.1 ± 21.6 | 114.4 ± 34.4 | 46.9 ± 13.0 | 25.4 | 0.0001 |
| Nematode abundance | 144.1 ± 111.0 | 80.0 ± 27.7 | 1577.2 ± 858.7 | 181.3 ± 98.3 | 60.0 ± 22.7 | 28.2 | 0.0001 |
| Microarthropod abundance | 18.7 ± 7.7 | 3.1 ± 5.4 | 11.7 ± 7.2 | 12.8 ± 11.2 | 3.5 ± 3.3 | 10.2 | 0.0001 |

*Notes*: Although non-transformed means and standard deviations are listed, all data were transformed to their standard normal deviates for the analysis in order to meet parametric assumptions. Upper part of table refers to multivariate tests, while the lower part lists the results from univariate *F* tests for each of the dependent variables. For all variables, the total degrees of freedom = 49. Abundance refers to number of individuals per sample (sample size defined in Methods).

*Table 2.* Results of *a posteriori* comparisons from MANOVA results, which examine the similarities and differences between ant cores and each experimental treatment.

| Dependent variable | Ant vs. control | Ant vs. food + water | Ant vs. food | Ant vs. water |
|---|---|---|---|---|
| Temperature (°C) | −0.26* | −0.30* | −0.34* | −0.26* |
| Moisture (%) | 0.00 | 0.08 | 0.03 | −0.06 |
| pH | −0.86* | −1.62* | −0.95* | −0.82* |
| Nitrogen (%) | 0.59* | −0.14 | −0.36 | 0.66* |
| Phosphorous (%) | 1.36* | −0.75 | −0.54 | 1.03* |
| Organic matter (ppm) | 0.51* | −0.13 | −0.05 | 0.56* |
| Bacteria abundance | 0.65* | −0.56* | −0.39 | 0.65* |
| Fungal abundance | 0.39 | 0.79 | 0.31 | 0.44 |
| Misc. eukaryote abundance | 1.44* | −0.65 | −0.36 | 1.36* |
| PLFA richness | 0.57* | 0.33 | 0.09 | 0.81* |
| PLFA abundance | 0.60* | −0.45 | −0.42 | 0.64* |
| Nematode abundance | 1.13* | −2.85* | −0.07 | 1.15* |
| Microarthropod abundance | 1.45* | 0.90* | 0.08 | 1.50* |

*Notes*: This is a partial listing of MANOVA unplanned comparisons for soil category using the Tukey test on mean differences. Abundance refers to number of individuals per sample (sample size defined in Methods). Mean differences are reported only for ant soil vs. all other soil types. The standard normal deviates were used for this analysis.
$*p < 0.05$.

food + water, and food cores generally similar (Table 2). This finding applied to N, P, and OM, miscellaneous eukaryotes, and PLFA richness and abundance. Exceptions to this result are as follows. Ant-nest cores were significantly different from all other soil treatments for soil temperature (cooler in ant nests vs. all other soil categories) and pH (more acidic in ant vs. non-ant soils). Soil moisture and fungal abundance did not differ across any of the five soil categories. The *a posteriori* results revealed that ant cores were similar to food-addition cores, but significantly different from food + water-addition cores for bacteria, nematode, and microarthropod abundance (Table 2).

Because there was collinearity among response variables, we performed a PCA (Tabachnick and Fidell 1996). The first two eigenvalues explained the majority of the variance in the data (58.8%). Axis one was composed of bacterial abundance, PLFA richness and abundance, and N, P, and OM. For axis two, nematode and fungal abundance and soil pH and temperature loaded high. The bivariate plot suggests two groups, similar to the MANOVA findings above: ant-nest soils group loosely with food- and food + water-addition soils, while the control and water-addition cores form a separate, tight cluster (Figure 2).

*Food web characteristics in post-treatment soils*

Based on our PLFA results, our post-treatment samples included all the major bacterial subgroups, such as methyl-, saturated-, unsaturated-, iso-, anteiso-, and branched-bacteria (Bossio and Scow 1998). One-way ANOVAs indicated

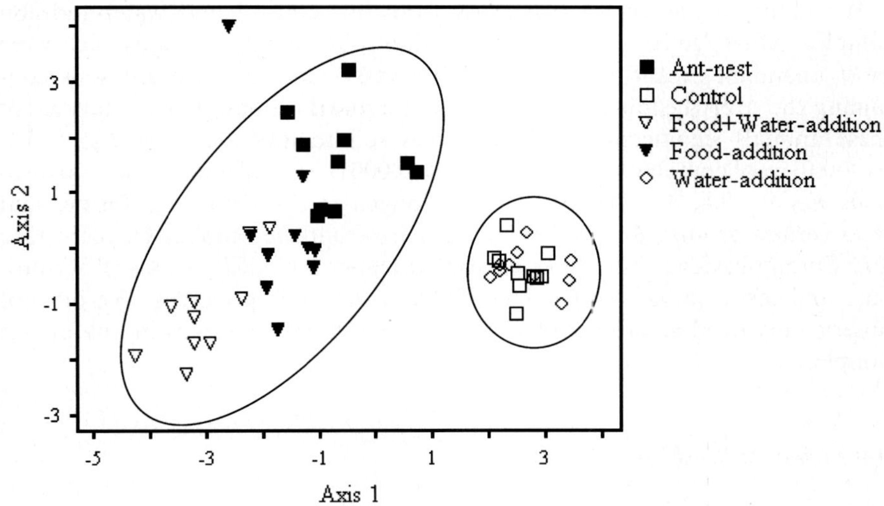


*Figure 2.* Scatter plot of factor scores for the 50 samples according to a PCA of 10 response variables.

that bacterial subgroups were significantly different across soil treatment: saturated $F_{49} = 11.8$, $p < 0.0001$; unsaturated $F_{49} = 5.6$, $p < 0.001$; iso $F_{49} = 29.4$, $p < 0.0001$; anteiso $F_{49} = 33.1$, $p < 0.0001$; methyl $F_{49} = 6.4$, $p < 0.0001$; and branched $F_{49} = 29.6$, $p < 0.0001$. In general, the food + water treatment had the most bacteria across bacterial subgroups, while ant and food cores resembled each other with the second highest amount; the control and water-addition cores had the fewest individuals across all bacterial subgroups.

Nematodes in all feeding groups were up to 10× more abundant in the food + water treatment compared to the other treatments. Ant and food soils were most similar to each other and had the next highest number of nematodes, while the control and water cores had the fewest nematodes. With the exception of plant parasites and predaceous nematodes, the number of individuals in each feeding guild was significantly different across treatments: bacterivores $F_{49} = 15.0$, $p < 0.0001$; fungivores $F_{49} = 25.9$, $p < 0.0001$; and omnivores $F_{43} = 3.6$, $p < 0.01$.

Fungivorous nematodes were numerically dominant across all soil treatments, accounting for 79% of all nematodes identified. Bacterivores were 14% of the remaining nematodes, omnivores were 3%, and predators and plant parasites were each <1% (3% of the total could not be identified). Of the fungivores, 9% were *Hexatylina* spp., 9% were *Tylenchus* spp., and the vast majority (82%) were *Aphelenchoides* and *Aphelenchus* spp. Bacterivores consisted of individuals from the orders Rhabditida and Araeolaimida. Omnivores were from the order Dorylaimida, and the few predators identified were from the order Mononchida.

We obtained few microarthropods from our control and water-addition samples. Mites, collembolans, and miscellaneous microarthropods each were most abundant in ant, food, and food + water samples, with ant soils containing the overwhelming majority of all microarthropods. The abundance of these animals significantly differed across soil treatments: mites $F_{49} = 5.1$, $p < 0.01$; collembolans $F_{49} = 6.7$, $p < 0.0001$; miscellaneous microarthropods $F_{49} = 10.4$, $p < 0.0001$. Mites belonged to the Opilioacariformes and Acariformes groups, and collembolans were from the families Onychiuridae and Entomobryidae. The miscellaneous category included various proturans, mite and insect larvae, and unidentifiable arthropod specimens. We did not observe any beetles, earthworms, or other macro-invertebrates in any of our samples.

### *Ant impact on fresh soil*

Ants affected implant soil in 2 months. Depauperate, nutrient-poor soils added to ant nests contained significantly more bacteria, nematodes, microarthropods, and other soil biota and had higher levels of N, P, and OM than controls (Table 1). When implant soil from ant nests was compared to ant soil from the pre-treatment samples, it had significantly more bacteria, nematodes, microarthropods, and other soil biota and had higher levels of P and OM (data not shown). The three exceptions to this trend were no change in soil nitrogen and higher soil moisture and pH in the pre-treatment nest-soils compared to the implant nest-soils.

## Discussion

We successfully mimicked many aspects of ant nests and their influence on belowground chemistry and biota through food additions. A surprising result was that *M. andrei* 'manipulated' the depauperate soil placed in their nests during our 2-month experiment by significantly increasing soil nutrients and organismal abundance and richness. This suggests that ant effects on soil food webs and nutrients can occur quickly, which may have important implications for restoration work as discussed below. Because the food + water effect was much greater than the ant effect for some variables, the reduced particle size of our food via a food processor could have increased the rate of decomposition and subsequent soil changes, which would explain this discrepancy between ant and food + water soils.

Although we predicted that food and water would interact to mimic ant-nest effects, food additions alone explain most of the variation in soil biota and nutrients at this site during early summer months. Our results indicate that soils from the ant cores most resemble soils from the food + water and food cores. Our water additions were effective only in combination with the food additions. This is further supported by the fact that the control implants (no additions) were most similar to the water-only treatment for the majority of the dependent variables. These results were consistently obtained for the sub-groupings within each taxon. Bacterial, nematode, and microarthropod sub-groups tended to be most similar between the ant, food + water and food cores, while the control and water soils had the lowest abundance for each of the sub-types. Although our moisture treatment was ineffective by itself, it is striking that the food + water treatment had the greatest abundance for the majority of the sub-groups.

There were visible fungal hyphae on top of the food and food + water cores when we retrieved them in the field; however, our MANOVA results indicate that fungi did not differ between treatments. Since large numbers of fungivorous nematodes, as obtained in our results, indicate high fungus production, absence of increased fungus biomass in those samples could be due to strong top–down effects of fungivores or to a failure in our PLFA analysis. PLFA markers are well developed for bacteria, but fungal markers are less well documented (White and Findlay 1988; Bossio and Scow 1995). Thus, our lack of findings for the fungi could be due to PLFA limitations and not to a real trend in nature.

Mites, collembolans and other microarthropods were significantly more abundant in ant soils than in all other cores, eventhough similarities were shared between the ant, food + water, and food treatments. There are two possible explanations for this result. First, there are many variables associated with ant nests that we could not mimic. For example, *M. andrei* probably affects the soil structure and/or behaves in ways that might facilitate colonization by these microarthropods. Second, because microarthropods were already present in greater numbers in pre-treatment ant soils than in non-ant soils, they probably colonized the ant cores more quickly than the treatment cores, which had a relatively depauperate microarthropod community before our experimental manipulation.

Our moisture treatment was ineffective, as evidenced by our MANOVA results. We may have overestimated the importance of soil moisture in this system, or we may have implemented the moisture treatment too early in the season when moisture was not a limiting factor. Had this experiment been carried out in the hottest summer months (July–September), perhaps we would have been able to capitalize on more dramatic soil moisture differences between ant and non-ant soil.

This research is the first mechanistic approach to the influence of ant nests on soil chemistry and biota. Our results align with previous studies showing that ants increase soil nutrients and the abundance of most soil taxa (Wagner et al. 1997; Laakso and Setälä 1998). Boulton et al. (2003) showed that *M. andrei* nests at this same site have higher concentrations of N, P, and OM and more abundant soil taxa. Although many studies have examined how ants affect soil chemistry, only a handful of studies have shown that ants positively affect soil food webs. Our results suggest that *M. andrei* exerts such a positive effect primarily via the addition of food, mostly seeds. Moreover, this ant effect can occur quickly – in just 2 months based on our findings. Because ants are widespread and are the most abundant eusocial insect with many long-lived species, they could substantially influence soil and belowground food webs in a number of ecosystems.

The results we report here have important implications for conservation and restoration. In terms of conservation, countless studies have suggested a variety of factors that negatively or positively associate with biota richness and abundance, although few attempt to unravel the mechanism behind such a relationship. Our findings show the relationship between a given variable and biota richness/abundance and then examine experimentally how this effect occurs. From a restoration perspective, native ant species could be crucial in improving soil quality for re-establishing indigenous animals and plants. For this reason, researchers have attempted to protect and/or restore ants to various wooded areas in Europe and Canada (e.g., Pavon 1950, 1960; Bradley 1972). Our research focuses on one species of ant on serpentine soil, so future work should address how this ant effect varies by season, habitat, and ant species.

## Acknowledgments

We would thank Dr Bruce Jaffee for his assistance in nematode identification and Dr Kate Scow for her PLFA analysis of the microbial community. Deasy Lontoh's assistance was invaluable in processing soil samples in the laboratory. The manuscript was greatly improved by comments from Drs Paul Stapp, Bruce Jaffee, Brian Inouye, Susan Harrison, Phil Ward, and two anonymous reviewers. This research was supported by the following grants to the senior author: a University of California, Davis Jastro-Shields Award, an N.S.F. Research Training Grant for Nonlinear Dynamics in Biology (NSF DBI-9602226), a US Environmental Protection Agency STAR Fellowship, and an award from the Consortium for Research at McLaughlin (supported by the Packard Foundation). We greatly appreciate input from the late Dr Gary Polis, who was instrumental in the design of our field experiment.

## References

Beare M., Parmelee R., Hendrix P. and Cheng W. 1992. Microbial and faunal interactions and effects on litter nitrogen and decomposition in agroecosystems. Ecol. Monogr. 62: 569–591.

Boag B. and Yeates G. 1998. Soil nematode biodiversity in terrestrial ecosystems. Biodiv. Conserv. 7: 617–630.

Bongers T. and Bongers M. 1998. Functional diversity of nematodes. Appl. Soil Ecol. 10: 239–251.

Bossio D. and Scow K. 1995. Impact of carbon and flooding on metabolic diversity of microbial communities in soils. Appl. Envir. Microbiol. 61: 4043–4050.

Bossio D. and Scow K. 1998. Impact of carbon and flooding on PFLA profiles and substrate utilization patterns of soil microbial communities. Microbial Ecol. 35: 265–278.

Boulton A.M., Jaffee B.A. and Scow K.M. 2003. Effects of a common harvester ant (*Messor andrei*) on richness and abundance of soil biota. Appl. Soil Ecol. 23: 257–265.

Brady N. and Weil R. 1996. The nature and properties of soils, 11th edn. Prentice-Hall, Inc., New York.

Bradley G. 1972. Transplanting *Formica obscuripes* and *Dolichoderus taschenbergi* (Hymenoptera: Formicidae) colonies in jack pine stands of southeastern Canada. Can. Entomol. 104: 245–249.

Brown M. 1999. Nest relocation and encounters between colonies of the seed-harvesting ant *Messor andrei*. Insectes Sociaux 46: 66–70.

Campbell C. and Biederbeck V. 1976. Soil bacterial changes as affected by growing season weather conditions: a field and laboratory study. Can. J. Soil Sci. 56: 293–310.

Cole B. 1994. Nest architecture in the western harvester ant, *Pogonomyrmex occidentalis* (Cresson). Insectes Sociaux 41: 401–410.

Dean W., Milton S. and Klotz S. 1997. The role of ant-mounds in maintaining small-scale patchiness in dry grasslands in Central Germany. Biodiv. Conserv. 6: 1292–1307.

Doran J. 1987. Microbial biomass and mineralizable nitrogen distributions in no-tillage and plowed soils. Biol. Fert. Soils 5: 68–75.

Folgarait P. 1998. Ant biodiversity and its relationship to ecosystem functioning: a review. Biodiv. Conserv. 7: 1221–1244.

Friese C. and Allen M. 1993. The interaction of harvester ants and vesicular-arbuscular mycorrhizal fungi in a patchy semi-arid environment: the effects of mound structure on fungal dispersion and establishment. Functional Ecol. 7: 13–20.

Hölldobler B. and Wilson E. 1990. The Ants. Harvard University Press, Massachusetts.

Jaffee B., Ferris H. and Scow K. 1998. Nematode-trapping fungi in organic and conventional cropping systems. Phytopathology 88: 344–349.

Laakso J. and Setälä H. 1997. Nest mounds of red wood ants (*Formica aquilonia*): hot spots for litter dwelling earthworms. Oecologia 111: 565–569.

Laakso J. and Setälä H. 1998. Composition and trophic structure of detrital food webs in ant nest mounds of *Formica aquilonia* and in the surrounding forest soil. Oikos 81: 266–278.

MacKay W. 1981. A comparison of the nest phenologies of three species of *Pogonomyrmex* harvester ants (Hymenoptera: Formicidae). Psyche 88: 25–74.

Mikola J., Yeates G., Barker G., Wardle D. and Bonner K. 2001. Effects of defoliation intensity on soil food-web properties in an experimental grassland. Oikos 92: 333–343.

Parmelee R., Bohlen P. and Edwards C. 1995. Analysis of nematode trophic structure in agrosystems: functional groups versus high resolution taxonomy. In: Collins H.P., Robertson G.P. and Klug M.J. (eds), The Significance and Regulation of Soil Biodiversity. Kluwer Academic Publishers, pp. 203–207.

Pavon M. 1950. Sugli inizi d'un esperimento practico di lotta biologica con *Formica rufa* L. contro la processionaria del pino (*ThaumetoPoea PityocamPa* Sch.). Atti Soc. Ital. Sci. Nat. 89: 195–201.

Pavon M. 1960. Les transplantations de *Formica lugubris* sur les Apennins de la province de Pavie. Italy Ministry of Agriculture, Collana Verde 7: 161–169.

Preisser E.L. 2003. Field evidence for a rapidly cascading underground food web. Ecology 84: 869–874.

Sokal R. and Rohlf F. 1995. Biometry, 3rd edn. W.H. Freeman and Company.

Strong D., Maron J. and Connors P. 1996. Top down from underground? The under-appreciated influence of subterranean food webs on above-ground ecology. In: Polis G.A. and Winemiller K.O. (eds), Food Webs: Integration of Patterns and Dynamics. Chapman and Hall, New York, pp. 170–175.

Tabachnick B. and Fidell L. 1996. Using Multivariate Statistics, 3rd edn. Harper Collins College Publishers.

University of California, Natural Reserve System (UC NRS). 2000. Natural history of the McLaughlin Reserve. UC Davis printing services.

Wagner D., Brown M. and Gordon D. 1997. Harvester ant nests, soil biota and soil chemistry. Oecologia 112: 232–236.

White D. and Findlay R. 1988. Biochemical markers for measurement of predation effects on the biomass, community structure, nutritional status, and metabolic activity of microbial biofilms. Hydrobiologia 159: 119–132.

Whitford W. 1996. The importance of the biodiversity of soil biota in arid ecosystems. Biodiv. Conserv. 5: 185–195.

Whitford W. and DiMarco R. 1995. Variability in soils and vegetation associated with harvester ant (*Pogonomyrmex rugosus*) nests on a Chihuahuan Desert watershed. Biol. Fert. Soils 20: 169–173.

Yeates G., Bongers T., deGoede R. and Georgieva S. 1993. Feeding habits in nematode families and genera – an outline for soil ecologists. J. Nematol. 25: 315–331.

Zak D.R., Holmes W.E., White D.C., Peacock A.D. and Tilman D. 2003. Plant diversity, soil microbial communities, and ecosystem function: are there any links? Ecology 84: 2042–2050.

Biodiversity and Conservation (2006) 15:69–91
DOI 10.1007/s10531-004-3106-5 

# Host specificity, alpha- and beta-diversity of phytophagous beetles in two tropical forests in Panama

FRODE ØDEGAARD
*Norwegian Institute for Nature Research (NINA), Tungasletta 2, NO-7485 Trondheim, Norway (e-mail: frode.odegaard@nina.no; phone: +47-73-80-15-55; fax: +47-73-80-14-01)*

Received 8 April 2004; accepted in revised form 31 August 2004

**Key words:** Biodiversity, Canopy, Effective specialization, Insect herbivores, Species turnover

**Abstract.** Species diversity, host specificity and species turnover among phytophagous beetles were studied in the canopy of two tropical lowland forests in Panama with the use of canopy cranes. A sharp rainfall gradient occurs between the two sites located 80 km apart. The wetter forest is located in San Lorenzo Protected Area on the Caribbean side of the isthmus, and the drier forest is a part of the Parque Natural Metropolitano close to Panama City on the Pacific slope. Host specificity was measured as effective specialization and recorded by probability methods based on abundance categories and feeding records from a total of 102 species of trees and lianas equally distributed between the two sites. The total material collected included more than 65,000 beetles of 2462 species, of which 306 species were shared between the two sites. The wet forest was 37% more species rich than the dry forest due to more saproxylic species and flower visitors. Saproxylic species and flower visitors were also more host-specific in the wet forest. Leaf chewers showed similar levels of species richness and host specificity in both forests. The effective number of specialized species per plant species was higher in the wet forest. Higher levels of local alpha- and beta-diversity as well as host specificity based on present data from a tropical wet forest, suggests higher number of species at regional levels, a result that may have consequences for ecological estimates of global species richness.

**Abbreviations:** PNM – Parque Natural Metropolitano, Panama Province, Panama; SLA – San Lorenzo Protected Area, Colon Province, Panama

## Introduction

Biodiversity is not only an issue of curiosity, but stands firm on the political agenda as a resource for humanity (Heywood 1995). Many species are under continuous threat as more natural ecosystems are changed, polluted, affected by climatic change, or exploited too heavily. To know the number of species, their ranges and ecology is therefore an important topic in conservation biology, for instance when estimating the rate of extinction and decline of species both locally and worldwide. More specifically, elucidation of the variation in host specificity and insect species richness along geographical gradients would be essential for refining our knowledge on the magnitude of local or even global species richness (May 1994; Mawdsley and Stork 1997; Stork 1997; Ødegaard 2000a) as well as ecosystem structure and dynamics, and eventually for thorough based decisions in nature management.

Ecological estimates of regional species diversity are almost entirely based on studies reporting insect species diversity from single tropical sites (e.g., Erwin 1982; Basset et al. 1996; Ødegaard 2000a; Novotny et al. 2002a). However, such alpha diversity studies are of limited value in terms of predicting species richness on larger spatial scales due to incomplete sampling caused by temporal and spatial limitations. It is even hard to predict regional species diversity based on data from very extensively studied areas because the influence from mass effects increases with sample size (Shmida and Wilson 1985; Novotny and Basset 2000).

Species composition and host specificity of phytophagous beetles may vary considerably between sites (Basset 1992). This variation can be explained by differences in resource availability or other determinants for insect species diversity, such as hostplant abundance, distribution and species richness (e.g., Southwood 1960, 1961; Neuvonen and Niemelä 1983; Condit et al. 2000), plant architecture (Lawton and Schröder 1977), chemical composition of plant tissue (e.g., Connor et al. 1980; Bernays and Chapman 1994), and interactions among animals e.g., competition for resources and enemy-free space and several abiotic factors (Strong et al. 1984). For these reasons, comparisons of insect communities may have a higher probability of being of general validity if replicates are taken in different types of forests.

Beta diversity is the extent to which the diversity of two or more spatial units differs (Magurran 2003). Originally, beta diversity measures the extent of difference between two or more areas relative to the total species richness (e.g., Whittaker 1960), but more commonly it is used for comparing similarity between sites through different indices based on abundance or presence/absence data (Magurran 2003). Beta diversity has been widely used also for estimation of regional species richness through application of the classical species–area relationship (Gleason 1922; Connor and McCoy 1979; Mawdsley 1996; Ricotta et al. 2002). The main factors explaining beta diversity is range and habitat restrictions (Harrison et al. 1992). While there is an increasing understanding of patterns and mechanisms responsible for species turnover in tropical plants (Condit et al. 2002), there are relatively few studies on beta diversity of tropical insects (Hespenheide 1994; Novotny and Missa 2000).

Beta diversity among phytophagous insects is obviously linked to species turnover among host plants. Similarity in tropical tree communities declines with distance between sites (Condit et al. 2002). If all species were monophagous, the species turnover among phytophagous insects would be similar or higher than that of their host plants. Host specialization among insects complicates the patterns of beta diversity, however. The effective specialization among tropical insect communities is probably as low as 5–10% (Basset et al. 1996; Ødegaard et al. 2000; Novotny et al. 2002a), and it is unknown how the large proportion of generalists among tropical insects affects species turnover.

Generalists tend to have wider geographical ranges than specialized species (Gaston 1991), but these relations are hard to study because geographical range of insect species in tropical forests is very poorly known for most groups, and even less knowledge exists about similarity among insect communities between sites (Basset 2001).

The aim of this study was to examine how different terrestrial insect communities are structured with regard to species richness and host specificity in two different tropical lowland forests in Panama. The model group for this study was phytophagous beetles which constitute a dominant component of biodiversity on Earth (Hammond 1992). Host specificity and taxonomic composition of beetle communities in tropical forests are therefore important parameters structuring terrestrial ecosystems and basic for development of a better knowledge of species richness and beta diversity.

The geography of Panama particularly lends itself to understanding patterns in species turnover. The sharp rainfall gradient between the two sites causes an almost complete turnover of plant species. In addition, population variability of insect assemblages is in some cases more prominent between dry and wet tropical forests than it is between temperate and tropical forests (Wolda 1978; Pimm 1991). It would be of fundamental ecological interest to test if related parameters such as host specificity and beta diversity of insect communities also differ between these forest types. The results may in turn give further implications about the validity of current ecological estimates of global species richness.

## Methods

### *Study sites*

The study was carried out at two Panamanian lowland forest sites which represent different tropical forest types. The first site is the Parque Natural Metropolitano (PNM), which consists of 265 ha dry tropical forest in Panama province, close to Panama City and 2 km from the Pacific coast (8°59′ N, 79°33′ W, ca. 30 m a.s.l.). The average annual temperature is 28 °C, and annual precipitation is 1740 mm. The dry season is distinct from December to April, when rainfall is usually less than 100 mm per month. This is a secondary forest that has escaped major human disturbance for about 90 years. The vegetation at this site is characterized by dominance of deciduous trees (30–35 m height) and lianas in the canopy (Wright et al. 2003).

The other site is located in an evergreen, wet forest in San Lorenzo Protected Area (SLA) (9°17′ N, 79°58′ W, ca. 130 m a.s.l.), Colon province, 4.4 km away from the Atlantic coast of the isthmus. The average annual temperature is 25.8 °C, and average annual precipitation is 3140 mm. There is a pronounced dry season from mid December to end of April that receives nearly 10% of the yearly precipitation. This forest is dominated by trees at 35–45 m height with lianas and epiphytes occurring regularly in the canopy. The San Lorenzo Protected Area includes 9600 ha of relatively old-growth tropical forest which has escaped anthropogenic disturbance for about 200 years (Wright et al. 2003).

Annual rainfall drops linearly from 3.1 to 1.7 m crossing the isthmus from north to south with little confounding elevational change. The study sites are located ca. 80 km apart in a contiguous protected forest along the Panama Canal with a total area of 370 km$^2$ (Wright and Colley 1994).

### *Canopy access*

The canopy was accessed by two canopy cranes erected at the sites. The crane in PNM is 44 m tall with an arm length of 52 m that gives access to ca. 0.8 ha of projected area. About 40 species of trees and ca. 35 species of climbers reach the middle or upper levels of the canopy which could be accessed from the crane gondola. The SLA-crane is 54 m tall with an arm length of 55 m. Hence, the projected area accessible for study was 0.88 ha. About 70 species of trees and ca. 10 species of lianas were easily accessible for study from the crane gondola in SLA.

### *Focal taxa*

The focal groups of this study were adult beetles of Buprestidae, Chrysomeloidea, and Curculionoidea, which make up nearly all herbivorous and a major part of saproxylic beetles in this forest. All the beetle material was identified to species level. Identifications were performed by the author or experts studying the different taxonomic groups. A species list from PNM is available in Ødegaard (2003). A large part of the material was deposited in the author's collection at the Norwegian Institute for Nature Research (NINA). Material of current taxonomic importance for the taxonomists has been deposited in their respective collections (see Acknowledgements), while a representative selection of the material were deposited in the synoptic insect collection at Smithsonian Tropical Research Institute (STRI) and the University of Panama.

For statistical treatment leaf chewers, saproxylic species, and flower visitors were distinguished as grouping variables. Leaf chewers were defined as all species feeding on green plant parts. Saproxylic species included species feeding on dead wood or wood associated fungi. Flower visitors included species attracted to flowers, presumably feeding on nectar or pollen. Fruit eaters and seed predators were included in this group as well. Taxonomically, the material was treated statistically at family level for small groups, and at subfamily level for Curculionidae, according to Alonzo-Zarazaga and Lyal (1999).

*Target plants*

The phytophagous beetle fauna of a total of 50 and 52 plant species were studied in PNM and SLA, respectively. Selection of plant species for study was limited by the crane perimeter, but as far as possible confamilial species at the two sites, and a representative proportion of trees and lianas at the sites were chosen. Additional criteria for the selection of target plant included size of plant biomass, and that the target plants as far as possible appeared without confounding epiphytes and lianas in order to minimize the influence from neighbouring plants. The number of plant taxa belonging to different life forms and taxonomic categories is indicated in Table 1.

Only two plant species were shared between the two sites; the lianas *Phryganocydia corymbosa* (Vent.) Bur., and *Cydista aequinoctalis* (L.) Miers of the family Bignoniaceae. The following genera among the study plants were common to the two sites: *Cecropia* (Cecropiaceae), *Cordia* (Boraginaceae), *Nectandra* (Lauraceae), and *Arrabidaea* (Bignoniaceae). A total of 14 plant families were shared between the two sites which include 82 and 58% of species in PNM and SLA, respectively (Table 2). Trees and lianas as prominent life forms of plants were considered separately as grouping variables for statistical treatment.

*Sampling programme*

The sampling procedure intended to survey a similar leaf and branch area of each plant species, and to maximize the number of microhabitats of each plant. Sampling was carried out from the crane gondola using a 1 $m^2$ beating sheet. Each sample was standardized by beating different parts of the tree or liana for 30 min by moving from different positions within the tree both along vertical and horizontal gradients. Each sampling position within the tree included beating of two or three branches before the material was collected by an aspirator. Movements between positions were repeated six to eight times within the 30 min period. Accordingly, appropriate statistical replication was based on equal beating time as a rough substitute measurement for leaf area

*Table 1.* Number of plant species distributed on life forms and taxonomic categories in PNM and SLA.

| | PNM | SLA |
|---|---|---|
| No. of plant species | 50 | 52 |
| Trees | 24 | 43 |
| Lianas | 26 | 9 |
| No. of plant genera | 43 | 48 |
| No. of plant families | 21 | 26 |
| No. of congeneric pairs | 6 | 5 |

*Table 2.* Plant families represented by the study plants shared between PNM and SLA, and the number of plant species studied in each family.

| Families | PNM | SLA |
|---|---|---|
| Anacardiaceae | 4 | 1 |
| Araliaceae | 1 | 1 |
| Arecaceae | 1 | 2 |
| Bignoniaceae | 12 | 6 |
| Boraginaceae | 1 | 1 |
| Combretaceae | 1 | 1 |
| Cecropiaceae | 2 | 1 |
| Fabaceae | 6 | 8 |
| Lauraceae | 3 | 1 |
| Moraceae | 4 | 2 |
| Rubiaceae | 1 | 1 |
| Sapindaceae | 3 | 3 |
| Sapotaceae | 1 | 1 |
| Tiliaceae | 1 | 1 |

(Ødegaard 2000b, 2004). Each tree crown was sampled regularly day and night once a month during 1 year. The range of height surveyed in the forest included canopy habitats from 10 to 40 m, but not the understorey. In addition, sampling was performed more frequently in periods of leaf flush or flowering in order to optimize insect species richness and the number of host observations. This sampling strategy was termed 'additional sampling' by Ødegaard (2000b). One individual plant of each species was sampled except for the big tree *Brosimum utile* in SLA. A total of six trees of this species were surveyed with similar methods in order to study the effect of sample size (i.e. number of tree individuals) (Ødegaard 2004).

The sampling effort was similar in PNM and SLA. The sampling period lasted from *primo* March 1995 to *medio* May 1996, and *primo* March 2001 to *medio* May 2002 for PNM and SLA, respectively. The beginning of the rainy season *(primo* May) was sampled two subsequent years at each site since this is a period of very high insect activity in Panama (Wolda 1980; Wolda et al. 1998).

### *Host observations*

Feeding observations were recorded at times of flowering, fruiting or leaf-flush. A host record was defined as at least one feeding observation. Generally, host specificity is overestimated if the number of specimens of an insect species is lower than the number of host species (Colwell and Futuyma 1971). Accordingly, the feeding observations of species recorded 50 times or more (h0), were distinguished from those encountered less than 50 times (h1). Rare species always constitute a major proportion of species in samples, and they may differ from common ones regarding patterns of host specificity (Price et al. 1995). Therefore, also host observations attended with lower level of confidence were included and determined by probability assessments based on abundance categories (Flowers and Janzen 1997; Ødegaard 2000b). These host occurrences were assigned to the following categories according to the number of individual records from the assumed host plant; h2: 10 or more records; h3: 5–9 records; h4: 2–4 record; h5: 1 record and additional evidence for host association based of collections or literature. Singletons recorded on a plant were treated as tourists unless additional evidence about host associations was available from collections or literature. Species of aerial drift (randomly distributed species) were treated as tourists although some of these were extremely common on the studied plants. These species are small in body size (often less than 2 mm), which seem to be rare among broad generalists. Frequently, they also belong to taxonomic groups which in general tend to be specialists (e.g., Anthonomini and Apioninae) *(personal observation)*. Species with proven host associations on trees other than the specific target trees were treated as aerial drift material when abundance was 20% less than of that of their host tree (Ødegaard 2004). Otherwise, aerial drift material was distinguished from polyphagous species through feeding records.

*Statistical methods*

Host specificity was measured as effective specialization (May 1990; Ødegaard et al. 2000). The principle behind effective specialization of a plant's insect fauna is to weight each insect species in accordance with its degree of specialization on other plants in the community. The monophagous species are given the heaviest weight, while broadly polyphagous species adds insignificant to the value. For a plant species $k$, in a community of $T$ plant species, the proportion of beetles effectively specialized on $k$, $f_k$, is given by

$$f_k = \sum_{i=1}^{T} (1/i) p_k(i), \qquad (1)$$

where $p_k(i)$ is the proportion of beetles associated with plant species $k$ and $i$ other plants. Knowing $f_k$, the number of insect species effectively specialized on each plant species ($k$) is given as $S_f$:

$$S_f = S_k f_k, \qquad (2)$$

where $S_k$ is the number of beetle species associated with plant species $k$. This is the parameter that implies the plants' relative contribution to the maintenance of insect species richness in the community.

The rarefied number of species present in samples was computed by Coleman's rarefaction (Coleman 1981; Colwell and Coddington 1994). Beta diversity between different insect communities was calculated both as presence/absence data (Jaccard and Sørensen index), and abundance data with Sørensen and Morisita-Horn statistics (Magurran 2003). Accumulation-curves and effective specialization-curves based on the observed number of species ($S_{obs}$) in each plant species, and similarity indices, were calculated with 50 randomizations by the programme Estimates (Colwell 1997). Comparisons of means were done by one sample *t*-tests using the program package SPSS 11.5.

## Results

A total of 35,479 beetles of 1165 species were recorded from PNM (Ødegaard 2003), while 30,352 beetles including 1603 species were recorded from SLA. The higher number of species from SLA (37%) was due to higher number of flower visitors and saproxylic species (Figure 1). These differences between the ecological guilds were seen also among taxonomic groups representing the guilds. Among the saproxylic species the difference was mostly due to Cerambycidae, Anthribidae (incl. Brentidae), Cryptorynchinae (incl. Molytinae), Conoderinae, and Scolytinae (incl. Platypodinae and Cossoninae). All these groups were nearly twice as species rich in SLA than in PNM. In contrast, flower visitors such as Bruchinae and Baridinae were more species rich at PNM. Large groups of leaf chewers like Chrysomelidae and Buprestidae (mostly leaf miners) had a similar number of species at the two sites (Figure 2).

### *Similarity between sites*

The total material included 2462 species of which 306 species were shared between the two sites. This proportion constitutes 12% of the species (Jaccard

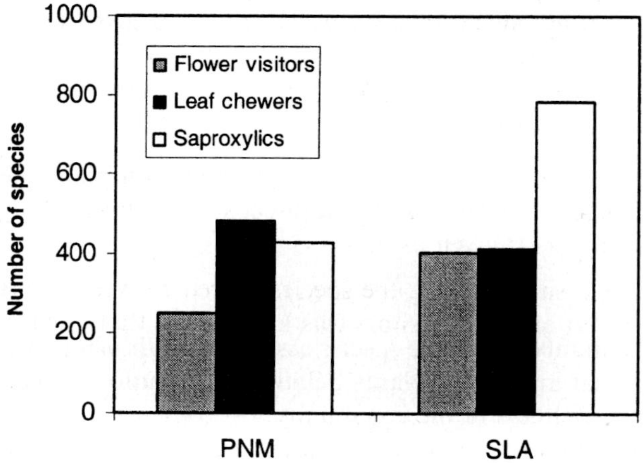


*Figure 1.* The total number of phytophagous beetle species belonging to different guilds at each of the two study sites.

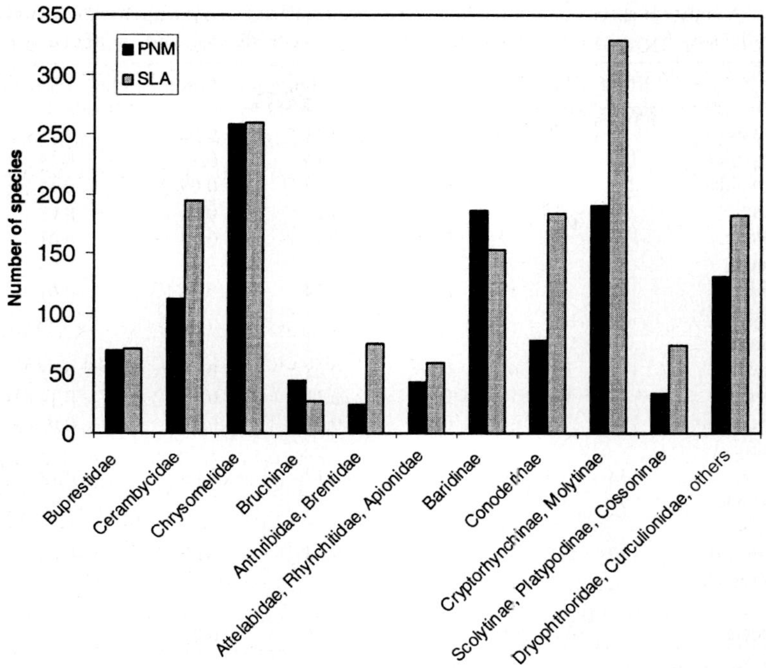


*Figure 2.* The number of beetle species belonging to different taxonomic groups at each of the two sites.

index). Abundance-based similarity indices for the total material were 0.08 (Sørensen Abundance index) and 0.07 (Morisita-Horn index) (Table 3). Presence/absence similarity indices on taxonomic subgroups indicated the lowest species turnover among Baridinae, Chrysomelidae, and Cerambycidae, and the highest among Anthribidae (incl. Brentidae) and Conoderinae. Abundance similarity indices indicated a relatively low turnover among Cerambycidae, Chrysomelidae, Scolytinae, while the Bruchinae and the remaining groups of Curculionidae showed a relatively high turnover (Morisita-Horn index, Table 3).

Among the three guilds of phytophagous beetles there were a higher proportion of leaf chewers in PNM than in SLA, and a higher proportion of saproxylic species in SLA than in PNM. The proportion of flower visitors was similar at the two sites. Among the species shared between the two sites, the relative proportion of flower visitors was higher than that at each of the sites. The relative proportion of leaf chewers and saproxylic species among the shared fauna was at an intermediate level relative to the sites (Figure 3a).

Regarding host relationships among the total beetle fauna, the relative proportion of tourists, generalists and specialists (up to family level of plants) were similar at the two sites. The fauna common to each site, was dominated by tourists and generalists that made up 47 and 36%, respectively. Specialists

*Table 3.* Number of species in taxonomic subgroups at each site, and the number of shared species between the sites along with similarity indices; ±, = presence/absence; abd. = abundance.

| Taxa | PNM | SLA | Shared | Jaccard ± | Sørensen ± | Sørensen abd. | Morisita-Horn |
|---|---|---|---|---|---|---|---|
| Buprestidae | 69 | 70 | 10 | 0.08 | 0.14 | 0.02 | 0 |
| Cerambycidae | 112 | 194 | 38 | 0.14 | 0.25 | 0.15 | 0.14 |
| Chrysomelidae[a] | 258 | 260 | 71 | 0.16 | 0.27 | 0.09 | 0.15 |
| Bruchinae | 44 | 26 | 7 | 0.11 | 0.2 | 0.05 | 0.03 |
| Anthribidae, Brentidae | 24 | 73 | 5 | 0.05 | 0.1 | 0.04 | 0.01 |
| Attelabidae, Rhynchitidae, Apionidae | 42 | 58 | 12 | 0.14 | 0.24 | 0.03 | 0.01 |
| Baridinae | 186 | 153 | 47 | 0.16 | 0.28 | 0.1 | 0.08 |
| Conoderinae | 77 | 184 | 14 | 0.06 | 0.11 | 0.06 | 0.11 |
| Cryptorhynchinae, Molytinae | 190 | 329 | 54 | 0.12 | 0.21 | 0.09 | 0.05 |
| Scolytinae, Platypodinae, Cossoninae | 33 | 73 | 10 | 0.09 | 0.17 | 0.05 | 0.13 |
| Dryophthoridae, Curculionidae, others[b] | 130 | 182 | 38 | 0.14 | 0.24 | 0.08 | 0.03 |
| Total species | 1165 | 1602 | 306 | 0.12 | 0.21 | 0.08 | 0.07 |

[a]Includes Megalopodidae and Orsodacnidae, but excludes Bruchinae.
[b]Includes Curculioninae, Entiminae, Mesoptiliinae.

up to family level of plants constituted only 17% of the shared fauna (Figure 3b).

## *The number of beetles per plant species*

Despite that more species were surveyed at SLA, the number of host-associated beetle species per plant species was not significantly different between the two sites, neither within life forms of plants, nor within insect guilds (Table 4). On average 49.1 ± 4.3 species were associated with plants in PNM, based on revised data from Ødegaard (2000b), while 59.4 ± 5.3 species were associated with plants in SLA. The number of species per plant varied from only a few to more than 150 species both within trees and lianas (Figure 4). Plotting the rates at which the host associated species accumulate when adding host plants to the samples clearly indicated that more species were recorded per plant in SLA, and that their accumulation rate was higher (Figure 5). Details on number of species associated with trees and lianas within flower visitors, leaf chewers, and wood eaters are given in Table 4.

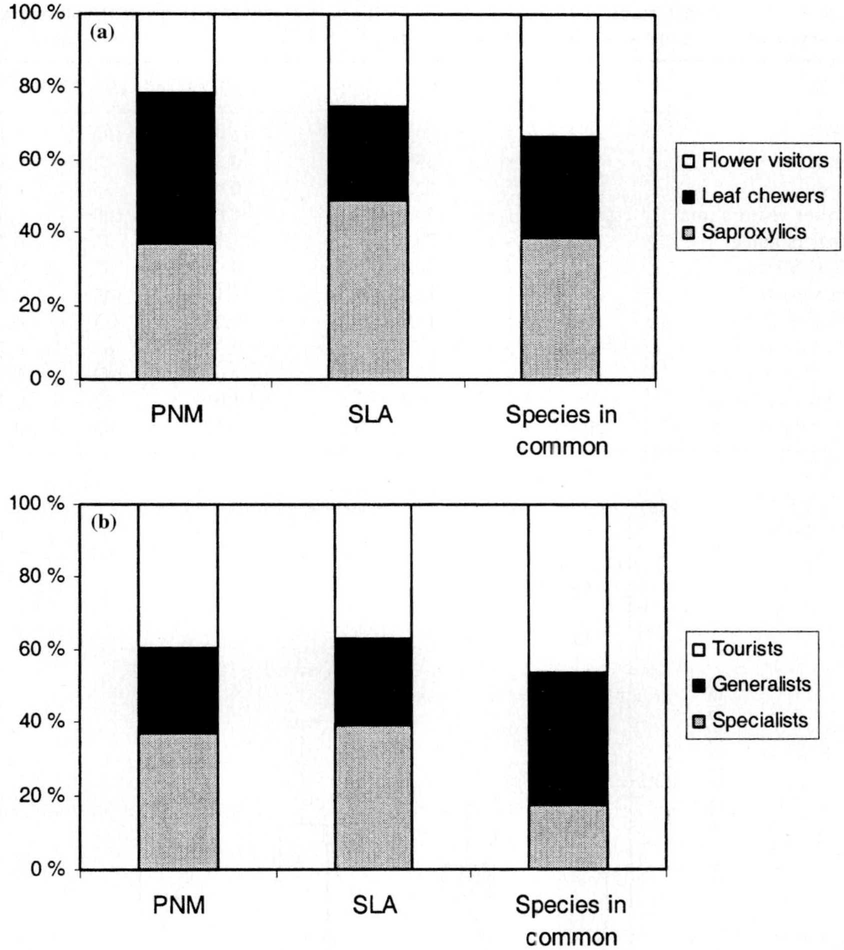


*Figure 3.* The relative proportion of species belonging to different guilds (a) and different categories of host associations (b) at each of the two sites and among the species shared between the sites. The specialists are defined as species associated within a plant family.

## *Host specificity*

A total of 3088 and 2453 host observations were recorded in SLA and PNM, respectively. Only 2.6 and 3.8% of the host observations were of the highest level of confidence in SLA and PNM, respectively. More than half of the data reflected host observations of the two lowest levels of confidence (Table 5). Hence, individual host observations should be treated with caution, although for the purpose of comparisons between sites and groups, they are useful as long as being consistent across entities compared.

*Table 4.* The average number of beetle species per plant for flower visitors, leaf chewers and saproxylic species on trees and lianas at the two sites.

| Species | PNM | SLA | *t*-test | df | *p* |
|---|---|---|---|---|---|
| S total | 49.1 ± 4.3 | 59.4 ± 5.3 | 1.499 | 100 | 0.137 |
| S lianas | 45.5 ± 5.8 | 45.1 ± 9.0 | −0.31 | 33 | 0.975 |
| S trees | 53.0 ± 6.5 | 62.4 ± 6.1 | 0.99 | 65 | 0.326 |
| S flower visitors total | 13.6 ± 2.1 | 20.7 ± 3.6 | 1.689 | 100 | 0.094 |
| S flowers lianas | 12.2 ± 2.6 | 20.9 ± 10.2 | 1.175 | 33 | 0.248 |
| S flowers trees | 15.1 ± 3.4 | 20.6 ± 3.8 | 0.969 | 65 | 0.336 |
| S chewers total | 18.2 ± 1.5 | 15.8 ± 1.5 | −1.164 | 100 | 0.247 |
| S chewers lianas | 19.2 ± 2.3 | 13.7 ± 3.1 | −1.273 | 33 | 0.212 |
| S chewers trees | 17.0 ± 1.9 | 16.2 ± 1.6 | −0.329 | 65 | 0.744 |
| S saproxylic total | 17.3 ± 1.9 | 23.0 ± 2.8 | 1.624 | 100 | 0.108 |
| S saproxylic lianas | 14.0 ± 1.9 | 10.4 ± 2.1 | −1.016 | 33 | 0.317 |
| S saproxylic trees | 20.9 ± 3.3 | 25.6 ± 3.3 | 0.928 | 65 | 0.357 |

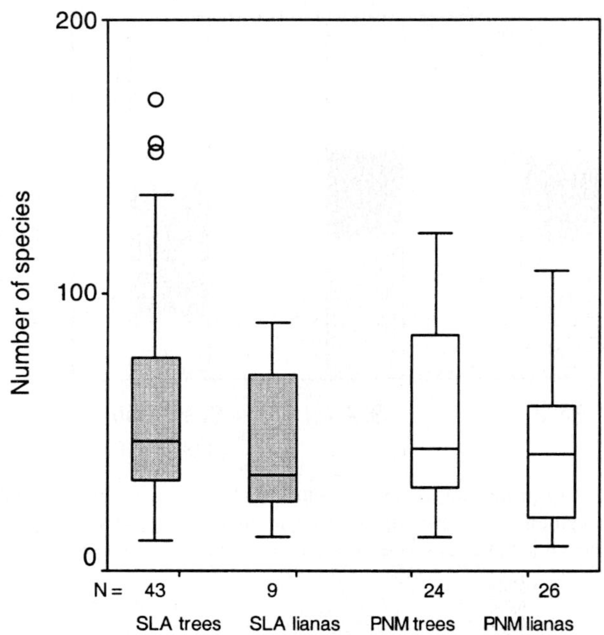


*Figure 4.* A Box and Wisker plot of the number of phytophagous beetle species per plant species of trees and lianas at each of the two sites.

Species associated with few host plants dominate among species at both sites. The proportion of monophagous insect species among the study plants was 47 and 54% in PNM and SLA, respectively. There was no obvious pattern regarding the proportion of beetles associated with two or more host plants across the sites, although there was a tendency that the proportion of generalists utilizing five or more plant species was higher in PNM (Figure 6).

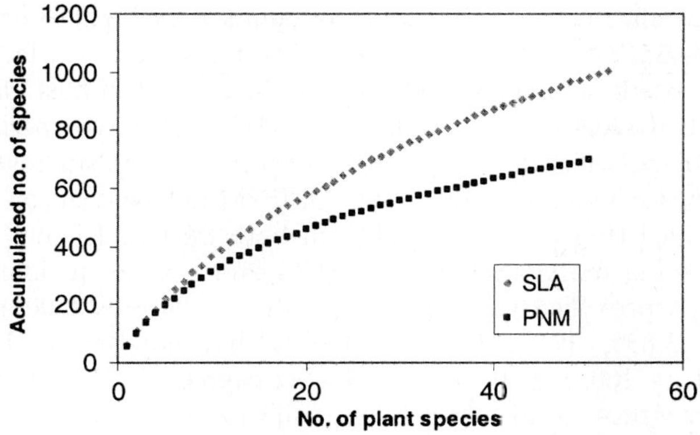


*Figure 5.* Rarefied accumulation curves of phytophagous beetle species with increased number of host plants at each of the two sites.

*Table 5.* Confidence level of host observations.

| | SLA | PNM[a] |
|---|---|---|
| h0: feeding (≥50 specimens) | 65 | 118 |
| h1: feeding ( < 50 specimens) | 173 | 212 |
| h2: > 10 individuals | 438 | 470 |
| h3: 5–9 individuals | 401 | 381 |
| h4: 2–4 individuals | 1057 | 1272[b] |
| h5: 1 individual | 954 | – |
| Total | 3088 | 2423 |

Feeding observations of species recorded 50 times or more were distinguished from those recorded less than 50 times in the total material.
[a]Revised data from Ødegaard (2000b).
[b]Includes also h5.

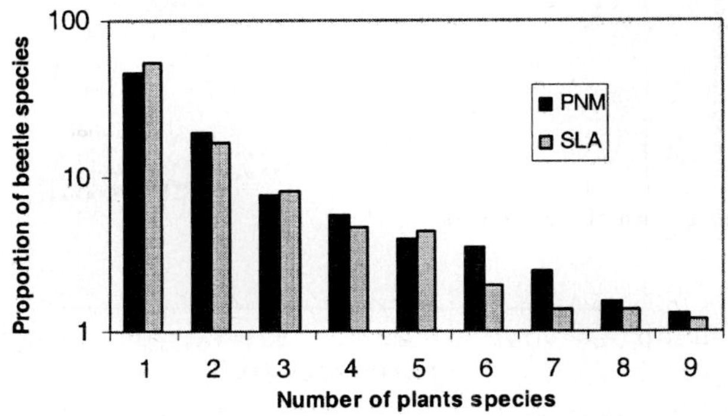


*Figure 6.* The relative proportion of phytophagous beetle species associated with one to nine plant species at each of the two sites.

The average effective specialization was significantly higher in SLA than in PNM ($t = 3.035$, df $= 100$, $p = 0.003$). The magnitude of effective specialization of the beetle assemblages depends on the number of host plants studied (Ødegaard et al. 2000). A calculation of average effective specialization of beetle assemblages on each of 1 to 50 plant species at each site reveals that the fauna in SLA was more specialized than in PNM independent on the number of plants studied (Figure 7). All guilds on both trees and lianas showed the tendency of being more specialized in SLA, and significant differences were found among saproxylic species ($t = 4.657$, df $= 100$, $p < 0.001$), and flower visitors ($t = 3.839$, df $= 100$, $p < 0.001$), but not among leaf chewers ($t = 0.69$, df $= 100$, $p = 0.492$). The higher degree of effective specialization among flower visitors in SLA was only significant in trees ($t = 3.2$, df $= 65$, $p = 0.002$) (Figure 8).

The number of beetle species effectively specialized on plants was higher in SLA than in PNM. ($t = 2.255$, df $= 100$, $p = 0.026$). This difference was due to flower visitors ($t = 2.536$, df $= 96$, $p = 0.013$) and saproxylic species ($t = 2.426$, df $= 100$, $p = 0.017$). The number of leaf chewers effectively specialized at the two sites was similar ($t = -0.798$, df $= 100$, $p = 0.427$) (Figure 9).

## Discussion

### *Alpha diversity*

Plant species diversity in tropical forests correlates strongly with annual precipitation (Gentry 1988; Wright 1992; Leigh 1999), and the number of

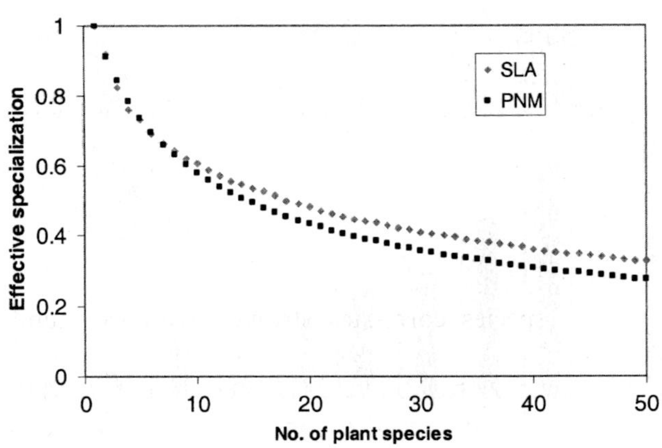


*Figure 7.* Effective specialization ($F_T$) as a function of the number of plant species at the two sites. The curves represent the average effective specialization of beetle assemblages on 1 to 50 plant species.

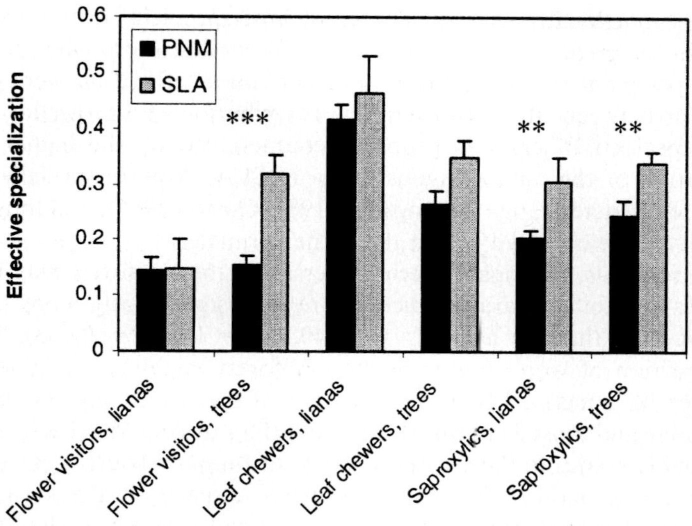


*Figure 8.* The average effective specialization per plant species (+SE) among different guilds on trees and lianas at each of the two sites. ** $p < 0.01$, *** $p < 0.005$.

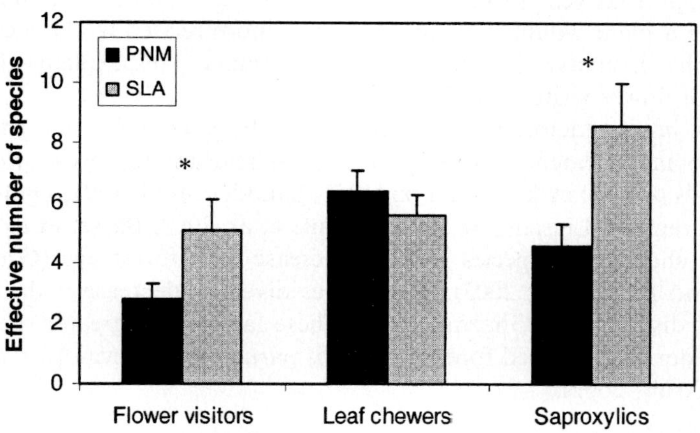


*Figure 9.* The average effective number of species per plant species (+SE) among different guilds at each of the two sites. * $p < 0.05$.

phytophagous insect species correlates strongly with the number of plant species in a community (e.g., Strong et al. 1984; Andow 1991). Thus, it is not surprising that the total number of phytophagous beetle species recorded from the wet forest was higher than from the dry forest. Similar patterns are also observed in butterflies (De Vries 1994).

A larger local species pool of insects also affects the average number of insects per plant species as the number of generalists among the insects will increase proportionally with the degree of host specificity in the insect community. In the present study, more flower visitors and saproxylic species were recorded per plant species in SLA. Species richness of leaf chewers per plant was similar between sites, however. The contribution of a larger local species pool to species richness was probably counteracted by the higher effective specialization of the phytophagous fauna in SLA. A more specialized fauna has a more restricted range (Strong et al. 1984; Gaston 1991), and thus, a lower probability to be collected within the crane perimeter.

Apart from this, the higher species diversity in the wet forest may be due to differences in habitat characteristics. There were significantly more saproxylic species on trees than on lianas ($t = 3.040$, df $= 100$, $p = 0.003$). Thus, the low proportion of wood borers in the dry forest might be explained by the dominance of lianas in PNM. The number of species among fungus feeders (Anthribidae and most Scolytinae) was more than twice in the wet forest, a fact that probably relates to the greater diversity of fungus resources in wet forests, and thus, greater habitat diversity for saproxylic species in the wet forest.

Regarding flower visitors, many species, especially within the Baridinae and Cerambycidae, are specialized wood borers as larvae. The adults are general flower visitors that may be attracted to plants in blossom from long distance. Consequently, many pollinator–flower interactions maintain 'loose niches' that may change from year to year (Roubik 1992). The number of flower visiting species on a plant would therefore be a trait more related to the local species pool (alpha diversity) than to specific host plants. This explains the higher number of flower visitors in SLA.

Human impact factors including forest age, fragmentation, and edge effects related to management history may also be relevant to species richness of arthropods (e.g., Hanski and Gilpin 1991; Saunders et al. 1991; Didham et al. 1998; Floren and Linsenmair 2001; Colville et al. 2002; Basset et al. 2004). It has been shown that species richness increase with forest age (Grove 2002; Floren and Linsenmair 2003), and species diversity decrease with increasing degree of disturbance (Ghazoul 2002). These facts imply greater species richness in older, undisturbed forests, which is particularly relevant for saproxylic beetles (Grove 2002).

### *Beta diversity*

The high species turnover among insects is not surprising regarding the almost complete change in plant composition. The figure compares well with Broadhead and Wolda (1985) who studied the diversity of Psocoptera in two tropical forests in Panama. They found even less overlap between sites (Jaccard index 0.03 opposed to 0.12 in this study), but their localities were more distant, and also separated by a significant difference in elevation. More relevant for comparison are studies of species turnover of Conoderinae and Buprestidae between La Selva, Costa Rica and Barro Colorado Island (BCI), Panama

(Hespenheide 1994). The faunas of these relatively distant sites were more similar than that of the same groups in the present study (for Conoderinae, Jaccard index: 0.23 opposed to 0.08 in the present study; for Buprestidae, Jaccard index: 0.14 opposed to 0.08 in the present study). In this case, the precipitation gradient is probably more important than geographical distance. This is supported by studies of the butterfly fauna of La Selva that is more similar to that of BCI (Jaccard index: 0.42), than that of the dry forests in Guanacaste, Costa Rica (Jaccard index: 0.27) (DeVries 1994; Hespenheide 1994).

The beetle studies from La Selva are mostly ground based in contrast to the present canopy study, which may indicate an additional stratum differences in species turnover when assuming higher beta diversity among specialists than generalists (Gaston 1991). Saproxylic species like Conoderinae and many Buprestidae, are likely to be more specialized in the canopy than on ground, as wood resources there are fresher and less dominated by fungi. In contrast, the understorey is dominated by wood resources of different decomposing stages that probably require a more generalized fauna of saproxylic insects (Price 1992; Bernays and Chapman 1994). These speculations, however, need further study.

Erwin (1983, 1991) suggested extremely high beta diversity among tropical insect communities. He recorded 1080 species of beetles in four neotropical forest types in the same area (at most 67 km apart). Only 1% of the species was shared between the sites. Among Lepidoptera species sampled in Beni, Bolivia (933 species and 1748 individuals), and in Paitza, Peru (1006 species and 1731 individuals; the two sites are 500 km apart), only 3.2% of the species were shared between sites (Erwin 1991). These arguments were used as indications of restricted distribution patterns and evidence for huge species richness. However, it is important to be aware of that calculations of beta-diversity of tropical forests tend to overestimate species turnover due to the dominance of rare species (Mawdsley 1996; Stork 1997) and small sample size (Chao et al. 2000). Simulations reveal that high species turnover will occur randomly given a small sample size and a large species pool (Chao et al. 2000). In the case of Erwin (1991), there were less than two specimens per species, an indication of a sample size representing a very small fraction of alpha diversity. On the other hand, comparisons of sites based on similar number of individuals but different species richness (as in the present study) may underestimate beta diversity because rare species shared between the sites may be unrecorded (Colwell and Coddington 1994).

It is obvious that species turnover increase across forest types where resources are completely different (Harrison et al. 1992). Likewise, abiotic factors like temperature are assumed to be more important than vegetation changes for species turnover in phytophagous insects as exemplified by geometrid moths in montane rainforests (Brehm et al. 2003). Additional insight would probably come from comparisons within similar forest types. However, it is difficult to find sites that are similar enough with regard to climate, habitat type, and other environmental factors, while also being geographically distant enough to be relevant (Bartlett et al. 1999).

It would also be important to study beta diversity in different regions, especially with respect to variation in geology and climate. For these reasons, species turnover among phytophagous insect communities in Panama may be relatively high compared with other tropical forests as beta diversity of trees in Panamanian forests is higher than in western Amazonian forests due to steeper climatic gradients (Condit et al. 2002).

### *Host specificity*

Although many hypotheses have been suggested for the evolution of host range (e.g., Basset 1992), very few studies, if any, have reported different levels of host specificity among comparable groups of herbivorous insects at different tropical sites. In many ways the patterns found in the present study fit well with current hypotheses. With basis in forests of different ages, the young forest should be dominated by pioneer plants investing more resources in growth than defence against herbivory (Coley 1983; Coley et al. 1985), a factor that promotes polyphagy (i.e. resource availability hypothesis). The dominance of deciduous species in PNM may have similar consequences (MacLean and Jensen 1985), although the outcome is not obvious since plant phenology is also seasonally predictable. Fluctuations in insect populations are likely to be higher in dry forests than in wet forests (Wolda 1978) which may be caused by unpredictable environments. Specialists may be more vulnerable than generalists if resources are unstable (e.g., Redfearn and Pimm 1988). In addition, dry forests are more isolated and vulnerable against human impact (Janzen 1988), a fact that probably affect specialists more than generalists due to smaller range (Ghazoul 2002). On the other hand, common species may be more heavily affected by fragmentation because rare species are better dispersers (Didham et al. 1998).

These hypotheses were mainly developed for leaf chewing insects, and it is unknown to what extent they apply for flower visitors and saproxylic species. The higher degree of specialization among flower visitor in SLA in the present study may be influenced by different phenology of the host plants. In seasonal forests (PNM), flowering events tend to synchronize in dry season, while they are more distributed over the year in wet forests (*personal observation*). Since flower visitors are mainly generalists and loosely associated with their flowers (Roubik 1992), synchronized flowering season may attract several seasonal insect populations simultaneously resulting in a lower effective specialization as the same insect species visits several flowers of different plant species. On the other hand, dispersed flowering events attract only those insect populations that happen to be in phase (adults) at the time of florescence. Thus, the high level of effective specialization among flower visitors in SLA may be an artefact caused by the current availability of resources.

Hypotheses for the evolution of host range may apply for saproxylic species as well as leaf chewers since the wood borers are dominated by species attracted to recently dead wood that may have certain levels of plant defence (Bernays and Chapman 1994). An additional hypothesis for explaining the higher degree of specialization among saproxylic species in SLA relates to the increased frequency of fungal diseases in wet forests. If insects escape from fungal attack trough specialization, this phenomenon works similar as increased predation pressure which is a major selection pressure towards specialization (Jermy 1988; Novotny et al. 1999).

The measured level of host specificity may be influenced by sample size because generalists and specialists accumulate in samples at different rates (Novotny et al. 2002b), a fact that may bias smaller samples towards higher specialization, as in SLA. This is probably of minor importance since sample sizes are not very different, and that the observed differences in host specificity only apply for saproxylic species and flower visitors. Another problem relates to the fact that effective specialization does not account for phylogeny of host plants. This is probably not critical given the present sample size and the procedure of selection of target plants.

### *Implications for regional species richness*

Global arthropod species richness was revised by Ødegaard (2000a) based on the present data set from PNM. He concluded that a working figure of 5 million species would be appropriate. Results from the present study suggest some consequences for ecological estimates of regional species richness. Higher alpha- and beta-diversities as well as more host specific faunas based on data from wet forests, all points in the same direction towards a higher number of species. More importantly, Ødegaard (2004) substantiated that the number of phytophagous insect species on a tropical tree might be an order of magnitude higher than that reported from extensive studies, because sample size, important microhabitats, and successional stages of trees linked to vertical stratification of insect assemblages are ignored or heavily underestimated in most studies (Basset et al. 2003). The conciliation in estimates of global arthropod species richness as indicated by Ødegaard (2000a) and Novotny et al. (2002a) might therefore be challenged by estimates based on data sets considering these confounding variables.

## Acknowledgements

I am grateful to the Canopy Crane Team at STRI including S.J. Wright, V. Horlyck, M. Ruiz, M. Samaniego, J. Herrera, E. Andrade, and O. Saldana for providing crane access during my stays in Panama. Several taxonomists kindly helped with identification of beetles: R.S. Anderson, L. Borowiec, W.E. Clark, R.W. Flowers, E. Giesbert, H.A. Hespenheide, F.T. Hovore, B. Jordal, J.M. Kingsolver, L.R. Kirkendall, D.G. Kissinger, C.W. O'Brian, C.L. Staines, H.P. Stockwell, K.H. Thunes, and D.M. Windsor. The warmest thanks go to Å. Viken for punching of data and fruitful discussions. The manuscript benefited from comments by Y. Basset, O.T. Sandlund, and Å. Viken. This study was financially supported by the Norwegian Research Council, the Smithsonian Institution Fellowship Programme, and the Norwegian Institute for Nature Research.

## References

Alonzo-Zarazaga M.A. and Lyal C.H.C. 1999. A world catalogue of families and genera of Curculionoidea (Insecta: Coleoptera) (Excepting Scolytidae and Platypodidae). Entomopraxis, S.C.P., Barcelona.

Andow D.A. 1991. Vegetational diversity and arthropod population response. Annu Rev. Entomol. 36: 561–586.

Barlett R., Pickering J., Gauld I. and Windsor D. 1999. Estimating global biodiversity: tropical beetles and wasps send different signals. Ecol. Entomol. 24: 118–121.

Basset Y. 1992. Host specificity of arboreal and free-living insect herbivores in rain forests. Biol. J. Linnean Soc. 47: 115–133.

Basset Y. 2001. Invertebrates in the canopy of tropical rain forests. How much do we really know? Plant Ecol. 153: 87–107.

Basset Y., Mavoungou J.F., Mikissa J.B., Missa O., Miller S.E., Kitching R.L. and Alonso A. 2004. Discriminatory power of different arthropod data sets for the biological monitoring of anthropogenic disturbance in tropical forests. Biodivers. Conserv. 13: 709–732.

Basset Y., Novotny V., Miller S.E. and Kitching R.L. (eds) 2003. Arthropods of Tropical Forests. Spatio-temporal Dynamics and Resource Use in the Canopy. Cambridge University Press, Cambridge.

Basset Y., Samuelson G.A., Allison A. and Miller S.E. 1996. How many species of host-specific insects feed on a species of tropical tree. Biol. J. Linnean Soc. 59: 201–216.

Bernays E.A. and Chapman R.F. 1994. Host-plant Selection by Phytophagous Insects. Chapman and Hall, New York.

Brehm G., Homeier J. and Fiedler K. 2003. Beta diversity of geometrid moths (Lepidoptera: Geometridae) in an Andean montane rainforest. Divers. Distrib. 9: 351–366.

Broadhead E. and Wolda H. 1985. The diversity of Psocoptera in two tropical forests in Panama. J. Animal Ecol. 54: 739–754.

Chao A., Hwang W.-H., Chen Y.-C. and Kuo C.Y. 2000. Estimating the number of shared species in two communities. Stat. Sinica. 10: 227–246.

Coleman B.D. 1981. On the random placement and species–area relations. Math. Biosci. 54: 191–215.

Coley P.D. 1983. Herbivory and defensive characteristics of tree species in a lowland tropical forest. Ecol. Monogr. 53: 209–233.

Coley P.D., Bryant J.P. and Chapin F.S. 1985. Resource availability and plant antiherbivore defence. Science 230: 895–899.

Colville J., Picker M.D. and Cowling R.M. 2002. Species turnover of monkey beetles (Scarabaeidae: Hopliini) along environmental and disturbance gradients in the Namaqualand region of the succulent Karoo, South Africa. Biodivers. Conserv. 11: 243–264.

Colwell R.K. 1997. Estimates: statistical estimation of species richness and shared species from samples. Version 5. User's guide and application published at: http://viceroy.eeb.uconn.edu/estimates.

Colwell R.K. and Coddington J.A. 1994. Estimating terrestrial biodiversity through extrapolation. Philos. Trans. Roy. Soc. Lond. B 345: 101–118.

Colwell R.K. and Futuyma D.J. 1971. On the measurement of niche breadth and overlap. Ecology 52: 567–576.

Condit R., Ashton P.S., Baker P., Bunyavejchewin S., Gunatilleke S., Gunatilleke N., Hubbell S.P., Foster R.B., Itoh A., LaFrankie J.V., Lee H.S., Losos E., Manokaran N., Sukumar R. and Yamakura T. 2000. Spatial patterns in the distribution of tropical tree species. Science 288: 1414–1418.

Condit R., Pitman N., Leigh E.G.J., Chave J., Terborgh J., Foster R.B., Nunez P.V., Aguilar S., Valencia R., Villa G., Muller-Landau H.C., Losos E. and Hubbell S.P. 2002. Beta-diversity in tropical forest trees. Science 295: 666–669.

Connor E.F., Feath S.H., Simberloff D. and Opler P.A. 1980. Taxonomic isolation and the accumulation of herbivorous insects: a comparison of introduced and native trees. Ecol. Entomol. 5: 205–211.

Connor E.F. and McCoy D. 1979. The statistics and biology of the species–area relationship. Am. Nat. 113: 791–833.

DeVries P.J. 1994. Patterns of butterfly diversity and promising topics in natural history and ecology. In: Mcdade L.A., Bawa K.S., Hespenheide H.A. and Hartshorn G.S. (eds), La Selva. Ecology and Natural History of a Neotropical Rain Forest. The University of Chicago Press, Chicago, pp. 187–194.

Didham R.K., Hammond P.M., Lawton J.H., Eggleton P. and Stork N.E. 1998. Beetle species responses to tropical forest fragmentation. Ecol. Monogr. 68: 295–323.

Erwin T.L. 1982. Tropical forests: their richness in Coleoptera and other arthropod species. Coleopts. Bull. 36(1): 74–75.

Erwin T.L. 1983. Beetles and other insects of the tropical forest canopies at Manaus, Brazil, sampled with insecticidal fogging techniques. In: Sutton S.L, Whitmore T.C. and Chadwick A.C. (eds), Tropical Rain Forest: Ecology and Management. Blackwell Scientific Publications, Oxford, pp. 59–75.

Erwin T.L. 1991. How many species are there? Revisited. Conserv. Biol. 5: 1–4.

Floren A. and Linsenmair E. 2001. The influence of anthropogenic disturbance on the structure of arboreal arthropod communities. Plant Ecol. 153: 153–167.

Floren A. and Linsenmair E. 2003. How do beetle assemblages respond to anthropogenic disturbance? In: Basset Y., Kitching R.L., Miller S. and Novotny V. (eds), Arthropods of Tropical Forests. Spatio-temporal Dynamics and Resource Use in the Canopy. Cambridge University Press, Cambridge, pp. 190–197.

Flowers R.W. and Janzen D.H. 1997. Feeding records of Costa Rican leaf beetles (Coleoptera: Chrysomelidae). Florida Entomol. 80: 334–366.

Gaston K.J. 1991. How large is a species' geographical range? Oikos 61: 434–438.

Gentry A.H. 1988. Changes in plant community diversity and floristic composition on environmental and geographical gradients. Ann. MO Bot. Gard. 75: 1–34.

Ghazoul J. 2002. Impact of logging on the richness and diversity of forest butterflies in a tropical dry forest in Thailand. Biodiv. Conserv. 11: 521–541.

Gleason H.A. 1922. On the relation between species and area. Ecology 3: 158–162.

Grove S.J. 2002. The influence of forest management history on the integrity of the saproxylic beetle fauna in an Australian lowland tropical rainforest. Biol. Conserv. 104: 149–171.

Hammond P.M. 1992. Species inventory. In: Groombridge B. (ed.), Global Biodiversity. Status of the Earth's Living Resources. Chapman & Hall, London, pp. 17–39.

Hanski I. and Gilpin M. 1991. Metapopulation dynamics: brief history and conceptual domain. Biol. J. Linnean Soc. 42: 3–16.

Harrison S., Ross S.J. and Lawton J.H. 1992. Beta diversity on geographic gradients in Britain. J. Animal Ecol. 61: 151–158.

Hespenheide H.A. 1994. An overview of faunal studies. In: Mcdade L.A., Bawa K.S., Hespenheide H.A. and Hartshorn G.S. (eds), La Selva. Ecology and Natural History of a Neotropical Rain Forest. The University of Chicago Press, Chicago.

Heywood V.H. (ed.) 1995. Global Biodiversity Assessment (GBA). UNEP, Cambridge University Press, Cambridge.

Janzen D.H. 1988. Tropical dry forests: the most endangered major tropical ecosystem. In: Wilson E.O. and Peter F.M. (eds), Biodiversity. National Academy Press, Washington, DC, pp. 130–137.

Jermy T. 1988. Can predation lead to narrow food specialization in phytophagous insects? Ecology 69: 902–904.

Lawton J.H. and Schröder D. 1977. Effects of plant type, size of geographical range and taxonomic isolation on the number of insect species associated with British plants. Nature 265: 137–140.

Leigh E.G.J. 1999. Tropical Forest Ecology. A View from Barro Colorado Island. Oxford University Press, New York.

MacLean S.F.J. and Jensen T.S. 1985. Food plant selection by insect herbivores in Alaskan arctic tundra: the role of plant life form. Oikos 44: 211–221.

Magurran A.E. 2003. Measuring Biological Diversity. Blackwell Publishing, Oxford.

Mawdsley N. 1996. The theory and practice of estimating regional species richness from local samples. In: Edwards D.S., Booth W.E. and Choy S.C. (eds), Tropical Rainforest Research-Current Issues. Kluwer Academic Publishers, Dordrecht, pp. 193–213.

Mawdsley N.A. and Stork N.E. 1997. Host-specificity and effective specialization of tropical canopy beetles. In: Stork N.E., Adis J. and Didham R.K. (eds), Canopy Arthropods. Chapman & Hall, London, pp. 104–130.

May R.M. 1990. How many species? Philos. Trans. Roy. Soc. Lond. B 330: 293–304.

May R.M. 1994. The effects of spatial scale on ecological questions and answers. In: Edwards P.J., May R.M. and Webb N.R. (eds), Large-Scale Ecology and Conservation Biology. Blackwell Scientific Publications, Oxford, pp. 1–17.

Neuvonen S. and Niemelä P. 1983. Species richness and faunal similarity of arboreal insect herbivores. Oikos 40: 452–459.

Novotny V. and Basset Y. 2000. Rare species in communities of tropical insect herbivores: pondering the mystery of singletons. Oikos 89: 564–572.

Novotny V., Basset Y., Auga J., Boen W., Dal C., Drozd P., Kasbal M., Isua B., Kutil R., Manumbor M. and Molem K. 1999. Predation risk for herbivorous insects on tropical vegetation: a search for enemy-free space and time. Aust. J. Ecol. 24: 477–483.

Novotny V., Basset Y., Miller S.E., Weiblen G.D., Bremer B., Cizek L. and Drozd P. 2002a. Low host specificity of herbivorous insects in a tropical forest. Nature 416: 841–844.

Novotny V., Basset Y., Miller S.E., Drozd P. and Cizek L. 2002b. Host specialization of leaf-chewing insects in a New Guinea rainforest. J. Animal Ecol. 71: 400–412.

Novotny V. and Missa O. 2000. Local versus regional species richness in tropical insects: one lowland site compared with the island of New Guinea. Ecol. Entomol. 25: 445–451.

Ødegaard F. 2000a. How many species of arthropods? Erwin's estimate revised. Biol. J. Linnean Soc. 71: 583–597.

Ødegaard F. 2000b. The relative importance of trees versus lianas as hosts for phytophagous beetles (Coleoptera) in tropical forests. J. Biogeogr. 27: 283–296.

Ødegaard F. 2003. Species richness, taxonomic composition and host specificity of phytophagous beetles in the canopy of a tropical dry forest in Panama. In: Basset Y., Novotny V., Miller S.E. and Kitching R.L. (eds), Arthropods of Tropical Forests. Spatio-temporal Dynamics and Resource Use in the Canopy. Cambridge University Press, Cambridge, pp. 220–236.

Ødegaard F. 2004. Species richness of phytophagous beetles in the tropical tree *Brosimum utile* (Moraceae): the effects of sampling strategy and the problem of tourists. Ecol. Entomol. 29: 76–88.

Ødegaard F., Diserud O.H., Engen S. and Aagaard K. 2000. The magnitude of local host specificity for phytophagous insects and its implications for estimates of global species richness. Conserv. Biol. 14: 1182–1186.

Pimm S.L. 1991. The Balance of Nature? Ecological Issues in the Conservation of Species and Communities. The University of Chicago Press, Chicago.

Price P.W. 1992. The resource-based organization of communities. Biotropica 24: 273–282.

Price P.W., Diniz I.R, Morais H.C. and Marques E.S.A. 1995. The abundance of insect herbivore species in the tropics: the high local richness of rare species. Biotropica 27: 468–478.

Redfearn A. and Pimm S.L. 1988. Population variability and polyphagy in herbivorous insect communities. Ecol. Monogr. 58: 39–55.

Ricotta C., Carranza M.L. and Avena G. 2002. Computing beta-diversity from species area curves. Basic Appl. Ecol. 3: 15–18.

Roubik D.W. 1992. Loose niches in tropical communities: why are there so many trees and so few bees? In: Hunter M.D., Ohgushi T. and Price P.W. (eds), Resource Distribution and Animal-Plant Interactions. Academic Press, San Diego, CA.

Saunders D.A., Hobbs J.R. and Margules C.R. 1991. Biological consequences of ecosystems fragmentation: a review. Conserv. Biol. 5: 18–32.

Shmida A. and Wilson M.V. 1985. Biological determinants of species diversity. J. Biogeogr. 12: 1–20.

Southwood T.R.E. 1960. The abundance of the Hawaiian trees and the number of their associated insect species. Proc. Hawaiian Entomol. Soc. 17: 299–303.

Southwood T.R.E. 1961. The number of species of insect associated with various trees. J. Animal Ecol. 30: 1–8.

Stork N.E. 1997. Measuring global biodiversity and its decline. In: Reaka-Kudla M.L., Wilson D.E. and Wilson E.O. (eds), Biodiversity II. Joseph Henry Press, Washington, DC, pp. 41–68.

Strong D.R., Lawton J.H. and Southwood S.R. 1984. Insects on Plants. Harvard University Press, Cambridge, MA.

Whittaker R.H. 1960. Vegetation of the Siskiyou Mountains, Oregon and California. Ecol. Monogr. 30: 279–338.

Wolda H. 1978. Fluctuations in abundance of tropical insects. Am. Nat. 112: 1017–1045.

Wolda H. 1980. Seasonality of tropical insects I. Leafhoppers (Homoptera) in Las Cumbres, Panama. J. Animal Ecol. 49: 277–290.

Wolda H., O'Brien C.W. and Stockwell H.P. 1998. Weevil diversity and seasonality in tropical Panama as deduced from light-trap catches (Coleoptera: Curculionidae). Smithsonian Contrib. Zool. 590: 1–79.

Wright S.J. 1992. Seasonal drought, soil fertility, and the species density of tropical forest plant communities. Trends Ecol. Evol. 7: 260–263.

Wright J. and Colley M. 1994. Accessing the Canopy. Assessment of Biological Diversity and Microclimate of the Tropical Forest Canopy: Phase I. United Nation Environmental Program, Nairobi, Kenya.

Wright S.J., Horlyck V., Basset Y., Barrios H., Bethancourt A., Bohlmann S.A., Gilbert G.S., Goldstein G., Graham E.A., Kitajima K., Lerdau M.T., Meinzer F.C., Ødegaard F., Reynolds D.R., Roubik D.W., Sakai S., Samaniego M., Sparks J.P., Van Bael S., Winter K. and Zotz G. 2003. Tropical Canopy Biology Program, Republic of Panama. In: Basset Y., Horlyck V. and Wright S.J. (eds), Studying Forest Canopies from Above: The International Canopy Crane Network. Smithsonian Tropical Research Institute and UNEP, Panama, pp. 137–155.

Biodiversity and Conservation (2006) 15:93–107
DOI 10.1007/s10531-004-3955-y 

# Vertical and temporal patterns of biodiversity of fruit-feeding butterflies in a tropical forest in Uganda

FREERK MOLLEMAN[1,*], ARJAN KOP[1], PAUL M. BRAKEFIELD[1], PHILIP J. DE VRIES[2] and BAS J. ZWAAN[1]
[1]*Institute of Biology, Leiden University, P.O. Box 9516, 2300 RA Leiden, The Netherlands;* [2]*Department of Biological Sciences, 2000 Lake Shore Drive, University of New Orleans, New Orleans, LA 70148, USA; *Author for correspondence (e-mail: freerkmolleman@hotmail.com; phone: +31-71-527-4882; fax: +31-71-527-4900)*

Received 31 March 2004; accepted in revised form 14 September 2004

**Key words:** Abundance, Canopy, Nymphalidae, Trap, Understory

**Abstract.** Quick surveys are often used by conservation biologists to assess biodiversity. In tropical forests, fruit-feeding butterflies are a convenient indicator group because they can be readily trapped and are comparatively easy to identify. However, studies carried out in Costa Rica and Ecuador have revealed that long-term sampling is needed to estimate biodiversity accurately. Furthermore, almost half of the biodiversity of fruit-feeding butterflies in the neotropics was found to be in the canopy. Short term sampling in the understory can, therefore, lead to inaccurate estimates of species richness and worse, to poorly informed conservation decisions. Comparable to the studies in South America, we performed a long-term trapping study of the same guild of butterflies in the understory and canopy of Kibale Forest in Uganda, to describe temporal and vertical patterns of biodiversity. We caught 32,308 individuals of 94 species over three years. About 14% of these species could be categorized as canopy specialists and 68% as understory specialists. Temporal variation was extensive and did not follow a clear seasonal pattern. This is the first study in an African forest with continuous sampling of fruit-feeding butterflies over multiple years and in both canopy and understory.

## Introduction

One of the great challenges in contemporary ecology is to elucidate the many spatial and temporal processes that affect patterns of biodiversity. For instance, why do we find a certain number of species in a habitat? Why does biodiversity change when the ecosystem is disturbed? and Why are some species typically rare, and others common? The first step towards understanding biodiversity is to describe it by taking appropriate samples on an appropriate temporal and spatial scale.

Quick surveys are often used to assess biodiversity and to compare it for different areas or for the effect of disturbance or management. Several studies have evaluated the value of different monitoring methods (Kremen 1994; Fagan and Kareiva 1997; Wood and Gillman 1998). Long-term studies are essential to assess temporal variation as well as changes due to climate change, disturbance and management, and to test the reliability of rapid inventories (Struhsaker

2002). Such studies are, however, rare because most funding schemes promote short projects and political problems that may arise in the country of study can cut projects short (Pennisi 1989; Hagen 1990; Mervis 1998).

Tropical forests encompass a high proportion of the biodiversity (Myers et al. 2000). Much of this biodiversity and its dynamics is unknown, and, in particular, the canopy is still a poorly studied habitat (Walter et al. 1998; Basset 2001; Mitchell 2001). Nearly all tropical forests lie within developing countries where there are few resources and opportunities for local researchers to work in the natural habitat, and most such work is carried out by Western biologists. The area of tropical forest has decreased dramatically in recent decades and will continue to decrease leaving isolated fragments within the borders of protected areas (FAO 1999; Jenkins 2003). Therefore, it is now crucial to study biodiversity in relation to conservation in tropical forests (Stork 2001). Particularly in Africa, little monitoring has taken place in tropical forests in protected areas, and management decisions are rarely based on scientific data representative for the area. The management is rarely evaluated in follow-up programs to examine its effectiveness in achieving the conservation objectives (Struhsaker 2002).

Which taxa are most appropriate for a biodiversity study depends on the properties of the specific area and the objectives of the study (Pearson 1994). The taxon has to be sufficiently diverse within the area of study to give the study statistical power, and at the same time, diversity should not be too high in order to keep the study manageable on the levels of sample size and identification. In other words, species accumulation curves should reasonably quickly level off. The taxon can be chosen from different trophic levels, each with their own scale of response to habitat changes. Another important criterion is that the taxon can be efficiently sampled and in a repeatable way, independent of the person who is performing the sampling. Apart from biodiversity in the area, the conservation status of the species in question should be considered: a larger number of widespread species may not compensate for a decrease in endemic or localized species (Fagan and Kareiva 1997).

To select indicator taxa, one has to quantify to what extent spatial patterns of species richness coincide among different groups. High congruence would be encouraging for rapid biodiversity assessment, but studies on temperate areas have revealed rather low congruence in species richness (but see Lund and Rahbek 2002). A biodiversity study on tropical forest in Uganda using woody plants, large moths, butterflies, birds, and small mammals, also revealed a very low congruence (Howard et al. 1998, 2000). The latter study suggested, however, that the combination of butterflies and birds could be used as a reliable indicator for conservation value of an area, in part, because of their complementarity. However, patterns of congruence and complementarity vary between biogeographical regions (Pearson and Carroll 1999; Cleary 2002), habitats (Ricketts et al. 2002) and conservation status (Moore et al. 2003).

In tropical forests, the diversity of Lepidoptera is particularly high, and especially butterflies are readily identifiable and sufficiently diverse to be widely used as indicators of biodiversity. Butterflies are sensitive to changes in the habitat and are, therefore, important candidates for monitoring (Pearson 1994). They are sensitive to floral diversity, vegetation structure (Brehm et al. 2003), structural components of the habitat (Hill et al. 2001) and climate change (Parmesan et al. 1999). Fruit-feeding butterflies can be monitored with standard traps baited with fermenting fruits, as butterflies generally do not recognize the traps as food sources after release, thus avoiding biassed sampling (Hughes et al. 1998). Butterflies of the fruit-feeding guild are all members of the Nymphalidae and in Africa include Charaxinae, Satyrinae and Nymphalinae (Larsen 1991).

Many studies have used butterflies to assess biodiversity in tropical forests and most of these studies have evaluated the effect of disturbance (Daily and Ehrlich 1995; Hill et al. 1995; Hamer et al. 1997, 2003; Lawton et al. 1998; Wood and Gillman 1998; DeVries et al. 1999; Hamer and Hill 2000; Willott et al. 2000; Lewis 2001; Ghazoul 2002; Summerville and Crist 2002; Cleary 2003; Stork et al. 2003) or natural gradients (Lewis et al. 1998; Hill et al. 2001; Pyrcz and Wojtusiak 2002) on butterfly communities. Studies in Costa Rica and Ecuador are particularly noteworthy for their long duration (>5 years) and the inclusion of the vertical dimension (DeVries 1988; DeVries et al. 1997, 1999; DeVries and Walla 2001; Engen et al. 2002; Walla et al. 2004). These studies have yielded the spatial and temporal scale needed to reliably estimate the biodiversity of fruit-feeding butterflies in neotropical forests using traps baited with fermenting fruit. An important component of biodiversity in these neotropical areas was found in the canopy, where a community with a different temporal pattern was detected (DeVries and Walla 2001).

Other studies on butterflies in the canopy of tropical forests are few in number and have generally used relatively short sampling periods. Systematic sampling with the use of fruit-baited traps was performed on Sabah (Malaysia) where a small community of canopy butterflies was found (Hill et al. 2001). In Northern Queensland (Australia) a limited canopy fauna was observed from canopy towers (Hill et al. 1992). In Vietnam, an important canopy component to butterfly diversity was noted, although, the method used did not allow for species richness analyses of the canopy (Spitzer et al. 1993). A combination of fruit-baited traps and standardized counts from a canopy walkway in Borneo revealed a lower abundance and species number of fruit-feeding butterflies in the canopy than in understory, whilst the nectar-feeding guild showed a reversed pattern (Schülze et al. 2001). Canopy traps in the Ivory Coast revealed a limited number of canopy specialists (Fermon et al. 2003). Few long-term trapping studies of fruit-feeding butterflies have been carried out in Africa. Owen (1971) performed trapping studies with a duration of more than one year in a garden in Freetown (Sierra Leone) and a garden in Kampala (Uganda). Windig et al. (1994) collected five species of *Bicyclus* in Malawi over a 3-year-period on a daily basis. Libert (1994) sampled butterflies at two forested hills in Cameroon for 7 years, using transect walks. Butterflies have only occasionally been used in biodiversity surveys on mainland Africa (Fuller et al. 1998; Howard et al. 2000; Rogo and Odulaja 2001; Fermon et al. 2003; Stork et al. 2003).

Here we present data from a three-year trapping study in canopy and understory at a single location in Kibale forest, Uganda, using a method adapted from DeVries and co-workers. This forest combines floral and faunal elements from both the West and East African biogeographical region and can therefore be considered a representative area for tropical Africa (Vonesh 2001). The abundance and biodiversity patterns of fruit-feeding butterflies are described, and indications are given for effective monitoring of fruit-feeding butterflies in Kibale Forest. This is the first study in Africa with continuous sampling of fruit-feeding butterflies over multiple years and in both canopy and understory.

## Material and methods

### *Field site*

This study was conducted from April 2000 to September 2003 at Makerere University Biological Field Station in Kibale Forest National Park, Western Uganda (0°35′ N 30°20′ E). The field station borders selectively logged moist evergreen forest at an altitude of around 1500 m and is therefore classified as a transition towards montane forest. The mean maximum temperature is 23.8 °C and the mean annual rainfall is 1749 mm and is bimodal in distribution (Rode et al. 2003).

### *Trapping method*

We used live-traps for butterflies (height 125 cm, diameter 35 cm) (DeVries et al. 1997), baited with two spoons of fermenting banana. The bait was prepared three days prior to baiting and not replaced unless lost. Replacement bait was taken from the original stock and was therefore at a similar stage of fermentation.

The traps were hung at 22 trap locations at a minimum of 100 m apart in closed canopy forest (Figure 1). Each trap location was baited with an understory trap, 40 cm from the ground, and a canopy trap, suspended from a high branch, between 20 and 30 m high. For each canopy trap, height of suspension, tree species, tree height, crown size and number of trees touching the tree in the canopy were noted.

During the study, canopy traps had to be replaced regularly due to broken ropes (monkeys occasionally ate them), broken branches, newly grown branches or fallen trees. In the latter case, both the understory and canopy traps were moved to a new nearby location. Occasionally, traps were regularly visited by predators (probably bats), and in such cases the trap location was also changed. The traps were baited once every four weeks and then scored for four consecutive days. On two days (21–22 November 2001), the data collection was

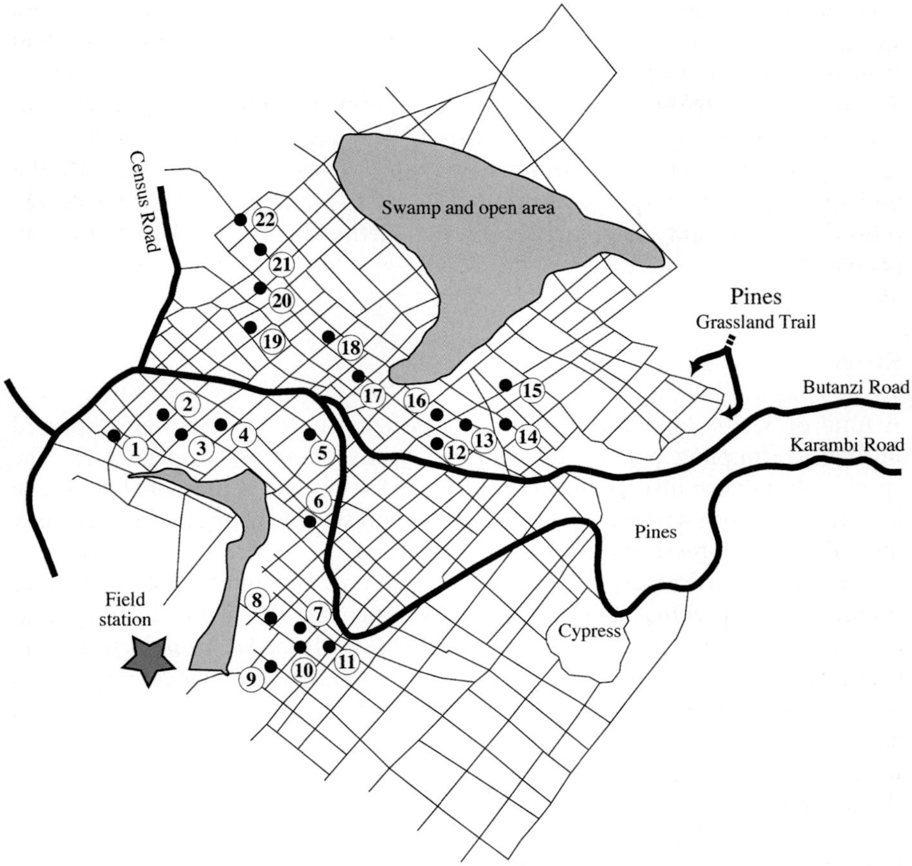


*Figure 1.* Map of the study site at Makerere University Biological Field-Station with trap locations and trail-system.

cut short due to adverse weather, and for two periods data were lost (January 2001, July 2002). Traps were excluded from analysis when traces of predation had been found, or the bait was lost. Individual abundance per period was calculated as mean number of individuals per trap per day to correct for the slight temporal variation in trap numbers.

When possible, butterflies were identified in the field and released, without marking. All other butterflies were collected in glassine envelopes for later identification. Selected and representative specimens have been donated to the Zoologisch Museum Amsterdam (The Netherlands). The analyses were limited to fruit-feeding butterflies, and excluded the facultativily fruit-feeding *Neptis*. Data on trapped *Neptis*, Acraeinae, Lycaenidae and Hesperidae will be published elsewhere. The females of *Bicyclus smithi*, *B. golo* and *B. istaris* are difficult to distinguish and, assuming that the sex-ratio of caught individuals was similar for these three species, numbers of females were assigned to these species in proportion to those found in males. The data for these species were then treated as if they were collected and identified as for all other species. When predation had taken place, the wings that could be found were identified and noted separately. Data analyses closely followed DeVries et al. (1997). To identify species level stratification, a two tailed $\chi^2$-test was performed using a null hypothesis of equal proportions (df = 1, $p$-critical = 0.05). Note the following important differences with the methods used by DeVries et al. (Table 1).

## Results

A total of 32,308 individual butterflies belonging to 94 species were captured during 42 sampling periods over 40 consecutive months (Appendix 1). The species abundance distribution ranged from five species represented by single individuals to one species, *Bicyclus smithi*, represented by approximately 9659 individuals (Figures 2 and 3).

Details on vertical aspects of species richness and individual abundance (number of individual butterflies) are summarized in Table 2. Individual abundances of the whole guild differed significantly between understory and canopy ($\chi^2$, df = 1 $F$ = 2159, $p < 0.001$). Individual abundances per species were typically low in the canopy with the exception of *Sevenia boisduvali*, which had two major peaks in abundance (Figure 4, Appendix 1). Five of the species were found in the canopy only, whilst nine were significantly more common in the canopy. Of all fruit-feeding butterflies caught in the canopy, 3050 (70%) were classified as belonging to canopy specialists (significantly more, using a $\chi^2$-test with a null hypothesis of equal proportions, or only in the canopy). Twenty species were found in understory only, whilst 54% of all species were significantly more common in understory. The vast majority of individuals captured

*Table 1.* Summary of important methodological differences between this study and those of DeVries et al.

| | Molleman et al. | DeVries et al. | Expected effect on measured abundance |
|---|---|---|---|
| Number of trap locations | 22 | 25 | – |
| Height of understory traps | 40–60 cm | 50–150 cm | + |
| Sampling periods | 4 days | 5 days | – |
| Rebaiting | Only when bait was lost | On day 3 | – |
| Recaptured individuals | Counted, predominantly field identifications | Not included, specimens were collected or marked | + |

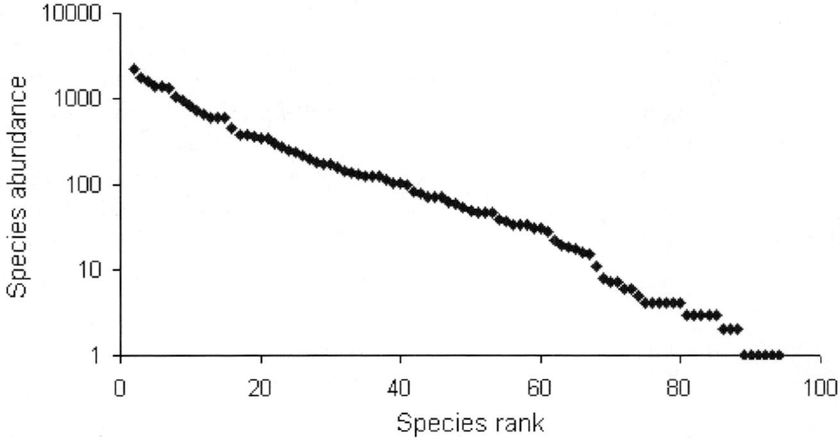


*Figure 2.* Rank-abundance distribution for total sample of fruit-feeding nymphalids form Kibale Forest at Kanyawara (94 species 32,308 individuals).

*Figure 3.* Species abundance distribution for total sample of fruit-feeding nymphalids form Kibale Forest at Kanyawara (94 species 32308 individuals) on a log base 3 scale.

*Table 2.* Species richness and abundance of the pooled Kibale Forest sample partitioned by vertical position.

| | Canopy | Understory | Both | Total |
|---|---|---|---|---|
| Species richness | 75 | 89 | | 94 |
| Specialist species ($\chi^2$) | 9 | 51 | | |
| Exclusively in 1 stratum | 5 | 12 | | |
| Stratum specialist species | 14 (15%) | 63 (67%) | 17 (18%) | 94 |
| Rare species (percentage per stratum) | 5 (36%) | 13 (21%) | 8 (47%) | 27 (28%) |
| Abundance | 4344 (13%) | 27,999 (87%) | | 32,308 |
| Abundance of stratum specialists | 3051 (70%) | 26,910 (96%) | | |

Stratum specialist species are those that are significantly ($\chi^2$-test) more abundant in a particular stratum or were exclusively caught in one of the strata. Rare species are those represented by a total of less than 10 individuals.

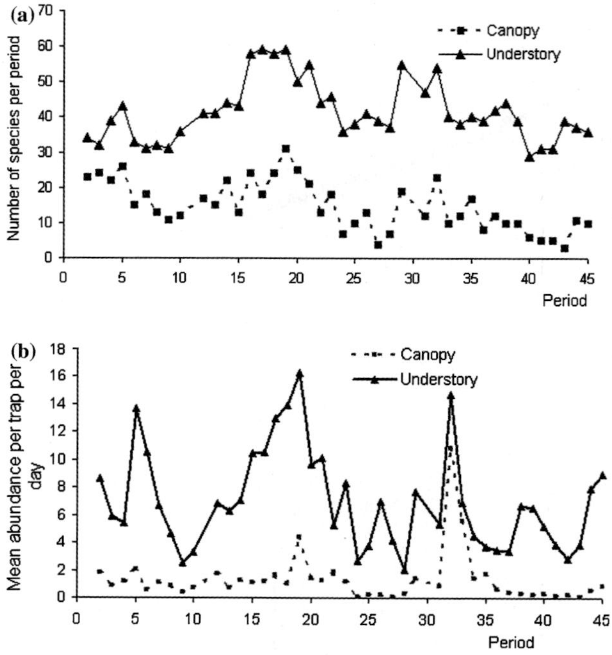


*Figure 4.* Temporal variation of the total Kibale Forest sample by vertical position, from May 2000 to September 2003 in 4 week periods (note that data for periods 11 and 30 are missing). A, species richness. B, individual abundance.

in the understory were classified as understory specialists. The canopy tended to host relatively more rare species than the understory.

Substantial temporal variation was found in species richness and abundance in both canopy and understory (Figure 4). However, this variation did not closely follow seasonal patterns: whilst rainfall shows a bimodal distribution (Rode et al. 2003), only three peaks in butterfly abundance were noted in these 40 months. These plots show the extent and unpredictability of temporal variation which affects the measurement of tropical butterfly diversity.

## Discussion

This is the first study providing estimates of vertical and temporal components of species diversity and individual abundance for fruit-feeding butterflies in Africa. Our data provide strong evidence that temporal and vertical patterns have to be taken into account when quick surveys on butterflies in African forests are considered. The canopy fauna represents an important component of the biodiversity, and temporal variation in abundance and species richness is extensive.

Our results reveal strong temporal variation in trap capture rates varying from 2.0 per trap per day to 16.3 in understory and 0.07–10.9 in canopy. This variation did not follow a clear seasonal pattern emphasizing that standardizing sampling on season or calendar month cannot avoid problems of precision that are associated with quick surveys. For a reliable survey of fruit-feeding butterfly communities with this sampling intensity, trapping should be performed for at least one complete year.

A small proportion of fruit-feeding nymphalids in Kibale Forest can be classified as canopy species and these generally have a low abundance. An exception is *Sevenia boisduvali* which had two abundance peaks during our sampling period (Appendix 1, Figure 4). Species that were typical understory dwellers were occasionally caught in the canopy, and together with the low abundance of true canopy species (70% of canopy abundance was due to canopy specialist vs. 96% of understory specialists in the understory), these may, in part, be responsible for the steep slope of the canopy species accumulation curve (Figure 5). A significant proportion (18%) of the species was regularly caught in both the canopy and understory. Perhaps surprising is the high proportion in the canopy of some species often considered to be understory specialists, in particular *Bicyclus mollitia* and *Euriphura chalcis* as well as some *Cymothoe* species (Appendix 1). The level of vertical stratification in fruit-feeding butterfly species may be affected by vegetation structure, and thus by forest disturbance (DeVries and Walla 2001; Fermon et al. 2003).

In concert with previous investigations in Neotropical forests (DeVries et al. 1997, 1999; DeVries and Walla 2001), this study confirms the utility of long-term, standardized sampling in diverse tropical butterfly communities. Our results demonstrate the significance of vertical and temporal patterns in butterfly biodiversity and invite detailed comparisons between different biogeographical regions and habitats, carefully accounting for differences in methodology like detailed in Table 1. The patterns of species diversity and

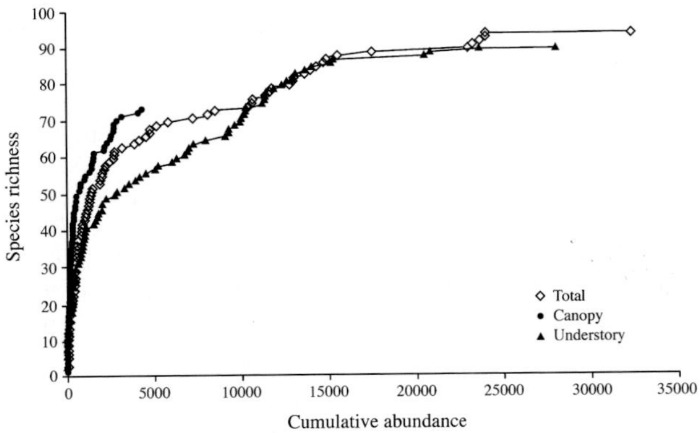


*Figure 5.* Species accumulation curves showing total species vs. individual abundance through time for total sample, and for canopy and understory separately.

abundance observed here in fruit-feeding nymphalid butterflies are likely to be due to underlying ecological and evolutionary factors, and implies that such patterns are inherent in other tropical insect communities.

## Acknowledgements

We thank the late Patrick Kagoro, and Boniface Balyeganira, Dennis Sebugwawo and Moses Musana for their great assistance with the fieldwork. We are grateful to Easter Mugwusi for accurate data entry. Alain Houle assisted with installing canopy traps and redrew the map using GPS co-ordinates. Russ Lande contributed advice on data analyses. Jacque Hecq, K. D. Dijkstra and Rob de Vos have been helpful with butterfly indentifications. This study was conducted with kind permission of the Uganda Wildlife Authority (U.W.A.) and the Ugandan National Council for Science and Technology (U.N.C.S.T.). The funding for this project was provided by the Netherlands Foundation for the Advancement of Tropical Research (WOTRO) (W80-82-238 to PMB) and the Schure-Beijerinck-Popping foundation (FM).

## Appendix 1

Number of individuals caught per species (estimated numbers in italics) with categorisation to stratum.

| Species | Abundance | % in canopy | $F$ ($\chi^2$-test) | Category |
|---|---|---|---|---|
| Charaxinae | | | | |
| *Charaxes etheocles* s.l. | 46 | 71.7 | 4.35 | c |
| *C. anticlea* | 3 | 0.0 | 1.50 | bu |
| *C. bipunctatus* | 129 | 40.3 | 2.42 | b |
| *C. brutus* | 7 | 42.9 | 0.07 | b |
| *C. candiope* | 30 | 33.3 | 1.67 | b |
| *C. castor* | 3 | 100.0 | 1.50 | bc |
| *C. cynthia* | 54 | 3.7 | 23.15 | u |
| *C. etesipe* | 2 | 0.0 | 1.00 | bu |
| *C. fulvescens* | 1405 | 1.3 | 666.96 | u |
| *C. numenes* | 48 | 12.5 | 13.50 | u |
| *C. paphianus* | 4 | 25.0 | 0.50 | b |
| *C. pleione* | 39 | 43.6 | 0.32 | b |
| *C. pollux* | 71 | 11.3 | 21.30 | u |
| *C. porthos* | 1 | 100.0 | 0.50 | bc |
| *C. protoclea* | 31 | 6.5 | 11.76 | u |
| *C. smaragdalis* | 3 | 33.3 | 0.17 | b |
| *C. tiridates* | 34 | 55.9 | 0.24 | b |
| *C. zelica* | 3 | 66.7 | 0.17 | bc |
| *C. zoolina* | 4 | 100.0 | 2.00 | bc |
| *Euxanthe crossleyi* | 15 | 40.0 | 0.30 | b |

*Appendix 1*. Continued.

| Species | Abundance | % in canopy | $F$ ($\chi^2$-test) | Category |
|---|---|---|---|---|
| Nymphalinae | | | | |
| *Charaxes porthos* | 1 | 100.0 | 0.50 | bc |
| *Cymothoe lurida* | 459 | 9.4 | 151.56 | u |
| *C. herminia* | 615 | 14.6 | 153.84 | u |
| *C. caenis* | 3 | 66.7 | 0.17 | b |
| *C. hobarti* | 144 | 36.1 | 5.56 | u |
| *Harma theobene* | 593 | 3.9 | 252.28 | u |
| *Pseudacraea eurytus* | 1 | 0.0 | 0.50 | bu |
| *P. clarckii* | 1 | 100.0 | 0.50 | bc |
| *P. lucretia* | 177 | 60.5 | 3.87 | c |
| *P. semire* | 1 | 0.0 | 0.50 | bu |
| *Catuna crithea* | 214 | 3.3 | 93.46 | u |
| *Aterica galene* | 350 | 1.7 | 163.21 | u |
| *Euphaedra alacris* | 1353 | 0.7 | 658.62 | u |
| *E. christyi* | 306 | 1.6 | 143.16 | u |
| *E. edwardsi* | 168 | 1.8 | 78.11 | u |
| *E. eusemoides* | 240 | 0.0 | 120.00 | u |
| *E. harpalyce* | 654 | 1.1 | 313.15 | u |
| *E. hollandi* | 19 | 5.3 | 7.61 | u |
| *E. kakamega* | 36 | 3.0 | 14.56 | u |
| *E. medon* | 1744 | 1.1 | 832.46 | u |
| *E. preussi* | 274 | 1.8 | 127.18 | u |
| *E. uganda* | 342 | 0.9 | 166.55 | u |
| *E. zaddachi* | 157 | 1.9 | 72.61 | u |
| *Euriphura chalcis* | 11 | 63.6 | 0.41 | b |
| *Neptidopsis ophione* | 122 | 32.0 | 7.93 | u |
| *Euriphene ribensis* | 71 | 2.8 | 31.61 | u |
| *E. saphirina* | 7 | 0.0 | 3.50 | u |
| *Bebearia absolon* | 28 | 0.0 | 14.00 | u |
| *B. sophus* | 251 | 0.8 | 121.53 | u |
| *Ariadne enotrea* | 81 | 32.1 | 5.19 | b |
| *A. pagenstecheri* | 6 | 16.7 | 1.33 | b |
| *Eurytela hiarbas* | 1432 | 58.8 | 22.17 | c |
| *E. dryope* | 6 | 50.0 | 0.00 | b |
| *Sevenia boisduvalli* | 1600 | 96.4 | 688.21 | c |
| *S. umbrina* | 113 | 92.9 | 41.63 | c |
| *S. occidentalium* | 47 | 97.9 | 21.54 | c |
| *Apaturopsis cleocharis* | 138 | 80.4 | 25.57 | c |
| *Junonia stygia* | 69 | 15.9 | 16.01 | u |
| *J. westermanni* | 17 | 29.4 | 1.44 | b |
| *Protogoniomorpha parhassus* | 8 | 25.0 | 1.00 | b |
| *P. temora* | 3 | 0.0 | 1.50 | bu |
| *Salamis cacta* | 62 | 6.5 | 23.52 | u |
| *Hypolimnas anthedon* | 4 | 25.0 | 0.50 | b |
| *H. monteironis* | 2 | 0.0 | 1.00 | bu |
| *H. salmacis* | 5 | 0.0 | 2.50 | bu |
| *Kallimoides rumia* | 374 | 1.3 | 177.13 | u |
| *Kamilla ansorgei* | 2 | 0.0 | 1.00 | bu |
| *Antanartia delius* | 124 | 67.7 | 7.81 | c |

*Appendix 1.* Continued.

| Species | Abundance | % in canopy | $F$ ($\chi^2$-test) | Category |
|---|---|---|---|---|
| *A. dimorphica* | 22 | 50.0 | 0.00 | b |
| *Lachnoptera anticlia* | 169 | 35.5 | 7.10 | u |
| *Phalanta eurytis* | 33 | 21.2 | 5.47 | u |
| *P. phalantha* | 201 | 85.1 | 49.46 | c |
| Satyrinae | | | | |
| *Melanitis ansorgei* | 4 | 0.0 | 2.00 | bu |
| *M. leda* | 99 | 3.0 | 43.68 | u |
| *Gnophodus betsimena* | 103 | 0.0 | 51.50 | u |
| *G. chelys* | 962 | 0.7 | 467.10 | u |
| *G. grogani* | 101 | 4.0 | 42.82 | u |
| *Bicyclus auricruda* | 613 | 6.0 | 236.97 | u |
| *B. buea* | 362 | 3.3 | 157.80 | u |
| *B. campinus* | 16 | 0.0 | 8.00 | u |
| *B. campus* | 4 | 0.0 | 2.00 | bu |
| *B. dentatus* | 78 | 5.1 | 31.41 | u |
| *B. funebris* | 18 | 0.0 | 9.00 | u |
| *B. golo* | 716 | 2.2 | 327.02 | u |
| *B. graueri* | 1075 | 0.7 | 523.59 | u |
| *B. istaris* | 59 | 0.0 | 29.71 | u |
| *B. mandanes* | 859 | 3.8 | 366.04 | u |
| *B. mesogena* | 376 | 2.1 | 172.34 | u |
| *B. mollitia* | 2205 | 18.8 | 429.96 | u |
| *B. safitza* | 4 | 0.0 | 2.00 | bu |
| *B. sambulos* | 46 | 2.2 | 21.04 | u |
| *B. sebetus* | 123 | 0.8 | 59.52 | u |
| *B. smithi* | 9659 | 1.4 | 4570.52 | u |
| *B. vulgaris* | 1 | 0.0 | 0.50 | bu |
| *Henotesia peitho* | 36 | 11.1 | 10.89 | u |

c, canopy species; u, understory species; b, both strata; bc, only in canopy, but NS; bu, only in understory, but NS.

## References


Basset Y. 2001. Invertebrates in the canopy of tropical rain forests: how much do we really know? Plant Ecol. 21: 87–107.

Brehm G., Homeier J. and Fiedler K. 2003. Beta diversity of geometrid moths (Lepidoptera: Geomtridae) in an Andean montane forest. Biodiv. Res. 9: 351–366.

Cleary D.F.R. 2002. Biodiversity and Environmental Change in the Rainforests of Borneo. Faculty of Natural Sciences, Mathematics and Informatics, Universiteit van Amsterdam, Amsterdam, pp. 193.

Cleary D.F.R. 2003. An examination of scale of assessment, logging and ENSO-induced fires on butterfly diversity in Borneo. Oecologia 135: 313–321.

Daily G.C. and Ehrlich P.R. 1995. Preservation of biodiversity in small rain-forest patches: rapid evaluations using butterfly trapping. Biodiv. Conserv. 21: 35–55.

DeVries P.J. 1988. Stratification of fruit-feeding Nymphalid butterflies in Costa Rican rainforest. J. Res. Lepidoptera 26: 98–108.

DeVries P.J., Murray D. and Lande R. 1997. Species diversity in vertical, horizontal, and temporal dimensions of a fruit-feeding butterfly community in an Ecuadorian rainforest. Biol. J. Linn. Soc. 62: 343–364.

DeVries P.J. and Walla T.R. 2001. Species diversity and community structure in neotropical fruit-feeding butterflies. Biol. J. Linn. Soc. 74: 1–15.

DeVries P.J., Walla T.R. and Greeney H.F. 1999. Species diversity in spatial and temporal dimensions of fruit-feeding butterflies from two Ecuadorian rainforests. Biol. J. Linn. Soc. 68: 333–353.

Engen S., Lande R., Walla T. and DeVries P.J. 2002. Analyzing spatial structure of communities using the two-dimensional Poisson lognormal species abundance model. Am. Nat. 160: 60–73.

Fagan W.F. and Kareiva P.M. 1997. Using compiled species lists to make biodiversity comparisons among regions: A test case using Oregon butterflies. Biol. Conserv. 80: 249–259.

FAO. 1999. State of the World's Forests 1999. Food and Agriculture Organization, Rome.

Fermon H., Waltert M. and Muhlenberg M. 2003. Movement and vertical stratification of fruit-feeding butterflies in a managed West African rainforest. J. Insect Conserv. 7: 7–19.

Fuller R.M., Groom G.B., Mugisha S., Ipulet P., Pomeroy D., Katende A., Bailey R. and Ogutu-Ohwayo R. 1998. The integration of field survey and remote sensing for biodiversity assessment: a case study in the tropical forests and wetlands of Sango Bay, Uganda. Biol. Conserv. 13: 379–391.

Ghazoul J. 2002. Impact of logging on the richness and diversity of forest butterflies in a tropical dry forest in Thailand. Biodiv. Conserv. 11: 521–541.

Hagen J.B. 1990. Problems in the institutionalization of tropical biology: the case of the Barro-Colorado-Island-Biological-Laboratory. Hist. Phil. Life Sci. 12: 225–247.

Hamer K.C. and Hill J.K. 2000. Scale-dependent effects of habitat disturbance on species richness in tropical forests. Conserv. Biol. 14: 1435–1440.

Hamer K.C., Hill J.K., Benedick S., Mustaffa N., Sherratt T.N., Maryati M. and Chey V.K. 2003. Ecology of butterflies in natural and selectively logged forests of northern Borneo: the importance of habitat heterogeneity. J. Appl. Ecol. 40: 150–162.

Hamer K.C., Hill J.K., Lace L.A. and Langan A.M. 1997. Ecological and biogeographical effects of forest disturbance on tropical butterflies of Sumba, Indonesia. J. Biogeogr. 9: 67–75.

Hill C.J., Gillison A.N. and Jones R.E. 1992. The spatial-distribution of rain-forest butterflies at 3 sites in North Queensland, Australia. J. Trop. Ecol. 8: 37–46.

Hill J.K., Hamer K.C., Lace L.A. and Banham W.M.T. 1995. Effects of selective logging on tropical forest butterflies on Buru, Indonesia. J. Appl. Ecol. 32: 754–760.

Hill J.K., Hamer K.C., Tangah J. and Dawood M. 2001. Ecology of tropical butterflies in rain-forest gaps. Oecologia 128: 294–302.

Howard P.C., Davenport T.R.B., Kigenyi F.W., Viskanic P., Baltzer M.C., Dickinson C.J., Lwanga J., Matthews R.A. and Mupada E. 2000. Protected area planning in the tropics: Uganda's national system of forest nature reserves. Conserv. Biol. 14: 858–875.

Howard P.C., Viskanic P., Davenport T.R.B., Kigenyi F.W., Baltzer M., Dickinson C.J., Lwanga J.S., Matthews R.A. and Balmford A. 1998. Complementarity and the use of indicator groups for reserve selection in Uganda. Nature 394: 472–475.

Hughes J.B., Daily G.C. and Ehrlich P.R. 1998. Use of fruit bait traps for monitoring of butterflies (Lepidoptera : Nymphalidae). Rev. Biol. Trop. 46: 697–704.

Jenkins M. 2003. Prospects for biodiversity. Science 302: 1175–1177.

Kremen C. 1994. Biological inventory using target taxa: a case-study of the butterflies of Madagascar. Ecol. Appl. 4: 407–422.

Larsen T.B. 1991. The butterflies of Kenya and their natural history. Oxford University Press, Oxford.

Lawton J.H., Bignell D.E., Bolton B., Bloemers G.F., Eggleton P., Hammond P.M., Hodda M., Holt R.D., Larsen T.B. and Mawdsley N.A. et al. 1998. Biodiversity inventories, indicator taxa and effects of habitat modification in tropical forest. Nature 391: 72–76.

Lewis O.T. 2001. Effect of experimental selective logging on tropical butterflies. Conserv. Biol. 15: 389–400.

Lewis O.T., Wilson R.J. and Harper M.C. 1998. Endemic butterflies on Grande Comore: habitat preferences and conservation priorities. Biol. Conserv. 85: 113–121.

Libert M. 1994. Monitoring of numbers and species composition of the butterfly community of 2 forested hills near Yaounde, Cameroun. Rev. Ecol-Terre Vie 25: 151–175.

Lund M.P. and Rahbek C. 2002. Cross-taxon congruence in complementarity and conservation of temperate biodiversity. Anim. Conserv. 5: 163–171.

Mervis J. 1998. Papua New Guinea – management dispute shutters tropical research institute. Science 279: 172–173.

Mitchell A. 2001. Canopy science: time to shape up. Plant Ecol. 153: 5–11.

Moore J.L., Balmford A., Brooks T., Burgess N.D., Hansen L.A., Rahbek C. and Williams P.H. 2003. Performance of sub-Saharan vertebrates as indicator groups for identifying priority areas for conservation. Conserv. Biol. 17: 207–218.

Myers N., Mittermeier R.A., Mittermeier C.G., da Fonseca G.A.B. and Kent J. 2000. Biodiversity hotspots for conservation priorities. Nature 403: 853–858.

Owen D.F. 1971. Tropical Butterflies: The Ecology and Behaviour of Butterflies in the Tropics with Special Reference to African Species. Clarendon Press, Oxford.

Parmesan C., Ryrholm N., Stefanescu C., Hill J.K., Thomas C.D., Descimon H., Huntley B., Kaila L., Kullberg J. and Tammaru T. et al. 1999. Poleward shifts in geographical ranges of butterfly species associated with regional warming. Nature 399: 579–583.

Pearson D.L. 1994. Selecting indicator taxa for the quantitative assessment of biodiversity. Philos. Trans. Roy. Soc. B – Biol. Sci. 345: 75–79.

Pearson D.L. and Carroll S.S. 1999. The influence of spatial scale on cross-taxon congruence patterns and prediction accuracy of species richness. J. Biogeogr. 26: 1079–1090.

Pennisi E. 1989. Panama lab overcomes political turmoil: business is booming at the Smithsonians-Tropical-Research-Institute despite increased conflicts surrounding Noriega regime. Scientist 3.

Pyrcz T.W. and Wojtusiak J. 2002. The vertical distribution of pronophiline butterflies (Nymphalidae, Satyrinae) along an elevational transect in Monte Zerpa (Cordillera de Merida, Venezuela) with remarks on their diversity and parapatric distribution. Global Ecol. Biogeogr. 11: 211–221.

Ricketts T.H., Daily G.C. and Ehrlich P.R. 2002. Does butterfly diversity predict moth diversity? Testing a popular indicator taxon at local scales Biol. Conserv. 103: 361–370.

Rode K.D., Chapman C.A., Chapman L.J. and McDowell L.R. 2003. Mineral resource availability and consumption by Colobus in Kibale National Park, Uganda. Int. J. Primatol. 24: 541–573.

Rogo L. and Odulaja A. 2001. Butterfly populations in two forest fragments at the Kenya coast. Afr. J. Ecol. 39: 266–275.

Schülze C.H., Linsenmair K.E. and Fiedler K. 2001. Understorey versus canopy: patterns of vertical stratification and diversity among Lepidoptera in a Bornean rain forest. Plant Ecol. 153: 133–152.

Spitzer K., Novotny V., Tonner M. and Leps J. 1993. Habitat preferences, distribution and seasonality of the butterflies (Lepidoptera, Papilionoidea) in a Montane tropical rain-forest, Vietnam. J. Biogeogr. 20: 109–121.

Stork N.E. 2001. The management implications of canopy research. Plant Ecol. 153: 313–317.

Stork N.E., Srivastava D.S., Watt A.D. and Larsen T.B. 2003. Butterfly diversity and silvicultural practice in lowland rainforests of Cameroon. Biodiv. Conserv. 12: 387–410.

Struhsaker T.T. 2002. Guidelines for Biological Monitoring and Research in Africa's Rainforest Protected Areas, A report to the Center for Applied Biodiversity Science. Conservation International, Durham, pp. 1–55.

Summerville K.S. and Crist T.O. 2002. Effects of timber harvest on forest Lepidoptera: community, guild, and species responses. Ecol. Appl. 12: 820–835.

Vonesh J. 2001. Natural history and biogeography of the Amphibians and Reptiles of Kibale National Park, Uganda. Contemporary Herpetology.

Walla T.R., Engen S., DeVries P.J. and Lande R. 2004. Modeling vertical beta-diversity in tropical butterfly communities. Oikos 107: 610–618.

Walter D.E., Seeman O., Rodgers D. and Kitching R.L. 1998. Mites in the mist: how unique is a rainforest canopy-knockdown fauna? Aust. J. Ecol. 23: 501–508.

Willott S.J., Lim D.C., Compton S.G. and Sutton S.L. 2000. Effects of selective logging on the butterflies of a Bornean rainforest. Conserv. Biol. 14: 1055–1065.

Windig J.J., Brakefield P.M., Reitsma N. and Wilson J.G.M. 1994. Seasonal polyphenism in the wild: survey of wing patterns in 5 species of *Bicyclus* butterflies in Malawi. Ecol. Entomol. 19: 285–298.

Wood B. and Gillman M.P. 1998. The effects of disturbance on forest butterflies using two methods of sampling in Trinidad. Biodiv. Conserv. 7: 597–616.

Biodiversity and Conservation (2006) 15:109–123
DOI 10.1007/s10531-004-4246-3


# Associations between weevils (Coleoptera: Curculionidea) and plants, and conservation values in two tussock grasslands, Otago, New Zealand

T.J. MURRAY[1,*], K.J.M. DICKINSON[2] and B.I.P. BARRATT[3]
*[1]Ensis, Private Bag 3020, Rotorua, New Zealand; [2]Department of Botany, University of Otago, P.O. Box 56, Dunedin, New Zealand; [3]AgResearch, Invermay Agricultural Centre, Private Bag 50034, Mosgiel, New Zealand; *Author for correspondence (e-mail: taramurray@ensisjv.com)*

Received 5 January 2003; accepted in revised form 20 September 2004

**Key words:** Land conservation values, MDS, Plant–weevil associations, PRIMER, Tussock grassland invertebrates

**Abstract.** Ecosystem level processes and species interactions have become important concepts in conservation and land management. Despite being New Zealand's greatest contributors to global diversity, native invertebrates have been largely overlooked in the assessment of land values, and their diversity has often been assumed to reflect native plant diversity without justification. Invertebrates can in fact affect plant species composition, and in ecosystems such as New Zealand's remaining indigenous and semi-modified tussock grasslands can do so in excess of more conspicuous vertebrate grazers. An understanding of the interactions between invertebrates and their plant hosts may be informative in assessing land conservation values, increase the efficiency of rapid inventory analyses and would be applicable across the production-conservation spectrum. This study considers the Curculionoidea, vascular plant, bryophyte and lichen communities of two semi-modified tussock grasslands in the Otago region of southern New Zealand. Quantitative plant and invertebrate sampling were carried out in January 2001. Data were analysed using ANalysis Of SIMilarity (ANOSIM) and Multi-Dimensional Scaling (MDS) ordination techniques. Vascular plant, bryophyte and lichen species richness was highest in the same site and plots as native weevil species richness, however the proportion of native vegetation in these locations was lower. Associations identified between Curculionoidea and vascular plants were dismissed due to the confounding effect of species frequency across samples. This appeared to have little influence on associations with bryophytes and/or lichens and these were given more weighting. The ecological implications of associations are considered and variability in weevil composition is discussed in reference to the tussock grassland environment. The importance of plant–invertebrate relationships to conservation and the uses and limitations of the PRIMER MDS ordination technique for determining associations are discussed.

## Introduction

Species interact both directly and indirectly through time and space. These interactions vary in form, frequency, intensity and complexity (Begon et al. 1990) and, in conjunction with abiotic factors, determine community structure and ecosystem function. Ecosystem and species interactions have become important concepts in sustainable management and conservation, as numbers of threatened species greatly outweigh conservation resources (Myers et al.

2000). By necessity therefore, biodiversity and ecosystem conservation are beginning to prevail over single species preservation (Ehrenfeld 2000), and previously overlooked organisms, like invertebrates, are beginning to receive greater consideration (McGuinness 2001).

Land conservation values, which might include the presence of rare, not specialised or unique species and communities or high total biodiversity, are frequently assessed using methods which assume that native invertebrate diversity reflects native plant diversity. Although such correlations have been demonstrated (Moeed and Meads 1985; Crisp et al. 1998; Harris and Burns 2000) there has been little evaluation of this as a general principle. In fact, although generally characterised by their respective plant communities, the differential grazing of phytophagous invertebrates can also affect plant-species composition (Patrick 1994) and even equal or exceed that of more conspicuous vertebrate grazers in, for example, tussock grasslands (White 1978; Andersen and Lonsdale 1990; Schowalter 2000). Nevertheless, there is pressure in ecological studies to use such methods which include 'rapid inventory techniques' because they are comparatively cheap and quick (Crisp et al. 1998). Understanding relationships between invertebrates and their host plants may increase the efficiency of rapid inventory techniques by allowing more accurate predictions of invertebrate community composition to be made from botanical information which is often already available or more easily collected. For example, in the case of host specific herbivores, for which the host plant has been determined, the absence of that plant can be informative. Furthermore, knowledge of the ecology of particular invertebrates could be useful in the assessment of ecosystem health by understanding trophic structure, identifying the causes of ecosystem degradation, or increasing the success of restoration attempts.

Invertebrates are New Zealand's major contributors to global diversity with less than 2% of species having been introduced (Patrick 1994) and endemism reaching approximately 96% (Watt 1975). The Otago region of southern New Zealand has a particularly diverse invertebrate fauna and is thought to represent a major centre of invertebrate endemism (Barratt and Patrick 1987; Barratt and Kuschel 1996). A significant proportion of this diversity occurs in the tussock grasslands characterising much of the montane inland areas. Although extensive, these areas are under-represented in Otago reserves (Allan 1978) and this is often due to their semi-modified floristic state with little regard for the invertebrate communities.

During a collaborative study by the Department of Conservation, New Zealand Forest Research Institute and Landcare Research Limited assessing the effects of fire on tussock-grassland ecosystems, spatial disparities in adult weevil communities (Coleoptera: Curculionidae) were noted (Barratt et al. 2003; Murray et al. 2003). This paper sets out to further analyse these disparities by relating them to the vegetation characteristics at each site and plot, in terms of species richness and percent native species, as such data are often more readily available and used to determine the conservation value of a site.

We also investigate potential weevil-plant and weevil-bryophyte/lichen species associations using PRIMER, a novel method of multidimensional scaling ordination designed specifically for examining community structure (Clark and Warwick 1994). This statistically flexible method is increasingly being used to analyse aquatic invertebrate communities (see Clarke 1993) but less widely in terrestrial settings.

## Methods

### *Study sites*

Sampling was conducted at Mt. Benger (45°35′ S, 169°15′ E) and Deep Stream (45°2′10″ S, 170°15′50″ E), in mid-January 2001. Both sites represent semi-natural *Chionochloa rigida* (Raoul) Zotov (1963) tall tussock grasslands with a diverse inter-tussock flora. They have a recent history of low density stocking (Gitay et al. 1992; Ross et al. 1997) and have remained un-burnt for at least 25 years. The Mt. Benger (MB) site is situated on the eastern slopes of the Old Man Range at 1167 m a.s.l., and Deep Stream (DS) at 500 m a.s.l., 10 km north of Lake Mahinerangi (see Murray et al. 2003). This study utilises three 20 m × 20 m plots at each site giving a total of 6 plots, MB1, MB2, MB3, DS1, DS2 and DS3.

### *Sampling*

A quantitative measure of vascular plant and weevil species abundance was achieved by removing 20 intact 0.1 $m^2$ inter-tussock turves (32 cm × 32 cm × 5 cm deep) per plot in a stratified design avoiding large *Aciphylla* plants and woody shrubs. Four additional turves per plot were taken to included tussocks (*Chionochloa rigida*) clipped to 100 mm above ground level. Individual turves were stored in paper sacks at 4 °C until they were examined to determine the local shoot frequency of vascular plant species (following the nomenclature of Allan Herbarium 2000) with the aid of a 25-cell grid placed over the turf surface. The presence/absence of bryophyte and lichen taxa were recorded and identified following the nomenclature of Beever et al. (1992) for mosses, Glenny (1998) for liverworts, and Galloway (1985) and Martin and Child (1972) for lichens.

Invertebrates were then collected by inverting whole turves and placing them individually in extraction funnels along with any material that may have come loose in storage or during vegetation recording. Turves were heated for 7 days by 150 W light bulbs mounted 400 mm above each turf and invertebrates were collected into 90% ethanol:10% glycerol. Curculionoidea were hand-separated from other invertebrates and stored in 70% ethanol. Bremner (1988) determined that between 97 and 100% of weevils present in a turf are recovered using these methods, and that cool-storage of turves for up to 3 weeks has no negative effect on this yield. Abundance of each weevil species/morpho-species, as identified by comparison to a reference collection held at AgResearch Invermay Agricultural Centre, was recorded per turf. Nomenclature follows Alonso-Zarazaga and Lyal (1999).

### *Analysis*

Non-metric Multi-Dimensional Scaling (MDS) sample ordinations (Kruskal and Wish 1978; Clarke and Warwick 1998) using the computer package PRIMER (Clarke and Warwick 1994) were employed to compare species compositions of each plot and site. Analysis of similarity (ANOSIM) permutation tests, which calculate the similarity coefficient *R*, were used to tests the null hypothesis of 'no difference' in community composition between sites and plots, providing statistical support for the primarily visual interpretation of these ordination results. Species ordinations were performed to identify potential ecological associations between weevils and vegetative taxa as taxa occurring frequently in the same samples appeared close together on these graphs. Analyses were based on similarity matrices produced using the Bray–Curtis similarity measure (Bray and Curtis 1957) to define a similarity coefficient between every pair of species (*species* ordinations) or turves (*sample* ordinations) and their varying abundances across all samples. All ordinations were run using one thousand iterations and were repeated multiple times. Abundance data were subject to 4th root transformation before ordination, to increase the weighting given to less common species while retaining abundance information. The resulting ordination graphs lack axes, instead each point, corresponding to a species or the composition of a turf in *species* and *sample* ordinations respectively, is placed in 2-dimensional space such that the relative distances between points match, in rank order, the corresponding pairwise similarities. The stress value indicates the adequacy of these distances (Clarke 1993), tending towards zero where the rank orders reach perfect agreement.

## Results

### *Community composition*

Mt. Benger supported greater species richness for all three taxonomic groups (Table 1) and almost twice the number of species recorded in only that site as opposed to in both sites. Vegetative species richness (vascular plants, bryophytes and lichens) was highest at Mt. Benger, however, the proportion of native vegetative species was slightly lower. Species richness was reasonably consistent between plots for all three groups at Deep Stream, as were the

*Table 1.* Number of vascular plants, bryophytes/lichens and weevil species collected in January 2001 at Mt. Benger (MB) and Deep Stream (DS). Values in parentheses indicate the number of species which are New Zealand native. The total proportion of vegetative species (vascular, bryophytes and lichens) that were native at each site and plot is also shown. Total number of species unique to each site and plot, in the context of this study is indicated.

| Site/plot | Vascular plants | Bryophytes and lichens | Weevil species | Native vegetation (%) | Unique species |
|---|---|---|---|---|---|
| MB | 57(44) | 32(22) | 17(17) | 74 | 39 |
| DS | 50(39) | 29(22) | 9(9) | 77 | 21 |
| MB1 | 38(27) | 14(10) | 8(8) | 71 | 14 |
| MB2 | 27(21) | 16(12) | 7(7) | 77 | 12 |
| MB3 | 40(29) | 20(13) | 16(16) | 70 | 28 |
| DS1 | 39(27) | 17(14) | 6(6) | 73 | 12 |
| DS2 | 36(27) | 17(14) | 7(7) | 77 | 10 |
| DS3 | 36(26) | 17(13) | 8(8) | 74 | 13 |

number of native species recorded, but varied considerably more between plots at Mt. Benger. Weevil species richness was twice as high in plot MB3 compared to both MB1 and MB2. Although the total number of native vascular, bryophyte and lichen species was also higher in MB3, the proportion of native vegetation was lower at 70%. Those plots with the highest proportion of native vegetation (MB2 and DS2) exhibited two of the lowest scores for both total unique species (those not recorded in at least one other plot) and weevil species richness.

MDS sample ordinations clearly indicated that the two sites differed in their species composition for all three taxonomic groups (Figure 1a, c, e) and this is supported by the associated ANOSIM *R*-values displayed in Table 2. All plots can be clearly separated based on the vascular plant ordination (Figure 1b), however, plots MB1 and MB2 showed a high degree of overlap between sample points for weevils and also bryophytes and lichens (Figure 1d, f) which is reflected in the non-significant *R*-values for these comparisons. The bryophyte and lichen species present at DS2 and DS3 were not sufficiently different (Figure 1d) to distinguish between the communities in these 2 plots.

### *Vascular plant–weevil associations*

An ordination of weevil and vascular plant abundance data from both sites indicated several instances where weevil and vascular plant species (or groups of species) frequently occurred in the same samples (Figure 2a). These particular associations, however, were not consistently evident in repeated species ordinations, or ordinations of each site independently, despite the low stress value (0.15) and high number of iterations. When the percentage of turves in which each weevil and plant species occurred was superimposed onto the

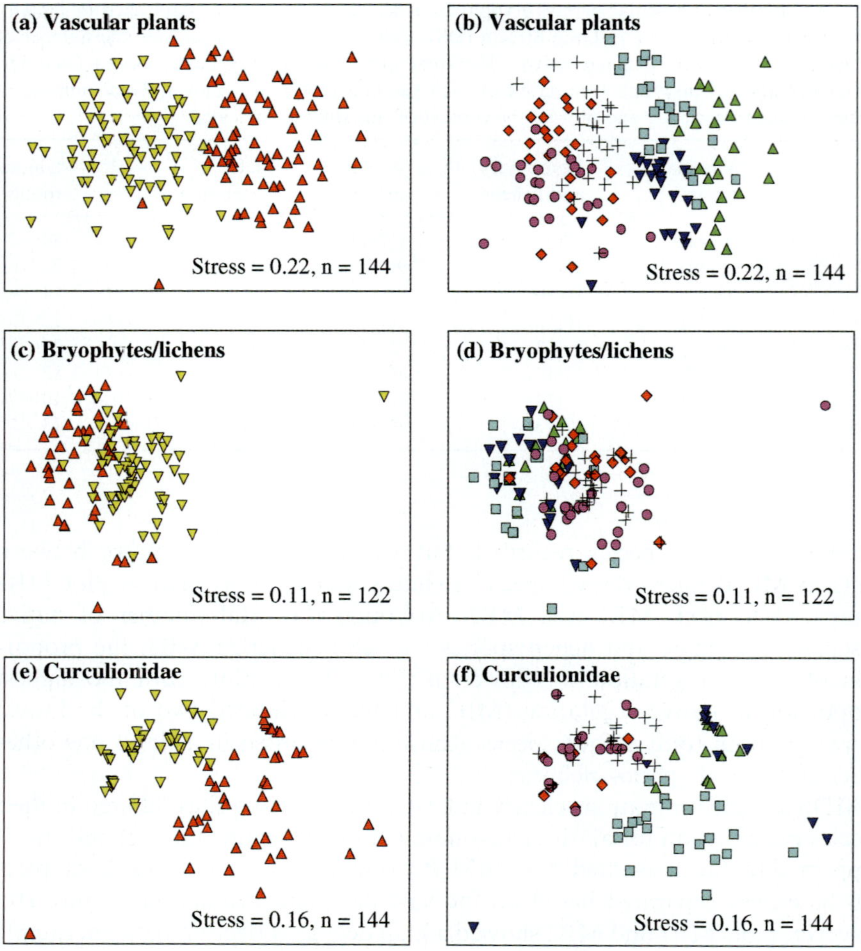


*Figure 1.* MDS sample ordinations of vascular plants, bryophytes/lichens and Curculionidae at the site (a, c, e) and plot (b, d, f) levels. Each point represents the species composition of an individual turf. (a, c, e) ▲Mt. Benger; ▼ Deep Stream; (b, d, f) ▲ MB1; ▼ MB2; ■ MB3; ◆ DS1; ● DS2; + DS3; *n* = number of turf samples.

ordination (Figure 2b) a pattern emerged. Generally, distances between species declined from the periphery to the centre of the ordination while the frequency of turves in which species occurred increased. Visual interpretation of repeated analyses indicated that each species remained at approximately the same position on this gradient from centre to periphery, however, at any point along this gradient directly adjacent species varied considerably from one repetition to the next, suggesting the potential associations in Figure 2a were an aberration of species frequency rather than representative of true associations between species.

*Table 2.* ANOSIM *R*-values and their associated significance for between-site and between plot differences in weevil, vascular plant and bryophyte/lichen community compositions (species presence and abundance) at Mt. Benger (MB) and Deep Stream (DS). Significance levels < 1.0 indicate statistical differences in species composition.

| Sites/plots compared | Curculionidae | | Vascular | | Bryophytes/ lichens | |
|---|---|---|---|---|---|---|
| | *R* | % sig | *R* | % sig | *R* | % sig |
| MB–DS | 0.56 | 0.1 | 0.55 | 0.1 | 0.38 | 0.1 |
| MB1-MB2 | 0.06 | 5.7 | 0.36 | 0.1 | 0.04 | 10.9 |
| MB1-MB3 | 0.66 | 0.1 | 0.40 | 0.1 | 0.16 | 0.1 |
| MB2-MB3 | 0.68 | 0.1 | 0.36 | 0.1 | 0.19 | 0.1 |
| DS1-DS2 | 0.16 | 0.4 | 0.28 | 0.1 | 0.12 | 0.4 |
| DS1-DS3 | 0.32 | 0.1 | 0.38 | 0.1 | 0.09 | 0.3 |
| DS2-DS3 | 0.01 | 26.8 | 0.46 | 0.1 | 0.14 | 0.2 |

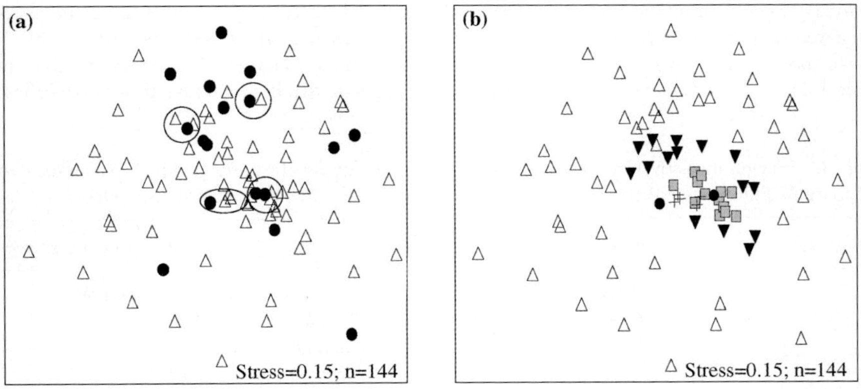


*Figure 2.* (a) Species ordination indicating possible associations between weevil (●) and vascular plant species (△). (b) Species ordination as in 2a with weevil/plant symbols replaced by symbols representing the frequency of turves in which each species occurred; △, < 10%; ▼, 10–20%; ■, 21–40%; ●, 41–60%; + > 60%.

## *Bryophyte/lichen–weevil associations*

Potential associations between weevil species and mosses, liverworts or lichens are summarised in Figure 3 and Table 3. Weevils showing no associations (i.e. those labelled directly on Figure 3), each occurred in < 3.5% of turves, re-iterating the possibility that species frequency across samples was driving the distances between points on ordinations. Species frequency, however, appeared to be less influential than for the weevil–vascular plant analysis, as repeated ordinations, and ordinations of each site independently, consistently produced these particular associations.

Group 1 associations (Figure 3, Table 3) were maintained in an ordination of Deep Stream data (not shown), but not at Mt. Benger where all three weevil

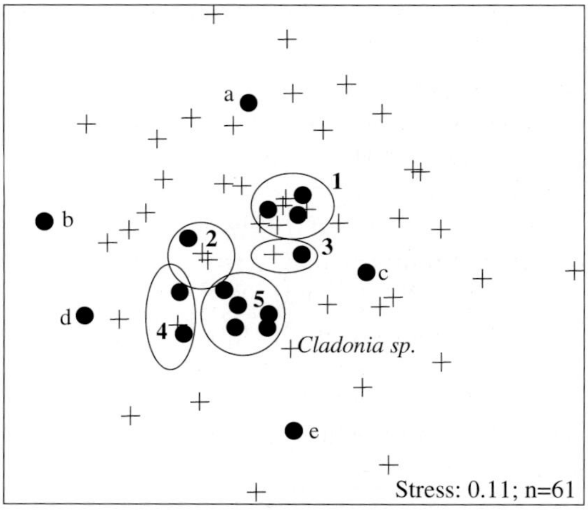


*Figure 3.* Species ordination of weevils (●) and bryophytes/lichens (+). Possible associations between weevils and bryophyte/lichen species are circled and the species involved indicated in Table 3. Weevil species labelled with letters are those that showed no associations, full names are given in Table 3. The lichen *Cladonia* sp. which forms part of group 5 for Mt. Benger data only is also labelled.

*Table 3.* Weevil, moss (M) liverwort (LW) and lichen (L) speceis forming groups 1–5 as indicated in Figure 3. Full names of the five weevil species showing no associations are also listed.

| Group/site | Weevil species | Associated bryophyte and lichen species |
|---|---|---|
| 1 | *Peristoreus insignis* | *Chiloscyphus chlorophyllus* (LW) |
| | *Baeosomus rugosus* | *Tortula truncata* (M) |
| | *Irenimus curvus* | *Cladina leptoclada* (L) |
| | | *Hypnum cupressiforme* (M) |
| | | *Leptodontium interruptum* (M) |
| 2 | *Baeosomus* cf.*angustus* | *Pseudocyphellaria pickeringii* (L) |
| | | *Ditrichum difficile* (M) |
| 3 | *Baeosomus* cf.*crassipes* | *Polytrichum juniperinum* (M) |
| 4 | *Baeosomus* sp. | |
| | *Gromilus* sp. 1 | *Psoroma histula* (L) |
| 5 | *Gromilus* sp. 2 | None |
| | *Nestrius* sp. 1 | |
| | *Nestrius* sp. 2 | |
| | *Baeosomus amplus* | |
| | *Crisus* sp. | |
| Remaining species | (a) *Catoptes sp.* | None |
| | (b) *Eugnomus durvillei* | |
| | (c) *Irenimus stolidus* | |
| | (d) *Catoptes dispar* | |
| | (e) cryptorhinchine sp. | |

species were considerably less common. In contrast, group 2 associations were only retained in an independent ordination of Mt. Benger data. These species occurred with low frequency at both sites (<4–15% of samples), although more so at Deep Stream. Group 3 represents an association between *Baeosomus* cf. *crassipes* (Broun) and *Polytrichum juniperinum* Hedw., the most frequently occurring weevil and moss species, respectively. This association was also apparent from analysis of Mt. Benger data alone, where *P. juniperinum* occurred in 93% of turves, compared to 63% at Deep Stream. In group 4, *Baeosomus* sp. and the lichen *Psoroma histula* Nyl. ex Cromb. were closely associated, while *Gromilus* sp. 1 was only loosely associated. When ordinated independently, Mt. Benger data showed a much stronger association between *Gromilus* sp. 1 and *P. histula*. Weevil species in group 5 were clustered together but were not closely associated to any particular bryophyte or lichen species. Four of these species, *Nestrius* sp. 1, *Nestrius* sp. 2, *Gromilus* sp. 2 and *Crisus* sp., were restricted to Mt. Benger, explaining the clustering. The latter two were, however, associated with the lichen *Cladonia* sp. (indicated in Figure 3), when Mt. Benger data were ordinated alone.

## Discussion

### *Community composition*

Significant differences were found in the species richness and biodiversity both between and within the two sites compared. The variability in species composition over the short distances between plots with no visible barriers within each site emphasises the complexity of the tussock grassland ecosystem. This highlights the importance of considering microhabitats and naturally discontinuous invertebrate and plant distributions in ecological studies on which conservation and management decisions are to be made. In the absence of certain micro-habitats, such as those maintained by an intact tussock canopy, some groups of species may be under-represented and others over-represented (Clarke and Warwick 2001). The tussock canopy protects the inter-tussock habitats from solar radiation and wind, thus affecting humidity, maintaining soil water balance and buffering temperatures (Yeates and Lee 1997). Such factors strongly influence invertebrate species composition, and plant–herbivore dynamics (Room 1990; Didham et al. 1998). Fragmentation and disturbance events including fire also cause changes in soil nutrients, structure and water balance and may have a measurable effect on the survivorship of weevil species with soil-dwelling immature stages as has been recorded for various soil nematodes (Yeates and Lee 1997). Some or all of these factors may explain the relative differences between sites and plots in the abundances of small, flightless weevil species (e.g. *Baeosomus* sp. and *Nestrius* sp.). Such species tend to be poor dispersers and as the two sites at least have independent disturbance histories which potentially altered canopy integrity in different ways and at different times, some species may have been lost from or failed to (re)colonise these sites.

If only one of the sites or plots studied here could be selected for the conservation of native biodiversity, the inclusion of invertebrate data in addition to botanical information could result in a different choice. For example, plot MB1 contains a greater proportion of native plants than MB3 and just two fewer species of both vascular and non-vascular plants. If there were resistance to the selection of MB3 due to factors such as opposition from adjacent landowners, MB1 may be selected more readily in compromise based on this botanical information alone. Plot MB3 however, contains twice the number of native weevil species, hence the relative biodiversity is actually greater and the inclusion of invertebrate data significantly strengthens the case for its conservation.

### *Associations*

Associations between bryophyte/lichen taxa and *Baeosomus* (Erirhinidae: Stenopelmini), the most speciose and abundant weevil genus found at the study sites (Murray et al. 2003), were of particular interest as the genus is thought to be composed of moss feeders (Kuschel 1964). Results suggested such associations for all five of the *Baeosomus* species recorded. Of the nine bryophyte and lichen species identified here as potential weevil host plants, only the moss *Polytrichum juniperinum* as a host of *Baeosomus* species including *B. amplus* (Broun) and *B. rugosus* (Broun) (Barratt and Patrick 1987; G. Kuschel, personal communication) is in agreement with the few other reports that exist. In contrast to the current analysis, few studies have observed even potential associations between weevils and liverworts or lichens although there is evidence that cryptogam herbivory is a characteristic of certain weevil species (Jolivet 1998). The associations between weevils and bryophyte/lichen taxa were considered more reliable than those with vascular plants. In the latter case, associations appeared to be confounded by the proportion of samples in which individual species occurred despite 4th root transformation, as this transformation only controls for abundance *within* samples. Essentially, weevil and vascular plant species that occurred close together on ordinations did so because they were similarly present in a large proportion of the total number of samples rather than necessarily having any specific ecological association. Using this ordination technique potential associations (plant and weevil species which are plotted close together on ordination graphs) representing less common species are most likely to reflect real relationships as this 'frequency' effect can be ruled out as a confounding factor. Some populations, however, such as those of *Eugnomus durvillei* Schönherr, *Crisus* sp., *Cryptorhynchine* sp. and *Catoptes* sp. may be so small that they do not utilise all available host plants, thus obscuring existing associations from detection by this method. Similarly, where weevil populations are primarily limited by factors other than food or habitats provided directly by their hosts (e.g. competitive rigor or altitudinal limits), associations with locally abundant or common plant species may be obscured.

In accordance with the literature (May 1977; Barratt and Patrick 1987; Bremner 1988; May 1993; Barratt and Kuschel 1996; Barratt et al. 1998), the lack of specific associations detected for the members of the Entiminae and Cyclominae recorded in this study, may indicate that they are generalist feeders. Such herbivores often exhibit more efficient dispersal capabilities than specialised species (Anderson 1988), making them more disturbance-tolerant. It has been shown (Barratt et al. 1998; Crisp et al. 1998; Harris and Burns 2000) that in the absence of intact habitat, semi-modified, fragmented and historically disturbed indigenous vegetation, such as that surveyed here, can support considerable indigenous invertebrate biodiversity. A rapid inventory of plants may not provide information on the likelihood of such generalists being represented in a particular grassland area.

Many authors (e.g. Lyal 1993; Colonnelli and Osella 1998) stress that the particular plants on which invertebrates are found cannot always be assumed to be their hosts. Laboratory rearing or gut content analysis may be required to prove associations beyond doubt (Košt'ál 1991; Lyal 1993), especially when considering species with subterranean habits (Marco et al. 1998). A case in point is the apparent association between *Peristoreus insignis* (Broun) and the liverwort *Chiloscyphus chlorophyllus* (Hook.f. and Taylor) Mitt., which is almost certainly a misleading result. Genera in the subfamily Curculioninae are usually considered to be host specific or oligophagous, restricted to areas where native plants still predominate (May 1993; Barratt et al. 1998), and *P. insignis* itself almost certainly feeds on plants in the family Asteraceae (G. Kuschel, personal communication). Furthermore, different weevil life stages may not necessarily feed on the same plant part or species (Lyal 1993; Simpson et al. 1996) and recent evidence suggests that maternal oviposition preferences can sometimes restrict host ranges more than feeding preferences (Singer 1986; Janz and Nylin 1997; Marco et al. 1998). Other factors are also influential, for example, during the winter months *Irenimus stolidus* Broun has been observed sheltering in the base of tussock tillers where temperatures remain higher than in the inter-tussock vegetation (Barratt and Kuschel 1996). Similarly the concealment afforded by the habitat of Cryptorhynchinae (e.g. leaf litter and dead wood) is thought to predominate as a factor determining host range (Lyal 1993). The apparent associations between *Baeosomus* spp., and bryophytes and lichens may result from a preference for the microhabitat provided by these life forms in general, and their suitability as refuge plants (Colonnelli and Osella 1998). The taxonomy and food potential of the specific bryophytes and lichens present may therefore be of little ecological consequence.

### *MDS ordination*

The PRIMER MDS ordination technique was developed to assess invertebrate communities in aquatic systems (Clarke and Warwick 1994) and has been successfully applied to floristic communities in terrestrial environments (e.g. Rose and Fairweather 1997). This form of MDS analysis was found here to be a useful tool in the identification of plant and weevil communities indicating that the technique could be employed for detecting similarities between sites. For example, it might be used to compare relative diversity or community composition at different sites or to validate replicate plots for ecological studies as having statistically similar species composition. Although this method can been used to detect associations between species and abiotic factors (Clarke and Warwick 1994), it may not be appropriate on its own to definitively detect ecological associations between species such as herbivorous insects and their plant hosts, especially in complex habitats and at varying spatial scales. It may however, provide a useful starting point in terms of indicating pairs or groups of species that frequently occur together.

## Conclusion

The effective management and conservation of indigenous biodiversity requires a fuller understanding of ecosystem function, including interactions between component species. An understanding of plant–invertebrate associations is especially important in New Zealand's remaining tussock grasslands as land transformation is occurring rapidly in this environment due to intensifying land use. Such knowledge would also be widely applicable across the production-conservation spectrum given that the invertebrate fauna span the functional groups essential for ecosystem processes, such as decomposition, to occur.

There is increasing pressure to conduct ecological studies using rapid inventory techniques, relying on plant biodiversity as a predictor of invertebrate biodiversity with insufficient justification. In contrast to aquatic systems, where invertebrate diversity and abundance is widely used as an indicator of ecosystem health (Wright et al. 1995; Saiz-Salinas 1997; O'Connor et al. 2000), terrestrial indicator species have rarely been determined, as terrestrial habitats are often far more diverse and species composition and abundance less consistent. Determining associations between terrestrial invertebrates and the climatic and floristic components of the ecosystems that they inhabit may increase the predictive power of such techniques. The current investigation determined several possible associations between weevils and non-vascular plants using an ordination approach, raising questions about their ecological relationships. It also illustrated the variability that exists in invertebrate species composition within the tussock grassland ecosystem even at small spatial scales, and the consequent importance of considering less obvious microhabitats when conducting ecological studies in such environments. Finally the study has provided an insight into methods that, once refined, could be used more effectively to assist in the determination of species associations within complex terrestrial communities.

## Acknowledgements

The writing of this contribution was supported and funded by the Miss E.L. Hellaby Indigenous Grasslands Research Trust and AgResearch Ltd. We thank Colin Ferguson (AgResearch Ltd.) and Rory Logan (AgResearch Ltd.) for assistance with fieldwork, John Steel (Botany Department, Otago University) for his identification of all mosses, liverworts and lichens and Ian Payton (Landcare Research Ltd.), Bill Lee (Landcare Research Ltd.) and Rory Logan for their help with the identification of vascular plants.

## References

Allan R.B. 1978. Scenic Reserves of Otago. Biological Survey of Reserves 4. Department of Lands and Survey, Wellington, New Zealand.

Allan Herbarium 2000. New Zealand plant names database. Landcare Research, New Zealand. Available http://nzflora.landcare.co.nz/plantnames.

Alonso-Zarazaga M.A. and Lyal C.H.C. 1999. A World Catalogue of Families and Genera of Curculionoidea (Insecta: Coleoptera) (Excepting Scolytidae and Platypodidae). Entomopraxis S.C.P, Barcelona, Spain.

Andersen R.S. 1988. The Curculionidae of the Queen Charlotte Island, British Columbia (Insecta; Coleoptera). Can. J. Zool. 66: 2406–2414.

Anderson A.N. and Lonsdale W.M. 1990. Herbivory by insects in Australia tropical savannas: a review. J. Biogeogr. 17: 433–444.

Barratt B.I.P. and Kuschel G. 1996. Broad-nosed weevils (Curculionidae: Brachycerinae: Entimini) of the Lammerlaw and Rock and Pillar ranges in Otago, with descriptions of four new species of *Irenimus*. N. Z. J. Zool. 23: 359–374.

Barratt B.I.P. and Patrick B. 1987. Insects of snow tussock grassland on the East Otago Plateau. N. Z. Entomol. 10: 69–77.

Barratt B.I.P., Evans A.A., Ferguson C.M., McNeill M.R., Proffitt J.R. and Barker G.M. 1998. Curculionoidea (Insecta: Coleoptera) of New Zealand agricultural grassland and lucerne as potential non-target hosts of the parasitoids *Microctonus aethiopoides* Loan and *Microctonus hyperodae* Loan (Hymenoptera: Braconidae). N. Z. J. Zool. 25: 47–63.

Barratt B.I.P., Ferguson C.M., Logan R.A.S., Murray T.J., Dickinson K.J.M. 2003. Impact of burning on Coleoptera, particularly Curculionoidea, in New Zealand native tussock grassland. In: Austin A.D., Mackay D.A. Cooper S.J.B. (eds), Invertebrate Biodiversity and Conservation. Records of the South Australia Museum, Monograph Series No. 7, Adelaide, pp. 171–178.

Beever J., Alison K.W. and Child J. 1992. The Mosses of New Zealand, 2nd edn. University of Otago Press, Dunedin, New Zealand.

Begon M., Harper J.L. and Townsend C.G. 1990. Ecology: Individuals, Populations and Communities, 2nd edn. Blackwell Scientific Press, Cambridge.

Bray R.J. and Curtis J.T. 1957. An ordination of the upland forest communities of Southern Wisconsin. Ecol. Monogr. 27: 325–349.

Bremner G. 1988. Ecological studies of the insect fauna of the East Otago Plateau. Ph.D. thesis, University of Otago, New Zealand.

Clarke K.R. 1993. Non-parametric multivariate analysis of changes in community structure. Aust. J. Ecol. 18: 177–143.

Clarke K.R. and Warwick R.M. 1994. Changes in Marine Communities: an Approach to Statistical Analysis and Interpretation. Plymouth Marine Laboratory, Plymouth.

Clarke K.R. and Warwick R.M. 1998. Quantifying structural redundancy in ecological communities. Oecologia 113: 278–289.

Clarke K.R. and Warwick R.M. 2001. A further biodiversity index applicable to species lists: variation in taxonomic distinctness. Mar. Ecol. Prog. Ser. 216: 265–278.

Colonnelli E. and Osella G. 1998. Host and refuge plants of weevils (Coleoptera: Curculionoidea). In: Colonnelli E., Louw S. and Osella G. (eds), Taxonomy, Ecology and Distribution of Curculionoides. Museo Regionale di Scienze Naturali, Torino, pp. 191–216.

Crisp P.N., Dickinson K.J.M. and Gibbs G.W. 1998. Does native invertebrate diversity reflect native plant diversity? A case study from New Zealand and implications for conservation Biol. Conserv. 83: 209–220.

Didham R.K., Hammond P.M., Lawton J.H., Eggleton P. and Stork N.E. 1998. Beetle species responses to tropical forest fragmentation. Ecol. Monogr. 68(3): 295–323.

Ehrenfeld J.G. 2000. Defining the limits of restoration: the need for realistic goals. Restor. Ecol. 8(1): 2–9.

Galloway D.J. 1985. The Flora of New Zealand: Lichens. P.D. Hasselberg New Zealand Government Printer, Wellington.

Gitay H., Lee W.G., Allen R.B. and Wilson J.B. 1992. Recovery of *Chionochloa rigida* tussocks from fires in South Island, New Zealand. N. Z. J. Environ. Manage. 35: 249–259.

Glenny D. 1998. A revised checklist of New Zealand liverworts and hornworts. Tuhinga 10: 119–149.

Harris R.J. and Burns B.R. 2000. Beetle assemblages of Kahikatea forest fragments in a pasture–dominated landscape. N. Z. J. Ecol. 24(1): 57–67.

Janz N. and Nylin S. 1997. The role of female search behaviour in determining host plant range in plant feeding insects: a test of the information processing hypothesis. Proc. R. Soc. Lond. Ser. B, Biol. Sci. 264: 701–707.

Jolivet P. 1998. Interrelationships Between Insects and Plants. CRC Press, Boca Raton.

Košt'ál M. 1991. Contribution to the knowledge of the host plants of Curculionidae (Coleoptera) from Middle Asia and Europe. Biologia 46(2): 169–172.

Kruskal J.B. and Wish M. 1978. Multidimensional Scaling. Sage Publications, Beverley Hills, California.

Kuschel G. 1964. Insects of Campbell Island: Coleoptera: Curculionidae of the subantarctic islands of New Zealand. In: Gressitt J.L. (ed.), Insects of Campbell Island, Pacific Insects Monograph Vol. 7, pp. 416–493.

Lyal C.H.C. 1993. Cryptorhynchinae (Insecta: Coleoptera: Curculioninae). Fauna of New Zealand 29. Manaaki Whenua Press, Lincoln, New Zealand.

Macro V., Lopez-Olguin J.F., Ortego F. and Castanera P. 1998. Restricted host range of the weevil *Aubeoymus mariaefranciscae*, a new sugar-beet pest. Entomologia Experimentalis et Applicata 87: 201–207.

Martin W. and Child J. 1972. The Lichens of New Zealand. A.H. and A.W. Reed, Wellington, New Zealand.

May B.M. 1977. Immature stages of Curculionidae: larvae of the soil-dwelling weevils of New Zealand. J. R. Soc. N. Z. 7(2): 189–228.

May B.M. 1993. Larvae of Curculionidae (Insecta: Coleoptera): a Systematic Overview. Fauna of New Zealand 28. Manaaki Whenua Press, Lincoln, New Zealand.

McGuinness C.A. 2001. The Conservation Requirements of New Zealand's Nationally Threatened Invertebrates. Threatened Species Occasional Publication 20. Department of Conservation, Wellington, New Zealand.

Moeed A. and Meads M.J. 1985. Seasonality of pitfall trapped invertebrates in three types of native forest, Orongorongo Valley, New Zealand. N. Z. J. Zool. 12: 17–53.

Murray T.J., Barratt B.I.P. and Dickinson K.J.M. 2003. Comparison of the weevil fauna (Coleoptera: Curculionoidea) in two tussock grassland sites in Otago, New Zealand. J. R. Soc. N. Z. 33(4): 703–714.

Myers N., Mittermeier R.A., da Fonseca G.A.B. and Kent J. 2000. Biodiversity hotspots for conservation priorities. Nature 403: 853–858.

O'Connor R.J., Walls T.E. and Hughes R.M. 2000. Using multiple taxonomic groups to index the ecological condition of lakes. Environ. Monit. Assess. 61(2): 207–228.

Patrick B.H. 1994. The importance of invertebrate biodiversity: an Otago Conservancy review. Conservation Advisory Science Notes 53.

Room P.M. 1990. Ecology of a simple plant-herbivore system: biological control of *Salvinia*. Trends Ecol. Evol. 5(3): 74–79.

Rose S. and Fairweather P.G. 1997. Changes in floristic composition of urban bushland invaded by *Pittosporum undulatum* in northern Sydney, Australia. Aust. J. Bot. 45(1): 123–149.

Ross D.J., Speir T.W., Tate K.R. and Feltham C.W. 1997. Burning in a New Zealand snow-tussock grassland: effects on soil microbial biomass and nitrogen and phosphorus availability. N. Z. J Ecol. 21(1): 63–71.

Saiz-Salinas J.I. 1997. Evaluation of adverse biological effects induced by pollution in the Bilbao estuary (Spain). Environ. Pollut. 96(3): 351–359.

Schowalter T.D. 2000. Insect Ecology: an Ecosystem Approach. Academic Press, London.

Simpson S.E., Nigg H.N., Coile N.C. and Adair R.A. 1996. *Diaprepes abberciatus* (Coleoptera: Curculionidae): host plant associations. Entomol. Soc. Am. 25(2): 333–349.

Singer M.C. 1986. The definition and measurement of oviposition preferences in plant feeding insects. In: Miller J.R. and Miller T.A. (eds), Insect-plant Interactions. Springer–Verlag, New York.

Watt J.C. 1975. The terrestrial insects. In: Kuschel G. (ed.), Biogeography and Ecology in New Zealand. Dr. Junk Publishers, The Hague, pp. 507–535.

White E.G. 1978. Energetics and consumption rates of alpine grasshoppers (Orthoptera: Acrididae) in New Zealand. Oecologia 33: 17–44.

Wright I.A., Cheesman B.C., Fairweather P.G. and Benson L.J. 1995. Measuring the impact of sewage effluent on the macroinvertebrate community of an upland stream: the effect of different levels of taxonomic resolution and quantification. Aust. J. Ecol. 20: 142–149.

Yeates G.W. and Lee W.G. 1997. Burning in New Zealand snow-tussock grassland: effects on vegetation and soil fauna. N. Z. J. Ecol. 21(1): 73–79.

Biodiversity and Conservation (2006) 15:125–141
DOI 10.1007/s10531-004-4247-2


# Effects of management intensity and season on arboreal ant diversity and abundance in coffee agroecosystems

STACY M. PHILPOTT[1,3,*], IVETTE PERFECTO[2] and JOHN VANDERMEER[1,2]

*[1]Department of Ecology and Evolutionary Biology, University of Michigan, 830 North University, Ann Arbor, MI 48109; [2]School of Natural Resources and the Environment, University of Michigan, 430 East University, Ann Arbor, MI 48109; [3]Current Address: Smithsonian Migratory Bird Center, National Zoological Park, 3001 Connecticut Ave. NW, Washington, DC 20008; *Author for correspondence (e-mail: philpotts@si.edu; phone: +1-202-673-0128, fax: +1-202-633-4206)*

Received 13 January 2004; accepted in revised form 20 September 2004

**Key words:** Agricultural intensification, Arboreal ants, Biodiversity, Chiapas, Coffee agroecosystem, Mexico

**Abstract.** Agricultural intensification decreases arthropod predator diversity, abundance and population stability, and may affect interactions between top predators and their arthropod prey – ultimately affecting ecosystem services. Coffee management intensification (reduction or removal of shade trees) reduces diversity of arthropod predators (ground-foraging ants). Because ants provide ecosystem services by controlling pests, influences of intensification on arboreal, coffee-foraging ant diversity and abundance are important. We here address how coffee intensification affects: (1) coffee-foraging ant diversity and abundance and (2) seasonal fluctuations in ant abundance. In each of four coffee sites of varying management intensity in Chiapas, Mexico, we sampled vegetation and using two methods, sampled ant diversity and abundance over two years. Sites significantly differed in vegetation and management intensity. Coffee-foraging ant diversity generally decreased with increasing management intensity (16–26% fewer species observed in the most intensively-managed site). Ant abundance was higher in the wet season. Management intensity, however, did not influence ant abundance or seasonal fluctuations in abundance. Our results highlight the importance of diverse agricultural systems in maintaining arthropod predator diversity, and point to one model system in which we may effectively test how diversity per se affects ecosystem services.

## Introduction

Conservation biologists strive to understand how habitat disturbance affects biodiversity in natural ecosystems (Didham et al. 1998; Kalif et al. 2001; Ricketts 2001; Tscharntke et al. 2002; Watt et al. 2002), agricultural habitats (Roth et al. 1994; Estrada and Coates-Estrada 2002; Ricketts et al. 2001; Siebert 2002), or across intensification gradients of agroecosystems (Perfecto et al. 1996; Greenberg et al. 1997; Kremen et al. 2002; Klein et al. 2002a). Yet, many examine how habitat modifications affect species richness or abundance without distinguishing between agricultural types (Aberg et al. 1995; Tilman 1999, but see Glor et al. 2001; Vandermeer and Carvajal 2001; Perfecto and Vandermeer 2002) or lack quantification of vegetation variables necessary to distinguish habitats (Ricketts et al. 2001; Rojas et al. 2001). Agriculture covers >75% of the earth's arable land (Young 1999) and biodiversity found therein provides ecosystem services (Balvanera et al. 2001; Klein et al. 2003).Thus quantifying effects of intensification and comparing conservation value of agricultural habitats is crucial to conservation.

Coffee (*Coffea arabica*) agroecosystems are highlighted for their conservation potential, but coffee management intensification eliminates biodiversity and may restrict ecosystem services. Coffee was traditionally grown under a diverse, dense shade canopy, but recent intensification includes reducing shade tree density and diversity and agrochemical use (Moguel and Toledo 1999; Mas and Dietsch 2003). With coffee intensification, diversity of predators such as ants decreases (Nestel and Dickschen 1990; Perfecto and Snelling 1995; Perfecto et al. 1996; Perfecto et al. 1997; Perfecto and Vandermeer 2002; Armbrecht and Perfecto 2003) yet no studies focus on arboreal (specifically coffee-foraging) ants (see Perfecto et al. 1996). Ants provide ecosystem services by preying on pests in agroecosystems including coffee (Way and Khoo 1992; Velez et al. 2000; Vandermeer et al. 2002) and ecosystem services may diminish as diversity is lost (Balvanera et al. 2001; Klein et al. 2003). Thus understanding losses of coffee-foraging ant diversity are particularly important.

Agroecosystem management and seasonality may also influence ant abundance. Theoretically, in vegetationally-diverse systems, predator populations are larger and more stable than in monocultures due to stable prey populations and other resources (Root 1973; Andow 1991). Empirical evidence shows some predators are more abundant in diverse (less intense) agricultural systems (Basedow 1991; Knops et al. 1999; Girma et al. 2000; Klein et al. 2002b) but management differences do not always affect ant abundance (Perfecto and Sediles 1992). Furthermore, some tropical insect populations are influenced by seasonal changes in temperature and rainfall (Tauber et al. 1998; Guedes et al. 2000) and ants are generally more abundant in the wet season (Alonso 1998; Kaspari and Weiser 2000). Thus although ant abundance may vary with management system and seasonality, abundance may fluctuate less in diverse systems.

In this study, we assess changes in diversity and abundance of coffee-foraging ants under the influences of coffee management intensification and seasonal changes investigating if: (1) Diversity of coffee-foraging ants declines with increasing management intensification; (2) Abundance of coffee-foraging ants increases with increasing management intensification; and (3) Abundance of coffee-foraging ants increases in the wet season and seasonal fluctuations in abundance are less under high-shade management.

## Methods

### *Site description and experimental design*

We set up sampling plots in four sites within three farms in the Soconusco region of SW Chiapas, Mexico: (1) Belen Rustic (TP; 15°15′ N, 92°22′ W); (2)

Belen Production (CPB; 15°15′ N, 92°23′ W); (3) Irlanda (CPI; 15°11′ N, 92°20′ W); and (4) Hamburgo (SM; 15°10′ N, 92°19′ W). All farms are between 950–1150 m elevation and receive ca. 4500 mm of rain per year. The sites represent a gradient of intensification based on density, diversity, and height of shade trees, and percent shade cover (Mas and Dietsch 2003; Perfecto et al. 2003). According to Moguel and Toledo's (1999) classification scheme, TP is a 'traditional polyculture', CPI and CPB are 'commercial polycultures', and SM is roughly a 'shaded monoculture'.

We sampled ants following an experimental design set up to study bird influences on coffee arthropod communities (I. Perfecto, unpublished data), thus ant sampling took place inside and outside of large bird exclosures. We set up 32 total exclosures (10 each in CPI and SM, 6 each in TP and CPB) with monofilament nylon netting (35 × 35 mm mesh) suspended from shade trees and covering roughly 10 × 8 m. Inside each exclosure we marked ten coffee plants for sampling and outside (< 10 m from exclosures) we marked ten control plants. Coffee plants in TP were larger, and nets were of set size, thus numbers of exclosure and control plants varied from seven to ten. On each sampling date we sampled a total of 200 plants each in CPI and SM, 120 in CPB, and 96 in TP. We maintained exclosures from Nov. 2000–Dec. 2002.

### *Ant sampling and diversity analysis*

We used two methods to sample ants: vacuum samples and tuna baits. Using vacuum samples, we sampled all marked plants four times in the dry (Nov. 2000, Feb. 2001, Nov. 2001, and Nov. 2002) and three times in the wet season (May 2001, Aug. 2001, and May 2002). On several days from 7:00–9:00 AM, for each marked coffee plant, we sampled two previously unused branches (> 6 leaves) with a 10 cm diameter reversed leaf-blower (D-vac) (WeedEater® Company, 1 Poulan Drive, Nashville, AR, 71852). Arthropods were vacuumed into mesh bags, placed in plastic bags and killed with ethyl acetate. We stored samples from sets of control or exclosure plants together and later identified ants. We standardized ant abundance as number of individuals per g of foliage sampled. We measured length and width of all vacuumed leaves and converted leaf area to a biomass estimate using an empirically generated equation (Biomass (g) = Leaf area (0.025)–0.08).

On marked plants, we baited for ants three times in the dry (Dec. 2000, Jan. 2002, and Dec. 2002) and twice in the wet season (May 01 and May 02) (Table 1). From 7:30–10:30 AM, we placed tuna baits (~5 g) 1 m above ground, collected, and identified all ants found after 30–45 min. In preliminary richness analyses the ant fauna in TP was further from reaching asymptotes than other sites so we sampled 200 extra plants under similar shade conditions in TP in Dec. 2002. We stored ants separately for each plant, but to compare with D-vac samples, we grouped sets of control and exclosure plants. In months where we used both methods, D-vac samples were collected first. We

*Table 1.* Vegetation characteristics sampled and totaled into a management index (MI) in four coffee sites (Belen Rustic (TP), Irlanda (CPI), Belen Production (CPB), and Hamburgo (SM)).

| Site | # tree species | % cover (Nov. 00) | % cover (Jan. 03) | % two species cover | % three species cover | Management Index | N |
|---|---|---|---|---|---|---|---|
| TP | 14.0 (a) | 71.7 ± 3.6 (a) | 75.3 ± 2.8 (a) | 10.3 ± 1.6 (a) | 1.6 ± 1.1 | 2.19 ± 0.19 (a) | 6 |
| CPI | 7.3 (b) | 65.0 ± 3.6 (a) | 66.4 ± 4.1 (a, b) | 10.6 ± 2.4 (a) | 0.2 ± 0.2 | 2.88 ± 0.14 (b) | 10 |
| CPB | 6.5 (b) | 42.3 ± 3.7 (b) | 59.0 ± 3.5 (b) | 0.7 ± 0.4 (b) | 0 | 3.60 ± 0.089 (c[a]) | 6 |
| SM | 3.0 (c) | 30.4 ± 2.1 (b) | 35.6 ± 3.2 (c) | 1.0 ± 0.4 (b) | 0 | 4.07 ± 0.06 (d) | 10 |
| $p$ | *** | *** | *** | ** | NS | ** | |

$^{**}p < 0.01$, $^{***}p < 0.001$, [a]CPB–SM, $p = 0.067$

Numbers show means (tree species richness, percent cover, proportion of shade points with two or three species' cover, and MI values) ± standard error for 35 × 35 m area (50 points) surrounding each of 32 bird exclosures. A higher MI shows more intensive coffee management. Letters show significant differences between sites based on Tukey's post-hoc tests.

recorded ant activity per bait for each species separately using an index where 0 = no ants, 1 = 1 to 2 ants, 2 = 3–10 ants, 3 ≥ 10 ants, and summed across species for total activity per plant. Mean activity was calculated as average total activity per plant for a set of control or exclosure plants.

We analyzed species richness data using EstimateS (Colwell and Coddington 1994; http://www/viceroy/eeb/uconn/.edu/estimates) for tuna and D-vac samples and for both data sets combined. We used sets of exclosure or control plants as samples for between site comparisons. Because sample sizes differed between sites we compared richness between sites with rarefaction (Gotelli and Colwell 2001). We used Coleman estimates, virtually indistinguishable from rarefaction (Colwell and Coddington 1994). To approximate total richness per site we used richness estimators (Chao2, Incidence-Based Coverage Estimator [ICE], and Michaelis–Menten Means [MMMeans]) that estimate total richness per site based on observed species richness plus the number of uniques (species found in only one sample) rather than singletons (species represented by only one individual) (Longino et al. 2002). EstimateS also calculated diversity indices (Fisher's alpha, 2) Shannon's index, and 3) Simpson's index (inverse of program results reported here). For all calculations we used presence/ absence data not abundance because ants are social insects (Longino et al. 2002). With simple linear regression we examined relationships between a Management Index (MI) (see below), diversity indices, and species richness.

### *Vegetation sampling*

We sampled shade tree diversity, percent shade cover, and structural diversity and summarized vegetation data using a management index (MI). In Nov. 2000, we counted the number of tree species in a 35 × 35 m area around each exclosure. In the same area, at fifty points (>5 m apart) we recorded: (1) foliage presence using a vertical tube densiometer and (2) tree species (one or more) directly above each point and calculated percent cover (% points with foliage) and structural diversity (proportion of points covered by two or three species). In Jan. 2003, we re-measured percent cover to account for changes in management during the time exclosures were maintained. We summarized vegetation variables per exclosure using the MI (Mas and Dietsch 2003) whereby raw data are converted to a scale from 0 to 1 and then summed. We divided values for each variable (% cover in Nov. 2000 and Jan. 2003, # shade tree species, and proportion of points with two or three species) by the highest overall value, and then subtracted this from 1. All values were summed for a total possible of 5, where 5 is most- and 0 is least-intensively managed.

### *Statistical analyses*

To differentiate between percent cover, structural diversity, tree richness, and MI we used ANOVA with site as a main factor. To assess differences in ant activity with sampling date and with season we used ANOVA. To determine seasonal fluctuations in ant abundance under differing management intensities, we calculated the coefficient of variation (CV) (the ratio of standard deviation in abundance for each date to mean abundance) for each site. For ant abundance, we used untransformed data for tuna baits and square-root transformed data for D-vac data to meet assumptions of normality. We used Tukey's post-hoc tests to make pair-wise comparisons where site differences were detected.

## Results

### *Vegetation and management system*

Sites significantly differed for each vegetation variable and MI where TP was generally the most shaded, CPI and CPB were intermediate, and SM was least shady (Table 1). Tree richness differed with site ($F_{3,28} = 63$, $p < 0.001$) and TP had more than double the number of tree species of CPI, CPB or SM ($p < 0.001$), and CPI and CPB had twice as many species as SM ($p < 0.001$). Tree richness in CPI and CPB did not differ ($p = 0.752$). In both Nov. 2000 ($F_{3,28} = 36.34$, $p < 0.001$) and Jan. 2003 ($F_{3,28} = 22.98$, $p < 0.001$) percent canopy cover differed between sites. In Nov. 2000, cover was higher in TP and CPI than in CPB and SM ($p < 0.001$), CPB tended to have higher cover than SM ($p = 0.080$), but cover in TP and CPI did not differ ($p = 0.507$). In Jan. 2003, percent canopy cover was higher in TP, CPI and CPB than in SM ($p < 0.001$), but CPI did not differ from TP ($p = 0.364$) or CPB ($p = 0.526$). Structural diversity differed between sites as well (two-species cover, $F_{3,28} = 11.33$, $p < 0.001$; three-species cover, $F_{3,28} = 2.333$, $p = 0.096$). TP and CPI had ten times more points covered with two species than CPB and SM ($p < 0.007$), and TP and CPI were the only sites with points covered by three species. MI values also differed with site ($F_{3,28} = 42.23$, $p < 0.001$). All sites significantly differed from one another ($p < 0.004$), and in order from highest to lowest shade were TP, CPI, CPB, and SM, but differences between TP and SM were only marginally significant ($p = 0.067$).

*Ant diversity and management system*

We collected a total of 81 ant species (67 with D-vac sampling and 57 with tuna baits) (Appendix 1). For both sampling methods the majority of ant species (D-vac, 61.2%; baits, 50%) were collected from fewer than ten samples. The most commonly encountered ants found in D-vac samples were *Brachymyrmex* sp. 1 (29.2% of samples), *Brachymyrmex heeri* (24.6%), *Nesomyrmex echinatinodis* (23.8%), and *Pseudomyrmex simplex* (23.6%). The most commonly encountered ants at tuna baits were *Brachymyrmex heeri* (29.3%), *Azteca instabilis* (28.7%), *P. simplex* (26.3%), and *N. echinatinodis* (23.9%).

Species richness differed with site and sampling method (Figure 1). In general, more species were captured with the D-vac (Figure 1), but rarefaction curves were closer to reaching asymptotes for baiting. For baits, ant richness was highest in TP, slightly lower in CPB, even lower in CPI, and lowest in SM (Figure 1a). For D-vac samples, CPB was substantially richer than other sites. TP had the second highest richness followed by CPI and SM, which had the lowest number of species (Figure 1b). For methods combined, CPB had slightly higher richness than TP (one species more) but both sites were much richer than CPI followed by SM (Figure 1c).

Estimates of total richness per site varied with sampling method and estimator but overall richness patterns were similar to observed results (Figure 2, Table 2). For tuna baiting, TP estimates were higher (8–39% more species) and SM estimates were lower (20–26% fewer species) than CPI and CPB, but estimates for CPI and CPB were within three species of one another (Table 2). Estimated richness was similar to observed richness (within 12 species) for all sites. For D-vac samples, however, richness estimates far exceeded observed numbers (up to 20 species more) – especially for TP where estimated richness was up to 61% higher than observed richness (Table 2), reflecting that we sampled less of the total community using this method. In general, however, estimated richness for TP, CPB, and CPI was higher than SM (11–45% more species). For samples combined, estimated richness for TP and CPB was similar (within 4 species). Estimated richness in TP and CPB was higher than for CPI (with 6 to 15 more species). SM was the least rich site with up to 10 fewer estimated species than CPI (Table 2).

Generally, diversity indices declined with increasing management, but results varied with the way the ant community was sampled (Table 2). For tuna bait data TP was more diverse than CPI and CPB which were more diverse than

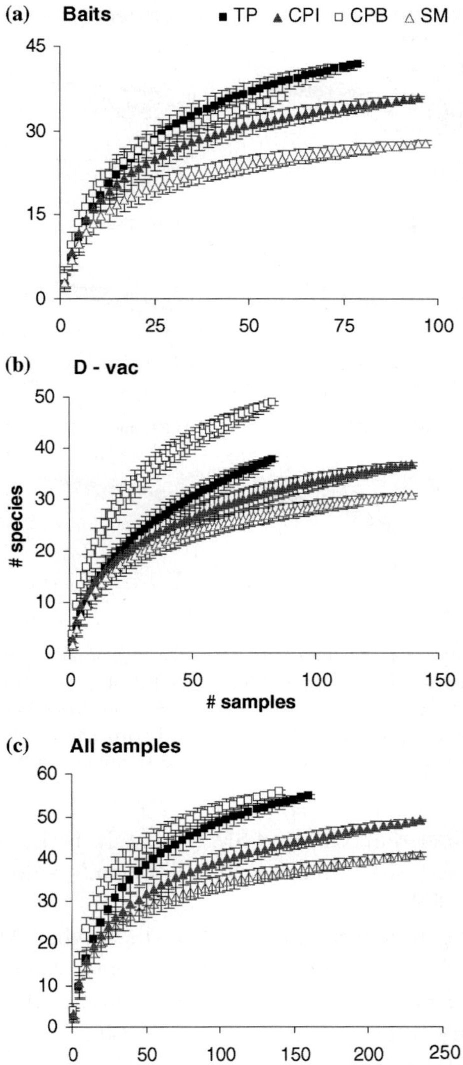


*Figure 1.* Arboreal ant species rarefaction curves in four coffee sites organized from most shady to least shady: Belen Rustic (TP), Irlanda (CPI), Belen Production (CPB), and Hamburgo (SM) in Chiapas, Mexico. Letters show rarefaction (Coleman) curves for (a) tuna baits, (b) D-vac samples, and (c) for all samples combined generated with EstimateS. Closed symbols show more shady sites, and open symbols show less shady sites.

SM. D-vac results were not as clear, but showed a general trend towards lower diversity in SM. For Fisher's alpha, TP was more diverse than all other sites, and SM was least diverse. Shannon Index values were higher in CPB than in TP and lowest in CPI. Simpson's values were the same for TP, CPB, and SM and lower in CPI. For all samples combined, general patterns showed TP and

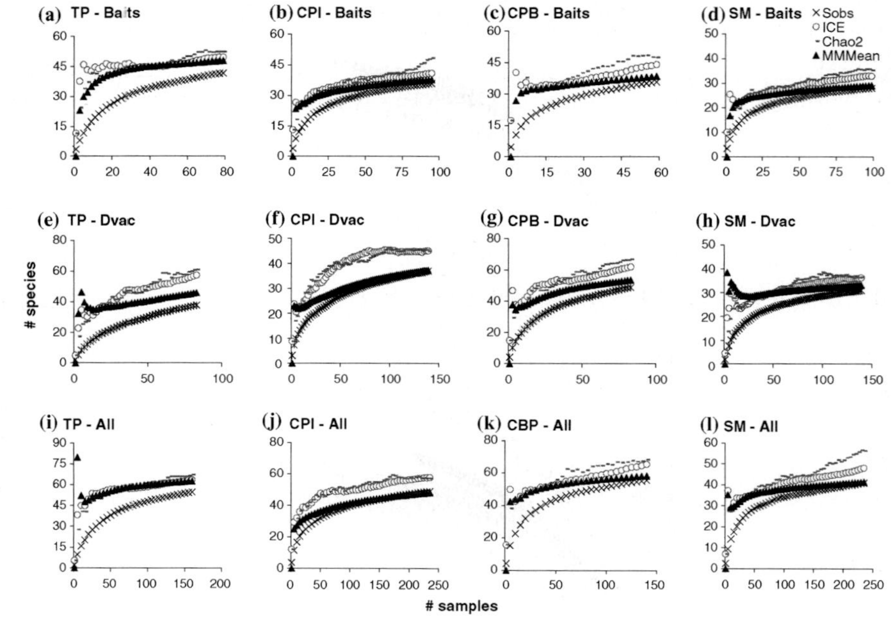


*Figure 2.* Ant species accumulation curves in four coffee sites of increasing management intensity (Belen Rustic (TP), Irlanda (CPI), Belen Production (CPB) and Hamburgo (SM)) for observed richness (SOBS), and for species richness estimators (ICE – Incidence-based Coverage Estimator; Chao2; MMMeans – Michaelis–Menten Means) created with EstimateS. Richness was assessed with tuna baits (a–d), D-vac samples (e–h), and for both sampling methods combined (i–l).

CPB to be more diverse than CPI and SM (Table 2). Fisher's alpha was highest for TP and lowest for SM. Shannon and Simpson's values were highest for CPB and lowest for CPI.

Overall, increasing management intensity (i.e. higher MI) correlated with decreasing observed and estimated ant species richness ($R^2 = 0.1389$, $y = -5.7718x + 65.893$, $F_{1,46} = 7.477$, $p = 0.009$), but did not correlate to changes in diversity index values ($R^2 = 0.0146$, $y = -0.8751x + 8.183$, $F_{1,34} = 0.507$, $p = 0.481$).

## *Ant abundance and activity by season, management system*

Ant abundance was somewhat higher in the wet season, but management intensity did not affect either ant abundance or seasonal fluctuations in abundance. Ant abundance at tuna baits was 21% higher ($F_{1,126} = 16.26$, $p < 0.001$) and with D-vac samples 19% (not significantly) higher in the wet season ($F_{1,126} = 0.144$, $p = 0.705$) (Figure 3). In three of five tuna sampling dates, and for all D-vac sampling dates, ant abundance differed with site (Figure 3, Table 3). To assess if site differences reflected changes in manage-

*Table 2.* Ant species richness (observed and estimated) and diversity indices for two sampling methods for four coffee sites (Belen Rustic (TP), Irlanda (CPI), (Belen Production (CPB), and Hamburgo (SM)).

| Site | SOBs | ICE | Chao2 | MM-Mean | Fisher's | Shannon | Simpson |
|---|---|---|---|---|---|---|---|
| *Tuna baiting* | | | | | | | |
| TP | 42.00 | 50.23 | 53.00 | 48.01 | 14.83 | 3.25 | 0.95 |
| CPI | 36.00 | 40.97 | 48.80 | 37.77 | 9.78 | 2.96 | 0.93 |
| CPB | 36.00 | 44.26 | 47.11 | 38.57 | 11.21 | 3.08 | 0.94 |
| SM | 28.00 | 32.91 | 35.20 | 29.07 | 7.34 | 2.79 | 0.92 |
| *D-vac Sampling* | | | | | | | |
| TP | 38.00 | 57.69 | 61.27 | 46.02 | 15.05 | 3.07 | 0.94 |
| CPI | 37.00 | 44.92 | 44.69 | 37.42 | 9.69 | 2.80 | 0.91 |
| CPB | 49.00 | 62.00 | 66.31 | 53.77 | 15.43 | 3.24 | 0.94 |
| SM | 31.00 | 36.33 | 36.44 | 33.18 | 8.99 | 2.94 | 0.94 |
| *All samples combined* | | | | | | | |
| TP | 55.00 | 63.83 | 68.00 | 62.75 | 17.09 | 3.40 | 0.95 |
| CPI | 49.00 | 57.89 | 57.07 | 48.14 | 11.48 | 3.00 | 0.93 |
| CPB | 56.00 | 65.70 | 67.08 | 58.58 | 14.98 | 3.49 | 0.96 |
| SM | 41.00 | 48.22 | 57.20 | 41.43 | 9.97 | 3.07 | 0.94 |

Numbers show observed richness (SOBs), estimated richness (Incidence-based Coverage Estimator (ICE), Chao2, and Michaelis-Menten Means (MM-Mean), and diversity indices (Fisher's, Shannon, and Simpson, all calculated with EstimateS.

ment intensity, we summed all Tukey's tests where 1) abundance was significantly lower in sites with lower MI, 2) abundance was significantly higher in sites with lower MI, and 3) where abundance did not differ. In 8 of 72 possible pair-wise comparisons, ant abundance was greater in the more intensively managed site, and in 10 of 72 comparisons abundance was greater in less intensively managed site. In most comparisons (54 of 72, 75 %); however, ant abundance did not differ with differences in management intensity.

Ant abundance did not fluctuate less with season under less intense management. For tuna baiting, the CV of abundance was higher in SM (0.289) than in CPI (0.194) or CPB (0.174), but was highest for TP (0.353), the site with the lowest MI. For D-vac samples, the CV was higher in CPB (0.688) than in CPI (0.525) or TP (0.439), but was lowest in SM (0.298), the site with the highest MI.

## Discussion

The four coffee sites selected differed for most vegetation variables and for the MI where TP was least intensively managed, CPI and CPB were intermediate,

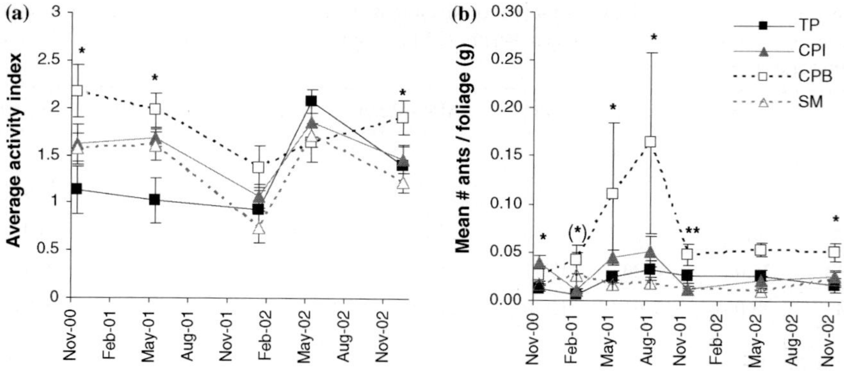


*Figure 3.* Effect of coffee management system and season on ant activity (a, tuna baits) and abundance (b, D-vac samples) in four coffee sites (Belen Rustic (TP), Irlanda (CPI), Belen Production (CPB) and Hamburgo (SM)) over a two year period. The dry season dates fall between November–April and wet season dates fall between May–October. Bars show standard error. Significant differences shown with asterisks are explained in the text and Table 3.

and SM was most intensively managed. Overall, increasing MI correlated with decreasing coffee-foraging ant richness. For tuna samples, higher management intensity reflected increased ant diversity. For D-vac sampling, ant diversity was still lowest under the highest management intensity, but differences between intermediate and high-shade sites were less clear, and diversity tended to be higher in CPB, a more intensely-managed site by some measures. Thus patterns of diversity loss depended somewhat on sampling method. Ant abundance was higher in the wet season, but abundance did not differ with management intensity, nor did abundance fluctuate less in less intensively managed sites.

Species richness estimators varied greatly in number of species estimated for a sampling method and site. In general, MMMeans and ICE were closest to reaching asymptotes and seemed to be the most reliable. MMMeans returned estimates closest to observed richness and was closest to reaching asymptotes, perhaps showing a minimum number of species per site/method. ICE returned higher estimates, but was also close to reaching asymptotes. These two estimators have shown high performance previously for tropical ants in Costa Rica (Longino et al. 2002) and for tropical trees (Chazdon et al. 1998). In contrast, Chao2 behavior was more erratic and sometimes not close to reaching asymptotes, even for where observed species accumulation curves were leveling off (Figure 2). For example, accumulation curves for SM for all methods combined was close to an asymptote, yet Chao2 estimates were sharply raising.

Overall, we found more ant species in D-vac samples than with tuna baits, likely due to differences in the ant communities sampled or due to interspecific competition. Tuna baits tend to attract generalist ants, perhaps from a larger foraging range including the ground and canopy, yet baits do not attract all

*Table 3.* ANOVA results comparing ant activity (tuna baiting) and abundance (# ants/foliage (g), D-vac sampling) between four coffee sites under differing management intensity.

| Date | df | F | P | Tukey's post-hoc |
|---|---|---|---|---|
| *Tuna baits* | | | | |
| Nov-00 | 3, 56 | 3.236 | 0.029 | TP < CPB (0.017) |
| May-01 | 3, 60 | 6.085 | 0.001 | TP < CPB (0.001), CPI (0.005), SM (0.026) |
| Jan-02 | 3, 60 | 1.959 | 0.130 | |
| May-02 | 3, 60 | 1.136 | 0.342 | |
| Dec-02 | 3, 60 | 2.969 | 0.039 | SM < CPB (0.025) |
| *D-vac sampling* | | | | |
| Nov-00 | 3, 60 | 5.06 | 0.003 | CPI > SM (0.038), TP (0.003) |
| Feb-01 | 3, 60 | 2.763 | 0.050 | TP < CPB (0.044) |
| May-01 | 3, 60 | 3.888 | 0.013 | SM < CPB (0.011) |
| Aug-01 | 3, 60 | 4.49 | 0.007 | SM < CPB (0.004) |
| Nov-01 | 3, 60 | 8.765 | <0.001 | CPB < CPI (0.001), SM (<0.001) |
| May-02 | 3, 60 | 12.317 | <0.001 | CPB < CPI (<0.001), SM (<0.001), TP (0.029); TP < SM (0.034) |
| Nov-02 | 3, 60 | 5.068 | 0.003 | CPB < SM (0.033), CPI (0.002) |

Results were calculated using untransformed data for tuna baiting and square-root transformed data for D-vac sampling. Tukey's post-hoc results (*p*-value in parenthesis) show significant site differences where sites in order from least to most intensively managed are: TP – Belen Rustic, CPI – Irlanda, CPB – Belen Production and SM – Hamburgo.

ants in the community. D-vac samples, on the other hand, will catch most ants foraging on coffee plants regardless of their diet preferences. Thus, the ant community sampled by tuna baits may be in part a subset of ants captured in D-vac samples, but may also include some ants more generally restricted to shade trees or the ground. Additionally, tuna baits attract some competitively dominant ants (such as *Solenopsis geminata*) that may have eliminated other ant species from baits before they were checked (Perfecto 1994).

Although ant diversity generally declined with increasing MI, this was not always the case, perhaps because habitat management is not the only factor controlling diversity. Ant richness in D-vac samples was exceptionally high in CPB, a more intensively managed site. Competitive interactions between ants may also strongly impact diversity within areas. For example, presence of *Solenopsis geminata* may restrict ant richness by excluding other ant species (Perfecto 1994). Surprisingly, CPB, the site with the highest overall richness, did not have any samples with *Solenopsis geminata.* It may be possible that absence of this ant in CPB allows other ants to exist there, thereby increasing richness. Additionally, although habitat characteristics may strongly affect species diversity (Collinge et al. 2003), regional factors such as distance from forest, (Ricketts et al. 2001; Perfecto and Vandermeer 2002) or landscape patterns such as extent of high-quality habitat and habitat arrangement may also be important (Vandermeer and Carvajal 2001; Steffan-Dewenter 2002; Weibull et al. 2003). Land-use history may also influence patterns of biodiversity loss such that species loss lags behind habitat destruction (Tilman et al. 1994). The landscape surrounding CPB includes fewer intensively managed coffee farms and more forest fragments than do CPI and SM (Pers. obsv.). CPB was recently converted to more intense management (< 5 years ago) and thus may be still undergoing species loss whereas CPI and SM have been under similar management for > 20 years (G. Ibarra-Nuñez, Pers. comm.). These three factors may explain why diversity in CPB was relatively high compared with CPI, a similarly managed farm. Richness and diversity in CPB was even slightly higher than in the least intensively managed site (TP). Although in this study we focus on coffee-foraging ants, all sites and in particular TP, may include arboreal ants more restricted to the shade tree layer that nonetheless may sometimes forage in the coffee plants. Given that accumulation curves were furthest from reaching asymptotes for TP, the true richness of arboreal ants foraging in the coffee layer may be much higher than our samples indicate. Furthermore, we did not sample nocturnally foraging ants that may account for a large part of the ant community. The inclusion of these ants may significantly alter the results found here.

In conclusion, coffee-foraging ant diversity, but not ant abundance, tended to decrease with increasing coffee management intensification. Many debate the relative importance of diversity and abundance in determining the function of biodiversity or ecosystem services (Balvanera et al. 2001; Kremen et al. 2002; Klein et al. 2003). Here, ant abundance was not changed, yet ant diversity (including both ground- and coffee-foraging ants) is affected by management intensity. Coffee may thus serve as a model system for investigating the interplay between diversity, abundance, and ecosystem services.

## Acknowledgements

We thank J.A. Garcia-Ballinas, P. Bichier, G. Lopez, J. Maldonado, B.E. Chilel, R. Velasquez, J.C. Mendez Lopez, A. Mendez Mendizabal, A. Hammond, F.B. Camposeco Silvestre, J.L. Cabrera Santos, L.H.L. Ramirez, A. Gonzalez, G. Dominguez, and S. Uno for field help. I. Armbrecht, J. Longino, B. McKay, R. Snelling, and P. Ward assisted in ant identification. I. Armbrecht and R. Burnham commented on the manuscript. The Peters and Edelman families and the ISMAM Cooperative allowed us to work on their farms. G. Ibarra-Núñez from El Colegio de la Frontera Sur in Tapachula provided logistical support. This research was supported by NSF Grant #DEB-9981526 to I. Perfecto, R. Greenberg, and G. Ibarra-Núñez, the Helen Olsen Brower Fellowship in Environmental Science of the University of Michigan, and an NSF Graduate Research Fellowship to S. Philpott.

*Appendix 1.* Ant species (organized by subfamilies) found in four coffee farms using tuna baits and D-vac sampling or for both methods combined over a 2-year sampling period.

| | All methods | | | | Baits | | | | D-vac | | | |
|---|---|---|---|---|---|---|---|---|---|---|---|---|
| Species | TP | CPI | CPB | SM | TP | CPI | CPB | SM | TP | CPI | CPB | SM |
| Dolichoderinae | | | | | | | | | | | | |
| *Azteca instabilis* | x | x | x | x | x | x | x | x | x | x | x | x |
| *Azteca* sp. 1 | | | x | | | | | | | | x | |
| *Azteca* sp. 2 | x | x | x | x | x | x | x | x | x | | x | |
| *Azteca* sp. 3 | | | x | x | | | | | | | x | x |
| *Dolichoderus debilis* | | x | | | | x | | | | | | |
| *Dolichoderus lutosus* | | | x | | | | x | | | | x | |
| *Dorymyrmex* sp. 1 | | | | x | | | | x | | | | |
| *Linepithema* sp. 1 | x | | | | x | | | | | | | |
| *Tapinoma* sp. 1 | | | x | | | | x | | | | | |
| *Tapinoma* sp. 2 | x | | x | | | | | | x | | x | |
| *Tapinoma* sp. 3 | x | | x | | | | | | x | | x | |
| *Technomyrmex* sp. 1 | x | | x | | x | | x | | | | | |
| *Technomyrmex* sp. 2 | x | x | x | x | | | x | x | x | x | x | x |
| Ecitoninae | | | | | | | | | | | | |
| *Labidus coecus* | | x | x | | | x | | | | | x | |
| Formicinae | | | | | | | | | | | | |
| *Brachymyrmex heeri* | x | x | x | x | x | x | x | x | x | x | x | x |
| *Brachymyrmex* sp. 1 | x | x | x | x | x | x | x | x | x | x | x | x |
| *Brachymyrmex* sp. 2 | x | x | x | x | | | | | x | x | x | x |
| *Brachymyrmex* sp. 3 | | x | x | | | | x | | | x | x | |
| *Camponotus abscisus* | | | | x | | | | | | | | |
| *Camponotus canescens* | x | x | x | | x | x | | | x | | x | |
| *Camponotus novogranadensis* | x | x | x | | x | x | x | | | | x | |
| *Camponotus senex* | x | | | | x | | | | x | | | |
| *Camponotus senex textor* | x | x | x | x | x | x | x | x | | x | x | x |
| *Camponotus sericeiventris* | x | | x | | x | | x | | | | | |
| *Camponotus striatus* | x | | x | | x | | x | | | | x | |
| *Camponotus* sp. 1 | x | x | x | | x | | x | | x | x | | |
| *Myrmelachista* sp. 1 | x | x | | | | | | | x | x | | |
| *Myrmelachista* sp. 2 | | x | | | | x | | | | x | | |
| *Myrmelachista* sp. 3 | x | x | x | x | x | x | | | x | x | x | x |
| *Myrmelachista* sp. 4 | x | | x | | | | | | x | | x | |

*Appendix 1.* Continued.

| | All methods | | | | Baits | | | | D-vac | | | |
|---|---|---|---|---|---|---|---|---|---|---|---|---|
| Species | TP | CPI | CPB | SM | TP | CPI | CPB | SM | TP | CPI | CPB | SM |
| *Paratrechina* sp. 1 | | | x | | | | | | | | x | |
| *Paratrechina* sp. 2 | | | x | | | | | | | | x | |
| Myrmicincae | | | | | | | | | | | | |
| *Cephalotes* sp. 1 | x | x | | | x | x | | x | x | | | |
| *Crematogaster* spp. | x | x | x | x | x | x | x | x | | | | |
| *Crematogaster carinata* | x | x | x | x | | | x | | x | x | x | x |
| *Crematogaster crinosa* | x | x | x | x | x | x | x | x | x | x | x | x |
| *Crematogaster formosa* | x | | | | x | | | | x | | | |
| *Crematogaster hirsuta* | | | | x | | | | | | | | x |
| *Crematogaster negrapilosa* | x | x | x | x | x | x | x | x | | | x | |
| *Crematogaster sumichrasti* | x | | x | x | x | | x | x | x | | x | x |
| *Crematogaster* sp. 1 | x | | x | | x | | x | | | | | |
| *Crematogaster* sp. 2 | x | | | | x | | | | x | | | |
| *Monomorium floricola* | x | x | x | x | | | | | x | x | x | x |
| *Monomorium pharoanis* | x | | x | | x | | | | | | x | |
| *Monomorium* sp. 1 | | | | x | | | | | | | | x |
| *Nesomyrmex echanatinodis* | x | x | x | x | x | x | x | x | x | x | x | x |
| *Nesomyrmex pittieri* | x | x | x | | x | x | x | | x | x | x | |
| *Pheidole indestincta* | x | x | x | x | x | x | x | x | x | x | x | x |
| *Pheidole punctatissima* | x | x | x | x | x | x | x | x | | x | x | |
| *Pheidole susannae* | | | x | | | | | | | | x | |
| *Pheidole* sp. 1 | x | | | | | | | | x | | | |
| *Pheidole* sp. 2 | x | x | x | x | x | x | x | x | x | x | x | x |
| *Pheidole* sp. 3 | x | x | x | x | | x | | x | x | x | x | x |
| *Pheidole* sp. 4 | | x | | | | x | | | | | | |
| *Pheidole* sp. 5 | x | x | x | | x | x | x | | | x | x | |
| *Pheidole* sp. 6 | x | | x | x | x | | x | x | | | | |
| *Pheidole* sp. 7 | | | x | | | | x | | | | x | |
| *Pheidole* sp. 8 | | x | | x | | | | | | x | | x |
| *Pheidole* sp. 9 | | | | x | | | | | | | | x |
| *Pheidole* sp. 10 | x | x | x | x | x | x | | | | x | x | x |
| *Procryptocerus scabriusculus* | x | x | x | x | x | x | x | x | x | x | x | x |
| *Pyramica* sp. 1 | | | x | | | | | | | | x | |
| *Solenopsis geminata* | x | x | | x | x | x | | x | x | x | | x |

*Appendix 1.* Continued.

| | All methods | | | | Baits | | | | D-vac | | | |
|---|---|---|---|---|---|---|---|---|---|---|---|---|
| Species | TP | CPI | CPB | SM | TP | CPI | CPB | SM | TP | CPI | CPB | SM |
| *Solenopsis* sp. 1 | x | x | x | x | x | x | x | x | | x | x | x |
| *Solenopsis* sp. 2 | x | x | | x | x | x | | | x | x | | x |
| *Solenopsis* sp. 3 | x | x | x | | | | | | x | x | x | |
| *Wasmannia auropunctata* | x | x | x | x | x | x | x | x | x | | x | |
| Poneromorph | | | | | | | | | | | | |
| *Gnamptogenys striatula* | | x | x | | | x | x | | | x | x | |
| *Hyperponera* sp. 1 | x | | | | x | | | | | | | |
| *Pachycondyla apicales* | x | x | x | x | | | | | x | x | x | x |
| Pseudomyrmecinae | | | | | | | | | | | | |
| *Pseudomyrmex ejectus* | x | x | x | x | x | x | x | x | x | x | x | x |
| *Pseudomyrmex elongatus* | x | x | x | x | x | x | x | x | | x | x | x |
| *Pseudomyrmex gracilis* | x | x | x | x | x | x | x | x | x | | x | |
| *Pseudomyrmex occulatus* | | | | x | | | | x | | | | |
| *Pseudomyrmex* PSW-06 | x | x | | | | | | | x | x | | |
| *Pseudomyrmex* PSW-53 | x | x | | x | x | x | | x | x | x | | x |
| *Pseudomyrmex simplex* | x | x | x | x | x | x | x | x | x | x | x | x |
| *Pseudomyrmex* sp. 1 | | x | | | | | | | | x | | |
| *Pseudomyrmex* sp. 2 | | x | | | | x | | | | | | |
| *Pseudomyrmex* sp. 3 | | | x | | | | | | | | x | |
| *Pseudomyrmex* sp. 4 | | x | | x | | | | | | x | | x |
| Total species observed | 55 | 49 | 56 | 41 | 42 | 36 | 36 | 28 | 38 | 37 | 49 | 31 |

## References


Aberg J., Jansson G., Swenson J.E. and Angelstam P. 1995. The effect of matrix on the occurrence of Hazel Grouse (*Bonasa bonasia*) in isolated habitat fragments. Oecologia 103: 265–269.

Alonso L.E. 1998. Spatial and temporal variation in the ant occupants of a facultative ant-plant. Biotropica 30: 201–213.

Andow D.A. 1991. Vegetational diversity and arthropod population response. Ann. Rev. Entomol. 36: 561–586.

Armbrecht I. and Perfecto I. 2003. Litter-twig dwelling ant species richness and predation potential within a forest fragment and neighboring coffee plantations of contrasting habitat quality in Mexico. Agric. Ecosyst. Environ. 97: 107–115.

Balvanera P., Daily G.C., Ehrlich P.R., Ricketts T.H., Bailey S.A., Kark S., Kremen C. and Pereira H. 2001. Conserving biodiversity and ecosystem services. Science 291: 2047.

Basedow T. 1991. Population-density and biomass of epigeal predatory arthropods, natural enemies of insect pests, in winter-wheat fields of areas with extremely different intensity of agricultural production. J. Plant Dis. Prot. 98: 371–377.

Chazdon R.L., Colwell R.K., Denslow J.S. and Guariguata M.R. 1998. Statistical methods for estimating species richness of woody regeneration in primary and secondary rain forests of northeastern Costa Rica. In: Dallmeier F. and Comiskey J.A. (eds), Forest Biodiversity Research, Monitoring and Modeling: Conceptual Background and Old World Case Studies. Parthenon Publishing, Paris, France, pp. 285–309.

Collinge S.K., Prudic K.L. and Oliver J.C. 2003. Effects of local habitat characteristics and landscape context on grassland butterfly diversity. Conserv. Biol. 17: 178–187.

Colwell R.K. and Coddington J.A. 1994. Estimating terrestrial biodiversity through extrapolation. Philos. Trans. R. Soc. London Ser. B 345: 101–118.

Didham R.K., Hammond P.M., Lawton J.H., Eggleton P. and Stork N.E. 1998. Beetle species responses to tropical forest fragmentation. Ecol. Monogr. 68: 295–323.

Estrada A. and Coates-Estrada R. 2002. Dung beetles in continuous forest, forest fragments and in an agricultural mosaic habitat island at Los Tuxtlas, Mexico. Biodiv. Conserv. 11: 1903–1918.

Girma H., Rao M.R. and Sithanantham S. 2000. Insect pests and beneficial arthropods population under different hedgerow intercropping systems in semiarid Kenya. Agroforest. Syst. 50: 279–292.

Glor R.E., Flecker A.S., Benard M.F. and Power A.G. 2001. Lizard diversity and agricultural disturbance in a Caribbean forest landscape. Biodiv. Conserv. 10: 711–723.

Gotelli N.J. and Colwell R.K. 2001. Quantifying biodiversity: procedures and pitfalls in the measurement and comparison of species richness. Ecol. Lett. 4: 379–391.

Greenberg R., Bichier P. and Sterling J. 1997. Bird populations in rustic and planted shade coffee plantations of Eastern Chiapas, Mexico. Biotropica 29: 501–514.

Guedes R.N.C., Zanuncio T.V., Zanuncio J.C. and Medeiros A.G.B. 2000. Species richness and fluctuation of defoliator Lepidoptera populations in Brazilian plantations of *Eucalyptus grandis* as affected by plant age and weather factors. Forest Ecol. Manag. 137: 179–184.

Kalif K.A.B., Azevedo-Ramos C., Moutinho P. and Malcher S.A.O. 2001. The effect of logging on the ground-foraging ant community in Eastern Amazonia. Stud. Neotropical Fauna Environ. 36: 215–219.

Kaspari M. and Weiser M.D. 2000. Ant activity along moisture gradients in a neotropical forest. Biotropica 32: 703–711.

Klein A., Steffan-Dewenter I. and Tscharntke T. 2003. Fruit set of highland coffee increases with the diversity of pollinating bees. Proc. R. Soc. London B 270: 955–961.

Klein A.M., Steffan-Dewenter I., Buchori D. and Tscharntke T. 2002. Effects of land-use intensity in tropical agroforestry systems on coffee flower-visiting and trap-nesting bees and wasps. Conserv. Biol. 16: 1003–1014.

Klein A.M., Steffan-Dewenter I. and Tscharntke T. 2002. Predator-prey ratios on cocoa along a land-use gradient in Indonesia. Biodiv. Conserv. 11: 683–693.

Knops J.M.H., Tilman D., Haddad N.M., Naeem S., Mitchell C.E., Haarstad J., Ritchie M.E., Howe K.M., Reich P.B., Siemann E. and Groth J. 1999. Effects of plant species richness on invasion dynamics, disease outbreaks, insect abundances and diversity. Ecol. Lett. 2:286–293.

Kremen C., Williams N.M. and Thorp R.W. 2002. Crop pollination from native bees at risk from agricultural intensification. Proc. Natl. Acad. Sci. USA 99: 16812–16816.

Longino J.T., Coddington J. and Colwell R.K. 2002. The ant fauna of a tropical rain forest: Estimating species richness three different ways. Ecology 83: 689–702.

Mas A.H. and Dietsch T.V. 2003. An index of management intensity for coffee agroecosystems to evaluate butterfly species richness. Ecol. Appl. 13: 1491–1501.

Moguel P. and Toledo V.M. 1999. Biodiversity conservation in traditional coffee systems of Mexico. Conserv. Biol. 13: 11–21.

Nestel D. and Dickschen F. 1990. The foraging kinetics of ground ant communities in different Mexican coffee agroecosystems. Oecologia 84: 58–63.

Perfecto I. 1994. Foraging behavior as a determinant of asymmetric competitive interaction between 2 ant species in a tropical agroecosystem. Oecologia 98: 184–192.

Perfecto I., Mas A., Dietsch T. and Vandermeer J. 2003. Conservation of biodiversity in coffee agroecosystems: a tri-taxa comparison in southern Mexico. Biodiv. Conserv. 12: 1239–1252.

Perfecto I., Rice R.A., Greenberg R. and VanderVoort M.E. 1996. Shade coffee: a disappearing refuge for biodiversity. Bioscience 46: 598–608.

Perfecto I. and Sediles A. 1992. Vegetational diversity, ants (Hymenoptera, Formicidae), and herbivorous pests in a neotropical agroecosystem. Environ. Entomol. 21: 61–67.

Perfecto I. and Snelling R. 1995. Biodiversity and the transformation of a tropical agroecosystem - ants in coffee plantations. Ecol. Appl. 5: 1084–1097.

Perfecto I. and Vandermeer J. 2002. Quality of agroecological matrix in a tropical montane landscape: ants in coffee plantations in southern Mexico. Conserv. Biol. 16: 174–182.

Perfecto I., Vandermeer J., Hanson P. and Cartin V. 1997. Arthropod biodiversity loss and the transformation of a tropical agroecosystem. Biodiv. Conserv. 6: 935–945.

Ricketts T.H. 2001. The matrix matters: Effective isolation in fragmented landscapes. Am. Nat. 158: 87–99.

Ricketts T.H., Daily G.C., Ehrlich P.R. and Fay J.P. 2001. Countryside biogeography of moths in a fragmented landscape: Biodiversity in native and agricultural habitats. Conserv. Biol. 15: 378–388.

Rojas L., Godoy C., Hanson P., Kleinn C. and Hilje L. 2001. Hopper (Homoptera: Auchenorrhyncha) diversity in shaded coffee systems of Turrialba, Costa Rica. Agroforest. Syst. 53: 171–177.

Root R.B. 1973. Organization of a plant-arthropod association in simple and diverse habitats – fauna of collards (*Brassica oleracea*). Ecol. Monogr. 43: 95–120.

Roth D.S., Perfecto I. and Rathcke B. 1994. The effects of management systems on ground-foraging ant diversity in Costa Rica. Ecol. Appl. 4: 423–436.

Siebert S.F. 2002. From shade- to sun-grown perennial crops in Sulawesi, Indonesia: implications for biodiversity conservation and soil fertility. Biodiv. Conserv. 11: 1889–1902.

Steffan-Dewenter I. 2002. Landscape context affects trap-nesting bees, wasps, and their natural enemies. Ecol. Entomol. 27: 631–637.

Tauber M.J., Tauber C.A., Nyrop J.P. and Villani M.G. 1998. Moisture, a vital but neglected factor in the seasonal ecology of insects: Hypotheses and tests of mechanisms. Environ. Entomol. 27: 523–530.

Tilman D. 1999. Global environmental impacts of agricultural expansion: The need for sustainable and efficient practices. Proc. Natl. Acad. Sci. USA 96: 5995–6000.

Tilman D., May R.M., Lehman C.L. and Nowak M.A. 1994. Habitat Destruction and the Extinction Debt. Nature 371: 65–66.

Tscharntke T., Steffan-Dewenter I., Kruess A. and Thies C. 2002. Characteristics of insect populations on habitat fragments: A mini review. Ecol. Res. 17: 229–239.

Vandermeer J. and Carvajal R. 2001. Metapopulation dynamics and the quality of the matrix. Am. Nat. 158: 211–220.

Vandermeer J., Perfecto I., Ibarra-Nunez G., Philpott S. and Garcia-Ballinas J.A. 2002. Ants (*Azteca* sp.) as potential biological control agents in organic shade coffee production in Southern Chiapas, Mexico: complication of indirect effects. Agroforest. Syst. 56: 271–276.

Velez M., Bustillo A.E. and Posada F.J. 2000. Predación sobre *Hypothenemus hampei*, (Ferrari) de las hormigas *Solenopsis* spp. , *Pheidole* spp., y *Dorymyrmex* spp. durante el secado del café. In: Vélez M., Bustillo A.E. and Posada F.J. (eds), Resúmenes XXVII Congreso. Sociedad Colombiana de Entomologia, Medellín, Colombia, pp. 17.

Watt A.D., Stork N.E. and Bolton B. 2002. The diversity and abundance of ants in relation to forest disturbance and plantation establishment in southern Cameroon. J. Appl. Ecol. 39: 18–30.

Way M.J. and Khoo K.C. 1992. Role of ants in pest-management. Ann. Rev. Entomol. 37: 479–503.

Weibull A.C., Ostman O. and Granqvist A. 2003. Species richness in agroecosystems: the effect of landscape, habitat and farm management. Biodiv. Conserv. 12: 1335–1355.

Young A. 1999. Is there really spare land? A critique of estimates of available cultivable land in developing countries. Environ. Dev. Sustainability 1: 3–18.

Biodiversity and Conservation (2006) 15:143–161
DOI 10.1007/s10531-004-4248-1 

# Influence of habitat fragmentation on the genetic variability in leaf litter ant populations in tropical rainforests of Sabah, Borneo

TOBIAS O. BICKEL[1,2,*], CARSTEN A. BRÜHL[1,3], JÜRGEN R. GADAU[2], BERT HÖLLDOBLER[2] and K. EDUARD LINSENMAIR[1]
*[1]Department of Animal Ecology and Tropical Biology, Biocenter 97074, Würzburg, Germany; [2]Department of Behavioral Physiology and Sociobiology, Bioenter, 97074, Würzburg, Germany; [3]Present address: Syngenta, Ecological Sciences, Jealott's Hill International Research Center, Bracknell, Berkshire, RG42 6EY, UK; *Author for correspondence: Department of Zoology, University of Otago, P.O. Box 56, Dunedin, New Zealand (e-mail: tobias-bickel@web.de; phone: +84-3-467-5244)*

Received 20 January 2004; accepted in revised form 26 September 2004

**Key words:** Biodiversity, Genetic distance, Genetic diversity, Habitat fragmentation, Leaf litter ants, *Odontomachus rixosus*, *Pheidole annexus*, Population subdivision, RAPD fingerprint, Tropical rainforest

**Abstract.** Two ant species, *Odontomachus rixosus* and *Pheidole annexus*, were studied in the tropical rainforests of Sabah, Malaysia, North Borneo, to analyze the impact of habitat fragmentation on the genetic diversity and population structure of ant populations using RAPD-fingerprinting. Ants were sampled in a contiguous (43,800 ha) and three patches of primary rainforests of varying size (4294, 146 and 20 ha) that were fragmented about 40 years ago.We found a decrease in genetic variability for both species in the fragmented populations compared to the contiguous. Genetic distances between populations resembled the geographical arrangement of populations and can be explained by an effect of isolation by distance. A high degree in population subdivision suggests a lack of meta-population dynamics due to a shortage of gene flow between populations, possibly the result of the high degree of habitat isolation by oil palm plantations. Although the results of this study are limited due to low replication this is the first data on genetic patterns of insect populations in fragmented rainforests and should be seen as starting point for future research. The value of small to medium sized protection areas for conservation needs to be carefully evaluated in the context of this study, as even relatively large areas (4294 ha) may not prevent the critical loss of genetic variability and guarantee long-term survival of organisms.

**Abbreviation:** RAPD – Random Amplified Polymorphic DNA

## Introduction

Fragmentation of formerly contiguous lowland tropical rainforests as a consequence of human use (e.g. agriculture, timber extraction) is recognized as a major threat to their biodiversity (Terborgh 1992; Dale et al. 1994; Fahrig and Merriam 1994; Laurance et al. 2000). Habitat fragmentation affects ecosystems in multiple ways, for example by the loss of biomass (Laurance et al. 1997), disturbance of important species interactions such as pollination (Aizen and

Feinsinger 1994), predation (Kareiva 1987; Oehler and Livaitis 1996) seed and nest predation (Wong et al. 1998) or changes in community composition (Brühl et al. 2003). Furthermore, the tree community, the fundamental matrix of a tropical rainforest, is substantially altered in smaller fragments (Benitez-Malvido 1998; Laurance et al. 1998).

Despite the dominance of arthropod species in tropical rainforests the majority of studies on the impact of habitat fragmentation have focused on vertebrates (but see below). However, insect species may be especially affected by the fragmentation processes as these can be more susceptible to ecosystem disturbance and often possess lower dispersal abilities (Didham et al. 1996). Additionally, effects of fragmentation on ecosystems and the genetic composition of populations should be detectable earlier as their generation times are generally shorter. Recent studies conducted on arthropods in fragmented rainforests have demonstrated a change in community composition, species reduction or a decline in abundances in fragmented habitats (Klein 1989: dung beetles, Fonseca de Souza and Brown 1994: termites, Brown and Hutchings 1997: butterflies, Didham 1997: leaf litter invertebrates, Davies and Margules 1998: carabid beetles, Gibbs 1998: carrion beetles, Kitching et al. 2000: moths).

Besides species diversity, genetic variability within species is also an important aspect of biodiversity. It is the raw material for evolutionary change, and consequently, the basis of evolution and adaptation leading to speciation and species diversity (Norse 1987; Hartl and Clark 1989; Borowsky 2000; Templeton et al. 2001). Beside that, a reduction in genetic variability is linked with a reduced fitness of populations (Reed and Frankham 2003). Following the theory of island biogeography (Mac Arthur and Wilson 1967) a loss of genetic diversity is expected in smaller islands or habitat fragments (Frankham 1996, 1997). Fragmentation of forest inevitably results in smaller habitat areas and subsequently reduces the effective population size which accelerates the accumulation of deleterious alleles and therefore increases the risk of extinction even for comparably large metapopulations. Additionally, a reduction in population size creates a higher level of inbreeding within populations (Hartl and Clark 1989) that may lead to the extinction of species (Frankham 1998; Saccheri et al. 1998; Elgar and Clode 2001). Despite its importance, there is little research conducted on the influence of habitat fragmentation on genetic diversity (Gibbs 1998; Knutsen et al. 2000; Vucetich et al. 2001) particularly in tropical forest ecosystems (but see: Cunningham and Craig 1998; White et al. 1999; Lacerda et al. 2001: mature trees; Sumner et al. 2004: skinks). To our knowledge there is no data available on insects from tropical ecosystems.

Though, studies on population genetics of insects from temperate regions show a higher degree of genetic differentiation in fragmented populations (Knutsen et al. 2000), focus on the identification of evolutionary significant units (Williams 2002) and investigate genetic variability in regard to previous bottleneck events (Debinski 1994; Clarke and O'Dwyer 2000; Williams 2002). Fragmented populations of the ant *Formica lugubris* showed no signs of inbreeding in fragments due to the high effective population size of this polygyne species (Gyllenstrand and Seppä 2003).

To analyze the impact of habitat fragmentation and isolation on intra-specific genetic variability, two leaf litter ant species (*Odontomachus rixosus* Smith 1857; *Pheidole annexus* Eguchi 2001) were sampled in the lowland dipterocarp rainforests of Sabah, Borneo, an exceptionally diverse rainforest in South East Asia (Whitmore 1984, 1998). Large parts of the primary rainforest have already been destroyed by timber extraction and conversion into oil palm plantations. Today oil palm plantations cover a huge area of Sabah (>one million hectares, Chung et al. 2000). Therefore, undisturbed contiguous primary rainforests persist only in central Sabah. Additional primary rainforest remnants remain in the form of isolated patches surrounded by large scale oil palm monocultures. These so called Virgin Jungle Reserves (VJR) have an average size of 1802 ha in Sabah comprising all together 90,386 ha (Laidlaw 1996; Brühl et al. 2003).

Ants are a key group in tropical ecosystems in terms of ecosystem functioning and animal biomass (Petal 1978; Hölldobler and Wilson 1990; Jones et al. 1994). Ant communities respond sensitively to habitat fragmentation by the extinction of specialist species in even large habitat patches (Boswell et al. 1998), the invasion of tramp species (Suarez et al. 1998) and by change in community composition and species richness (Bestelmeyer and Wiens 1996; Carvalho and Vasconcelos 1999; Majer and Delabie 1999; Brühl et al. 2003). A significant decline in species diversity, alteration of community composition and reduced density of leaf litter ants has previously been demonstrated for the forest fragments of this study (Brühl et al. 2003).

In this study we performed a genetic analysis by RAPD fingerprinting (Randomly Amplified Polymorphic DNA, Williams et al. 1990), a suitable method for the genetic analysis of insect populations (Haymer 1994; Harry et al. 1998). RAPD fingerprinting has commonly been used for conservation studies on a wide range of organisms (Fritsch and Riesenberg 1996: for an overview, Dhar et al. 1997: birds, Gibbs 1998: amphibians, Maki and Horie 1999: plants). Genetic diversity was assessed by nucleotide diversity, heterozygosity and the proportion of polymorphic loci. Furthermore, *F*-statistic and Nei's genetic distance were calculated to analyze population structure and differentiation.

## Materials and methods

### *The ant species*

Two leaf litter ant species, *O. rixosus* (*Ponerinae*) and *P. annexus* (*Myrmicinae*), were selected for this study because, (1) both species are regarded as primary forest specialists (Brühl 2001; K. Eguchi personal communication), (2) are readily identifiable in the field, and (3) could be sufficiently sampled (at least

10 colonies per habitat) by tuna baiting to collect enough material for a thorough statistical analysis. Both species are common and widespread in South East Asian rainforest habitats (Brown 1976; Eguchi 2001).

*Pheidole annexus* is a tiny ant (2 mm, worker) from the subfamily *Myrmicinae* with a monogynous (single queen per colony) colony structure and predominantly nests in small rotting twigs on the ground (Eguchi 2001). In eight sampled twig nests from different locations (Sepilok, Kebun Cina, Labuk see below) an average (±SD) of 62 (±48) workers and 33 (±38) majors was counted. The largest nest had 170 workers, the smallest only 17. A queen could not be detected in any of the nests and alate individuals were only found in two separate nests (39 males and 5 females). Therefore it is assumed that the species occupies multiple nests per colony. Brood (larvae and pupae) was abundant in all nests. The mean determined foraging range of *P. annexus* was 47 cm (±43 cm SD) estimated from 20 nest observations in Sabah and the maximum measured foraging range was 180 cm (T.B.: personal observations).

In contrast *O. rixosus* (subfamily: *Ponerinae*) is a larger species (12 mm, worker) with a polygynous colony structure (multiple queens per colony) and nests in the soil, frequently under rotten wood (Ito et al. 1996). The colonies of *O. rixosus* in Java contained 40–350 workers (Ito et al. 1996), but two excavated nests investigated in Danum Valley appeared to be much larger with an estimated number of 1000 workers per colony and multiple queens. There are no data on foraging ranges for *O. rixosus* available, but as this slow moving species is only found if the nest is in the near vicinity of the bait in the time-frame of the experiment (not more than 5 m), it is likely that the foraging distance is not much larger.

### *Study sites*

The study sites are located in the lowland dipterocarp rainforests of Sabah, Borneo, Malaysia. Samples were collected from February to March 2001 in four forests of different size: Danum Valley Conservation Area (large, 43,800 ha, considered as contiguous for the purpose of this study), Sepilok Virgin Forest Reserve (medium, 4294 ha), Kebun Cina Forest Reserve (small, 146 ha) and Labuk Road Forest Reserve (very small, 20 ha) (Figure 1; a detailed description of the study sites can be found in Brühl et al. (2003)). The latter three are located in the vicinity of Sandakan and are all protected primary forest reserves. Sepilok was slightly disturbed by selective logging ca. 50 years ago. At present this forest is recognized as regenerated and the research plots were located in an area far from the historical logging sites as well. For the very small fragment (Labuk) only the core area (20 ha), that can be described as primary forest was sampled. The edges of Labuk were heavily disturbed by human interference and parts of the reserve were destroyed by a forest fire in 1997 (outside of the 20 ha core area). Danum Valley (large) is about 100 km away from the three smaller fragments and is surrounded by

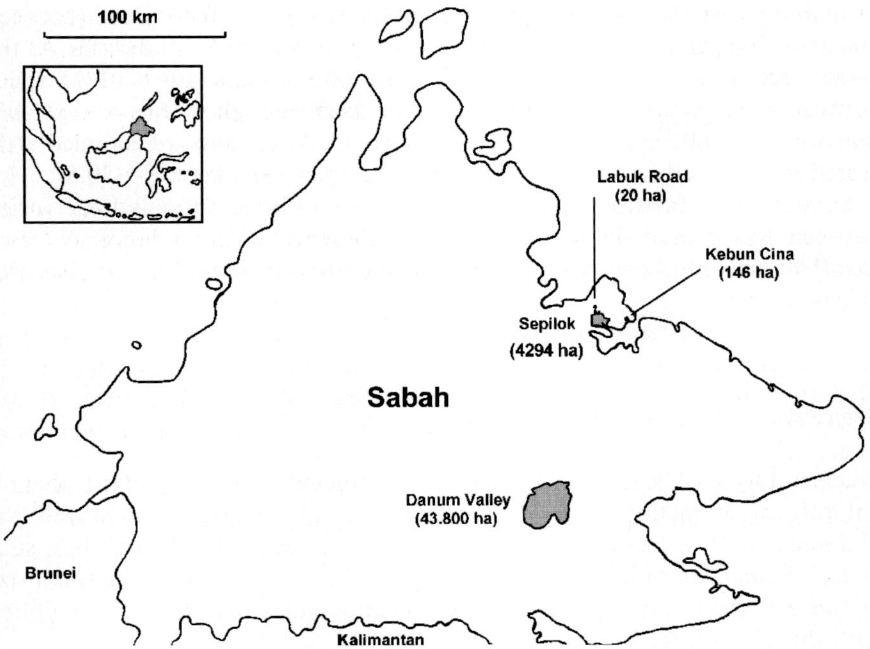


*Figure 1.* The location and habitat area of the four forests in Sabah, Borneo. The distances (by air) between habitats are as following: Danum – all fragments approximately 100 km; Kebun Cina – Sepilok/Labuk 15 km; Sepilok – Labuk 5 km.

secondary forest. Distances between the small and the medium and very small fragment is 15 km. The very small fragment is 5 km away from the medium sized one. These three fragments are surrounded by a matrix of agricultural land and urban development. The closest contiguous forested areas (primary or secondary forests) are at least 50 km away. The landscape between these contiguous forests and the three smaller fragments is dominated by oil palm plantations, a highly unsuitable habitat for forest dwelling leaf litter species in general. The two focal species of this study were never collected in oil palm plantation although this habitat was sampled for over 2 years (Brühl 2001). Therefore, the populations in the fragments are probably isolated in space and time as fragmentation started about 1960 with the beginning of the timber boom in Sabah.

### *Sampling*

In Danum Valley (large) we sampled along an existing transect system, in Sepilok (medium) and Kebun Cina (small) existing trail systems were utilized. Labuk (very small) was baited in the remaining intact core area. Tuna in oil was the bait of choice because of its ability to attract both species and its availability and standardized quality (Bestelmeyer et al. 2000). Baits (between 800 and 2000 per forest) were placed on the forest floor at 5 m intervals. As the target species occurred only at a small proportion of baits (10–0.001%), space between sampled colonies is regarded to be large enough to ensure that only individuals of different colonies were sampled. Collected ants were immediately stored in 100% ethanol. In the laboratory samples were kept at −20 °C.

Sample sizes (number of colonies) for the different populations varied between species and habitats because of differences in abundance. *O. rixosus*: (Danum (22), Sepilok (19), Kebun Cina (16) and Labuk (3); *P. annexus*: (12/24/23/22).

### *Molecular methods – DNA extraction*

Nuclear DNA of single individuals was extracted with a standard phenol/chloroform extraction method (Sambrook et al. 1989). We removed the abdomen of *O. rixosus* workers because it contained a PCR inhibiting substance, *P. annexus* individuals were processed entirely because of the small size of this ant species and there were no indications that the PCR was inhibited with this procedure.

### *RAPD fingerprinting*

As ants in general (and other eusocial insects) show a relatively low genetic variability (Crozier 1977) a primer screen was conducted for each species to find RAPD primers that produce a distinct, reproducible and easy to score banding pattern. For these screens 520 primers (Operon Technologies, Set A01 - Z20) were tested with four randomly chosen individuals of *O. rixosus* and *P. annexus*, respectively. PCR reactions were performed in a Biometra® Thermocycler with a total reaction volume of 12.5 $\mu l$ per sample containing: 20 ng template DNA, 1.25 $\mu l$ 10 × Buffer (MBI Fermentas), 2 mM $MgCl_2$, 100 $\mu M$ of each dNTP, 1.25 U *Taq* polymerase (MBI Fermentas), and 0.6 $\mu M$ primer (Operon Technologies). Amplification was performed with the following cycle parameters: 5 cycles of 94 °C/1 min, 35 °C/1 min, and 72 °C/2 min, another 32 cycles of 94 °C/10 s, 35 °C/30 s, and 72 °C/30 s.

The PCR products were resolved by horizontal gel electrophoresis at 120 V for 4.5 h in 1% Synergel (Diversified Biotech, Newton Center, MA) and 0.6% agarose in a 0.5 × TBE buffer. DNA was stained with ethidium bromide for 25 min, destained in distilled water for 20 min, and the banding patterns visualized under a UV-light source. The gel was photographed and a digitized version was saved for further analysis.

*Statistical analysis – Genetic diversity and population structure*

Genetic diversity was estimated as nucleotide diversity ($\pi$) according to Borowsky (2000). Nucleotide diversity is defined as the mean expected number of base mutations between two randomly drawn alleles from a population (Nei and Li 1979; Borowsky 2000). To reduce any bias possibly caused by differing numbers of sampled colonies in the populations a $C^{++}$ computer program was used to calculate nucleotide diversity from randomly chosen pairs of samples. After 50 replication steps the mean and variance of $\pi$ was calculated. For the *O. rixosus* population of the very small (Labuk) habitat with only 3 sampled colonies $\pi$ was calculated for all possible permutations between pairs of colonies. Mean nucleotide diversity between populations was regarded as statistically significant different if the variances were non-overlapping (Borowsky, personal communication).

All other population genetic parameters were calculated with the TFPGA software (Miller 1997). Heterozygosity (allelic diversity) was estimated with the approach of Lynch and Milligan which estimates heterozygosity based on a Taylor expansion (Lynch and Milligan 1994). Heterozygosity is the proportion of individuals in a population that have two different alleles in one gene locus and are therefore heterozygous for this gene. Here, heterozygosity is defined as the probability that a pair of randomly chosen individuals, one will show a 1-allele (present band) and one a 0-allele (missing band) on a locus. To test for differences in mean heterozygosity between populations, data was arcsine transformed and an independent sample *t*-test was used (Archie 1985).

The genetic population structure was analyzed by calculating the *F*-statistic (Wright 1951). For pairs of populations $F_{ST}$ values (fixation indices) were determined from the allele frequency data (Weir and Cockerham 1984) to get a measurement of genetic differentiation (population subdivision) between populations. The $F_{ST}$ values are equivalent to the loss of genetic diversity (heterozygosity) in populations due to genetic drift (Hartl and Clark 1989). Therefore, population subdivision is an indirect measurement of gene flow between populations which can be calculated from the $F_{ST}$ values as the number of effective migrants per generation (Waples 1987).

Additionally, a Markov Chain Monte Carlo simulation (2000 permutations) approach of Fisher's Exact Test was performed to determine if significant differences in marker distributions between populations exist (Raymond and Rousset 1995).

To measure genetic distance (genetic relationship) Nei's unbiased genetic distance (Nei 1978) was calculated. A cluster analysis (UPGMA) was performed on the basis of these genetic distance measurements. To get a measure of confidence in the resulting tree chart a bootstrap method (1000 permutations) which evaluates the relative strength of individual nodes was applied. A mantel statistic was used to investigate a possible relationship between genetic and geographic distance in the TFPGA program. All other general statistical tests were performed with SPPS (SPSS for Windows 10.5).

## Results

### *Primer screen*

In *O. rixosus* 402 and in *P. annexus* 412 (about 80%) of the 520 tested primers generated only few monomorphic bands or failed at all. Only 118 primers in *O. rixosus* and 108 in *P. annexus* produced useful markers for our genetic analysis.

### *RAPD fingerprint*

For *O. rixosus* 31 primers that produced multiple polymorphisms in the primer screen were selected for fingerprinting. These primers produced 105 polymorphic genetic markers. In *P. annexus* 23 selected primers generated 187 polymorphic markers. For both species only a small number of markers were population specific (5 in *O. rixosus*, 9 in *P. annexus*). Because of the small number of samples of *O. rixosus* colonies obtained from the very small habitat the following values for this habitat should be taken with caution and are only presented for completeness. This caution does not apply to the *P. annexus* population of the very small fragment. There was a gradual decrease in the proportion of polymorphic markers in *O. rixosus* populations from the largest to the smallest population (Table 1). *P. annexus* had the highest proportion of polymorphic markers in the large habitat and exhibited the smallest percentage in the medium sized fragment, with the two other fragments falling between these two (Table 1).

### *Genetic diversity*

*Odontomachus rixosus* populations showed a decrease in heterozygosity and nucleotide diversity with decreasing habitat size (Table 1). Heterozygosity and nucleotide diversity in the medium populations was reduced to 85% ($t$ = 0.887, $p$ = 0.376) and 86% (significant, non-overlapping variances) of the values reached in the large habitat.

The *P. annexus* populations displayed the highest genetic variability in the large habitat for both measurements (Table 1). The degree of heterozygosity in the next most diverse habitat (small) reached 84% ($t$ = 2.610, $p$ = 0.009) of the large habitat. Nucleotide diversity was reduced to 70% from the large habitat to the very small habitat (non-overlapping variances). For both indices

*Table 1.* Measurements of genetic variability in the four *O. rixosus* and *P. annexus* populations as determined by nucleotide diversity $\pi$ ($*10^3$ $\pm$ variance; var.), heterozygosity $H_j$($\pm$ var.) and the proportion of polymorphic loci (% polym. loci).

| | Measurement | Large | Medium | Small | Very small |
|---|---|---|---|---|---|
| *O. rixosus* | $\pi \pm$ var. * $10^3$ | 13.294 ± 0.010 | 11.359 ± 0.006 | 11.042 ± 0.012 | *7.460 ± 0.002* |
| | $H_j \pm$ var. | 0.167 ± 0.032 | 0.145 ± 0.030 | 0.137 ± 0.031 | *0.046 ± 0.018* |
| | % polym. loci | 64.7 | 57.1 | 49.5 | *11.4* |
| *P. annexus* | $\pi \pm$ var. * $10^3$ | 15.789 ± 0.017 | 7.671 ± 0.002 | 10.242 ± 0.022 | 11.166 ± 0.010 |
| | $H_j \pm$ var. | 0.315 ± 0.033 | 0.189 ± 0.038 | 0.265 ± 0.036 | 0.261 ± 0.040 |
| | % polym. loci | 85.6 | 59.4 | 80.2 | 74.3 |

All values of nucleotide diversity $\pi$ were significantly different from each other (non-overlapping variances). Heterozygosity in the large population of *P. annexus* was significantly different from the fragments (*t*-test: $p = 0.009$, $p = 0.009$, $p = 0.006$). Differences between the large *O. rixosus* population and medium and small were not significant (*t*-test: $p = 0.37$, $p = 0.243$). Note that values for *O. rixosus* in the very small habitat are only shown for completeness (*italic*).

the medium sized population showed the lowest degree of genetic diversity of all populations. The ratio of mean heterozygosity of the three fragments compared to the large population was 0.76 for *P. annexus* and 0.84 for *O. rixosus* (very small excluded).

A comparison between the two species revealed a similar degree of nucleotide diversity in most habitats with the exception of the large population of *P. annexus* which was more variable than the corresponding one of *O. rixosus*. Contrary to this, the heterozygosity values were higher in *P. annexus* for all habitats. In the same way, there was a much higher proportion of polymorphic loci in *P. annexus*.

We performed a regression analysis to investigate possible relationships between habitat area and genetic diversity. Both, the heterozygosity and the nucleotide diversity displayed a linear trend for both species with an increase in genetic variability with increasing habitat size (very small fragment excluded for *O. rixosus*). The habitat area explained a high proportion of the increase in genetic variability for heterozygosity (*O. rixosus*: $r = 0.981$; *P. annexus*: $r = 0.672$) and nucleotide diversity (*O. rixosus*:$r = 0.732$; *P. annexus*: $r = 0.856$) – though the relationships were not significant.

### *Population structure*

Cluster analysis of Nei's unbiased genetic distance, visualized in tree diagrams, revealed a pattern expected under an isolation by distance model for *P. annexus* with the three fragments (medium, small, very small) forming a separate cluster with the large habitat as an outgroup (Figure 2). This matches the geographical distribution of habitats as can be seen in a comparison between the tree diagrams and distances between habitats from the map

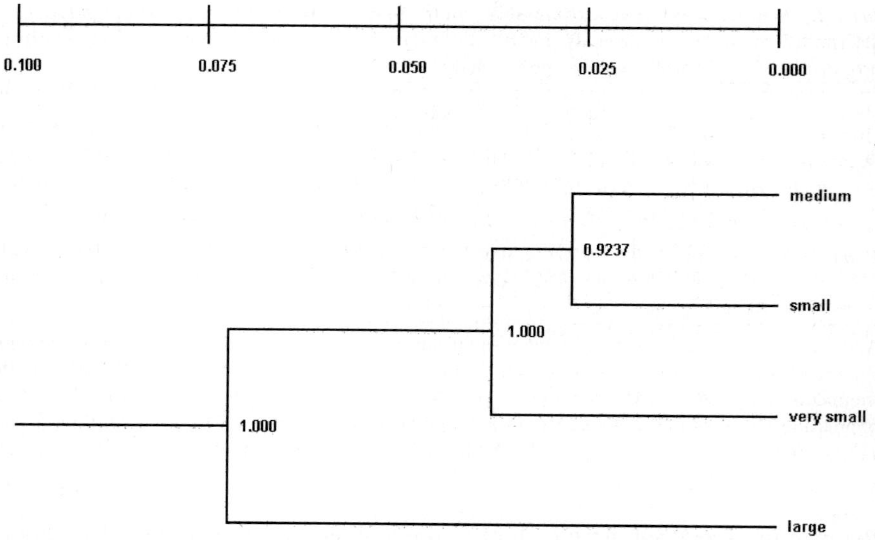


*Figure 2.* Cluster analysis of Nei's unbiased genetic distance (1978) between populations of *P. annexus* using the UPGMA cluster algorithm. Bootstrap proportions (1000 iterations) are shown beside nodes.

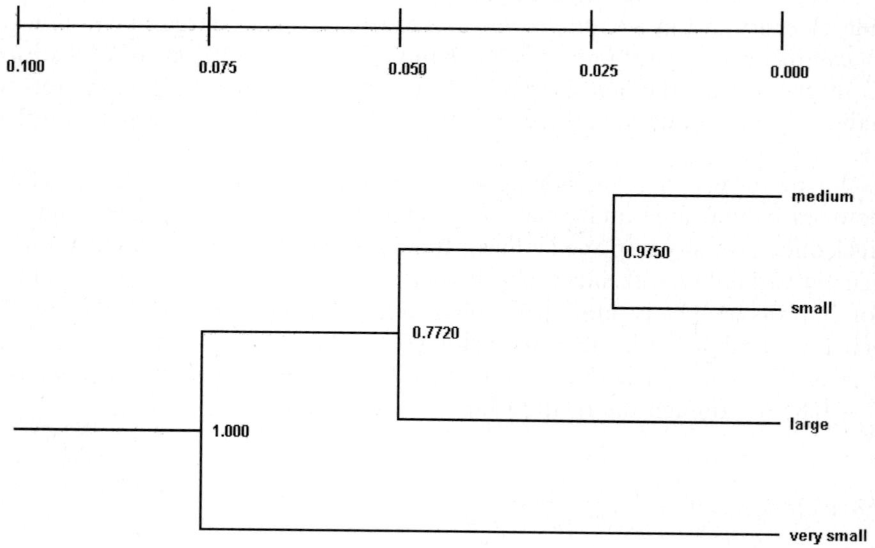


*Figure 3.* Cluster analysis of Nei's unbiased genetic distance (1978) between populations of *O. rixosus* using the UPGMA cluster algorithm. Bootstrap proportions (1000 iterations) are shown beside nodes.

(Figure 1). The situation was similar for the *O. rixosus* data (Figure 3) though the very small habitat does not fall in the same cluster as the other fragments.

*Table 2.* Population differentiation and gene flow between pairs of *O. rixosus* populations.

| | Large | Medium | Small | Very small |
|---|---|---|---|---|
| Large | | 0.233 ± 0.042 *** | 0.171 ± 0.031 * | 0.321 ± 0.059 |
| Medium | 0.823 | | 0.112 ± 0.020 | 0.226 ± 0.065 |
| Small | 1.212 | 1.982 | | 0.279 ± 0.066 |
| Very small | 0.529 | 0.856 | 0.646 | |

$F_{ST}$ values ± SD are given above the diagonal, gene flow in migrants per generation below. All $F_{ST}$ values were significant different from zero (SD non-overlapping with zero). All values for the very small habitat must be taken with caution (*italic*). A Fisher's Exact Test (2000 permutations) revealed significant differences in the marker distribution between the large and medium/small habitats as indicated by an asterisk (*$p < 0.05$, ***$p < 0.0001$).

*Table 3.* $F_{ST}$ values for pairs of all populations of *P. annexus* ± SD above diagonal, below the calculated gene flow.

| | Large | Medium | Small | Very small |
|---|---|---|---|---|
| Large | | 0.213 ± 0.022 ** | 0.127 ± 0.016 ** | 0.134 ± 0.020 ** |
| Medium | 0.922 | | 0.085 ± 0.012 | 0.119 ± 0.015 |
| Small | 1.712 | 2.695 | | 0.086 ± 0.012 |
| Very small | 1.619 | 1.851 | 2.647 | |

All $F_{ST}$ values were significant different from Fisher's Exact test revealed significant differences in the marker distribution between the large habitat and all fragments (**$p < 0.001$).

A Mantel test between geographic and genetic distance showed a high coefficient of correlation for *P. annexus* ($r = 0.92$, $p = 0.15$) and a low correlation for *O. rixosus*: $r = 0.30$, $p = 0.51$. The very small fragment was not included in the Mantel test for *O. rixosus*.

Both species showed a similar genetic distance between the medium, small and very small habitats (ca. 0.025) but the genetic distance from those to the large habitat is more pronounced in *P. annexus* (ca. 0.075) than in *O. rixosus* (ca. 0.050).

Measurements for population differentiation and gene flow indicated a restricted genetic exchange between the fragments in both species (Tables 2 and 3). All $F_{ST}$ values were significantly different from zero and exceptionally high, indicating a low gene flow between the large habitat and the fragments, and a restricted gene flow between the three fragments alone. These results were confirmed by a Monte Carlo simulation of Fisher's Exact Test that showed highly significant differences in the marker frequencies between the large and the three smaller habitats, but not between the three smaller fragments alone (Tables 2 and 3).

## Discussion

This study is a first attempt to tackle the important question: Does forest fragmentation influence the population dynamics of tropical insects? It should be seen as a starting point for future research in this area. A lack of suitable remaining forest fragments in Sabah limited our conclusions.

Despite limited replication, we found a measurable effect on the genetic diversity and population structure of two leaf litter ant species in the dipterocarp rain forests of Sabah, probably caused by habitat fragmentation due to deforestation. Therefore, the reduction of biodiversity in leaf litter ant populations in these fragmented forests is not only detectable on community level, measured by species diversity and density in the same area (Brühl et al. 2003), but also on the level of single populations, measured as a decline in genetic diversity. This result is surprising and alarming, as the effect was already measurable after only 40 years of fragmentation and isolation even in quite large habitat fragments comprising more than 4000 ha. The reduction in genetic diversity was supported by several measurements of genetic variability (nucleotide diversity, heterozygosity, and the proportion of polymorphic loci). Nucleotide diversity was significantly smaller in the fragments relative to the large habitat for both species. Additionally, heterozygosity was significantly reduced in one species, *P. annexus*, a finding similar to that for rainforest skinks in fragmented rainforests of Austalia (Sumner et al. 2004). This decrease in genetic diversity in smaller habitats demonstrates their sensitivity to population density fluctuations.

The lack of heterozygosity in fragmented forests in both species can be interpreted as an effect of inbreeding and genetic drift, effects that are certainly to be expected in small populations. Both phenomena should not be analyzed separately (Hartl and Clark 1989; Templeton and Read 1994) because of reciprocal interactions. Therefore, the fixation of alleles is also an indication for inbreeding. Unfortunately, the inbreeding coefficient $F_{IS}$ cannot be calculated from RAPD data unless data from more than one generation exist (Lynch and Milligan 1994). Increased inbreeding in fragmented populations is expected to be associated with a loss in the ability to adapt to changing environmental parameters because genetic variation is lost (Hartl and Clark 1989; Templeton and Read 1994).

Besides the deficiency of evolutionary flexibility, inbreeding in Formicidae can lead to a serious inbreeding depression as diploid males are produced due to the haplo-diploid sex determination system in ants. Diploid males are usually sterile and a burden for the colony (Giraud et al. 2000). Additionally, inbreeding can also be directly correlated with a higher susceptibility of extinction for populations (Frankham 1998; Saccheri et al. 1998; but see Elgar and Clode 2001).

The reduction of genetic diversity in the rainforest patches was of the same magnitude as in island populations of comparable sizes. For example, in several insect island populations (drosophilid species and beetles) the ratio of heterozygosity from island and continental populations was 0.79 (Frankham 1997). The ratios of the mean heterozygosities in the forest fragments compared to Danum Valley were comparable (0.84 for *O. rixosus* and 0.76 for *P. annexus*). This expected lower genetic variability is caused by the typically lower effective population sizes in island populations. Therefore, the reduction in genetic variability in our fragmented populations matched the theoretical expectations of isolated island populations.

It is unlikely that these results are the consequence of a sampling artifact, as a comparable number of samples from each population were taken from an equivalent area (with the exception of the very small (Labuk) population of *O. rixosus*). The reason for the low number of sampled colonies in Labuk is not the result of a lower sampling effort as the highest number of baits (about 2000) was applied in this habitat, but a result of the low density of *O. rixosus*. Additionally, sampling of *P. annexus* indicated that sample size is not correlated with genetic diversity in our analysis. Sample size in the Danum Valley population of *P. annexus* was lower than in the fragmented habitats (12 compared to more than 20 in the fragments). Nevertheless, the Danum Valley population of *P. annexus* had a higher genetic diversity compared to the fragmented populations.

Although a regression analysis showed a linear trend between habitat size and genetic diversity (nucleotide diversity and heterozygosity) the results were not significant. This is mainly due to the low number of replicates and the wide gap in habitat area between the large (> 40,000 ha) and the smaller fragments (< 5000 ha). However, the general trend of decreasing genetic variability in smaller fragments with two measurements for genetic diversity and the smaller proportion of polymorphic loci in the fragments make a linear relationship possible. It must be emphasized that this low number of replicates is the result of the high logging activity in the area with the result that there are no other fragments left or at least none that can be defined as primary forests. So the number of replicates is not a matter of a poor planning process of this study, but the fact that field studies must deal with the remainder of what is available, even if this is not an ideal situation.

### *Population structure*

The-isolation-by-distance-population-model predicts a correlation of genetic distance (genetic dissimilarity between populations) with geographic distance due to lack in dispersal and gene flow over long distances. Cluster analysis (UPGMA) of the genetic distance measures revealed the same pattern for the two species supporting an isolation by distance model. The Danum Valley population (large) with the highest distance to any other populations (100 km) forms the out group in the phenograms, the three smaller fragments cluster together (two in *O. rixosus*). The high correlation between geographic and genetic distance for *P. annexus* in a Mantel test gave further support towards the isolation-by-distance scenario.

However, genetic distances between the three fragments were higher than expected under an isolation by distance model alone. For example, Labuk (very small) is genetically furthest apart from the other fragments, yet Labuk is only 5 km distant to Sepilok (medium). This may be a result of the high degree of genetic isolation of Labuk and a resulting high allele fixation by inbreeding and genetic drift due to small effective population size (Hartl and Clark 1989). Finally, the high degree of disturbance in the Labuk population compared to the other fragments and harsh microclimatic conditions due to extensive edge effects could hinder new immigration and foster the isolation effect.

Analysis of population subdivision (*F*-statistic) supported findings of high genetic isolation in fragmented populations and emphasized the low gene flow between habitats. The values for population subdivision were exceptionally high (ca 5–10 times higher) if compared to other studies on fragmented insect populations (Knutsen et al. 2000; Gyllenstrand and Seppä 2003). For both species there is only a very restricted gene flow detectable between Danum Valley (large) and the fragments (Tables 2 and 3). Fisher's Exact Test was able to detect significant and highly significant differences in marker frequencies between the contiguous habitat and the fragments supporting the findings in the *F*-statistics. Between the fragmented populations a slightly higher gene flow between approximately 2.0 and 2.7 effective migrants per generation occurred. To hinder the effect of genetic drift one migrant per generation would theoretically be sufficient (Hartl and Clark 1989). But experimental findings suggest that even 10 migrants are not sufficient to stop genetic differentiation of populations (Allendorf and Phelbs 1981; Vucetich and Waite 2000). Considering this, gene flow between populations seems limited and very restricted. From this point of view there is no meta population dynamic – with gene flow between subpopulations – present in both species. In both species, the lack of gene flow may lead to an increase in genetic differentiation between the populations and foster an additional loss of genetic variability due to inbreeding effects in the future. The low migration rates can be explained by the quality of the matrix habitat (Fahrig and Merriam 1994) preventing recolonization (Templeton et al. 2001). Such conditions occur in the case of our study, as the intervening matrix habitat is dominated by urban and agricultural landscapes (mostly oil palm plantation) which are uninhabitable for both species (Brühl 2001). Although the dispersal abilities of the two species are unknown, it is unlikely that founding queens of either species are able to fly the whole distance from Danum Valley to the Sandakan area. Additionally, as there are no suitable habitat patches (stepping stones) between the two localities hence stepwise migration between these habitats is impossible.

The genetic population structure in this study is comparable with another study on isolated ant populations. Island populations in the Florida Key's showed a 10 time higher genetic distance from an island to a continental population than the genetic distance from the large habitat to the fragments in our study, although the geographic distance (220 km) was in the same range (100 km) in both studies (Gadau et al. 1996). The difference can probably be explained by the longer isolation time of the Florida Keys compared to the isolation time in our fragmentation study, although population substructuring (estimated as $F_{ST}$) was comparable. This combination of low genetic distance combined with a high differentiation (population subdivision) can be explained by similar founding populations (situation before fragmentation) and a low gene flow, fostering inbreeding and genetic drift resulting in a high population subdivision. This means that historically there was a sufficient gene flow between the forests prior to fragmentation. The same result of low genetic distance but high population subdivision was found in a study on fish populations inhabiting small isolated water bodies (Schug et al. 1998). These fish populations derived from a similar founding population but subsequently had no more genetic exchange.

This study shows that even a relatively large protection area like Sepilok (more than 4000 ha) may not be sufficient to maintain the original genetic diversity of tropical leaf litter ant populations. As most Virgin Jungle Reserves (VJR) – with a mean size of approximately 2000 ha (Chung et al. 2000) – are smaller than Sepilok, most protection areas in Sabah are not large enough to prevent genetic erosion with all its consequences including the higher susceptibility to extinction. Whether these results are representative for other organisms in different regions remains speculative. But, if even such small organisms as the ants under study are already measurably affected after such a short period of isolation it is more than likely that larger organisms with smaller population densities are even more severely affected. If these results are confirmed for other organisms, the conservation value of small forest fragments has to be critically evaluated.

## Acknowledgements

This study was supported by a DAAD PhD scholarship to C. A. B. (D-98-19438). J.G. was funded by the DFG (GA 661, SFB-554 C-5). Thanks are due to the staff of the Forest Research Centre (FRC) Sandakan and especially to Chey vun Khen and Arthur Chung for their help and support while we stayed in Sepilok. Thank goes to the people in the field center of the Danum Valley Conservation Area for the possibility to carry out fieldwork there. A special 'thank you' for Thomas Eltz as well, for lively discussions and critical comments on the study.

## References

Aizen M.A. and Feinsinger P. 1994. Forest fragmentation, pollination and plant reproduction in a Chaco dry forest, Argentina. Ecology 75: 330–351.

Allendorf F.W. and Phelbs S.R. 1981. Use of allelic frequencies to describe population structure. Can. J. Fish. Aquat. Sci. 38: 1507–1514.

Archie J.W. 1985. Statistical analysis of heterozygosity data: independent sample comparisons. Evolution 39: 623–637.

Benitez-Malvido J. 1998. Impact of forest fragmentation on seedling abundance in a tropical rain forest. Conserv. Biol. 12: 380–389.

Bestelmeyer B.T., Agosti D., Alonso L.E., Brandao C.R.F., Brown W.L.Jr., Delabie J.H.C. and Silvestre R. 2000. Field techniques for the study of ground dwelling ants. In: Agosti D., Majer J.D., Alonso E.L. and Schultz T.R. (eds), Ants: Standard Methods for Measuring and Monitoring Biodiversity. Smithonian Institute Press, Washington, pp. 122–144.
Bestelmeyer B.T. and Wiens J.A. 1996. The effects of land use on the structure of ground-foraging ant communities in the Argentine chaco. Ecol. Appl. 6: 1225–1240.
Borowsky R.L. 2000. Estimating nucleotide diversity from random amplified polymorphic DNA and amplified fragment length polymorphism data. Mol. Phylogenet. Evol. 18: 143–148.
Boswell G.P., Britton F.N. and Franks N.R. 1998. Habitat fragmentation, percolation theory and the conservation of a keystone species. Proc. R. Soc. Lon. Ser. B Biol. Sci. 265: 1921–1925.
Brown K.S.J. and Hutchings R.W. 1997. Disturbance, fragmentation, and the dynamics of diversity in Amazonian forest butterflies. In: Laurance W.F. and Bierregard O.R. (eds), Tropical Forest Remnants: Ecology, Management, and Conservation of Fragmented Communities. The University of Chicago Press, Chicago, pp. 91–110.
Brown W.L.J. 1976. Contributions toward a reclassification of the Formicidae. Part VI. Ponerinae, Tribe Ponerini, Subtribe Odontomachiti. Section A. Introduction, subtribal characters, Genus *Odontomachus*. Stud. Entomol. 19: 67–171.
Brühl C.A. 2001.Leaf litter ant communities in tropical lowland rainforests in Sabah, Malaysia: effects of forest disturbance and fragmentation. PhD Thesis, University of Würzburg, Würzburg, http://opus.bibliothek.uni-wuerzburg. de/opus/volltexte/2002/104/pdf/dissbruehl.pdf.
Brühl C.A., Eltz T. and Linsenmair K.E. 2003. Size does matter – effects of tropical rainforest fragmentation on the leaf litter ant community in Sabah, Malaysia. Biodivers. Conserv. 12: 1371–1389.
Carvalho K.S. and Vasconcelos H.L. 1999. Forest fragmentation in central Amazonia and its effects on litter-dwelling ants. Biol. Conserv. 91: 151–157.
Chung A.Y.C., Eggleton P., Speight R.M., Hammond P.M. and Chey V.K. 2000. The diversity of beetle assemblages in different habitat types in Sabah, Malaysia. Bull. Entomol. Res. 90: 475–496.
Clarke G.M. and O'Dwyer C. 2000. Genetic variability and population structure of the endangered golden sun moth, *Synemon plana*. Biol. Conserv. 92: 317–381.
Crozier R.H. 1977. Evolutionary genetics of the Hymenoptera. Ann. Rev. Entomol. 22: 263–268.
Cunningham M. and Craig M. 1998. Genetic effects of forest fragmentation on a rainforest restricted lizard (Scinidae: *Gnypetoscinus queenslandiae*). Biol. Conserv. 83: 19–30.
Dale V.H., Pearson M.S., Offerman L.H. and O'Neill R.V. 1994. Relating patterns of land-use change to faunal biodiversity in the central Amazon. Conserv. Biol. 8: 1027–1036.
Davies K.F. and Margules C.R. 1998. Effects of habitat fragmentation on carabid beetles: experimental evidence. J. Anim. Ecol. 67: 460–471.
Debinski D.M. 1994. Genetic diversity assessment in a metapopulation of the butterfly *Euphyryas gilleti*. Biol. Conserv. 70: 25–31.
Dhar A.K., Pokras K.D., Garcia C.D., Evers D.C. and Gordon Z.J. 1997. Analysis of genetic diversity in common loon Gavia immer using RAPD and mitochondrial RFLP techniques. Mol. Ecol. 6: 581–586.
Didham R.K. 1997. The influence of edge effects and forest fragmentation on leaf litter invertebrates in Central Amazonia. In: Laurance W.F. and Bierregard O.R. (eds), Tropical Forest Remnants: Ecology, Management, and Conservation of Fragmented Communities. The University of Chicago Press, Chicago, pp. 55–69.
Didham R.K., Ghazoul J., Stork N.E. and Davis A.J. 1996. Insects in fragmented forests: a functional approach. Trends Ecol. Evol. 11: 255–261.
Eguchi K. 2001. A revision of the Bornean species of the ant genus *Pheidole* (Insecta: Hymenoptera: Formicidae: Myrmicinae). Tropics Monograph Series Nr. 2: 1–154.
Elgar M.A. and Clode D. 2001. Inbreeding and extinction in island populations: a cautionary note. Conserv. Biol. 15: 284–286.

Fahrig L. and Merriam G. 1994. Conservation of fragmented populations. Conserv. Biol. 8: 50–59.

Fonseca de Souza O.F. and Brown V.K. 1994. Effects of habitat fragmentation on Amazonian termite communities. J. Tropic. Ecol. 10: 197–206.

Frankham R. 1996. Realtionship of genetic variation to population size in wildlife. Conserv. Biol. 10: 1500–1508.

Frankham R. 1997. Do island populations have less genetic variation than mainland populations? Heredity 78: 311–327.

Frankham R. 1998. Inbreeding and extinction: island populations. Conserv. Biol. 12: 665–675.

Fritsch P. and Riesenberg L.H. 1996. The use of RAPD in conservation genetics. In: Smith T.B. and Wayne R.K. (eds), Molecular Genetic Approaches in Conservation. Oxford University Press, Oxford, pp. 54–73.

Gadau J., Heinze J., Hölldobler B. and Schmid M. 1996. Population and colony structure of the carpenter ant *Camponotus floridanus*. Mol. Ecol. 5: 785–792.

Gibbs J.P. 1998. Genetic structure of redback salamander *Phletodon cinereus* populations in continuous and fragmented forests. Biol. Conserv. 86: 77–81.

Giraud T., Blatrix R., Poteaux C., Solignac M. and Jaisson P. 2000. Population structure and mating biology of the polygynous ponerine ant *Gnamptogenys striatula* in Brazil. Mol. Ecol. 9: 1835–1841.

Gyllenstrand N. and Seppä P. 2003. Conservation genetics of the wood ant, *Formica lugubris*, in a fragmented landscape. Mol. Ecol. 12: 2931–2940.

Harry M., Robin S. and Lachaise D. 1998. Use of polymorphic genetic markers (RAPDs) in evolutionary and applied entomology. Ann. Soc. Entomol. France 34: 9–32.

Hartl D.L. and Clark A.G. 1989. Principles of Population Genetics. 2nd ed. Sinauer Associates Inc. Publishers, Sunderland, Massachusetts.

Haymer D.S. 1994. Arbitrary (RAPD) primer sequences used in insect studies. Insect Mol. Biol. 3: 191–194.

Hölldobler B. and Wilson E.O. 1990. The ants. Belknap press of Cambridge University, Cambridge.

Ito F., Yusoff N.R. and Idris A.H. 1996. Colony composition and queen behavior in polygynous colonies of the Oriental ponerine ant *Odontomachus rixosus* (Hymenoptera Foricidae). Insectes Socieaux 43: 77–86.

Jones C.G., Lawton J.H. and Shachak M. 1994. Organisms as ecosystem engineers. Oikos 69: 373–386.

Kareiva P. 1987. Habitat fragmentation and the stabillity of predator–prey interactions. Nature 26: 388–390.

Kitching R.L., Orr A.G., Thalib L., Mitchell H., Hopkins M.S. and Graham A.W. 2000. Moth assemblages as indicators of environmental quality in remnants of upland Australian rain forest. J. Appl. Ecol. 37: 284–297.

Klein B.C. 1989. Effects of forest fragmentation on dung and carrion beetle communities in central Amazonia. Ecology 70: 1715–1725.

Knutsen H., Rukke B.A., Jorde P.E. and Ims R.A. 2000. Genetic differentiation among populations of the beetle *Bolitophagus reticulatus* (Coleoptera: Tenebrionidae) in a fragmented and a continuous landscape. Heredity 84: 667–676.

Lacerda D.R., Acedo M.D.P., Lemos Filho J.P. and Lovato M.B. 2001. Genetic diversity and structure of natural populations of *Plathymenia reticulata* (Mimosoideae), a tropical tree from the Brazilian Cerrado. Mol. Ecol. 10: 1143–1152.

Laidlaw R.K. 1996. A comparison between populations of primates, squirrels, tree shrews and other mammals inhabiting virgin, logged, fragmented and plantation forests in Malaysia. In: Lee S.S., May D.Y., Gauld I.D. and Bishop J. (eds), Preserving Ecological Systems. The Agenda for Long-term Research and Development. Praeger Publishers, NY, pp. 93–113.

Laurance W.F., Ferreira L.V., Rankin-de Merona J.M., Laurance S.G., Hutchings R.W. and Lovejoy T.E. 1998. Effects of forest fragmentation on recruitment patterns in Amazonian tree communities. Conserv. Biol. 12: 460–464.

Laurance W.F., Laurance S.G., Ferreira L.V., Rankin-de Merona J.M., Gascon C. and Lovejoy T.E. 1997. Biomass collapse in Amazonian forest fragments. Science 278: 1117–1118.

Laurance W.F., Vasconcelos H.L. and Lovejoy T.E. 2000. Forest loss and fragmentation in the Amazon: implications for wildlife conservation. Oryx 34: 39–45.

Lynch M. and Milligan B.G. 1994. Analysis of population genetic structure with RAPD markers. Mol. Ecol. 3: 91–99.

Mac Arthur R.H. and Wilson E.O. 1967. The Theory of Island Biogeography. Princeton University Press, Princeton, New Jersey.

Majer J.D. and Delabie J.H.C. 1999. Impact of tree isolatin on arboreal and ground ant communities in cleared pasture in the Atlantic rain forest region of Bahia, Brazil. Insectes Socieaux 46: 281–290.

Maki M. and Horie S. 1999. Random amplified polymorphic DNA (RAPD) markers reveal less genetic variation in the endangered plant *Cerastium fischerianum* var. *molle* than in the widespread conspecific *C. fischerianum*var. *fischerianum*(Caryophyllaceae). Mol. Ecol. 8: 145–150.

Miller M.P. 1997. TFPGA (Tools for Population Genetics Analysis) 1.3. A windows program for the analysis of allozyme and molecular population genetic data, in, Computer program distributed by the author.

Nei M. 1978. Estimation of average heterozygosity and genetic distance from a small number of individuals. Genetics 89: 583–590.

Nei M. and Li W.-H. 1979. Mathematical model for studying genetic variation in terms of restriction endonucleases. Proc. Natl. Acad. Sci. USA 76: 5269–5273.

Norse E.A. 1987. Habitat diversity and genetic variability: are they necesarry ecosystem properties?. In: Draggan S., Cohrssen J.J. and Morrison R.E. (eds), Preserving Ecological Systems. The Agenda for Long-term Research and Development. Praeger Publishers, NY, London, pp. 93–113.

Oehler J.D. and Livaitis J.A. 1996. The role of spatial scale in understanding responses of medium sized carnivores to forest fragmentation. Can. J. Zool. 74: 2070–2079.

Petal J. 1978. The role of ants in ecosystems. In: Brian M.V. (ed), Production ecology of ants and termites (International Biology Programme, no. 13). Cambridge University Press, pp. 293–325.

Raymond M.L. and Rousset F. 1995. An exact test for population differentiation. Evolution 49: 1280–1283.

Reed D.H. and Frankham R. 2003. Correlation between fitness and genetic diversity. Conserv. Biol. 17: 230–237.

Saccheri I., Kuusaari M., Kankare M., Vikman P., Fortelius W. and Hanski I. 1998. Inbreeding and extinction in a butterfly metapopulation. Nature 392: 491–494.

Sambrook J., Fritsch E.F. and Maniatis T. (eds) 1989. Molecular Cloning: A Laboratory Manual. 2nd edn. Cold Spring Harbor Laboratory Press, Cold Spring Harbor.

Schug M., Downhower J.F., Brown L.P., Sears D.B. and Fuerst P.A. 1998. Isolation and genetic diversity of *Gambusia hubbsi* populations in blueholes on Andros Island, Bahamas. Heredity 80: 336–346.

Suarez A.V., Bolger D.T. and Case T.J. 1998. Effects of fragmentation and invasion on native ant communities in coastal southern California. Ecology 79: 2041–2056.

Sumner J., Jessop T., Paetkau D. and Moritz C. 2004. Limited effect of anthropogenic habitat fragmentation on molecular diversity in a rain forest skink, *Gnypetoscinus queenslandiae*. Mol. Ecol. 13: 259–269.

Templeton A.R. and Read B. 1994. Inbreeding: one word, several meanings, much confusion. In: Loeschcke V., Tomiuk J. and Jain S.K. (eds), Conservation Genetics. Birkhäuser Verlag, Basel, Boston, Berlin, pp. 91–105.

Templeton A.R., Robertson R.J., Brisson J. and Strasburg J. 2001. Disrupting evolutionary processes: the effect of habitat fragmentation on collared lizards in the Missouri Ozarks. Proc. Natl. Acad. Sci. USA 98: 5426–5432.

Terborgh J. 1992. Maintenance of diversity in tropical forests. Biotropica 24: 283–292.

Vucetich J.A. and Waite T.A. 2000. Is one-migrant-per-generation sufficient for the genetic management of fluctuating populations? Anim. Conserv. 3: 261–266.
Vucetich L.-M., Vucetich J.A., Waite T.A. and Peterson R.O. 2001. Genetic (RAPD) diversity in *Peromyscus maniculatus* populations in naturally fragmented landscapes. Mol. Ecol. 10: 35–40.
Waples R. 1987. A multispecies approach to the analysis of gene flow in marine shore fishes. Evolution 41: 385–400.
Weir B.S. and Cockerham C.C. 1984. Estimating F-statistics for the analysis of population structure. Evolution 36: 1358–1370.
White G.M., Boshier D.H. and Powell W. 1999. Genetic variation within a fragmented population of *Swietenia humilis* Zucc. Mol. Ecol. 8: 1899–1909.
Whitmore T.C. 1984. Tropical Rain Forests of the Far East. 2nd ed. Clarendon Press, Oxford.
Whitmore T.C. 1998. An Introduction to Tropical Rain Forests. 2nd ed. Oxford University Press, Oxford.
Williams B.I. 2002. Conservation genetics, extinction, and taxonomic status: a case history of the regal fritillary. Conserv. Biol. 16: 148–157.
Wong T.C.M., Sodhi N.S. and Turner I.M. 1998. Artificial nest and seed predation experiments in tropical lowland rainforest remnants of Singapore. Biol. Conserv. 85: 97–104.
Wright S. 1951. The genetical structure of populations. Ann. Eugenet. 15: 323–354.

Biodiversity and Conservation (2006) 15:163–176
DOI 10.1007/s10531-004-5016-y 

# Comparing relative model fit of several species-accumulation functions to local Papilionoidea and Hesperioidea butterfly inventories of Mediterranean habitats

ALBERTO JIMÉNEZ-VALVERDE[1,2,*], SILVIA JIMÉNEZ MENDOZA[1], JOSÉ MARTÍN CANO[1] and MIGUEL L. MUNGUIRA[1]
*[1]Departamento de Biología (Zoología), Universidad Autónoma de Madrid, Cantoblanco, E-28049, Madrid, Spain; [2]Present address: Museo Nacional de Ciencias Naturales, Departamento de Biodiversidad y Biología Evolutiva. C/ José Gutiérrez Abascal 2, E-28006, Madrid, Spain; *Author for correspondence (e-mail: mcnaj651@mncn.csic.es; phone: +34-914111328 ext. 1212)*

Received 16 March 2004; accepted in revised form 12 October 2004

**Key words:** Accumulation curves, Accumulation models, Local scale, Mediterranean habitats, Papilionoidea & Hesperioidea, Species-richness inventories

**Abstract.** When compiling an inventory of hyperdiverse taxa, it is impossible to record the total number of species during fieldwork. To ensure the accuracy of species-richness data it is necessary to assess the reliability of inventories. Accumulation curves are an easy method for doing this and are extensively described in the literature. In this study, we compare the relative fit of various models of species-accumulation functions for six local butterfly inventories, evaluating them by a consideration of the values of the fit, coefficient of determination and sum-of-squares, and the residual patterns and Akaike's Information Criterion. In general, complex functions, such as the Weibull or Chapman-Richards, performed better than simpler and more widely used models (e.g., the Clench and negative exponential models). The performance of models varied among sampling plots, indicating the influence of factors such as land use and community structure. Thus, although the application of more complex models should replace the use of simple ones, further research into the factors affecting model fit of accumulation functions is necessary.

## Introduction

Species richness is an essential criterion for land use and conservation planning. It is an easy parameter to interpret, and it has gradually replaced the traditional diversity indices based on species-number and evenness (Gaston 1996). However, reliable inventory data for hyperdiverse taxa, such as arthropods, are difficult to obtain due to their intrinsic attributes. For example, the impossibility of recording the total number of species during a survey is a serious methodological problem (Colwell and Coddington 1994). Therefore, we need a reliable method to assess the completeness of inventories in order to ensure that we are working with accurate data.

In a species-accumulation curve, the cumulative number of species is plotted against a cumulative measure of sampling effort. As sampling effort increases, the rate at which new species are added to the inventory declines asymptotically. Theoretically, the asymptotic value represents the total species richness that would be achieved with an infinite survey using the methods and over the time period of the actual study. Once a species-accumulation model has been fitted, accumulation curves should (1) allow formality to be conferred on the species inventories, thereby facilitating evaluation of their completeness, (2) provide a better basis for planning sampling work once the effort required to obtain reliable inventories has been estimated, and (3) facilitate the extrapolation of the total number of species present in a given area (Lamas et al. 1991; Soberón and Llorente 1993; Colwell and Coddington 1994; Gotelli and Colwell 2001). Several other methods have been proposed for estimating the number of species in an assemblage. Non-parametric estimators use the relative abundance of rare species to estimate the number of species not seen (Heltshe and Forrester 1983; Smith and van Belle 1984; Chao 1987; Colwell and Coddington 1994). Although widely used, some authors (Chiarucci et al. 2003; Petersen and Meier 2003; Petersen et al. 2003) have argued that these estimators are inaccurate and difficult to interpret, and Chiarucci et al. (2003) proposed that other estimators should be developed. Fitting a lognormal abundance distribution and estimating the unsampled proportion of the curve (e.g., Fagan and Kareiva 1997; Longino et al. 2002) is another approximation that involves several dubious assumptions and practices (Colwell and Coddington 1994). Moreover, these methods only give an estimate of species richness, and provide no information of use in planning fieldwork.

Species-accumulation curves have been widely used to assess completeness of different taxonomic group inventories. The Clench and negative exponential functions are the most frequently used models (e.g., butterflies: Fagan and Kareiva 1997; Jiménez-Valverde et al. 2004; Soberón and Llorente 1993; Araneae: Jiménez-Valverde and Lobo 2004; bats: Moreno and Halffter 2000; Coleoptera: Hortal et al. 2001; Diptera: Petersen et al. 2003). However, in general, these works did not compare the relative model fit to their data and assessed the performance of the chosen function solely by the coefficient of determination ($R^2$). Several authors have pointed out that this is a misleading method (Keating and Quinn 1998; Motulsky and Christopoulos 2003). To our knowledge, only Flather (1996) has compared different species-accumulation models, using regional-scale bird data, concluding that the Weibull model was the most appropriate for quantifying and comparing species-accumulation curves. However, model performance may be expected to depend on numerous factors, including community structure (Keating and Quinn 1998) and spatial scale (Tjørve 2003), among others.

In this work we compare the relative model fit and predictive validity of various species-accumulation models in six annual butterfly inventories carried out at a local scale in sampling plots with different land uses in a Mediterranean area. The overall objective was to derive some recommendations about the most appropriate type of function for modelling the inventory process and for obtaining reliable estimates of species richness.

Butterflies (Papilionoidea and Hesperioidea) fulfil many of the requirements associated with ecological indicator taxa (Pearson 1994), and this characteristic is strengthened by their close relationship with the vegetation layer that is a consequence of their phytophagy and host-specific behaviour. Thus, butterflies could reveal trends in plants and therefore, in many other terrestrial arthropod groups (Martín Cano et al. 1996). This justifies their usefulness for assessing areas of interest in environmental conservation. Moreover, butterflies could have the potential to be an umbrella group for biodiversity conservation (New 1997). However, because of the bias in the geographic distribution of our chorological knowledge (Dennis and Thomas 2000; García-Barros and Munguira 1999; Martín and Gurrea 1999), under-recording is a common problem even in regions with a well-known fauna (Spain: García-Barros et al. 2000; Portugal: García-Pereira et al. 1999; Great Britain: Dennis et al. 1999; France: Dennis et al. 2002 and Dennis and Shreeve 2003). Thus, there is a pressing need to make extensive field inventories and evaluate them in order to obtain accurate species-richness data.

## Methods

### *Study site*

The study was carried out in Cabañeros National Park (Spain), which is situated between the northwest of Ciudad Real Province and the southeast of Toledo Province (latitude 39°23′47″ north and longitude 4°29′14″ west). Cabañeros is part of the Montes de Toledo mountain range. The Park has a moderate Mediterranean climate, it is situated in the Mesomediterranean bioclimatic stage and has a dry-subhumid tendency (mixed oak forests of *Quercus pyrenaica, Q. suber* and *Q. ilex ballota*) (Rivas-Martínez 1987).

Six 1-km$^2$ sampling plots with different land uses were selected:

- Sampling plot 1: Forest. An oak forest located on a shady rocky slope. These environmental conditions preserve the best Mediterranean forests of the area (Vaquero de la Cruzc 1997; Costa Tenorio et al. 1998).
- Sampling plot 2: Pine plantation. A *Pinus pinaster* plantation dating from the late 1960s. The shrub layer mainly consists of *Erica arborea* inside the plantation and *Cistus ladanifer* shoots in the firebreaks.
- Sampling plot 3: Shrubs. Plot with a dense, well-developed shrub layer dominated by *C. ladanifer*.
- Sampling plot 4: Crops. Plot dominated by cultivated land.
- Sampling plot 5: Grasslands. An extensive area of pastures with extensive sheep grazing.
- Sampling plot 6: Mixed use. Plot with no single dominant use.

The percentages of each habitat in each sampling plot are shown in Table 1.

*Table 1.* Description of the six sampling plots.

| Sampling plot | Habitats and relative proportions |
|---|---|
| 1 | Mediterranean oak forest (97%); screes (3%) |
| 2 | *Pinus pinaster* plantation (90%); firebreaks (10%) |
| 3 | Shrub (60%); grasslands (25%); crops (15%) |
| 4 | Crops (75%); shrub (10%); grasslands (15%) |
| 5 | Grasslands (75%); holm oak shrub (25%) |
| 6 | Shrub (35%); grasslands (25%); crops (25%); fallow land (15%) |

### *The biological assemblage of the study*

The butterfly fauna of Cabañeros National Park has recently been studied by Jiménez-Valverde et al. (2002, 2004). The traditional land use of the area, which can be appreciated throughout the territory, results in a few common and widespread species typical of pastures and grasslands being highly dominant among the fauna. The phenology of the butterfly assemblage is a typically Mediterranean, with a decrease in the number of species in the hot and dry summer and a subsequent recovery due to the emergence of bivoltine species that have already flown in spring.

### *Fieldwork*

The linear transect method developed by Pollard (1977, 1982) was employed (see Pollard and Yates 1993, for a thorough description of the method). This procedure is quick and easy to use. The six sampling plots were visited weekly from mid-March to September 2002. The six transects were 1 km long, and aerial photographs were used to divide them into sections of lengths proportional to the amount each land-use type in each plot. Transects were walked at a constant pace and butterflies that appeared 5 m ahead and 2.5 m on either side of the observer were counted. Butterflies were identified in flight wherever possible, while those that were more difficult to identify were captured in order to study their morphological characters in the laboratory. Counts were initially carried out between 1100 and 1500 h, when butterfly activity was greatest, but as the summer went on, the daily sampling period was progressively changed in order to avoid the hottest hours in the middle of the day, since butterflies are less active and shelter in the vegetation at these times. No sampling was undertaken on days with climatologically adverse conditions (Pollard and Yates 1993).

### *Statistical analysis*

When building accumulation curves to infer recommendations about sampling planning for other researchers, effort is expressed as sampling units ($n$)

(Moreno and Halffter 2001), which, in our case, were 'sampling weeks'. The order in which they were added was randomized 100 times and the mean number of accumulated species, for each $n$ value between 1 and the total number of sampling weeks, was calculated. This randomization process is detailed in Colwell and Coddington (1994). Calculations were performed with EstimateS 6.0 software (Colwell 2000).

Seven species-accumulation models that are widely used in the literature were examined (Table 2; see Tjørve 2003, for sources), all of which were asymptotic except the Power function. Models were fitted by non-linear regression using the STATISTICA program (Statsoft 2001). The simpler models (those with only two parameters) were fitted using the Simplex & Quasi-Newton algorithm. However, this method was not able to fit the more complex functions so Hooke-Jeeves or Rosenbrock algorithms were used (Statsoft 2001). Evaluation of non-sensical best-fit parameter values and visual assessment of model-fitting were used to reject non-useful functions. The coefficient of determination ($R^2$) and sum-of-squares ($SS$) were used to determine overall model performance. $R^2$ will usually be higher and $SS$ lower in those models the more parameters are included, simply because these models are more flexible. The model with the lowest $SS$ and the fewest parameters will be chosen over more complex functions. Otherwise, graphical examination of residuals and the Akaike's Information Criterion (AIC) enabled the definitive selection of the models that best fit the data. AIC is a method based on information theory that addresses the questions of which model is most likely to have generated the data and how much more likely that model is compared with the others under consideration (Motulsky and Christopoulos 2003). AIC is defined by the equation:

$$\mathrm{AIC} = N * \ln\left(\frac{SS}{N}\right) + 2^{*}K + \left(\frac{2^{*}K^{*}(K+1)}{N-K-1}\right)$$

where $N$ is the number of data points, $SS$ is the sum-of-squares and $K$ is the number of parameters plus one; the data are most likely to come from the model with the lowest AIC value (Motulsky and Christopoulos 2003). The

*Table 2.* Candidate function models for species-accumulation curves. The dependent variable ($f(x)$) is the number of species, and the independent variable ($x$) is the number of sampling units.

| Function | | Number of parameters | Asymptote |
|---|---|---|---|
| Power | $a^{*}x^{b}$ | 2 | No |
| Negative exponential | $a/b^{*}(1-\exp(-b^{*}x))$ | 2 | Yes ($a/b$) |
| Clench | $a^{*}(x/(b+x))$ | 2 | Yes ($a/b$) |
| Weibull | $a^{*}(1-\exp(-b^{*}x^{c}))$ | 3 | Yes ($a$) |
| Morgan–Mercer–Flodin | $a^{*}x^{c}/(b+x^{c})$ | 3 | Yes ($a$) |
| Chapman-Richards | $a^{*}(1-\exp(-b^{*}x))^{c}$ | 3 | Yes ($a$) |
| Beta-P | $a^{*}(1-(1+(x/c)^{d})^{-b})$ | 4 | Yes ($a$) |

probability of choosing the correct model is given by the equation:

$$\text{probability} = \frac{e^{-0.5*\Delta \text{AIC}}}{1 + e^{-0.5*\Delta \text{AIC}}}$$

**Results**

The number of samples, individuals and species per sampling plot are shown in Table 3. Non-sensical best-fit parameter values, whereby the value of the asymptote (the predicted number of species) was lower than the actual number observed, implied that the negative exponential model should be rejected (Table 3). Moreover, visual inspection of the fitted function (Figure 1) gave no grounds for favouring its selection. The Power function did not give a good fit to the data (Figure 2a), since this and the negative exponential model had the lowest $R^2$ and the largest *SS* values (Table 4). Only in sampling plot 5 did this function fit the data well (Figure 2b), exhibiting a high $R^2$ and a low *SS* that were quite similar to those of the other models. Thus, the first two criteria for discarding non-useful models led us to reject those based on the negative exponential and the Power distributions.

Visually, the rest of the models fitted the data quite well (Figure 3). As the *SS* values of the simpler models were always higher than those of the more complex ones (Table 4), we compared residuals and employed the AIC criterion with all of them.

In general, all the models exhibited lack of fit (Table 4 and Figure 4), although the magnitude varied among models and sampling plots. The Clench model function had the largest values of *SS* and exhibited the most systematic pattern in the residuals, followed by the Chapman-Richards and Weibull models. The Morgan–Mercer–Flodin and the Beta-P model functions

*Table 3.* Number of samples, species and individuals recorded for each sampling plot and predicted number of species (value of the asymptote) of the six asymptotic species accumulation models.

| Sampling plot | 1 | 2 | 3 | 4 | 5 | 6 |
|---|---|---|---|---|---|---|
| Number of samples | 24 | 22 | 25 | 23 | 23 | 24 |
| Number of species | 40 | 20 | 27 | 21 | 21 | 27 |
| Number of individuals | 1294 | 361 | 1121 | 688 | 378 | 619 |
| Functions | | | | | | |
| Negative exponential | 39.4 | 19.7 | 26.4 | 20.3 | 19.9 | 26.3 |
| Clench | 49.5 | 25.1 | 32.9 | 24.1 | 24.6 | 32.9 |
| Weibull | 44.3 | 23.3 | 31.1 | 22.0 | 36.8 | 37.5 |
| Morgan–Mercer–Flodin | 54.2 | 29.1 | 38.9 | 25.5 | 61.2 | 50.9 |
| Chapman-Richards | 42.7 | 22.1 | 29.5 | 21.5 | 27.2 | 32.1 |
| Beta-P | 51.9 | 118.6 | 36.2 | 23.8 | 68.2 | 64.5 |

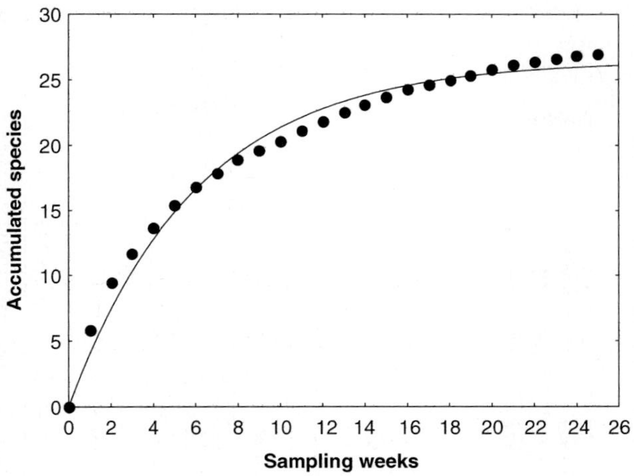


*Figure 1.* Negative exponential function fitted to the species-accumulation curve for sampling plot 3. Visual inspection and comparison with other functions (Figure 4) suggested that the model is unsuitable.

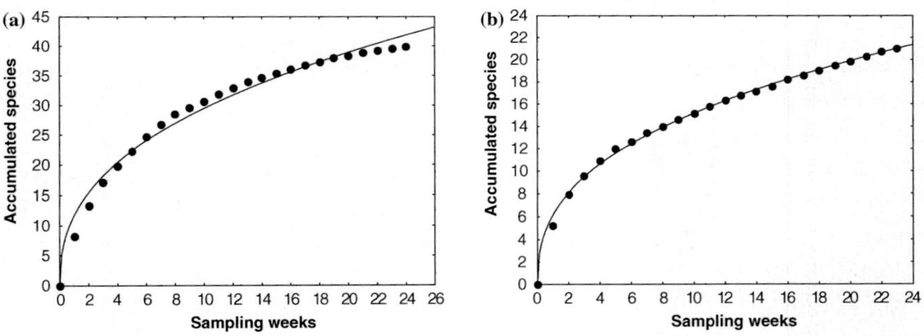


*Figure 2.* Power function fitted to the data of sampling plots 1 (a) and 5 (b). This model did not produce a good fit (a) except in inventories where the accumulation curve ended while still rising (b).

gave the best fit to the data. There was a marked lack of fit in all the models in sampling plot 5, and to a lesser extent in sampling plots 2 and 3. In general, the lack of fit was most pronounced in the lower sections of the accumulation curves.

Mean AIC values differed significantly from one function to another (Kruskal–Wallis test: $H$ (4, $N$ = 30) = 13.12, $p < 0.05$), the Clench model having the highest value (Table 4). The mean AIC also differed significantly between sampling plots (Kruskal–Wallis test: $H$ (5, $N$ = 30) = 11.88, $p < 0.05$), being highest in sampling plot 5 (Table 4).

Although the Clench model extrapolations of species richness were reasonable (Table 3), it had the highest AIC values of all the functions and of all six

*Table 4.* Coefficients of determination ($R^2$) and sum-of-squares ($SS$) for the different species-accumulation models in the six sampling plots.

| Function | 1 | 2 | 3 | 4 | 5 | 6 | Mean AIC |
|---|---|---|---|---|---|---|---|
| Power | $R^2$ = 0.984 | $R^2$ = 0.989 | $R^2$ = 0.988 | $R^2$ = 0.977 | $R^2$ = 0.997 | $R^2$ = 0.995 | – |
| | $SS$ = 41.04 | $SS$ = 6.67 | $SS$ = 14.58 | $SS$ = 14.51 | $SS$ = 1.41 | $SS$ = 4.73 | |
| Negative exponential | $R^2$ = 0.991 | $R^2$ = 0.988 | $R^2$ = 0.984 | $R^2$ = 0.985 | $R^2$ = 0.958 | $R^2$ = 0.975 | – |
| | $SS$ = 24.53 | $SS$ = 7.07 | $SS$ = 18.96 | $SS$ = 9.33 | $SS$ = 26.15 | $SS$ = 28.25 | |
| Clench | $R^2$ = 0.999 | $R^2$ = 0.997 | $R^2$ = 0.997 | $R^2$ = 0.999 | $R^2$ = 0.982 | $R^2$ = 0.991 | – 40.16 (±10.51) |
| | $SS$ = 2.81 | $SS$ = 1.31 | $SS$ = 3.66 | $SS$ = 0.56 | $SS$ = 11.18 | $SS$ = 10.03 | |
| | AIC = – 44.24 | AIC = – 54.72 | AIC = – 40.88 | AIC = – 78.03 | AIC = – 9.33 | AIC = – 13.74 | |
| Weibull | $R^2$ = 0.999 | $R^2$ = 0.999 | $R^2$ = 0.999 | $R^2$ = 0.999 | $R^2$ = 0.998 | $R^2$ = 0.999 | – 85.40 (±7.20) |
| | $SS$ = 0.49 | $SS$ = 0.50 | $SS$ = 0.51 | $SS$ = 0.18 | $SS$ = 1.09 | $SS$ = 0.18 | |
| | AIC = −83.35 | AIC = −72.89 | AIC = −87.19 | AIC = −101.64 | AIC = −59.99 | AIC = −107.32 | |
| Morgan–Mercer–Flodin | $R^2$ = 0.999 | $R^2$ = 0.999 | $R^2$ = 0.999 | $R^2$ = 0.999 | $R^2$ = 0.998 | $R^2$ = 0.999 | – 90.27 (± 6.84) |
| | $SS$ = 0.45 | $SS$ = 0.27 | $SS$ = 0.38 | $SS$ = 0.12 | $SS$ = 0.98 | $SS$ = 0.22 | |
| | AIC = −85.10 | AIC = −86.28 | AIC = −94.95 | AIC = −110.47 | AIC = −62.31 | AIC = −102.51 | |
| Chapman-Richards | $R^2$ = 0.999 | $R^2$ = 0.998 | $R^2$ = 0.999 | $R^2$ = 0.999 | $R^2$ = 0.997 | $R^2$ = 0.999 | – 75.14 (±6.90) |
| | $SS$ = 0.88 | $SS$ = 0.73 | $SS$ = 0.79 | $SS$ = 0.38 | $SS$ = 1.42 | $SS$ = 0.22 | |
| | AIC = −69.24 | AIC = −64.64 | AIC = −76.47 | AIC = −84.09 | AIC = −53.87 | AIC = −102.51 | |
| Beta-P | $R^2$ = 0.999 | $R^2$ = 0.999 | $R^2$ = 0.999 | $R^2$ = 0.999 | $R^2$ = 0.998 | $R^2$ = 0.999 | – 90.87 (±7.34) |
| | $SS$ = 0.41 | $SS$ = 0.12 | $SS$ = 0.39 | $SS$ = 0.10 | $SS$ = 0.96 | $SS$ = 0.23 | |
| | AIC = −84.54 | AIC = −100.61 | AIC = −90.66 | AIC = −111.80 | AIC = −59.43 | AIC = −98.21 | |
| Mean AIC | −73.29 (±7.83) | −75.83 (±8.07) | −78.03 (±9.78) | −97.21 (±6.88) | −48.99 (±10.01) | −84.86 (±17.84) | – |

The AIC value is shown for the Weibull, Morgan–Mercer–Flodin, Chapman-Richards and Beta-P models (see text). Also, mean AIC values (±SE) for each function and sampling plot. The model with the highest mean AIC was the Clench model, and the sampling plot with the highest mean AIC was plot 5.

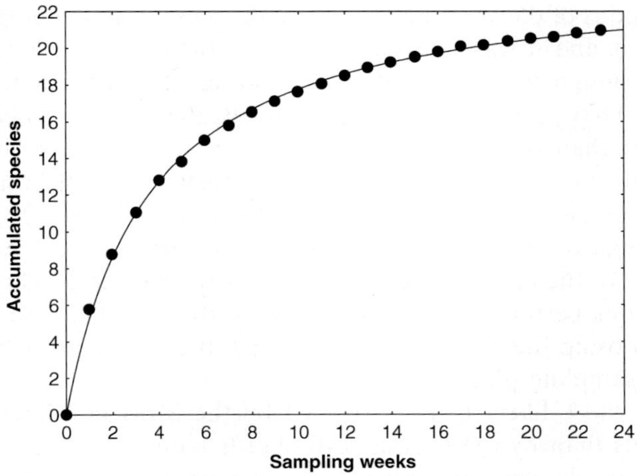


*Figure 3.* Example of the Clench model function fitted to the data of sampling plot 4. Compared with the negative exponential and Power functions, all other models fitted the data from the six sampling plots quite well.

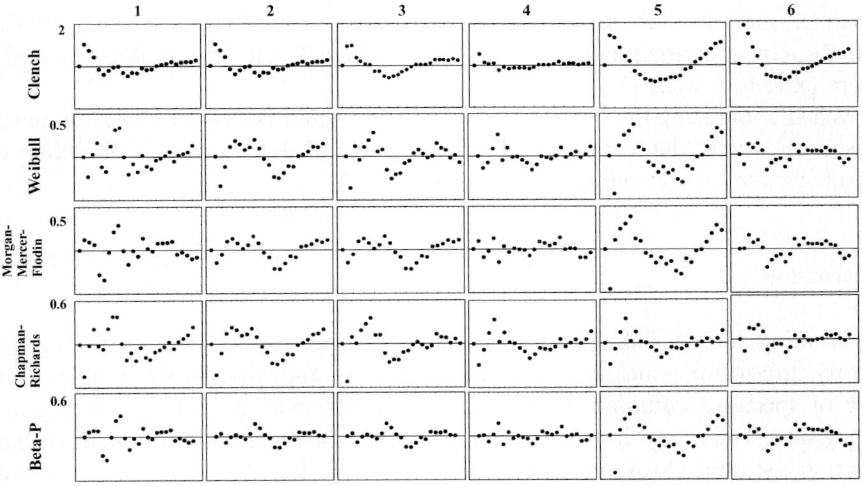


*Figure 4.* Residuals of the models plotted against the number of sampling weeks in the six sampling plots (horizontal lines = 0). The Clench model function had the most systematic pattern in the residuals. All the models had a greater lack of fit in sampling plot 5.

sampling sites (Table 4), implying that data were better explained by any model other than the Clench model. The probabilities of choosing the correct model (that is, any one other than the Clench model) were maximum (0.95–1.00). Thus, the Clench model does not seem to be a good option for modelling our data on butterfly species accumulation.

Extrapolations of the Weibull function make sense from a biological point of view (Table 3), and it was a better model than the Chapman-Richards model (Table 4), although the probabilities of choosing the right model were quite low (0.0001–0.08). In sampling plots 5 and 6, the Weibull model was only slightly better than the Beta-P; however, as the probability of choosing the correct model was $< 0.5$, these results are inconclusive. Only in sampling plot 6 did the Weibull model perform better than the Morgan–Mercer–Flodin model, although the probability of choosing the correct model was extremely low ( $< 0.1$). In the remaining cases, the Morgan–Mercer–Flodin and Beta-P model functions performed better than the Weibull model, with high probabilities of choosing the right model (0.65–0.99; the lowest probabilities corresponding to sampling plots 1 and 5).

Data were more likely to be explained by the Morgan–Mercer–Flodin or Beta-P models than by the Chapman-Richards model in sampling plots 1–5. Probabilities of choosing the correct model were high (0.94–0.99), with the lowest values corresponding to sampling plot 5. For plot 6, the Morgan–Mercer–Flodin and Chapman-Richard models had the same AIC value, and thus a probability of choosing the best model of 0.5. Both were superior to the Beta-P function, although they had quite low probabilities of choosing the correct model ( $< 0.2$) Extrapolations of species richness with the Chapman-Richard models were reasonable, while those of the Morgan–Mercer–Flodin function in sampling plots 5 and 6 and of the Beta-P function in plots 2, 5 and 6 were extremely high (Table 3).

When comparing the Morgan–Mercer–Flodin and Beta-P models, and in the cases in which the former performed better (Table 4), the probability of choosing the correct model was less than 0.5.

## Discussion

Estimating species richness and assessing the accuracy of inventories are currently important issues in conservation biology and biodiversity studies. The use of species-accumulation functions is an easy way to achieve these goals. They have been used in many studies and are an essential component of biodiversity studies (Moreno and Halffter 2000; Willot 2001). However, little is known about the performance of different models in different ecological situations (variation with spatial scale, habitat structure, community structure, etc.). Researchers have usually applied one widely accepted model, without considering whether any other would fit their data better. In this work, we have seen that model performance varies among local-scale sampling plots. Moreover, there was a consistent lack of fit for each model in the same sampling units. Thus, there are certain factors that systematically affect the fitting of models, and to determine what they are requires much more accurate field data and studies based on simulated data. Until then, comparative tests must be undertaken before selecting the definite model for our data. However, in spite of this systematic lack of fit, we can find species-accumulation functions that generally have significantly greater lack of fit than others.

If the sampling zone is relatively small or the taxonomic group is sufficiently well known, or both, then all species have a high probability of being recorded. In these cases, the negative exponential model is recommended (Soberón and Llorente 1993). However, these conditions are not usually found when working with hyperdiverse groups, such as arthropods. Palmer (1990) argued that if an estimator produces a lower value than the observed species richness, it should be considered a bad estimator. In this work, the negative exponential model gave such a result, as was also found by León-Cortes et al. (1998) for sphinx moth inventories and by Peterson and Slade (1998) for two model data sets. Conclusions are contradictory in the species-area literature (Tjørve 2003).

The non-asymptotic Power model did not fit the data well, except for sampling plot 5. The species-accumulation curve ends while it is still rising, so it could be fitted equally well by non-asymptotic and asymptotic models (Flather 1996). However, the Power model did not perform any better than other more flexible asymptotic functions.

The Clench model is recommended for large study sites and for protocols in which the longer the time spent in the field, and thus the greater the experience with the sampling method and taxonomic group, the higher is the probability of adding new species to the inventory (Soberón and Llorente 1993). However, Keating and Quinn (1998), employing simulated data, demonstrated that the Clench model was not appropriate for large and heterogeneous communities, as Soberón and Llorente (1993) proposed. Moreover, as the communities that best fitted the Clench model were of the broken-stick distribution type (moderately large and very even structure; Magurran 1988), and these communities are rarely found in nature, Keating and Quinn (1998) did not recommend the use of this species-accumulation model. We have found little evidence for the suitability of the Clench model, which provides poorer fits than any of the four other functions (Chapman-Richards, Morgan–Mercer–Flodin, Weibull and Beta-P). Of these functions, Morgan–Mercer–Flodin and Beta-P seem to perform best, the latter being slightly better than the former. However, as the predicted species richness values were sometimes too high, caution is required when using these two functions. Instead, the Weibull or Chapman-Richards models are preferred.

Flather (1996) found that the Weibull function was the best model for his data (he did not test the Morgan–Mercer–Flodin model). It was superior to the Beta-P, since the latter failed to converge in 36% of cases. Both are very flexible models and are very effective at fitting data sets closely (Tjørve 2003). Though its performance in our study was slightly worse than that of the Morgan–Mercer–Flodin and Beta-P models, its reasonable species richness predictions led us to conclude that the Weibull model is the best option for modelling the species accumulation process in our butterfly inventories.

Much more research is needed if we are to understand the behaviour of different species-accumulation models and the influence of different factors (e.g., spatial scale of the study, taxonomic group, community structure, habitat type and land use) on them. Unfortunately, we have not been able to evaluate the reliability of the model at showing how well the estimated asymptote matches the true asymptote of the data. We cannot know this unless all the taxa in an area have been sampled. Thus, even if the models fit the data well, they still may not give a good estimate of asymptotic species richness, as strikingly high values indicate. Virtual and real data sets where the true species richness is known are needed to measure the predictive power with respect to this characteristic. However, we may at least conclude that the traditional Clench model, widely used because of its simplicity, is not the best choice for modelling butterfly species-accumulation at the local scale, and that more complex and flexible functions are therefore definitely a better option.

## Acknowledgements

We are grateful to Jorge M. Lobo and Joaquín Hortal for their valuable suggestions. This study was partially supported by the European Project *Biodiversity Assessment Tools* (EVK2-CT-1999–00041) and a PhD grant from the Museo Nacional de Ciencias Naturales/C.S.I.C./Comunidad de Madrid.

## References

Chao A. 1987. Estimating the population size for capture–recapture data with unequal catchability. Biometrics 43: 783–791.

Chiarucci A., Enrigut N.J., Perry G.L.W., Miller B.P. and Lamont B.B. 2003. Performance of nonparametric species richness estimators in a high diversity plant community. Div. Distribut. 9: 283–295.

Colwell R.K. 2000. EstimateS: Statistical Estimation of Species Richness and Shared Species from Samples (Software and User's Guide), Version 6.0. Available at http://viceroy.eeb.uconn.edu/estimates.

Colwell R.K. and Coddington J.A. 1994. Estimating terrestrial biodiversity through extrapolation. Philos.Trans. R. Soc. Ser. B 345: 101–118.

Costa Tenorio M., Morla Juaristi C. and Sainz Ollero H. (eds) 1998. Los bosques ibéricos. Una interpretación geobotánica. Geoplaneta, Barcelona.

Dennis R.L.H. and Shreeve T.G. 2003. Gains and losses of French butterflies: tests of predictions, under-recording and regional extinction from data in a new atlas. Biol. Conserv. 110: 131–139.

Dennis R.L.H., Shreeve T.G., Sparks T.H. and Lhonore J.E. 2002. A comparison of geographical and neighbourhood models for improving atlas databases. The case of the French butterfly atlas. Biol. Conserv. 108: 143–159.

Dennis R.L.H., Sparks T.H. and Hardy P.B. 1999. Bias in butterfly distribution maps: the effects of sampling effort. J. Insect Conserv. 3: 33–42.

Dennis R.L.H. and Thomas C.D. 2000. Bias in butterfly distribution maps: the influence of hot spots and recorder's home range. J. Insect Conserv. 4: 73–77.

Fagan W.F. and Kareiva P.M. 1997. Using compiled species lists to make biodiversity comparisons among regions: a test case using Oregon butterflies. Biol. Conserv. 80: 249–259.

Flather C.H. 1996. Fitting species-accumulation functions and assessing regional land use impacts on avian diversity. J. Biogeogr. 23: 155–168.

García-Barros E., García-Pereira P. and Munguira M.L. 2000. The geographic distribution and state of butterfly faunistic studies in Iberia (Lepidoptera Papilionoidea Hesperioidea). Belg. J Entomol. 2: 111–124.

García-Barros E. and Munguira M.L. 1999. Faunística de mariposas diurnas en España peninsular. Áreas poco estudiadas: una evaluación en el umbral del siglo XXI (Lepidoptera: Papilionidae and Hesperiidae). SHILAP Revista de Lepidopterología 27(106): 189–202.

García-Pereira P., García-Barros E. and Munguira M.L. 1999. Evaluación del conocimiento de la fauna de mariposas de Portugal continental. SHILAP Revista de Lepidopterología 27(106): 225–231.

Gaston K.J. (ed.) 1996. Biodiversity A Biology of Numbers and Difference. Blackwell Science, Oxford.

Gotelli N.J. and Colwell R.K. 2001. Quantifying biodiversity: procedures and pitfalls in the measurement and comparison of species richness. Ecol. Lett. 4: 379–391.

Heltshe J.F. and Forrester N.E. 1983. Estimating species richness using the jackknife procedure. Biometrics 39: 1–11.

Hortal J., Lobo J.M. and Martín-Piera F. 2001. Forecasting insect species richness scores in poorly surveyed territories: the case of the Portuguese dung beetles (Col. Scarabaeinae). Biodiv. Conserv. 10: 1343–1367.

Jiménez-Valverde A. and Lobo J.M. 2004. Determining a combined sampling procedure for a reliable estimation of Araneidae and Thomisidae assemblages (Arachnida: Araneae). J. Arachnol. 32(3): in press.

Jiménez-Valverde A., Martín Cano J. and Munguira M.L. 2002. Fauna de mariposas del Parque Nacional de Cabañeros y su entorno (Lepidoptera: Papilionoidea, Hesperioidea). SHILAP Revista de Lepidopterología 30(120): 271–279.

Jiménez-Valverde A., Martín Cano J. and Munguira M.L. 2004. Patrones de diversidad de la fauna de mariposas del Parque Nacional de Cabañeros y su entorno (Ciudad Real, España central) (Lepidoptera: Papilionoidea, Hesperioidea). Anim. Biodiv. Conserv. 27(2): 15–24.

Keating K.A. and Quinn J.F. 1998.Estimating species richness: the Michaelis-Menten model revisited. Oikos 81(2): 411–416.

Lamas G., Robbins R.K. and Harvey D.J. 1991. A preliminary survey of the butterfly fauna of Pakitza, Parque Nacional del Manu, Peru, with an estimate of its species richness. Publicaciones del Museo de Historia Natural, Universidad Nacional Mayor de San Marcos, Serie A Zoología 40: 1–19.

León-Cortés J.L., Soberón-Mainero J. and Llorente-Bousquets J. 1998. Assessing completeness of Mexican sphinx moth inventories through species accumulation functions. Div. Distribut. 4: 37–44.

Longino J.T., Coddington J. and Colwell R.K. 2002. The ant fauna of a tropical rain forest: estimating species richness three different ways. Ecology 83: 689–702.

Magurran A.E. 1988. Ecological Diversity and Its Measurement. Princeton University Press, New Jersey.

Martín J. and Gurrea P. 1999. Áreas de especiación en España y Portugal. Boletín de la Asociación española de Entomología 23(12): 83–103.

Martín Cano J., Ferrín J.M., García-Barros E., García-Ocejo A., Gurrea P., Luciañez M.J., Munguira M.L., Perez Barroeta F., Ruiz Ortega M., Sanz Benito M.J., Simon J.C. and Viejo J.L. 1996. Las comunidades de insectos del Parque Regional de la Cuenca Alta del Manzanares (centro de España): estado de conservación. Graellsia 51: 101–111.

Moreno C.E. and Halffter G. 2000. Assessing the completeness of bat biodiversity inventories using species accumulation curves. J. Appl. Ecol. 37: 149–158.

Moreno C.E. and Halffter G. 2001. On the measure of sampling effort used in species accumulation curves. J. Appl. Ecol. 38: 487–490.

Motulsky H.J. and Christopoulos A. 2003. Fitting Models to Biological Data Using Linear and Nonlinear Regression. A practical guide to curve fitting, Available at http://www.graphpad.com. GraphPad Software Inc., San Diego CA.

New T.R. 1997. Are Lepidoptera an effective 'umbrella group' for biodiversity conservation? J. Insect Conserv. 1: 5–12.

Palmer M.W. 1990. The estimation of species richness by extrapolation. Ecology 71(3): 1195–1198.

Pearson D.L. 1994. Selecting indicator taxa for the quantitative assessment of biodiversity. Philos. Trans. R. Soc. Ser. B 345: 75–79.

Petersen F.T. and Meier R. 2003. Testing species-richness estimation methods on single-sample collection data using the Danish Diptera. Biodiv. Conserv. 12: 667–686.

Petersen F.T., Meier R. and Larsen M.N. 2003. Testing species richness estimation methods using museum label data on the Danish Asilidae. Biodiv. Conserv. 12: 687–701.

Peterson A.T. and Slade N.A. 1998. Extrapolating inventory results into biodiversity estimates and the importance of stopping rules. Div. Distribut. 4: 95–105.

Pollard E. 1977. A method for assessing changes in the abundance of butterflies. Biol. Conserv. 12: 115–134.

Pollard E. 1982. Monitoring butterfly abundance in relation to the management of a nature reserve. Biol. Conserv. 24: 317–328.

Pollard E. and Yates T.J. 1993. Monitoring Butterflies for Ecology and Conservation. Chapman & Hall, London.

Rivas Martínez S. 1987. Mapas de series de vegetación de España 1:400.000 y Memoria. ICONA, Ministerio de Agricultura, Pesca y Alimentación, Spain.

Smith E.P. and van Belle G. 1984. Nonparametric estimation of species richness. Biometrics 40: 119–129.

Soberón J. and Llorente J. 1993. The use of species accumulation functions for the prediction of species richness. Conserv. Biol. 7(3): 480–488.

Statsoft. 2001. STATISTICA (Data Analysis Software System and User's Manual). Version 6. StatSoft Inc., Tulsa, OK.

Tjørve E. 2003. Shapes and functions of species-area curves: a review of possible models. J. Biogeogr. 30: 827–835.

Vaquero de la Cruz J. 1997. Flora vascular y vegetación. In: García Canseco V. et al.(eds), Parque Nacional de Cabañeros. Ecohábitat, Ciudad Real, Spain pp. 95–154.

Willott S.J. 2001. Species accumulation curves and the measure of sampling effort. J. Appl. Ecol. 38: 484–486.

Biodiversity and Conservation (2006) 15:177–189
DOI 10.1007/s10531-004-6903-y 

# The habitat requirement of the Genji-firefly *Luciola cruciata* (Coleoptera : Lampyridae), a representative endemic species of Japanese rural landscapes

MARI TAKEDA[1], TATSUYA AMANO[1,*], KAZUHIRO KATOH[2] and HIROYOSHI HIGUCHI[1]

*[1]Laboratory of Biodiversity Science, School of Agricultural and Life Sciences, The University of Tokyo, 1-1-1 Yayoi, Bunkyo-ku, Tokyo 113-8657, Japan; [2]Laboratory of Agricultural Field Ecology, School of Agricultural and Life Sciences, The University of Tokyo, 1051 Hata-machi, Hanamigawa-ku, Chiba City, Chiba 262-0018, Japan; *Author for correspondence (e-mail: tatsu@es.a.u-tokyo.ac.jp; phone: +81-3-5841-7542; fax: +81-3-5841-7542)*

Received 10 June 2004; accepted in revised form 11 November 2004

**Key words:** Abundance, Agricultural landscape, Aquatic insects, Ditch modification, Water quality

**Abstract.** Raising public interest in nature through conserving species of high social interest is crucial in achieving effective conservation of biodiversity. In Japan, the Genji-firefly *Luciola cruciata* (Coleoptera Lampyridae) in biodiversity rich agricultural landscapes called the satoyama has always attracted exceptional public interest. This study provides rare information on environmental factors associated with the abundance of the Genji-firefly. Stepwise backward multiple regression revealed that firefly abundance increased with increasing pH, DO and prey abundance while decreasing with water depth and the proportion of artificially modified ditch length. These factors are thought to be influential mainly to the larval and pre-pupal periods of the firefly. The implications of the results for the conservation of the Genji-firefly are discussed, with reference to the relationship between Genji-firefly conservation and extensive biodiversity conservation in the satoyama.

## Introduction

Today, with increasing interests for the conservation of wild fauna and flora, people have become aware not only of the importance of natural environments but also of semi-natural environments (Buckley et al. 1997; Endels et al. 2002). Typical examples of such semi-natural environments are agricultural landscapes found in rural areas worldwide (Burel 1996; Elphick 2000). Hedgerows in Europe for instance, have been studied intensively for their roles as refuges or corridors for the conservation of biodiversity in rural areas (Burel 1996; Burel et al. 1998).

In Japan, agricultural landscapes called the satoyama provide a variety of habitat types for wildlife, helping to maintain rich biodiversity in the Japanese countryside (Kobori and Primack 2003). The satoyama consists of a mosaic of forests, grasslands, rice-fields, ponds, creeks and irrigation ditches that have historically provided resources for agricultural life (Kobori and Primack 2003).

Today, however, the reduced rice production policy by the government (Moore 1990), together with modernized agricultural schemes, urban developments and changes in forestry, have altered the rural landscapes drastically (Fukamachi et al. 2001), causing a corresponding decline in aquatic plants [e.g., *Sparganium japonicum* (Typhales : Sparganiaceae), Kato 2001], birds [e.g., *Butastur indicus* (Falconiformes : Accipitridae), Fujioka and Yoshida 2001], and insects [e.g., *Japonica saepestriata* (Lepidopter : Lycaenidae) and *Antigius attilia* (Lepidopter : Lycaenidae), Kato 2001, *Oligoaeschna pryeri* (Odonata : Aeshnidae) and *Rhyothemis fuliginosa* (Odonata : Libellulidae), Washitani 2001].

The Genji-firefly, with its unique luminescence, has been a representative insect of Japanese satoyama. Japanese people's admiration for the insect is exceptional to the extent that it has had a prominent influence on the Japanese culture (Kobori and Primack 2003). Nevertheless, the Genji is no exception to today's general trend and its population has declined in many areas of Japan (Ohba 1988). There are various reasons to believe that the conservation of the Genji-firefly is significant for the conservation of satoyama as a whole. Firstly, the Genji-firefly is an object of exceptionally high social interest mainly due to its aesthetic value. It has demonstrated to date that it can attract a large number of people to be involved in its conservation (Ohba 1988). Many authors have argued that raising public interest in nature through conserving such species of high interest is a crucial feature in successful conservation projects (e.g., Harrison and Burgess 2000; Primack et al. 2000; Suh and Samways 2001). Secondly, throughout its life cycle, the Genji utilises the diverse spatial environment typical to the satoyama; in the larval period, it uses water in the irrigation ditches; in the pupal period, it uses the soil surrounding the ditches; and finally in the adult phase, it makes use of the vegetation and space around and above the ditches (Ohba 1988). Therefore, it can be said that they are relatively sensitive to environmental changes in the habitat (Yuma 2000).

Due to their popularity, a considerable amount of conservation efforts has already been dedicated to this species. The Genji is a designated national natural treasure in at least 10 districts, and conservation activities are found in almost all prefectures of Japan, where various environmental conditions are typically used as general guidelines to distinguish favourable habitats from non-favourable ones (Ohba 1988). However, for the coexistence of human activities and the effective conservation of the Genji-firefly, accurate understanding of the abundance and distribution of this species is indispensable. Surprisingly, the number of scientific data on the habitat requirement of the Genji-firefly that are likely to lead to practical conservation measures is minimal. In the only study on the habitat requirement of the Genji-firefly thus far, Shibue et al. (1995) do not take into account a number of potentially important factors, such as prey abundance, water quality and artificial alteration conditions of irrigation ditches. Thus, the aim of this study was to identify the habitat characteristics associated with the abundance and distributional pattern of the Genji-firefly in the study area, based on which a more robust predictive model can be constructed and applied to practical conservation activities.

## Material and methods

### *Study material*

After hatching from eggs laid on mosses and vegetation on the walls of irrigation ditches, the Genji larvae drop into the irrigation ditches where they sink to the bottom. They remain under water for the majority of the life cycle (usually approximately 10 months), preying on pleurocerid snails (Kato 2001), most commonly Kawanina snails, *Semisulcospira libertina*, to which it is highly specific (Ohba 1988). Around April, fully-grown larvae emerge from the water by climbing up the ditch walls to reach suitable soil. They burrow underground and develop into pupae; emergence of the luminous adults occurs in June. They then fly above the rice fields and irrigation ditches before mating on suitable nearby lower vegetation, after which the females oviposit on mosses and lower vegetation. The adults only consume water in the form of dew on leaves (Yuma 2000).

### *Study sites*

Study sites were in Ichikai-town (36°33′ N, 140°07′ E), Tochigi prefecture in the North Kanto region of Japan. In Ichikai, where many satoyama landscapes remain, anthropogenic developments are causing changes to the long used habitat of the Genji firefly.

Within the study area, 44 yatos with various environmental conditions were selected as study sites. Yatos are defined as networks of valleys with flat bottoms where rice fields are typically developed, surrounded by coppice (Washitani 2001). Today, in Ichikai, there are some yatos with relatively high firefly abundance, while a decline in abundance has been observed at others (M. Komori, personal communication).

The 44 yatos had rice field components of at least 200 m in length (on Digital Map 25,000 of Mito by the Geographic Survey Institute, Japan) to ensure independence of each study site. Here, the average migration distance of adult Genji-fireflies, known to be approximately 100 m (Ohba 1988) was taken into consideration. Yatos unsuitable for adult Genji observation due to unfavourable conditions such as strong artificial illumination, and/or the presence of vision hindering obstacles were eliminated. Both the firefly abundance survey and the environmental factors survey were carried out at each of the 44 sites.

*Abundance survey of the Genji-firefly*

Past observations have shown that the number of adult fireflies in the study area reaches its peak at the end of June on average. Since the weather condition in the survey year was an average one, the abundance survey of adult Genji-firefly was conducted on 17th, 19th, and 21st June 2002. The number of illuminating adult fireflies was recorded from a fixed observation point between the generally accepted peak hours of 8 pm to 8:30 pm (Ohba 1988) by over 120 pre-trained local volunteers. Pairs of volunteers were allocated to each study site, for which they were responsible for all three survey dates. The observation point at each site was selected so that viewing conditions were as similar as possible among sites, and observations were made for a uniform duration of 1 min within a uniform 180° field of view, divided into three directions, right, centre, and left, for practical means. In the study area, it was not difficult to distinguish individual fireflies flying in near proximity, such that double counts were unlikely. Thus, it seems plausible to assume that overestimation of firefly abundance has been avoided. On the other hand, to avoid underestimation, the maximum number of Genji-fireflies observed over the three survey dates was used for analysis.

*Environmental factors survey*

Based on the life cycle of the Genji firefly, aquatic and terrestrial environmental factors in and around the irrigation ditches that are likely to be relevant to firefly abundance were selected and investigated (Tables 1 and 2). Generally, for each of the irrigation ditches surveyed, the structural and ecological features were uniform throughout the length of the ditch, and drainage openings that could potentially affect water qualities were not found in any of the surveyed ditches. Those environmental factors that were assumed to be consistent throughout the entire length of a single ditch were measured at a fixed point. Other factors were surveyed every 10 m for 200 m along each irrigation ditch. All environmental factors were surveyed four times in total unless when data were not available. The four study periods were: February 2002, March–April 2002, end of May–July 2002, September–October 2002, which almost covers the entirety of a single life-cycle of the Genji-firefly. Since seasonal variations observed were minimal, the mean of four seasonal measurements were used for statistical analysis.

*Statistical analyses*

A multiple regression analysis was performed for the dependent variable, firefly abundance, with surveyed environmental factors as potential explanatory variables using SYSTAT 8.0 (SPSS 1998). The firefly abundance data, as well

*Table 1.* Environmental factors investigated and incorporated into the analysis, with definitions, survey methods, and survey style.

| Variable | Definition | Survey method | Survey style |
|---|---|---|---|
| Ditch modification% [a] | proportion of ditch length modified artificially | observation | 5 m interval along ditch length |
| Ditch base material [b] | prevalent material forming base of ditch | observation | single fixed point |
| Silt% [a] | proportion of silt in sediment | observation | along 1 m ditch length upstream of single fixed point |
| Litter% [a] | proportion of litter in sediment | observation | along 1 m ditch length upstream of single fixed point |
| Flow continuity [a] | proportion of ditch length with flowing water | observation | 5 m interval along ditch length |
| Ditch width (cm) [c] | width of ditch | measured by measure | single fixed point |
| Water depth (cm) [c] | depth from water surface to base of ditch | measured by measure | single fixed point[e] |
| Base current velocity (cm/s) [c] | water current velocity at base of ditch | measured by electric meter | single fixed point[e] |
| Vegetation height (cm) [c] | Height of vegetation adjacent to ditch | measured by measure | single fixed point |
| Vegetation cover [a] | proportion of vegetation cover adjacent to ditch | observation | single fixed point |
| PH | pH | measured by pH meter | single fixed point[f] |
| DO (ppm) | DO (dissolved oxygen) | measured by Oxymeter | single fixed point[f] |
| EC (mS/m) | EC (electric conductivity) | measured by electric meter | single fixed point[f] |
| Water temperature (°C) | water temperature | measured by thermometer | single fixed point |
| S. libertina [c] | abundance of *S. libertine* | soil samples taken from ditch bases using a sieve | mean of five samples taken within 1.5 m of single fixed point |
| Alternative land use % [a] | proportion of ditch length in alternative land use | observation | 5 m interval along ditch length |
| Ditch coverage [d] | whether vegetation covers only water edges or water surface as well | observation | 5 m interval along ditch length |

[a] Arcsin transformed.
[b] Included in analysis as three dummy variables: unexposed base material (1) or not (0), exposed natural material (1) or not (0), and concrete (1) or not (0).
[c] Log transformed.
[d] Dummy variable coding: 0 = no vegetation cover, 1 = vegetation covers only water edges, 2 = vegetation also covers water surface.
[e] Mean of 3 points; left, centre and right for ditch width above 1 m.
[f] Mean of data for ripple and pool when both existed within 1 m of single fixed point.

*Table 2.* Environmental elements expected to affect the abundance of the Genji-firefly, specific environmental factors relevant to each element, how each affect firefly abundance, and the life stage it affects.

| Environmental elements | Variables | Role/effect | Relevant life stage |
|---|---|---|---|
| Artificial ditch alteration conditions | ditch modification % | availability of climbable & burrowable ditch wall | pre-pupae |
| Ditch hydrological conditions | ditch base material | larval habitat | larvae |
| | Silt % | larval habitat | larvae |
| | litter % | larval habitat | larvae |
| | flow continuity | stability of habitat | larvae |
| | ditch width | water quantity/velocity | larvae |
| | Water depth | water quantity/velocity | larvae |
| | base current velocity | water quantity/velocity | larvae |
| Water quality | canal coverage | availability of source of shade | larvae |
| | vegetation height | source of shade | larvae |
| | vegetation cover | source of shade | larvae |
| | pH | water quality | larvae |
| | DO (dissolved oxygen) | water quality | larvae |
| | EC (electric conductivity) | water quality | larvae |
| | water temperature | water quality | larvae |
| Prey abundance | *S. libertina* | prey abundance for larvae | larvae |
| | litter % | prey food | larvae |
| | vegetation height | prey food | larvae |
| | vegetation cover | prey food | larvae |
| Environment surrounding ditch | alternative land use % | continuity of habitat | adult |
| | canal coverage | disturbance of flight | adult |
| | vegetation height | resting site | adult |
| | vegetation cover | resting site | adult |

as those explanatory variables that showed non-normal distributions were log transformed, and data expressed as proportions were arcsin transformed prior to the analysis. Intercorrelations of explanatory variables were examined prior to the regression analysis and strongly correlated variables (defined as Pearson correlation $r > 0.8$) were eliminated from the analysis to avoid multicollinearity. Explanatory variables were selected based on the stepwise backward elimination method and p(elimination) $>0.05$.

## Results

Correlations between variables were low to moderate in most cases (Table 3), though five combinations of the variables showed correlations of 0.50 to 0.76. However, since correlations above 0.8 were not found between any of the variables, all variables were included as explanatory variables in the regression analysis.

*Table 3.* Pearson correlations between each explanatory variable surveyed.

| | Ditch modification % | Unexposed base material | Exposed natural base material | Concrete base | Silt % | Litter % | Flow continuity | Ditch width | Water depth | Base current velocity |
|---|---|---|---|---|---|---|---|---|---|---|
| Ditch modification % | 1 | | | | | | | | | |
| Unexposed base material [a] | −0.425 | 1 | | | | | | | | |
| Exposed natural base material [a] | −0.049 | −0.478 | 1 | | | | | | | |
| Concrete base [a] | 0.507 | −0.736 | −0.243 | 1 | | | | | | |
| Silt% | 0.122 | −0.024 | −0.203 | 0.184 | 1 | | | | | |
| Litter% | −0.332 | 0.212 | −0.002 | −0.232 | 0.193 | 1 | | | | |
| Flow continuity | 0.070 | 0.040 | 0.087 | −0.111 | 0.204 | 0.172 | 1 | | | |
| Ditch width | 0.193 | −0.079 | 0.309 | −0.151 | −0.272 | −0.252 | −0.227 | 1 | | |
| Water depth | 0.078 | −0.011 | 0.109 | −0.071 | −0.317 | −0.203 | 0.047 | 0.512 | 1 | |
| Base current velocity | 0.130 | 0.045 | 0.002 | −0.050 | −0.368 | −0.488 | −0.035 | 0.241 | 0.590 | 1 |
| Vegetation height | −0.512 | 0.395 | 0.082 | −0.499 | −0.190 | 0.150 | −0.119 | 0.198 | −0.005 | 0.062 |
| Vegetation cover | −0.765 | 0.395 | 0.030 | −0.459 | 0.008 | 0.292 | 0.027 | −0.176 | 0.042 | 0.003 |
| PH | 0.217 | −0.149 | 0.240 | −0.020 | −0.144 | −0.124 | −0.076 | 0.383 | −0.074 | −0.152 |
| DO | 0.120 | −0.177 | 0.288 | −0.026 | −0.165 | 0.066 | 0.224 | −0.009 | 0.145 | 0.219 |
| EC | 0.259 | −0.238 | 0.106 | 0.181 | −0.005 | −0.035 | −0.002 | 0.307 | 0.142 | −0.117 |
| Water temperature | 0.332 | −0.379 | 0.006 | 0.414 | 0.068 | −0.270 | −0.130 | 0.278 | −0.001 | −0.010 |
| *S. libertina* | −0.399 | 0.235 | −0.101 | −0.181 | 0.092 | 0.130 | 0.078 | −0.464 | −0.142 | 0.018 |
| Alternative land use % | −0.189 | 0.220 | −0.062 | −0.195 | 0.149 | 0.261 | 0.054 | −0.239 | −0.322 | −0.440 |
| Ditch coverage [a] | −0.009 | −0.023 | 0.011 | 0.017 | −0.063 | −0.026 | 0.125 | −0.022 | 0.124 | 0.041 |

*Table 3*. Continued.

| | Vegetation height | Vegetation cover | PH | DO | EC | Water temperature | *S. libertina* | Alternative land use % | Ditch coverage |
|---|---|---|---|---|---|---|---|---|---|
| | Vegetation height | Vegetation cover | PH | DO | EC | Water temperature | *S. libertina* | Alternative land use % | Ditch coverage |
| Vegetation height | 1 | | | | | | | | |
| Vegetation cover | 0.497 | 1 | | | | | | | |
| PH | −0.008 | −0.291 | 1 | | | | | | |
| DO | 0.107 | −0.034 | −0.043 | 1 | | | | | |
| EC | −0.159 | −0.238 | 0.434 | −0.146 | 1 | | | | |
| Water temperature | −0.220 | −0.352 | 0.395 | −0.238 | 0.220 | 1 | | | |
| *S. libertina* | 0.029 | 0.447 | −0.154 | −0.093 | −0.436 | −0.040 | 1 | | |
| Alternative land use % | 0.075 | 0.164 | −0.036 | −0.107 | −0.007 | −0.199 | 0.008 | 1 | |
| Ditch coverage [a] | −0.133 | −0.064 | −0.252 | 0.028 | −0.053 | 0.131 | 0.071 | 0.202 | 1 |

[a] Dummy variables.

*Table 4.* Results of backward stepwise multiple regression analysis on Genji-firefly abundance.

| Explanatory variables adopted | Partial regression coefficients | Standard error | Standardized partial regression coefficients | Tolerance | $t$ | $p$ |
|---|---|---|---|---|---|---|
| Constant | −2.593 | 0.750 | | | | |
| PH | 0.269 | 0.082 | 0.329 | 0.934 | 3.286 | 0.002 |
| DO | 0.251 | 0.055 | 0.454 | 0.962 | 4.602 | <0.001 |
| Water depth | −0.422 | 0.140 | −0.299 | 0.953 | −3.022 | 0.004 |
| *S. libertina* | 0.316 | 0.083 | 0.405 | 0.821 | 3.790 | 0.001 |
| Ditch modification % | −0.245 | 0.078 | −0.339 | 0.807 | −3.146 | 0.003 |

Shown are only significant variables. $F_{5,\,38} = 13.762$, $p < 0.001$, $R^2 = 0.644$, $R^2_{adj} = 0.597$.

The number of Genji-fireflies in a yato varied from 0 to 71 and the mean was 17.34 ± 2.49 (± SE). According to the regression analysis, five variables, PH, DO, water depth, *S. libertina* abundance, and the proportion of artificially modified ditch length were found to be significantly affecting firefly abundance; 60% of the variation in firefly abundance was explained by these variables (Table 4). PH, DO, and *S. libertina* abundance showed positive relationships with firefly abundance, while the number of fireflies decreased with increasing water depth and modified ditch%. All of the five variables selected as significant variables had tolerance values above 0.8, indicating the absence of strong collinearity among those variables.

## Discussion

It has been reported that not only aquatic conditions, but also terrestrial conditions, especially bank-side structure and vegetation, have an impact on the abundance of aquatic insects with terrestrial adult phases, such as caddisflies and dragonflies (Samways and Steytler 1996; Collier et al. 1997; Harrison and Harris 2002; Iwata et al. 2003). The results of this study clearly supported this idea by showing that the abundance of Genji-fireflies was also affected by both aquatic (e.g., pH, DO) and terrestrial factors (e.g., the proportion of artificially modified ditch length).

The negative impact of increased proportions of artificially modified ditch length on fireflies is understandable since it is crucial for the firefly larvae to be able to find ditch walls suited for pre-pupal landing after emergence from the water, followed by underground burrowing (Ohba 1988; The Japanese Firefly Society 1996). Further, the relatively high correlation of the proportion of artificially modified ditch length to vegetation height ($r = -0.512$) and vegetation cover ($r = -0.765$, Table 3) indicated the possibility that bank side vegetation positively affects firefly abundance through its role in resting, mating and oviposition, as has been reported in other species of aquatic insects (Ormerod et al. 1990; Samways and Steytler 1996).

Water quality is thought to affect firefly abundance as the direct determinant of larval habitat quality. Though many studies have found that the sensitivity to acid varies among aquatic insects (Ward 1992), from highly tolerant dragonflies (e.g., Pollard and Berrill 1992) to particularly sensitive mayflies (e.g., Courtney and Clements 1998), the positive influence of relatively high pH (range: 6.07–8.46) on firefly abundance indicated that the Genji-firefly larvae were relatively sensitive to water acidification. The positive relationship between DO (range: 6.26–9.12) and firefly abundance is consistent with the generally accepted knowledge that DO is an environmental variable of considerable importance to many aquatic insects (Ward 1992; Williams and Feltmate 1992).

Prey abundance has often been considered as one of the most powerful determinants of firefly abundance (Ohba 1988; The Japanese Firefly Society 1996). Indeed, *S. libertina* abundance was found to affect the Genji-firefly's abundance significantly. However, when compared with the other selected explanatory variables, the standardized partial regression coefficient of prey abundance indicated that its impact on firefly abundance was not much stronger, or even weaker. Since *S. libertina* abundance was not highly correlated to pH or DO (Table 3), while those variables strongly affected firefly abundance, *S. libertina* in the study area is thought to be more tolerant with regard to water quality than the firefly. That is, in the study area, the two species differ in their habitat requirements, with the prey being more tolerant than the predator, causing factors other than prey abundance to have a stronger impact on the abundance of the Genji-firefly.

The results of this study also appeared to suggest water depth (a maximum of 21 cm) to have a significantly negative impact on firefly abundance. However, Genji-fireflies have been found to inhabit streams as deep as 2 m (Ohba 1988). Thus, it is more likely that the apparent negative impact of water depth on firefly abundance is an indirect one through other factors not investigated in this study. Indeed, the three deepest ditches were characterised by the presence of heavy traffic roads running along the yato, together with street lights, as well as relatively large numbers of houses, which are all indicative of increased human activities. These surrounding environmental conditions are expected to affect the behaviour of adult fireflies, especially their mating behaviour. Investigating the influence of such factors on firefly abundance will be needed in future studies. In fact, for other aquatic insects, it has been argued that the adult terrestrial stage can be critical in regulating population size (Werneke and Zwick 1992; Enders and Wagner 1996).

### *Conservation recommendations*

The present regression model represents a basic model explaining Genji-firefly abundance distribution, upon which a more extensive and robust predictive model can be developed. Based on this fundamental model, the following potential conservation recommendations are provided. When development programmes and other human activities with potential impacts on the environment are considered in the study area, special attention needs to be paid to the artificial ditch alteration conditions, water quality and prey abundance. As much change is in progress in Ichikai town, including a ditch modification programme in planning, the information provided by this study is thought to become of immediate use.

Though much care should be taken when the results of this study is applied to other habitats of the Genji-firefly, the biological basis for most of the variables selected in this study is likely to confirm, at least, the qualitative generality of the conclusion. Therefore, when considering the recovery of the Genji-firefly abundance at other habitats, the determinants identified here could serve as practical guidelines. For example, increasing the abundance of *S. libertina*, which had previously been regarded as a factor of highest and sometimes sole priority in firefly abundance recovery, may be insufficient on its own unless other environmental factors are improved.

Although the model needs to be further improved, this study represents the first extensive attempt to examine the habitat requirements of the Genji-firefly and provide reliable information about the key habitat variables for this insect of high public interest. Further, habitat requirements of the Genji-firefly identified here also apply to a wide range of organisms in agricultural landscapes. Water quality, for instance, is known to have a marked impact on the abundance of many aquatic insects (e.g., Ward 1992; Courtney and Clements 1998; Suh and Samways 2001), while ditch modification in Japanese rice fields has been reported to affect negatively the abundance of various frogs and fishes, through which the distribution of predatory birds, such as egrets and herons, is also changed (Fujioka and Lane 1997; Lane and Fujioka 1998). Therefore, we believe that raising public interest towards the conservation of the Genji-firefly in the satoyama would be a crucial first step for the extensive conservation of biodiversity in the satoyama.

**Acknowledgements**

We would like to thank G. Fujita, Dr T. Miyashita and Dr S. Takatsuki for valuable comments on the draft, S. Nishihara for advice and introductory materials. K. Ushiyama, Y. Kunitake, M. Eda, Y. Wu, T. Ishizaki, Y. Yamaura and K. Itoh provided indispensable help in the abundance survey. We are equally grateful to the teachers, parents, and pupils at Kokai Chuo Primary School, Ichikai town and the volunteers of the town of Ichikai for participating in the survey. Our greatest gratitude goes to the two families, Mr & Mrs Mizuo Komori (Ichikai Town Cultural Asset Preservation Council member) and Mr & Mrs Norio Takatoku for their kindness toward our study in Ichikai. This work was supported by a grant from the Nippon Life Insurance Foundation to Higuchi, H.

## References


Buckley G.P., Howell R., Watt T.A., Ferris-Kaan R. and Anderson M.A. 1997. Vegetation succession following ride edge management in lowland plantations and woods. 1. The influence of site factors and management practices. Biol. Conserv. 82: 289–304.

Burel F. 1996. Hedgerows and their role in agricultural landscapes. Crit. Rev. Plant Sci. 15: 169–190.

Burel F., Baudry J., Butet A., Clergeau P., Delettre Y. and LeCoeur D. 1998. Comparative biodiversity along a gradient of agricultural landscapes. Acta Oecologica 19: 47–60.

Collier K.J., Smith B.J. and Baillie B.R. 1997. Summer light-trap catches of adult Trichoptera in hill-country catchments of contrasting land use, Waikato, New Zealand. New Zeal. J. Mar. Freshwater Res. 31: 623–634.

Courtney L.A. and Clements W.H. 1998. Effects of acidic pH on benthic macroinvertebrate communities in stream microcosms. Hydrobiologia 379: 135–145.

Elphick C.S. 2000. Functional equivalency between rice fields and seminatural wetland habitats. Conserv. Biol. 14: 181–191.

Endels P., Jacquemyn H., Brys R., Hermy M. and Blust G.D. 2002. Temporal changes (1986–1999) in populations of primrose (*Primula vulgaris* Huds.) in an agricultural landscape and implications for conservation. Biol. Conserv. 105: 11–25.

Enders G. and Wagner R. 1996. Mortality of *Apatania fimbriata* (Insecta: Trichoptera) during embryonic, larval and adult life stages. Freshwater Biol. 36: 93–104.

Fujioka M. and Lane S.J. 1997. The impact of changing irrigation practices in rice fields on frog populations of the Kanto Plain, central Japan. Ecol. Res. 12: 101–108.

Fujioka M. and Yoshida H. 2001. The potential and problems of agricultural ecosystems for birds in Japan. Global Environ. Res. 5: 151–162.

Fukamachi K., Oku H. and Nakashizuka T. 2001. The change of a satoyama landscape and its causality in Kamiseya, Kyoto Prefecture, Japan between 1970 and 1995. Landscape Ecol. 16: 703–717.

Harrison C. and Burgess J. 2000. Valuing nature in context: the contribution of common good approaches. Biodiv. Conserv. 9: 1115–1130.

Harrison S.S.C. and Harris I.T. 2002. The effects of bankside management on chalk stream invertebrate communities. Freshwater Biol. 47: 2233–2245.

Iwata T., Nakano S. and Inoue M. 2003. Impacts of past riparian deforestation on stream communities in a tropical rain forest in Borneo. Ecol. Appl. 13: 461–473.

The Japanese Firefly Society 1996. A Handbook on Fireflies. The Japanese Firefly Society, Tokyo, Japan. (in Japanese).

Kato M. 2001. 'SATOYAMA' and biodiversity conservation: 'SATOYAMA' as important insect habitats. Global Environ. Res. 5: 135–149.

Kobori H. and Primack R.B. 2003. Participatory conservation approaches for *satoyama*, the traditional forest and agricultural landscape of Japan. Ambio 32: 307–311.

Lane S.J. and Fujioka M. 1998. The impact of changes in irrigation practices on the distribution of foraging egrets and herons (Ardeidae) in the rice fields of central Japan. Biol. Conserv. 83: 221–230.

Moore R.H. 1990. Japanese Agriculture: Patterns of Rural Development. Westview Press, Boulder.

Ohba N. 1988. Japanese Insects Ser.12 The Genji-firefly. Bunichi Sogo Press, Tokyo, Japan. (in Japanese).

Ormerod S.J., Weatherley N.S. and Merrett W.J. 1990. The influence of conifer plantations on the distribution of the golden ringed dragonfly *Cordulegaster boltoni* (Odonata) in upland Wales. Biol. Conserv. 53: 241–251.

Pollard J.B. and Berrill M. 1992. The distribution of dragonfly nymphs across a pH gradient in south-central Ontario lakes. Can. J. Zool. 70: 878–885.

Primack R., Kobori H. and Mori S. 2000. Dragonfly pond restoration promotes conservation awareness in Japan. Conserv. Biol. 14: 1553–1554.

Samways M.J. and Steytler N.S. 1996. Dragonfly (Odonata) distribution patterns in urban and forest landscapes, and recommendations for riparian management. Biol. Conserv. 78: 279–288.
Shibue K., Ohba N. and Fujii E. 1995. Analysis of habitat factor which influence on population of *Luciola cruciata* at Nobi district in the Miura peninsula. J. Jpn. Inst. Landscape Architecture 58: 121–124. (In Japanese with English summary).
SPSS Inc. 1998. SYSTAT 8.0. SPSS Inc.,Chicago, Illinois.
Suh A.N. and Samways M.J. 2001. Development of a dragonfly awareness trail in an African botanical garden. Biol. Conserv. 100: 345–353.
Ward J.V. 1992. Aquatic Insect Ecology, 1. Biology and Habitat. John Wiley & Sons, Inc., New York.
Washitani I. 2001. Traditional sustainable ecosystem 'SATOYAMA' and biodiversity crisis in Japan: conservation ecological perspectives. Global Environ. Res. 5: 119–134.
Werneke U. and Zwick P. 1992. Mortality of the terrestrial adult and aquatic nymphal life stages of *Baetis vernus* and *Baetis rhodani* in the Breitenbach, Germany (Insecta: Ephemeroptera). Freshwater Biol. 28: 249–255.
Williams D.D. and Feltmate B.W. 1992. Aquatic Insects. CAB International, Oxon.
Yuma M. 2000. Firefly facts by Yuma. In: Water and Culture Research Group (ed.), Hotaru-DAS: Survey on Aquatic Fireflies in relation to the Nearby Freshwater by Residents of the Lake Biwa Region. Shin'yo-sha Ltd, Tokyo, Japan, pp. 36–44 (In Japanese with English Summary).

Biodiversity and Conservation (2006) 15:191–216
DOI 10.1007/s10531-004-6904-x 

# Diversity of Chrysomelidae (Coleoptera) in Galicia, Northwest Spain: estimating the completeness of the regional inventory

ANDRÉS BASELGA* and FRANCISCO NOVOA
*Departamento de Biología Animal, Facultad de Biología, Universidad de Santiago de Compostela, 15706 Santiago de Compostela, Spain; *Author for correspondence: Address: Museo Nac. de Ciencias Naturales, Dept. de Biodiversidad y Biología Evolutiva, c/José Gutiérrez Abascal, 2, 28006 Madrid, Spain (e-mail: baselga@usc.es)*

Received 10 June 2004; accepted in revised form 11 November 2004

**Key words:** Chrysomelidae, Cumulative models, Galicia, Iberian peninsula, Inventory, Species richness

**Abstract.** The diversity of Chrysomelidae (Coleoptera) in Galicia, Northwest Spain was examined. A long-term sampling was conducted during 1996–2001 and 267 species were collected, but including bibliographic citations a total of 276 species were recorded. As a result of this study the regional inventory has grown from 83 taxa cited before 1998 to the current 276 species. Species accumulation models were used to measure the inventory completeness and estimate the actual species richness of Chrysomelidae occurring in Galicia. Estimates were generated by analyzing both the rarefaction curve from the long-term sampling and the cumulative number of species recorded from Galicia since 1866. Values of total richness predicted by these different methods range between 290 and 323 species. Therefore, it seems that between 85 and 95% of the leaf beetle fauna was recorded and thus the inventory has reached an acceptable level of completeness.

## Introduction

The family Chrysomelidae is a highly diverse phytophagous group which represents an important proportion of the Coleoptera, and thus the whole diversity of terrestrial communities. Estimations for Iberian and world Coleoptera indicate that leaf beetles reach about 10% of species within the order (Martin-Piera and Lobo 2000). At local scale the same pattern was observed in Galicia, where Chrysomelidae represent about 15% of the recorded beetles in Natural Park of Fragas del Eume (Baselga and Novoa 2004), though this percentage should be slightly reduced if some incompletely studied families were added to the inventory. Therefore, leaf beetle inventories and reliable estimations of its completeness are interesting tools for assessing biodiversity patterns, and thus optimize the conservation effort.

With this purpose, many papers were focused on the need to determine the degree of completion of faunistic inventories and thus estimate the true species richness for a wide range of taxonomic groups. Two different methods can be applied to this problem. The use of randomized sample accumulation curves was firstly developed for standardized samplings (Soberón and Llorente 1993;

Colwell and Coddington 1994; Carlton and Robison 1998; Moreno and Halffter 2000; Summerville et al. 2001; Noguera et al. 2002) but also was applied to non-standardized data from museum collections or taxonomic databases (Soberón et al. 2000; Hortal et al. 2001, 2004; Petersen et al. 2003; Martín-Piera and Lobo 2003; Meier and Dikow 2004). A second way to determine de degree of completion of species inventories in a region is the growth over time of the cumulative species number as a function of the year of description (Medellín and Soberón 1999; Cabrero-Sañudo and Lobo 2003).

Knowledge about the Iberian leaf beetle diversity has been notably increased since last quarter of 20th century, with many papers dealing on faunistics of mountain areas (Daccordi and Petitpierre 1977; Petitpierre 1981, 1994, 1997; Petitpierre and Gómez-Zurita 1998; García-Ocejo et al. 1992; García-Ocejo and Gurrea 1995) or greater regions (Petitpierre 1980, 1983, 1988, 1999, 2000; Biondi 1991; Bastazo et al. 1993; Doguet et al. 1996) Most of these studies are focused on Mediterranean regions that occupy the greatest part of the Iberian peninsula, whereas Eurosiberian areas located in northern regions are still poorly known. This was the case of Galicia, northwest Spain (Figure 1), with only 83 species cited before 1998 (most of them recorded by López Seoane 1866; Heyden 1870; Chapman and Champion 1907; Iglesias 1928). Galicia is mostly located within Eurosiberian region, but some southeastern areas are included in Mediterranean region (Izco 1987; Rivas-Martínez 1987).

We carried out an extensive study of Chrysomelidae fauna from Galicia and some faunistic and taxonomic results were partially published (see Appendix A for references). The purpose of this paper is (i) to describe the regional species richness of this highly diverse group, and (ii) to estimate the completeness of

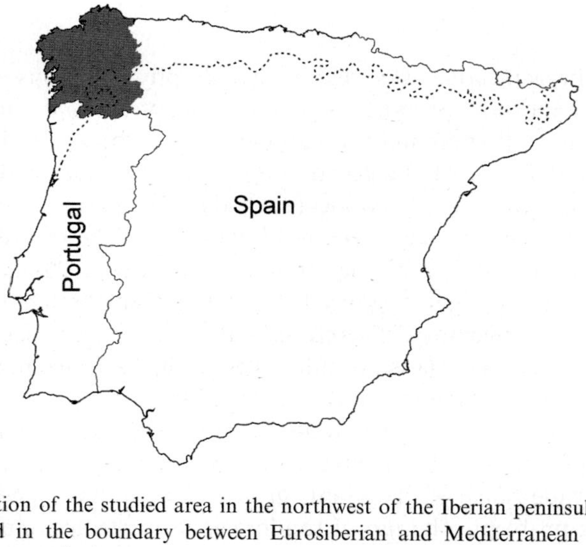


*Figure 1.* Location of the studied area in the northwest of the Iberian peninsula. Galicia (shadowed) is located in the boundary between Eurosiberian and Mediterranean phytogeographic regions (discontinuous line).

the inventory, assessing both the rarefaction curve generated from one long-term sampling (1996–2001) which was the first attempt to get a significant picture of the regional leaf beetle fauna, and the cumulative number of species recorded since the first citations in 1866.

## Materials and methods

Long-term sampling was carried out by Baselga, between 1996 and 2001, studying 191 localities in Galicia that were visited at least once (Figure 2). Among them, eight areas were selected and intensively sampled: coastal dunes and associated marshes (A), agricultural landscape near Santiago de Compostela (B), low altitude Atlantic mixed forest in Fragas del Eume Natural Park (C), medium altitude mountain ranges of Dorsal Gallega (D) and Larouco (E), the temperate valley of the Sil river (G) and high mountain ranges of Ancares (F) and Eixo-Segundera (H). Other localities visited during the field study are noted as X in Appendix A, whereas locations extracted from bibliographic sources are noted as Y. Areas E, G and H are located in the Mediterranean phytogeographic region, all the others in the Eurosiberian region (Izco 1987; Rivas-Martínez 1987). A total of 8602 specimens were collected. Another important source of material was the sampling carried out by G. Cerviño (1190 specimens) between 1991 and 1994. All this material is

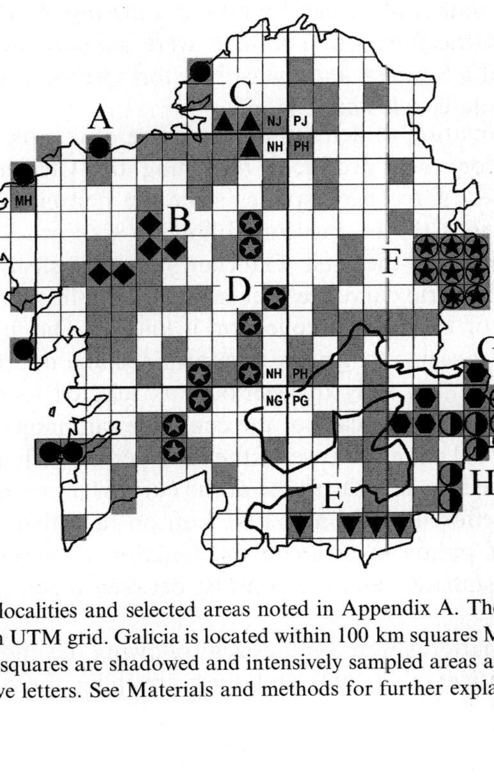


*Figure 2.* Sampling localities and selected areas noted in Appendix A. The squared pattern corresponds to the 10 km UTM grid. Galicia is located within 100 km squares MH, NG, NH, NJ, PG, PH and PJ. Sampled squares are shadowed and intensively sampled areas are noted with different symbols and respective letters. See Materials and methods for further explanation.

deposited in Baselga collection in the Departamento de Biología Animal, Universidad de Santiago de Compostela, Spain.

To assess the completeness of the inventory both asymptotic model and non-parametric estimators (ICE, Chao 2 and jackknife of first order) were used (Colwell and Coddington 1994). Estimations were produced by two different approaches: (i) the rarefaction curve from long-term sampling and (ii) the cumulative number of recorded species since the first citations (López Seoane 1866) to the present.

The rarefaction curve and the non-parametric estimators were generated with EstimateS 6.0 software (Colwell 2000), randomizing the sample order 100 times. Database records were used as a sampling-effort surrogate (Soberón et al. 2000; Hortal et al. 2001; Martin-Piera and Lobo 2003). Our database comprised 3332 records for 10098 specimens and includes the results of the collections by G. Cerviño (1991–1994) and Baselga (1996–2001), along with some other specimens collected by Novoa and collaborators since 1973. Each record is comprised of the following fields: species name, locality, date, host plant, number of specimens and collector. Any difference in any database field value give rise to a new database record, so increments of the number of records provide correlative increments of the sampling effort (Martin-Piera and Lobo 2003). Thereafter, the asymptotic Clench function was fitted to the smoothed curve (Soberón and Llorente 1993; Hortal et al. 2004):

$$S_{(e)} = ae/(1 + be)$$

where $S_{(e)}$ is the number of species found per sampling-effort unit ($e$); $a$ and $b$, the parameters of the function. The later were adjusted to the data of each curve by means of a Simplex and Quasi Newton method (StatSoft 2001). The predicted asymptote is calculated as $a/b$.

The second estimation to determine the degree of completeness of Galician leaf beetle inventory was produced by fitting the Clench function to the cumulative number of recorded species since the first citations in the region (López Seoane 1866) to the present, following Cabrero-Sañudo and Lobo (2003) but considering the first Galician record instead of the year of description. This historic curve was generated taking into account the year of published papers or the year of collection if known. The final section (1991–2001) of this curve seems to get an asymptotic shape due to the increase in sampling effort and thus only this period was selected to adjust the Clench function. We considered the case of the complete sampling with two different collectors (1991–2001) and the case of the final period with an single collector (1996–2001). In the first case the shape of the curve is more irregular due to the lack of sampling effort in 1994 and 1995 but, on the other hand, it is used a higher number of points to estimate the function than in the second case. Therefore, both estimates (and the contrast between them) are interesting for different reasons.

Biogeographic patterns were synthesized following the chorotypes proposed by Vigna Taglianti et al. (1992) and then grouping them into four major categories: Iberian elements (Ibe), for species endemic from the Iberian Peninsula; Mediterranean elements (Med), for species widespread in the Mediterranean countries; Eurosiberian elements (Eur), for species widespread in Europe, or Europe and the Siberian range; and finally the wide range elements (WR), for species widespread in all or a great part of the Palaearctic region, and reaching parts of both Eurosiberian and Mediterranean areas. These major divisions are established in order to make clear the contrast between Eurosiberian (septentrional) and Mediterranean (meridional) contributions to Galician fauna which is located across the Eurosiberian–Mediterranean boundary. The other two categories are neutral regarding this aspect, because almost all Iberian elements are present in both sides of Eurosiberian–Mediterranean limit, as well as WR elements reach both regions.

**Results**

A total of 276 species have been recorded from Galicia, including the bibliographic citations (Appendix A) and 267 of them were recorded during the field study. The complete inventory (276 taxa) represents between 34% (Petitpierre 2000) and 44% (Vela and Bastazo 1999) of the total Iberian Chrysomelidae diversity, since there are two different estimations of actual number of Iberian leaf beetles. The Galician fauna of leaf beetles is comprised of about the same proportion of Eurosiberian (Eur: 19.2%) and Mediterranean (Med: 19.6%) elements. Species of wide range (WR) reach near half of the fauna (45.7%) and Iberian endemisms (Ibe) represent the 15.2%.

Two new species were described from Galicia: *Aphthona sandrae* (Baselga and Novoa 2002a) and *Psylliodes cervinoi* (Baselga and Novoa 2003) which are only known from their type localities. Eight taxa were recorded from the Iberian peninsula for the first time: *Oulema erichsonii* (Suffrian), *Phyllotreta exclamationis* (Thunberg), *Phyllotreta ganglbaueri* Heikertinger, *Longitarsus australis* (Mulsant and Rey), *Longitarsus fulgens* (Foudras), *Chaetocnema confusa* (Boheman), *Psylliodes vindobonensis* Heikertinger and *Cassida subreticulata* Suffrian (Baselga and Novoa 1998, 1999a, b, 2000c, 2001a, b, 2002a, b). Taking into account present new records (22 species) and previous papers, 193 species (69.9% of known richness) have been newly recorded for Galicia since 1998.

The species accumulation curve (Figure 3) generated from the field study (excluding bibliographic records) nearly reaches the asymptote (292) predicted by the Clench function. Moreover, the observed richness ($S = 267$) is not far from the non-parametric estimators ICE (297), Chao 2 (300) and Jackknife 1 (311). Therefore between 85.9 and 91.4% of the estimated number of species living in Galicia seems to be detected in our field study (Table 1), but we should expect that between 24 and 44 taxa will be added to the inventory in the future.

On the other hand, the cumulative number of species recorded per year (Figure 4a) shows a slow growth since 1866 to 1990 and thereafter a great rise

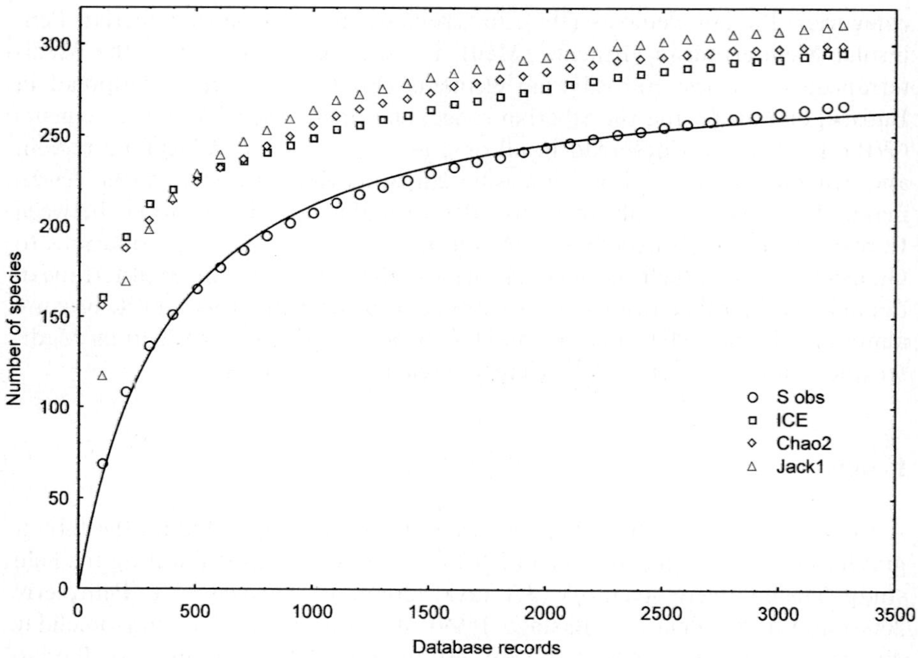


*Figure 3.* Species accumulation curve generated from the field study with the fitted Clench function and the non-parametric estimators ICE, Chao2 and first-order jackknife.

*Table 1.* Number of species recorded (*S* obs), estimates and percentage of the estimated value recorded for the long-term sampling, the whole historic inventory and the citations published before 1998.

| Method | *S* obs | Estimate | % |
|---|---|---|---|
| *Rarefaction curve* | | | |
| Clench asymptote | 267 | 292 | 91.4 |
| ICE | 267 | 297 | 89.9 |
| Chao 2 | 267 | 300 | 89.0 |
| Jackknife 1 | 267 | 311 | 85.9 |
| *Historic curves* | | | |
| Clench for 1991–2001 | 276 | 322 | 85.7 |
| Clench for 1996–2001 | 276 | 290 | 95.2 |
| Clench for citations published before 1998 | 83 | 128 | 64.8 |

in the rate of addition of new records due to the significant increase of sampling effort since 1991 to 2001. In this final section the historic curve becomes asymptotic. The fitted Clench function estimates an asymptote of 322 species for the curve between 1991 and 2001 (Figure 4b), and an asymptote of 290 species if the function is adjusted to the section between 1996 and 2001

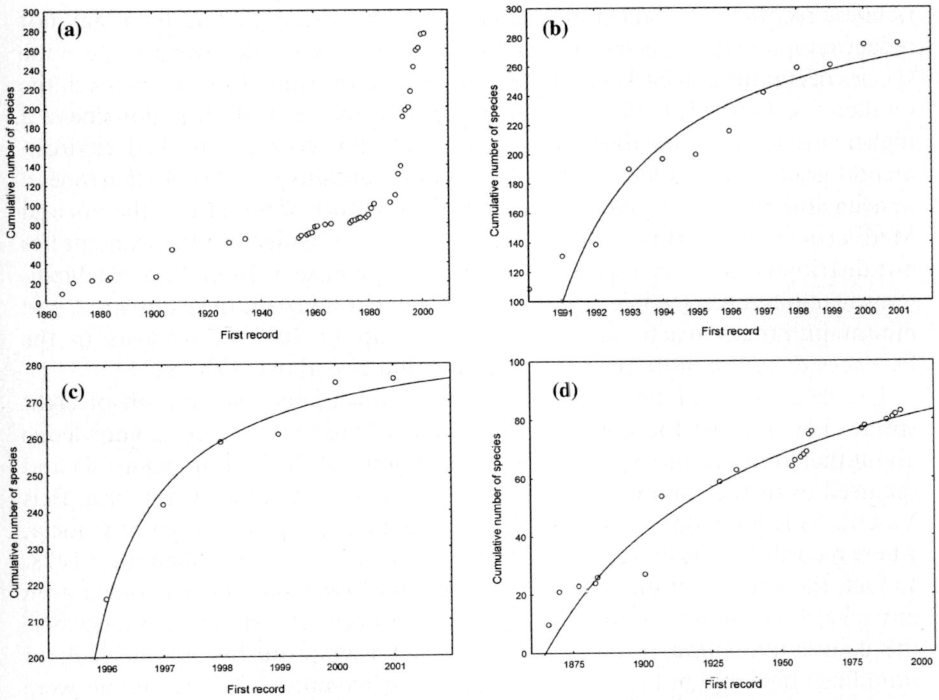


*Figure 4.* Cumulative number of species recorded per year. (a) Historic curve (1866–2001). (b) Fitted Clench function considering the final period with two collectors (1991–2001). (c) Fitted Clench function considering the final period with a single collector (1996–2001). (d) Cumulative number of citations published before 1998. The curve reaches an asymptote due to the lack of sampling effort and the unrealistic estimation is 128 species.

(Figure 4c). Since the recorded species richness is 276, the estimated completeness of the inventory reaches about 85.7–95.2% (Table 1).

## Discussion

The current state of the inventory is considered reasonably complete, because the different estimation methods agree in their values. The Clench function and the non-parametric estimators computed from the sampling rarefaction curve, as well as the historic curve of the cumulative number of recorded species predict a percentage of undetected species between 5 and 15% (14–44 species). The fauna of Chrysomelidae from Galicia (29574 km$^2$) comprises 276 recorded species, but the estimated total richness reaches between 290 and 323. If these values are compared with those of other Iberian regions, Galician diversity is quite similar to the 339 species cited from Aragón (47,6974 km$^2$), though this number might be an underestimation (Vives 2000), but it is far from the 384 taxa recorded from Cataluña (32091 km$^2$) (Petitpierre 1994). The reduced species richness recorded in Galicia in comparison with Aragón or Cataluña may not reflect sampling deficiencies but real differences in leaf beetle diversity, since the species accumulation curves and the non-parametric estimators nearly reach the predicted values (85.9–95.1% of the estimated richness). Both regions have a higher species richness than Galicia probably due to their marked environmental gradients between the high Pyrenees mountains and the Mediterranean or even arid plains and valleys, that allow the presence of both Eurosiberian and Mediterranean elements, as well as the presence of endemic Pyrenean species not distributed in other regions of the Iberian peninsula. In Galicia the Mediterranean phytogeographic region is restricted to a small southeastern area, and mountain ranges reach maximum altitudes up to 2000 m, whereas in the Pyrenees there are more than 200 summits that rise above 3000 m.

The description of two new species from Galicia and the citation of eight species new for the Iberian peninsula, indicate the previous poor knowledge about the faunas occurring in the northern regions of the Iberian peninsula and the need of further sampling effort in such areas (i.e. Asturias, Cantabria, País Vasco). This poor state of knowledge was particularly pronounced in Galicia, where records for the two thirds of the known species were published since 1998. In fact, the scarcity of published records before 1998 would have avoided even any reliable estimation of the number of species occurring in the region, because the cumulative curve (Figure 4d) reaches an asymptote due to the lack of sampling effort and not to the saturation of the inventory. Only 83 species were cited before 1998 and using these data the estimation generated by the Clench function is 128 species. This fact shows the importance of publishing faunistic inventories, which are the base for research on biodiversity and biogeography.

In summary, the study presented herein points out the need of extensive field work to collect highly diverse groups in still poorly surveyed territories, as well how this knowledge from non-standardized but extensive samplings can be used to get estimates of actual species richness. In the case of Galician leaf beetles, the long-term sampling has showed the previous poor knowledge of Iberian Eurosiberian-type faunas with the discovery of new taxa and the addition of many new records to the Galician and Iberian inventories. Also, the use of different methods in assessing the completeness of the inventory allows us to consider that estimates are accurate, because all non-parametric estimators, sample-based and historic curves predict values of total species richness within a narrow range. The recorded richness represent about 85–95% of the predicted values. Thus, the inventory seems to be reasonably complete and therefore valuable to describe biodiversity patterns of Iberian and European Chrysomelidae.

**Acknowledgements**

We are grateful to E. Petitpierre (Universitat de les Illes Balears, Palma de Mallorca) for valuable help on many taxonomic and bibliographic topics, G. Cerviño (A. Coruña) for the loan of his interesting collection of

Chrysomelidae and S. Barallobre (Santiago de Compostela) for her collaboration in different expeditions to collect beetles. We also thank J.M. Lobo, J. Hortal and A. Jiménez-Valverde (Museo Nacional de Ciencias Naturales, Madrid) for helpful advice on statistical methods and valuable comments that greatly improved this paper. This study was supported by Xunta de Galicia with projects PGIDT01PXI20004PR and PGIDT01MAM20001PR, as well as Ph.D. and postdoctoral grants to A.B.

## Appendix A

See Materials and methods section for explanation of terms.

| Species | First record | Published records | Main areas | Dist. |
|---|---|---|---|---|
| *Donacia reticulata* (Gyllenhal) | 1866 | López Seoane (1866) | Y | WR |
| *Donacia galaica* Báguena | 1959 | Báguena (1959), Petitpierre (2000) and Baselga and Novoa (2002b) | B, H, X, Y | Ibe |
| *Donacia marginata* Hoppe | 1960 | Báguena (1960a) and Baselga and Novoa (1999a, 2000a) | A, D, G, X, Y | WR |
| *Donacia bicolora* Zschach | 1960 | Báguena (1960a) | X, Y | WR |
| *Donacia vulgaris* Zschach | 1992 | New record | X | WR |
| *Donacia simplex* Fabricius | 1960 | Báguena (1960a) | Y | WR |
| *Plateumaris sericea* (Linné) | 1960 | Báguena (1960a) and Baselga and Novoa (2000a, b, 2002b) | A, B, C, D, E, F, H, X, Y | WR |
| *Oulema erichsonii* (Suffrian) | 1990 | Baselga and Novoa (1999b, 2000b, 2002b) | E, F, G, X | Eur |
| *Oulema gallaeciana* (Heyden) | 1870 | Heyden (1870), Iglesias (1928), Vives and González (1994) and Baselga and Novoa (1999a, 2000a, b, 2002b) | A, B, C, D, E, F, H, X, Y | Eur |
| *Oulema hoffmannseggi* (Lacordaire) | 1993 | Baselga and Novoa (2000c) | X | Med |
| *Oulema melanopus* (Linné) | 1993 | Baselga and Novoa (1999a, 2000a, b) | A, B, C, D, E, F, X | WR |
| *Oulema duftschmidi* (Redtenbacher) | 1993 | Baselga and Novoa (2000a, b, 2002b) | A, B, C, D, E, F, H, X | WR |
| *Oulema rufocyanea* (Suffrian) | 2000 | Baselga and Novoa (2002b) | H | Eur |

*Appendix A* Continued.

| Species | First record | Published records | Main areas | Dist. |
|---|---|---|---|---|
| *Crioceris paracenthesis* (Linné) | 1997 | Baselga and Novoa (2000a) | A | Med |
| *Crioceris asparagi* (Linné) | 1928 | Iglesias (1928), González de Andrés (1934) and Baselga and Novoa (1999a, 2000a) | A, X, Y | WR |
| *Lilioceris lilii* (Scopoli) ssp. *laeviuscula* (Weise) | 1907 | Chapman and Champion (1907) and Baselga and Novoa (2002b) | X, Y | WR |
| *Labidostomis lusitanica* (Germar) | 1907 | Chapman and Champion (1907), Iglesias (1928), Baselga and Novoa (1999a, 2000b, 2002b) and Petitpierre (2000) | E, F, G, X, Y | Med |
| *Labidostomis taxicornis* (Fabricius) | 1974 | Baselga and Novoa (2001a) | X | Med |
| *Lachnaia cylindrica* (Lacordaire) | 1998 | Baselga and Novoa (2000b) | F | Med |
| *Lachnaia hirta* (Fabricius) | 1998 | Baselga and Novoa (1999a, 2000b, 2002b) | D, F, G, X | Med |
| *Lachnaia pubescens* (Dufour) | 1907 | Chapman and Champion (1907), Iglesias (1928) and Baselga and Novoa (1999a, 2000b, 2002b) | E, F, G, H, X, Y | Med |
| *Lachnaia tristigma* (Lacordaire) | 1907 | Chapman and Champion (1907) and Baselga and Novoa (2002b) | D, E, X, Y | Med |
| *Tituboea biguttata* (Olivier) | 1994 | Baselga and Novoa (2002b) | G | Med |
| *Tituboea sexmaculata* (Fabricius) | 2000 | New record | G | Med |
| *Clytra atraphaxidis* Pallas | 1907 | Chapman and Champion (1907) and Baselga and Novoa (2002b) | Y | WR |
| *Clytra quadripunctata* (Linné) ssp. *quadripunctata* (Linné) | 1998 | Baselga and Novoa (2000b) | F | Eur |
| *Clytra espanoli* Daccordi and Petitpierre | 1907 | Chapman and Champion (1907), Iglesias (1928) and Baselga and Novoa (1999a, 2000b, 2002b) | B, C, F, H, X, Y | Ibe |
| *Smaragdina concolor* (Fabricius) | 1907 | Chapman and Champion (1907) and Baselga and Novoa (1999a, 2000b, 2002b) | B, C, D, E, F, G, H, X, Y | Med |
| *Smaragdina reyi* (Brisout) | 1960 | Báguena (1960b), Cobos (1969), Baselga and Novoa (2000b, 2002b) and Petitpierre (2000) | B, C, D, E, F, G, H, X, Y | Ibe |

*Appendix A* Continued.

| Species | First record | Published records | Main areas | Dist. |
|---|---|---|---|---|
| *Coptocephala brevicornis* (Lefévre) | 1960 | Báguena (1960c) and Baselga and Novoa (1999a, 2000a, 2002b) | A, G, X, Y | Ibe |
| *Coptocephala scopolina* (Linné) | 1981 | Baselga and Novoa (2000a, 2002b) | A, G, H, X | Med |
| *Stylosomus ilicicola* Suffrian | 1961 | Codina Padilla (1961a) | Y | Med |
| *Stylosomus rugithorax* Abeille | 1907 | Baselga and Novoa (2000b, 2002b) | B, C, D, E, F, H, X | Ibe |
| *Pachybrachis pteromelas* Graëlls | 2000 | New record | G | Ibe |
| *Pachybrachis hippophaes* Suffrian | 1907 | Chapman and Champion (1907) and Baselga and Novoa (2002b) | Y | Eur |
| *Pachybrachis azureus* Suffrian | 1907 | Chapman and Champion (1907) and Baselga and Novoa (1999b, 2000b, 2002b) | D, F, G, H, X, Y | Med |
| *Cryptocephalus excisus* Seidlitz | 1998 | Baselga and Novoa (2000b) | F | Ibe |
| *Cryptocephalus lusitanicus* Suffrian | 1907 | Chapman and Champion (1907) and Baselga and Novoa (2000b, 2002b) | E, F, G, X, Y | Ibe |
| *Cryptocephalus pominorum* Burlini | 1907 | Chapman and Champion (1907), Bourdonné (1994) and Baselga and Novoa (2000b, 2002b) | F, G, X, Y | Ibe |
| *Cryptocephalus pexicollis* Suffrian | 1955 | Burlini (1955) | Y | Med |
| *Cryptocephalus obliteratifer* Pic | 1907 | Chapman and Champion (1907) and Baselga and Novoa (2002b) | G, Y | Med |
| *Cryptocephalus bipunctatus* (Linné) | 1980 | Chapman and Champion (1907) and Baselga and Novoa (2000b, 2002b) | C, F, G, X, Y | Eur |
| *Cryptocephalus rugicollis* Olivier | 2000 | Baselga and Novoa (2002b) | G | Med |
| *Cryptocephalus aureolus* Suffrian | 1982 | Baselga and Novoa (1999a, 2000a, b, d, 2002b) | A, B, C, D, F, X | Eur |
| *Cryptocephalus globicollis* Suffrian | 1907 | Chapman and Champion (1907) and Baselga and Novoa (2002b) | X, Y | Med |
| *Cryptocephalus cantabricus* Franz | 1958 | Franz (1958) and Baselga and Novoa (2000b, d, 2002b) | B, C, E, F, G, H, X, Y | Ibe |
| *Cryptocephalus violaceus* Laicharting | 1907 | Chapman and Champion (1907), Petitpierre (2000) and Baselga and Novoa (2002b) | Y | Eur |

*Appendix A* Continued.

| Species | First record | Published records | Main areas | Dist. |
|---|---|---|---|---|
| *Cryptocephalus tibialis* Brisout | 1955 | Burlini (1955) and Baselga and Novoa (1999a, 2000b, 2002b) | E, F, G, H, X, Y | Eur |
| *Cryptocephalus parvulus* Müller | 2000 | New record | X | WR |
| *Cryptocephalus androgyne* Marseul ssp. *pelleti* Marseul | 1993 | New record | C, X | Eur |
| *Cryptocephalus cynarae* Suffrian | 1866 | López Seoane (1866), Chapman and Champion (1907) and Baselga and Novoa (1999a, 2002b) | D, G, X, Y | Ibe |
| *Cryptocephalus moraei* (Linné) | 1907 | Chapman and Champion (1907) and Baselga and Novoa (2002b) | B, D, X, Y | Eur |
| *Cryptocephalus crassus* Olivier | 1907 | Chapman and Champion (1907) and Baselga and Novoa (2002b) | G, Y | Med |
| *Cryptocephalus octoguttatus* (Linné) | 1992 | Baselga and Novoa (1999a, 2000b, 2002b) | C, D, E, F, G, H, X | Med |
| *Cryptocephalus vittatus* Fabricius | 1870 | Heyden (1870), Chapman and Champion (1907), Iglesias (1928), Burlini (1955) and Baselga and Novoa (1999a, 2000a, b, 2002b) | A, B, C, D, E, F, G, X, Y | WR |
| *Cryptocephalus celtibericus* Suffrian | 1877 | López Seoane (1877) and Baselga and Novoa (2000b, 2002b) | F, G, Y | Med |
| *Cryptocephalus bilineatus* (Linné) | 1870 | Heyden (1870), Plaza Infante (1979) and Baselga and Novoa (2000b, 2002b) | C, F, H, X, Y | WR |
| *Cryptocephalus mystacatus* Suffrian | 1907 | Chapman and Champion (1907) and Baselga and Novoa (2000b, 2002b) | D, E, F, G, H, X, Y | Ibe |
| *Cryptocephalus labiatus* (Linné) | 1991 | Baselga and Novoa (1999b, 2000b) | C, F, X | Eur |
| *Cryptocephalus pygmaeus* Fabricius | 1907 | Chapman and Champion (1907) and Baselga and Novoa (2000a, b, 2002b) | A, G, X, Y | WR |
| *Cryptocephalus fulvus* (Goeze) | 1994 | Baselga and Novoa (2000a, 2002b) | A, G, H, X | WR |
| *Cryptocephalus macellus* Suffrian | 1979 | Plaza Infante (1979) | Y | Eur |
| *Cryptocephalus pusillus* Fabricius | 1991 | New record | C, X | Eur |
| *Cryptocephalus rufipes* (Goeze) | 1991 | Baselga and Novoa (2000a, b, 2002b) | A, B, C, D, E, F, H, X | WR |

*Appendix A* Continued.

| Species | First record | Published records | Main areas | Dist. |
|---|---|---|---|---|
| *Oomorphus concolor* (Sturm) | 1998 | Baselga and Novoa (1999b, 2000b) | C, F | Eur |
| *Bromius obscurus* (Linné) | 2000 | Baselga and Novoa (2001a, 2002b) | H | WR |
| *Timarcha calceata* Pérez Arcas | 1975 | Baselga and Novoa (1999a, 2000b) | F, X | Ibe |
| *Timarcha geniculata* (Germar) | 1977 | Baselga and Novoa (2002b) | F, H, Y | Ibe |
| *Timarcha asturiensis* Kraatz | 1883 | Marseul (1883), Bechyné (1948) and Baselga and Novoa (2000a) | A, Y | Ibe |
| *Timarcha chloropus* (Germar) | 1884 | Fairmaire (1884), Bechyné (1948), Vives and González (1998) and Baselga and Novoa (1999a, 2000a) | A, D, X, Y | Ibe |
| *Timarcha gougeleti* Fairmaire | 1877 | López Seoane (1877), Fairmaire (1884), Bechyné (1948), Vives and González (1998) and Baselga and Novoa (1999a, 2000a) | A, B, C, X, Y | Ibe |
| *Timarcha trapezicollis* Fairmaire | 1884 | Fairmaire (1884) | Y | Ibe |
| *Leptinotarsa decemlineata* Say | 1993 | Baselga and Novoa (1999a, 2000a, 2002b) | A, B, H, X | Intr. |
| *Chrysolina herbacea* (Duftschmid) | 1866 | Baselga and Novoa (1999a, 2000a, b, 2002b) | A, B, D, E, F, G, H, X | WR |
| *Chrysolina fastuosa* (Scopoli) | 1996 | Baselga and Novoa (1999a, 2002b) | B, H, X | WR |
| *Chrysolina polita* (Linné) | 1982 | Baselga and Novoa (1999a, 2000a, b) | A, D, F, X | Eur |
| *Chrysolina bankii* (Fabricius) | 1928 | Iglesias (1928) and Baselga and Novoa (1999a, 2000a, b) | A, B, C, D, F, X, Y | Med |
| *Chrysolina americana* (Linné) | 1907 | Chapman and Champion (1907), Iglesias (1928) and Baselga and Novoa (1999a, 2002b) | G, X, Y | Med |
| *Chrysolina rufoaenea* (Suffrian) | 1991 | New record | D, E, X | Eur |
| *Chrysolina haemoptera* (Linné) | 1866 | López Seoane (1866), Iglesias (1928) and Baselga and Novoa (1999a, 2000a, b, 2002b) | A, C, D, H, X, Y | Eur |
| *Chrysolina mactata* (Fairmaire) | 1866 | López Seoane (1866), López Seoane (1877) and Baselga and Novoa (1998, 1999a, 2000b) | C. F, X, Y | Ibe |

*Appendix A* Continued.

| Species | First record | Published records | Main areas | Dist. |
|---|---|---|---|---|
| *Chrysolina gypsophilae* (Küster) | 1928 | Baselga and Novoa (1999a, 2000a) | A, G | WR |
| *Chrysolina latecincta* (Demaison) ssp. *decipiens* (Franz) | 1988 | Petitpierre (1988) and Baselga and Novoa (2002b) | X, Y | Eur |
| *Chrysolina affinis* (Fabricius) ssp. *rufofemorata* (Heyden) | 1973 | New record | X | Med |
| *Chrysolina grossa* (Fabricius) ssp. *chloromaura* (Olivier) | 1866 | López Seoane (1866), Iglesias (1928), Vives and González (1998) and Baselga and Novoa (1999a, 2000a) | A, D, X, Y | Med |
| *Chrysolina lucida* (Olivier) ssp. *torresi* Bechyné | 1991 | Baselga and Novoa (1999a, 2000a) | A, X | Eur |
| *Chrysolina diluta* (Germar) | 1907 | Chapman and Champion (1907), Codina Padilla (1961b) and Baselga and Novoa (1999a, 2000a, 2002b) | A, X, Y | Ibe |
| *Chrysolina varians* (Schaller) | 1907 | Baselga and Novoa (2000b) | F | Eur |
| *Chrysolina brunsvicensis* (Gravenhorst) | 1996 | Baselga and Novoa (1999b, 2000b) | B, D, F, X | Eur |
| *Chrysolina quadrigemina* (Suffrian) | 1991 | Baselga and Novoa (2000a, b, 2002b) | A, B, F, G, X | WR |
| *Chrysolina interstincta* (Suffrian) ssp. *graellsi* (Pérez Arcas) | 1992 | Baselga and Novoa (2000c, 2002b) | B, E, H, X | Med |
| *Oreina alpestris* (Schummel) | 1981 | Baselga and Novoa (2000b, c) | F, X | Eur |
| *Oreina ganglbaueri* (Jakob) | 1998 | New record | F | Ibe |
| *Gastrophysa polygoni* (Linné) | 1990 | Baselga and Novoa (2000a) | A, B, G, X | WR |
| *Gastrophysa unicolor* (Marsham) | 1866 | López Seoane (1866) and Baselga and Novoa (1999a, 2000a, b, 2002b) | A, B, C, D, E, F, G, H, X, Y | Ibe |
| *Colaspidema atrum* (Olivier) | 1934 | González de Andrés (1934) | Y | WR |
| *Colaspidema dufouri* (Pérez Arcas) | 1993 | Baselga and Novoa (2000c) | X | Ibe |
| *Phaedon tumidulus* (Germar) | 1901 | Bedel (1901) and Baselga and Novoa (2000a, b) | A, F | WR |
| *Phaedon cochleariae* (Fabricius) | 1999 | Baselga and Novoa (2000a, 2002b) | A, H | WR |
| *Phaedon armoraciae* (Linné) | 1998 | New record | E, X | WR |

*Appendix A* Continued.

| Species | First record | Published records | Main areas | Dist. |
|---|---|---|---|---|
| *Cyrtonus corruscans* Vuillefroy | 1994 | New record | X | Ibe |
| *Cyrtonus cupreovirens* Pérez Arcas | 1998 | New record | D | Ibe |
| *Cyrtonus franzi* Cobos | 1954 | Cobos 1954 | Y | Ibe |
| *Hydrothassa glabra* (Herbst) | 1993 | Baselga and Novoa (2000a, b, 2002b) | A, B, D, F, H, X | Eur |
| *Hydrothassa fairmairei* (Brisout) | 1996 | Baselga and Novoa (2000a, b) | A, B, C, D, F, X | Ibe |
| *Prasocuris junci* (Brahm) | 1982 | Baselga and Novoa (2000a, b) | A, B, F, G, X | WR |
| *Prasocuris phellandrii* (Linné) | 2001 | New record | X | WR |
| *Plagiodera versicolora* (Laicharting) | 1907 | Chapman and Champion (1907), González de Andrés (1934), Vives and González (1998) and Baselga and Novoa (2000b, 2002b) | B, C, D, E, F, H, X, Y | WR |
| *Plagiosterna aenea* (Linné) | 1991 | Baselga and Novoa (1999a, 2000b) | B, F, G, X | WR |
| *Chrysomela populi* Linné | 1928 | Iglesias (1928), González de Andrés (1934) and Baselga and Novoa (1999a, 2000b) | B, F, X, Y | WR |
| *Gonioctena aegrota* Fabricius | 2000 | New record | G | Med |
| *Gonioctena leprieuri* (Pic) | 1957 | Bechyné (1957) and Baselga and Novoa (2002b) | F, G, H | Ibe |
| *Gonioctena olivacea* (Forster) | 1991 | Baselga and Novoa (1999a, 2000a, b, 2002b) | A, B, C, D, E, F, G, H, X | WR |
| *Phratora vulgatissima* (Linné) | 1991 | Baselga and Novoa (1999b) | B, C, D, X | WR |
| *Phratora tibialis* (Suffrian) | 1988 | Vives and González (1988) | Y | Eur |
| *Phratora laticollis* (Suffrian) | 1998 | Baselga and Novoa (2000b) | F | WR |
| *Phratora vitellinae* (Linné) | 1998 | Baselga and Novoa (2000b) | F | Eur |
| *Entomoscelis adonidis* (Pallas) | 1974 | New record | X | WR |
| *Galerucella lineola* (Fabricius) | 1993 | New record | B, C, D, E, X | WR |
| *Galerucella calmariensis* (Linné) | 1993 | Baselga and Novoa (2000a) | A, B, G, X | WR |

*Appendix A* Continued.

| Species | First record | Published records | Main areas | Dist. |
|---|---|---|---|---|
| *Xanthogaleruca luteola* (Muller) | 1998 | Baselga and Novoa (1999a, 2000b) | F, G, X | WR |
| *Galeruca macchoi* (Joannis) | 1981 | Baselga and Novoa (2000b, 2002b) | C, F, H, X | Ibe |
| *Galeruca luctuosa* (Joannis) | 1934 | Laboissière (1934) and Baselga and Novoa (1999a, 2002b) | B, G, X | Eur |
| *Galeruca angusta* Küster | 1995 | Baselga and Novoa (1999a, 2000a, c) | A, X | Med |
| *Lochmaea scutellata* (Chevrolat) | 1934 | Laboissière (1934) and Baselga and Novoa (1999b, 2000b, 2002b) | B, C, E, F, H, X | Ibe |
| *Lochmaea suturalis* (Thomson) | 1995 | Baselga and Novoa (1999b, 2000a, b, 2002b) | A, C, D, E, F, H, X | Eur |
| *Exosoma lusitanica* (Linné) | 1907 | Chapman and Champion (1907), Iglesias (1928) and Baselga and Novoa (1999a, 2000a, b, 2002b) | A, B, C, E, F, G, H, X, Y | WR |
| *Calomicrus circumfusus* (Marsham) | 1866 | López Seoane (1866), Chapman and Champion (1907) and Baselga and Novoa (2000b, 2002b) | B, C, D, E, F, G, H, X, Y | Eur |
| *Calomicrus suturalis* (Joannis) | 1998 | Baselga and Novoa (2000b, 2002b) | F, G | Eur |
| *Luperus flavus* Rosenhauer | 1998 | Baselga and Novoa (2000b, 2002b) | E, F, G, H | Ibe |
| *Luperus sulphuripes* Graëlls | 1866 | López Seoane (1866) and Baselga and Novoa (1999a, 2000b, 2002b) | B, C, D, E, F, G, H, X, Y | Ibe |
| *Agelastica alni* (Linné) | 1928 | Iglesias (1928) and Baselga and Novoa 1999a, 2000b, 2002b | B, E, F, G, X, Y | WR |
| *Sermylassa halensis* (Linné) | 1998 | Baselga and Novoa (2000b) | F, X | Eur |
| *Leptomona erythrocephala* (Olivier) | 1907 | Heyden (1870) and Baselga and Novoa (2000a, b) | A, B, C, D, F, G, X, Y | Med |
| *Phyllotreta variipennis* (Boieldieu) | 1993 | Baselga and Novoa (2001a) | B, G | Med |
| *Phyllotreta vittula* (Redtenbacher) | 1993 | Baselga and Novoa (1999b) | B, X | WR |
| *Phyllotreta nemorum* (Linné) | 1993 | Baselga and Novoa (2000b) | B, C, D, E, F, X | WR |
| *Phyllotreta undulata* Kutschera | 1997 | Baselga and Novoa (2000a, b) | A, B, C, D, F, X | WR |

*Appendix A* Continued.

| Species | First record | Published records | Main areas | Dist. |
|---|---|---|---|---|
| *Phyllotreta tetrastigma* (Comolli) | 1996 | Baselga and Novoa (1998, 2000a, b, 2002b) | A, B, C, D, E, F, H | WR |
| *Phyllotreta striolata* (Fabricius) | 1870 | Heyden (1870), González de Andrés (1934) and Baselga and Novoa (2000a, b) | A, B, D, F, X, Y | WR |
| *Phyllotreta ochripes* (Curtis) | 1993 | Baselga and Novoa (2000c) | G | Eur |
| *Phyllotreta exclamationis* (Thunberg) | 1997 | Baselga and Novoa (1998, 2000a, b) | A, B, C, D, E, F, X | Eur |
| *Phyllotreta atra* (Fabricius) | 1870 | Heyden (1870) and Baselga and Novoa (2000a, 2002b) | A, B, C, D, G, H, X, Y | WR |
| *Phyllotreta cruciferae* (Goeze) | 1870 | Heyden (1870), González de Andrés (1934) and Baselga and Novoa (2000a) | A, B, E, X, Y | WR |
| *Phyllotreta foudrasi* Brisout | 1993 | New record | E, G | Med |
| *Phyllotreta temperei* Doguet | 1993 | Baselga and Novoa (2001a) | X | Ibe |
| *Phyllotreta consobrina* (Curtis) | 1993 | Baselga and Novoa (1999a, 2000a, b, 2002b) | A, B, C, D, E, F, G, X | WR |
| *Phyllotreta nigripes* (Fabricius) | 1993 | Baselga and Novoa (2000b, 2002b) | B, C, D, F, H, X | WR |
| *Phyllotreta ganglbaueri* Heikertinger | 2000 | Baselga and Novoa (2001b) | X | Eur |
| *Phyllotreta procera* (Redtenbacher) | 1993 | Baselga and Novoa (2000a) | A, C, G, X | WR |
| *Aphthona lutescens* (Gyllenhal) | 1996 | Baselga and Novoa (2000a) | A, B, D, X | WR |
| *Aphthona nigriceps* (Redtenbacher) | 1997 | Baselga and Novoa (2000a) | A, X | WR |
| *Aphthona punctiventris* Mulsant and Rey | 1993 | Baselga and Novoa (1998, 2000a) | A, C, G, X | Med |
| *Aphthona occitana* Doguet | 1997 | Baselga and Novoa (1999b) | B | Ibe |
| *Aphthona atrocaerulea* (Stephens) | 1996 | Baselga and Novoa (2000a, 2002b) | A, E, H | WR |
| *Aphthona melancholica* Weise | 1993 | Baselga and Novoa (1999b, 2000b, c) | C, F, X | Ibe |
| *Aphthona sandrae* Baselga and Novoa | 1994 | Baselga and Novoa (2002a) | C, X | Ibe |
| *Aphthona nonstriata* (Goeze) | 1995 | Baselga and Novoa (2000a, 2002b) | A, B, H, X | WR |
| *Aphthona herbigrada* (Curtis) | 1997 | Baselga and Novoa (2001a) | X | WR |
| *Longitarsus pellucidus* (Foudras) | 1994 | Baselga and Novoa (2002b) | G | WR |

*Appendix A* Continued.

| Species | First record | Published records | Main areas | Dist. |
|---|---|---|---|---|
| *Longitarsus codinai* Madar and Madar | 1996 | Baselga and Novoa (2000a) | A, X | Med |
| *Longitarsus ochroleucus* (Marsham) | 1997 | Baselga and Novoa (2000a, 2002b) | A, G, X | WR |
| *Longitarsus candidulus* (Foudras) | 1997 | Baselga and Novoa (2000a) | A, G | Med |
| *Longitarsus leonardii* Doguet | 2000 | Baselga and Novoa (2001a) | X | Ibe |
| *Longitarsus flavicornis* (Stephens) | 1994 | Baselga and Novoa (2000a, b, 2002b) | A, C, D, F, G, X | WR |
| *Longitarsus succineus* (Foudras) | 1991 | Baselga and Novoa (2000a) | A, C, X | WR |
| *Longitarsus aeruginosus* (Foudras) | 1996 | Baselga and Novoa (2000a, b) | A, B, C, F, X | WR |
| *Longitarsus rubiginosus* (Foudras) | 1997 | Baselga and Novoa (1999b) | X | Eur |
| *Longitarsus tabidus* (Fabricius) | 1992 | Baselga and Novoa (2000a, b, 2002b) | A, B, E, F, G, X | WR |
| *Longitarsus australis* (Mulsant and Rey) | 1997 | Baselga and Novoa (1998, 2000a) | A, X | Med |
| *Longitarsus foudrasi* Weise | 2000 | Baselga and Novoa (2001a, 2002b) | G | WR |
| *Longitarsus nigrofasciatus* (Goeze) | 1992 | Baselga and Novoa (2000a, b, 2002b) | A, B, E, F, G, H, X | WR |
| *Longitarsus suturatus* (Foudras) | 1993 | Baselga and Novoa (2000a, 2002b) | A, E, G, H, X | Med |
| *Longitarsus rutilus* (Illiger) | 1993 | Baselga and Novoa (1999b, 2000a, b) | A, B, C, E, F, G, X | WR |
| *Longitarsus lycopi* (Foudras) | 1993 | Baselga and Novoa (2000a, b, 2002b) | A, B, F, G, H, X | WR |
| *Longitarsus ordinatus* (Foudras) | 1993 | Baselga and Novoa (2000b, 2002b) | E, F, G, H, X | Med |
| *Longitarsus ferrugineus* (Foudras) | 1993 | Baselga and Novoa (2000a, b) | A, F, G, X | WR |
| *Longitarsus albineus* (Foudras) | 1993 | Baselga and Novoa (2000b) | G | WR |
| *Longitarsus membranaceus* (Foudras) | 1993 | Baselga and Novoa (2000a, b, 2002b) | A, C, D, E, F, G, X | WR |
| *Longitarsus melanocephalus* (De Geer) | 1870 | Heyden (1870) and Baselga and Novoa (2000a, b, 2002b) | A, B, C, D, E, F, G, H, X, Y | WR |
| *Longitarsus kutscherae* (Rye) | 1997 | Baselga and Novoa (2001a) | B | WR |
| *Longitarsus curtus* (Allard) | 1991 | Baselga and Novoa (2000c) | C, G, X | Eur |
| *Longitarsus exsoletus* (Linné) | 1997 | Baselga and Novoa (2000a, b) | A, B, D, F, X | WR |

*Appendix A* Continued.

| Species | First record | Published records | Main areas | Dist. |
|---|---|---|---|---|
| *Longitarsus cerinthes* (Schrank) | 1997 | Baselga and Novoa (2000a, b, 2002b) | A, B, F, G, X | WR |
| *Longitarsus scutellaris* (Mulsant and Rey) | 1993 | New record | X | Eur |
| *Longitarsus pratensis* (Panzer) | 1990 | Leonardi and Doguet (1990) and Baselga and Novoa (2000a, b, 2002b) | A, B, C, D, E, F, G, H, X, Y | WR |
| *Longitarsus reichei* (Allard) | 1870 | Heyden (1870), Jolivet (1953), Petitpierre (1999) and Baselga and Novoa (2000b) | B, F, X, Y | Eur |
| *Longitarsus ballotae* (Marsham) | 1870 | Heyden (1870) | G, Y | WR |
| *Longitarsus gracilis* Kutschera | 1993 | Baselga and Novoa (2000a, b, 2002b) | A, B, F, H, X | WR |
| *Longitarsus atricillus* (Linné) | 1991 | Baselga and Novoa (2000b, 2002b) | B, C, D, E, F, G, X | WR |
| *Longitarsus bedeli* Uhagón | 1993 | Baselga and Novoa (2002b) | H | Ibe |
| *Longitarsus lateripunctatus* Rosenhauer | 1993 | Baselga and Novoa (2000a, b) | A, B, F, G, X | WR |
| *Longitarsus dorsalis* (Fabricius) | 1993 | Baselga and Novoa (2000b, 2002b) | F, G, H, X | WR |
| *Longitarsus luridus* (Scopoli) | 1991 | Baselga and Novoa (2000a, b, 2002b) | A, B, C, D, E, F, G, H, X | WR |
| *Longitarsus fulgens* (Foudras) | 2000 | Baselga and Novoa (2001a) | C | Eur |
| *Longitarsus echii* (Koch) | 1993 | New record | E, X | WR |
| *Longitarsus aeneus* Kutschera | 1993 | Baselga and Novoa (2000a, b) | A, F, X | Med |
| *Longitarsus niger* (Koch) | 1993 | New record | X | Eur |
| *Longitarsus parvulus* (Paykull) | 1870 | Heyden (1870) and Baselga and Novoa (2000a, b) | A, D, F, X, Y | WR |
| *Longitarsus obliteratoides* Gruev | 1997 | Baselga and Novoa (2002b) | G | Med |
| *Longitarsus bergeali* Doguet and Gruev | 1998 | Baselga and Novoa (2000b) | F | Eur |
| *Longitarsus anchusae* (Paykull) | 1993 | Baselga and Novoa (2000b) | E, F, X | WR |
| *Altica ampelophaga* Guérin | 1981 | Baselga and Novoa (1999a, 2000a, b, 2002b) | A, C, D, E, F, H, X. | Med. |
| *Altica brevicollis* Foudras | 1990 | Baselga and Novoa (1999b, 2000b) | C, F, X | WR |
| *Altica quercetorum* Foudras | 1993 | Lombardero et al. (1993) and Baselga and Novoa (1999a) | X, Y | Eur. |

*Appendix A* Continued.

| Species | First record | Published records | Main areas | Dist. |
|---|---|---|---|---|
| *Altica tamaricis* Schrank ssp. *franzi* Král | 1966 | Král (1966) | Y | WR |
| *Altica oleracea* (Linné) | 1981 | Baselga and Novoa (2000a, b, 2002b) | A, B, D, E, F, G, H, X | WR |
| *Altica palustris* (Weise) | 1993 | Baselga and Novoa (2000b) | C, G, X | WR |
| *Altica carduorum* Guérin | 1990 | Baselga and Novoa (2000c) | X. | WR. |
| *Altica ericeti* (Allard) | 1981 | Baselga and Novoa (1999a, 2000b, 2002b) | B, C, D, E, F, G, H, X | Eur |
| *Altica longicollis* (Allard) | 1980 | Baselga and Novoa (1999b, 2000b) | F, X | Eur |
| *Hermaeophaga cicatrix* (Illiger) | 1999 | Baselga and Novoa (2001a) | C | WR |
| *Batophila aerata* (Marsham) | 2000 | Baselga and Novoa (2001a, 2002b) | G | WR |
| *Arrhenocoela lineata* (Rossi) | 1991 | Baselga and Novoa (2000b, 2002b) | C, D, F, G, X | Med |
| *Ochrosis ventralis* (Illiger) | 1997 | Baselga and Novoa (2000a, b, 2002b) | A, F, G | WR |
| *Neocrepidodera transversa* (Marsham) | 1997 | Baselga and Novoa (2000b, 2002b) | B, E, F, H, X | Eur |
| *Neocrepidodera impressa* (Fabricius) | 1976 | Baselga and Novoa (2000a) | A, X | Med |
| *Neocrepidodera ferruginea* (Scopoli) | 1996 | Baselga and Novoa (1999a, 2000b) | B, D, F, X | WR |
| *Neocrepidodera crassicornis* (Faldermann) ssp. *hispanica* (J. Daniel) | 1996 | Baselga and Novoa (2000a, b) | A, B, F, X | Med |
| *Crepidodera fulvicornis* (Fabricius) | 1996 | Baselga and Novoa (2000a, 2002b) | A, B, C, D, E, H, X | WR |
| *Crepidodera aureola* (Foudras) | 1993 | Baselga and Novoa (2000a, b, 2002b) | A, B, C, D, E, F, G, H, X | Med |
| *Crepidodera aurata* (Marsham) | 1981 | Baselga and Novoa (2000b, 2002b) | F, G, X | WR |
| *Epitrix pubescens* (Koch) | 1907 | Chapman and Champion (1907) and Baselga and Novoa (2000a, b) | A, F, X, Y | WR |
| *Epitrix intermedia* Foudras | 1996 | Baselga and Novoa (1999b, 2000b) | F, X | Eur |
| *Podagrica fuscicornis* (Linné) | 1997 | New record | B, X | WR |
| *Podagrica fuscipes* (Fabricius) | 1993 | Baselga and Novoa (2000a, b, 2002b) | A, B, C, E, F, G, H, X | Eur |
| *Mantura chrysanthemi* (Koch) | 1993 | New record | X | WR |

*Appendix A* Continued.

| Species | First record | Published records | Main areas | Dist. |
|---|---|---|---|---|
| *Mantura lutea* (Allard) | 1998 | Baselga and Novoa (2000a) | A | Med |
| *Mantura rustica* (Linné) | 1997 | Baselga and Novoa (2000a) | A, D | WR |
| *Chaetocnema chlorophana* (Duftschmid) | 1993 | Baselga and Novoa (2000a, 2002b) | A, E, G, X | WR |
| *Chaetocnema concinna* (Marsham) | 1992 | Baselga and Novoa (1999b, 2000a) | A, B, D, X | WR |
| *Chaetocnema laevicollis* (Thomson) | 1992 | Baselga and Novoa (1999b, 2000a, b) | A, B, C, E, F, X | WR |
| *Chaetocnema tibialis* (Illiger) | 1993 | Baselga and Novoa (2000a, b, 2002b) | A, E, F, G, X | WR |
| *Chaetocnema depressa* (Boieldieu) | 2000 | Baselga and Novoa (2001a, 2002b) | G | Med |
| *Chaetocnema aridula* (Gyllenhal) | 1996 | Baselga and Novoa (2000b) | D, F, X | WR |
| *Chaetocnema confusa* (Boheman) | 1996 | Baselga and Novoa (2000c, 2002b) | H, X | Eur |
| *Chaetocnema arida* Foudras | 1991 | Baselga and Novoa (2000a, b, 2002b) | A, B, C, D, E, F, G, H, X | WR |
| *Chaetocnema paganettii* Heikertinger | 1964 | Baselga and Novoa (2002b) | H | Med |
| *Chaetocnema hortensis* (Geoffroy) | 1870 | Heyden (1870) and Baselga and Novoa (2000a, b, 2002b) | A, B, C, D, E, F, H, X, Y | WR |
| *Oedionychus cinctus* (Fabricius) | 1928 | Iglesias (1928), Biondi (1991), Bastazo et al. (1993) and Baselga and Novoa (1998, 1999a, 2000a, b, 2002b) | A, B, C, D, E, F, G, H, X, Y | Ibe |
| *Sphaeroderma testaceum* (Fabricius) | 1990 | Baselga and Novoa (2000b) | C, D, F, X | Eur |
| *Sphaeroderma rubidum* (Graëlls) | 1991 | Baselga and Novoa (2000a, b, 2002b) | A, C, F, G, H, X | WR |
| *Apteropeda globosa* (Illiger) | 1991 | Baselga and Novoa (1999b, 2000b) | C, F, X | Eur |
| *Apteropeda orbiculata* (Marsham) | 1991 | Biondi (1991) and Baselga and Novoa (2000a, b) | A, B, C, D, F, X, Y | Eur |
| *Apteropeda ovulum* (Illiger) | 1996 | Baselga and Novoa 1999a | D, X | Med |
| *Dibolia occultans* (Koch) | 1992 | Baselga and Novoa (2000a, b, 2002b) | A, D, E, F, H, X | WR |
| *Psylliodes affinis* (Paykull) | 1993 | Baselga and Novoa (1999b, 2000b, 2002b) | B, C, D, F, G, X | WR |
| *Psylliodes marcidus* (Illiger) | 1997 | Baselga and Novoa (2000a) | A | WR |

*Appendix A* Continued.

| Species | First record | Published records | Main areas | Dist. |
|---|---|---|---|---|
| *Psylliodes pallidipennis* Rosenhauer | 1997 | Baselga and Novoa (2000a) | A | Med |
| *Psylliodes chrysocephalus* (Linné) | 1993 | Baselga and Novoa (2000a, b) | A, B, C, E, F, X | WR |
| *Psylliodes napi* (Fabricius) | 1934 | González de Andrés (1934) and Baselga and Novoa (2000a, b) | C, F, X, Y | WR |
| *Psylliodes laticollis* Kutschera | 1997 | Baselga and Novoa (2000a) | A, D | WR |
| *Psylliodes toelgi* Heikertinger | 1997 | Baselga and Novoa (2002b) | G | Eur |
| *Psylliodes cupreus* (Koch) | 1993 | Baselga and Novoa (2000a, b, 2002b) | A, B, E, F, H, X | WR |
| *Psylliodes fusiformis* (Illiger) | 1997 | Baselga and Novoa (2000a) | A | Med |
| *Psylliodes vindobonensis* Heikertinger | 1982 | Baselga and Novoa (2002b) | H | Eur |
| *Psylliodes hispanus* Heikertinger | 1993 | Baselga and Novoa (2001a) | C, X | Ibe |
| *Psylliodes dulcamarae* (Koch) | 1994 | Baselga and Novoa (2001a, 2002b) | G, X | WR |
| *Psylliodes cucullatus* (Illiger) ssp. *heydeni* Weise | 1991 | Baselga and Novoa (1999b, 2000a, b) | A, B, C, D, F, X | WR |
| *Psylliodes gougeleti* All. | 1866 | López Seoane (1866, 1877) and Baselga and Novoa (2000a, b, 2002b) | A, B, C, D, E, F, H, X, Y | Med |
| *Psylliodes cervinoi* Baselga and Novoa | 1993 | Baselga and Novoa (2003) | H | Ibe |
| *Dicladispa testacea* (Linné) | 1907 | Chapman and Champion (1907) and Baselga and Novoa (1999a, 2000a, b, 2002b) | A, B, F, G, X, Y | Med |
| *Hispa atra* Linné | 1907 | Chapman and Champion (1907) and Baselga and Novoa (2000b, 2002b) | B, C, D, E, F, H, X, Y | WR |
| *Cassida viridis* Linné | 1993 | Baselga and Novoa (1999a, 2000a, 2002b) | A, B, D, E, G, H, X | WR |
| *Cassida hemisphaerica* Herbst | 1997 | Baselga and Novoa (2000a, b, 2002b) | A, F, H, X | WR |
| *Cassida nebulosa* Linné | 1998 | Baselga and Novoa (2000b) | F | WR |
| *Cassida flaveola* Thunberg | 1993 | Baselga and Novoa (1999b, 2000a, b) | A, B, F, X | WR |
| *Cassida seladonia* Gyllenhal | 1991 | Baselga and Novoa (2000b) | B, F, X | WR |

*Appendix A* Continued.

| Species | First record | Published records | Main areas | Dist. |
|---|---|---|---|---|
| *Cassida vibex* Linné | 1991 | Baselga and Novoa (2000b) | B, D, F, X | WR |
| *Cassida rubiginosa* Müller | 1990 | Baselga and Novoa (2000a, b, 2002b) | A, B, C, D, E, F, H | WR |
| *Cassida inquinata* Brullé | 1866 | López Seoane (1866) and Baselga and Novoa (2000a) | B, Y | WR |
| *Cassida hexastigma* Suffrian | 1997 | Bordy (2000) | A, G | Med |
| *Cassida denticollis* Suffrian | 1993 | Baselga and Novoa (1999b) | B, C, D, X | WR |
| *Cassida subreticulata* Suffrian | 2000 | Baselga and Novoa (1999a, 2001a) | D, X | WR |
| *Cassida vittata* Villers | 1997 | Baselga and Novoa (2000a) | A, B | WR |

## References

Báguena L. 1959. Cuatro novedades y un comentario sobre Coleopteros de España. Eos 35: 209–214.

Báguena L. 1960a. Los Donaciinae ibéricos de la Colección del Instituto Español de Entomología. Graellsia 18: 3–14.

Báguena L. 1960b. Clave práctica para la determinación de los Clytrini ibéricos. Graellsia 18: 123–145.

Báguena L. 1960c. Notas críticas y descriptivas sobre algunos Clytrini. Graellsia 18: 15–28.

Baselga A. and Novoa F. 1998. Citas nuevas o interesantes de Chrysomelidae (Coleoptera) del Noroeste peninsular. Bol. Asoc. Española Entomol. 22(3–4): 246–248.

Baselga A. and Novoa F. 1999a. La colección de Chrysomelidae (Coleoptera) del Museo de Historia Natural Luis Iglesias (Universidad de Santiago de Compostela, España). Nova Acta Científica Compostelana (Bioloxía) 9: 307–312.

Baselga A. and Novoa F. 1999b. Citas interesantes de Alticinae (Coleoptera, Chrysomelidae) de la Península Ibérica. Bol. Asoc. Española Entomol. 23(12): 53–57.

Baselga A. and Novoa F. 2000a. Los Chrysomelidae (Coleoptera) de los sistemas dunares de Galicia (Noroeste de España). Bol. Real Soc. Española Hist. Nat. (Sec. Biol.) 96(12): 113–124.

Baselga A. and Novoa F. 2000b. Los Chrysomelidae (Coleoptera) de la Sierra de Ancares, Noroeste de la España (Coleoptera). Nouvelle Rev. d'Entomol. (N.S.) 17(2): 165–180.

Baselga A. and Novoa F. 2000c. Citas nuevas o interesantes de Chrysomelidae (Coleoptera) del Norte de la Península Ibérica. Bol. Asoc. Española Entomol. 24(3–4): 15–22.

Baselga A. and Novoa F. 2000d. *Crypytocephalus cantabricus* Franz, a poorly known endemic species from the northwest of the Iberian Peninsula (Coleoptera: Chrysomelidae). Koleopterologische Rundschau 70: 191–195.

Baselga A. and Novoa F. 2001a. Citas nuevas de Chrysomelidae (Coleoptera) de Galicia (Noroeste de la Península Ibérica). Bol. Asoc. Española Entomol. 25(1–2): 125–131.

Baselga A. and Novoa F. 2001b. *Phyllotreta ganglbaueri* Heikertinger, 1909, especie nueva para la Península Ibérica (Coleoptera, Chrysomelidae). Nouvelle Rev. d'Entomol. (N.S.) 18(1): 89–90.

Baselga A. and Novoa F. 2002a. New species of *Aphthona* (Coleoptera: Chrysomelidae: Alticinae) and key to Iberian species of *Aphthona hammarstroemi* group. Can. Entomol. 134(1): 1–7.

Baselga A. and Novoa F. 2002b. Los Chrysomelidae (Coleoptera) de las sierras orientales de Ourense (Galicia, noroeste de la Península Ibérica). Bol. Asoc. Española Entomol. 26(3–4): 57–73.

Baselga A. and Novoa F. 2003. A new species of *Psylliodes* (Coleoptera: Chrysomelidae) and key to the wingless species from the Iberian Peninsula. Ann. Entomol. Soc. Am. 96(6): 689–692.

Baselga A. and Novoa F. 2004. Coleópteros del Parque Natural de las Fragas del Eume (Galicia, noroeste de la Península Ibérica), II: Scarabaeoidea, Buprestoidea, Byrrhoidea, Elateroidea, Bostrichoidea, Lymexyloidea, Cleroidea, Cucujoidea, Tenebrionoidea, Chrysomeloidea y Curculionoidea. Bol. Asoc. Española Entomol. 28(1–2): 121–143.

Bastazo G., Vela J.M. and Petitpierre E. 1993. Datos faunísticos sobre Alticinae ibéricos. Bol. Asoc. Española Entomol. 17(1): 45–69.

Bechyné J. 1948. Contribution à la connaiscence du genre *Timarcha* Latr. 12: Etudes phylogenetiques et zoogeographiques (Col. Phyt. Chrys.). Sborn. Narod. Mus. Praze 4B: 1–62.

Bechyné J. 1957. Notes sur le *Phytodecta (Spartoxena) variabilis* Oliv. de l'Espagne (Col. Phytophaga). Eos 33: 263–267.

Bedel L. 1901. Faune des Coléoptères du Bassin de la Seine, V. Phythophaga. Société Entomologique de France, Paris.

Biondi M. 1991. Note faunistiche, tassonomiche ed ecologiche su alcune specie di Chrysomelidae Alticinae della Penisolla Iberica (Col.). Eos 66(2): 161–172.

Bordy B. 2000. Coléoptères Chrysomelidae. Vol. 3, Cassidinae. Faune de France 80. Federation Française des Sociétés de Sciences Naturelles, Paris.

Bourdonné J.C. 1994. Les *Cryptocephalus* rouges ouest méditerranéens du groupe de *C. bimaculatus* Fabricius (Coleoptera, Chrysomelidae, Cryptocephalinae). Nouvelle Rev. d'Entomol. (N.S.) 11(2): 185–209.

Burlini M. 1955. Revisione dei *Cryptocephalus* Italiani e della maggior parte specie di Europa. Mem. Soc. Entomol. Ital. 34: 5–287.

Cabrero-Sañudo F.J. and Lobo L.M. 2003. Estimating the number of species not yet described and their characteristics: the case of Western Palaearctic dung beetle species (Coleoptera, Scarabaeoidea). Biodiv. Conserv. 12: 147–166.

Carlton C.E. and Robison H.W. 1998. Diversity of litter-dwelling beetles in the Ouachita Highlands of Arkansas, USA (Insecta: Coleoptera). Biodiv. Conserv. 7: 1589–1605.

Chapman T.A. and Champion G.C. 1907. Entomology in NW Spain (Galicia and León). Trans. Ent. Soc. London 1907: 147–171.

Cobos A. 1954. Nota sobre el género Cyrtonus Latr. (Coleoptera Chrysomelidae) y diagnosis de formas nuevas ibero-marroquíes. Tamuda 2: 318–324.

Cobos A. 1969. Nuevos materiales para el estudio del Género *Gynandrophthalma* Lac. Graellsia 24: 67–77.

Codina Padilla F. 1961a. Apuntes sobre Chrysomelidae ibero-marroquíes (Coleoptera). Graellsia 19: 45–57.

Codina Padilla F. 1961b. Nota sobre las *Chrysomela* ibero-marroquies de subgenero *Polysticta* Hope (Col. Chrys.). Bull. Soc. Sci. Nat. Maroc 41: 69–83.

Colwell R.K. 2000. EstimateS: statistical estimation of species richness and shared species from samples. Version 6. User's guide and application published at: http://viceroy.eeb.uconn.edu/estimates.

Colwell R.K. and Coddington J.A. 1994. Estimating terrestrial biodiversity through extrapolation. Philos. Trans. R. Soc. London B 345: 101–118.

Daccordi M. and Petitpierre E. 1977. Coleópteros Crisomélidos de la Sierra de Cazorla (Jaén) y descripción de una nueva especie de *Clytra* Laich. (Coleoptera, Chrysomelidae). Misc. Zool. 4(1): 225–236.

Doguet S., Bastazo G., Bergeal M. and Vela J.M. 1996. Contribution a l'étude des Chrysomelidae d'Andalousie (Coleoptera). Nouvelle Rev. d'Entomol. (N.S.) 13(4): 315–323.

Fairmaire L. 1884. Liste complementaire des espéces du G. *Timarcha*. An. R. Soc. Esp. Hist. Nat. 13: 75–112.

Franz H. 1958. Neue und ungenugend bekannte Coleopterenarten aus Spanien. Eos 34: 117–130.

García-Ocejo A. and Gurrea P. 1995. Los crisomélidos (Coleoptera: Chrysomelidae) de la Sierra de Guadarrama (España Central). Análisis biogeográfico. Bol. Asoc. Española Entomol. 19(3–4): 51–68.

García-Ocejo A., Gurrea P. and Petitpierre E. 1992. Chrysomelidae (Coleoptera) de la sierra de Gredos (Sistema Central): datos faunísticos, ecológicos y fenológicos. Misc. Zool. 16: 81–92.

González de Andrés C. 1934. Insectos perjudiciales a las plantas cultivadas en Galicia. An. Asoc. Española Prog. Ciencias 1(4): 829–837.

Heyden L. von 1870. Entomologische Reise nach dem Sudlichen Spanien, der Sierra Guadarrama und Sierra Morena, Portugal und den Cantabrischen Gebirgen. Nicolaische Verlagbuckhandlung, Berlin.

Hortal J., García-Pereira P. and García-Barros E. 2004. Butterfly species richness in mainland Portugal: predictive models of geographic distribution patterns. Ecography 27: 68–82.

Hortal J., Lobo J.M. and Martín-Piera F. 2001. Forecasting insect species richness scores in poorly surveyed territories: the case of the Portuguese dung beetles (Col. Scarabaeinae). Biodiv. Conserv. 10: 1343–1367.

Iglesias L. 1928. Notas entomolóxicas. Insectos Galiza 1(53): 89–94.

Izco J. 1987. Galicia. In: Peinado-Lorca M. and Rivas-Martínez S. (eds), La vegetación de España. Universidad de Alcalá de Henares, Alcalá de Henares.

Jolivet P. 1953. Les Chrysomeloidea des Îles Baléars. Mém. Inst. R. Sci. Nat. Belg. 2(58): 3–152.

Král J. 1966. Zur Kenntnis der palaearktischen *Altica*-Arten. IV. Entomol. Blätter 62: 159–168.

Laboissière V. 1934. Galerucinae de la faune française (Coléoptères). Ann. Soc. Entomol. Fr. 103: 1–108.

Leonardi C. and Doguet S. 1990. Studio critico sui *Longitarsus* del gruppo *pratensis* (Panzer) (Col. Chrysomelidae). Atti Soc. Ital. Sci. Nat. Museo Civico Storia Nat. 131(2): 13–74.

Lombardero J., Rodríguez-Fernandez R.J. and Fernández de Ana-Magan F.J. 1993. *Altica quercetorum* (Foudras, 1860) (Col. Chrys.) plaga de las "Carballeiras" gallegas. Actas I Congreso Forestal Español 3: 307–312.

López Seoane V. 1866. Reseña de la Historia Natural de Galicia. Imprenta de Soto Freire, Lugo.

López Seoane V. 1877. Notas para la fauna gallega. Imprenta de El Eco Ferrolano, Ferrol.

Marseul S. 1883. Monographie de Chrysomelides de L'Ancien Monde. L'Abeille 21: 1–108.

Martín-Piera F. and Lobo J.M. 2000. Diagnóstico sobre el conocimiento sistemático y biogeográfico de tres órdenes de insectos hiperdiversos en España: Coleoptera, Hymenoptera y Lepidoptera. In: Martín-Piera F., Morrone J.J. and Melic A. (eds), Hacia un Proyecto CYTED para el Inventario y Estimación de la Diversidad Entomológica en Iberoamérica: prIBES-2000. Monografías Tercer Milenio, Vol. 1. SEA, Zaragoza, pp. 287–308.

Martín-Piera F. and Lobo J.M. 2003. Database records as a sampling effort surrogate to predict spatial distribution of insects in either poorly or unevenly surveyed areas. Acta Entomol. Ibérica Macaronésica 1: 23–35.

Medellín R.A. and Soberón J. 1999. Predictions of mammal diversity of four land masses. Conserv. Biol. 13: 143–149.

Meier R. and Dikow T. 2004. Significance of taxonomic databases from taxonomic revisions for estimating and mapping the global species diversity of invertebrates and repatriating reliable specimen data. Conserv. Biol. 18(2): 478–488.

Moreno C.E., and Halffter G. 2000. Assessing completeness of bat diversity inventories using species accumulation curves. J. Appl. Ecol. 37: 149–158.

Noguera F.A., Zaragoza-Caballero S., Chemsak J.A., Rodríguez-Palafox A., Ramírez E., González-Soriano E. and Ayala R. 2002. Diversity of the family Cerambycidae (Coleoptera) of the tropical dry forest of Mexico, I. Sierra de Huautla, Morelos. Ann. Entomol. Soc. Am. 95(5): 617–627.

Petersen F.T., Meier R. and Larsen M.N. 2003. Testing species richness estimation methods using museum label data on the Danish Asilidae. Biodiv. Conserv. 12: 687–701.

Petitpierre E. 1980. Catàleg dels Coleòpters Crisomèlids de Catalunya, I. Cryptocephalinae. Butll. Inst. Catalana d'Hist. Nat. (Sec. Zool.) 45: 65–76.

Petitpierre E. 1981. Chrysomelidae (Col.) de la Sierra de Albarracín. Bol. Asoc. Española Entomol. 4: 7–18.

Petitpierre E. 1983. Catàleg dels Coleòpters Crisomèlids de Catalunya, II. Zeugophorinae, Donaciinae, Criocerinae, Clytrinae, Lamprosomatinae i Eumolpinae. Butll. Inst. Catalana d'Hist. Nat. (Sec. Zool.) 49: 87–96.

Petitpierre E. 1988. Catàleg dels Coleòpters Crisomèlids de Catalunya, III. Chrysomelinae i Galerucinae. Butll. Inst. Cat. Hist. Nat. (Sec. Zool.) 55: 79–100.

Petitpierre E. 1994. Estudi faunístic i ecològic dels Coleòpters Crisomèlids de la Vall d'Aran. Butll. Inst. Catalana d'Hist. Nat. 62: 77–108.

Petitpierre E. 1997. Los Chrysomelidae (Coleoptera) del Moncayo (Aragón). Zapateri, Rev. Aragonesa Entomol. 7: 273–280.

Petitpierre E. 1999. Catàleg dels Coleòpters crisomèlids de Catalunya, IV. Alticinae. Butll. Inst. Catalana d'Hist. Nat. (Sec. Zool.) 67: 91–129.

Petitpierre E. 2000. Coleoptera, Chrysomelidae I. In: Ramos M.A. et al. (eds), Fauna Ibérica, Vol. 13. Museo Nacional de Ciencias Naturales, Madrid.

Petitpierre E. and Gómez-Zurita J. 1998. Los Chrysomelidae de León; NO de España (Coleoptera). Nouvelle Rev. d'Entomol. (N.S.) 15(1): 13–26.

Plaza Infante E. 1979. Contribución al conocimiento de los *Cryptocephalus* ibéricos. Bol. Asoc. Española Entomol. 3: 5–10.

Rivas-Martínez S. 1987. Memoria del Mapa de Series de Vegetación de España. Ministerio de Agricultura, Pesca y Alimentación, Madrid.

Soberón J. and Llorente J. 1993. The use of species accumulation functions for the prediction of species richness. Conserv. Biol. 7: 480–488.

Soberón J.M., Llorente J.B. and Oñate L. 2000. The use of specimen-label databases for conservation purposes: an example using Mexican Papilionid and Pierid butterflies. Biodiv. Conserv. 9: 1441–1466.

StatSoft. 2001. STATISTICA 6.0 (data analysis software system). StatSoft, Inc., Tulsa, OK.

Summerville K.S., Metzler E.H. and Crist T.O. 2001. Diversity of Lepidoptera in Ohio Forests at local and regional scales: how heterogeneous is the fauna? Ann. Entomol. Soc. Am. 94(4): 583–591.

Vela J.M. and Bastazo G. 1999. Ecological and biogeographical aspects of the Andalusian Leaf Beetle endemisms. In: Cox M.L. (ed.), Advances in Chrysomelidae Biology 1. Backhuys Publishers, Leiden, pp. 137–158.

Vigna Taglianti A., Audisio P.A., Belfiore C., et al. 1992. Riflessione di gruppo sui corotipi fondamentali della fauna W-paleartica ed in particolare italiana. Biogeographia 16: 159–179.

Vives E. 2000. Listado preliminar de los crisomélidos de Aragón (Coleoptera: Chrysomelidae). Cat. entomofauna aragon. 21: 11–18.

Vives E. and González C.F. 1994. Coleópteros Crisomélidos de la colección Español Auqué (1ª parte). (Coleoptera: Chrysomelidae). Zapateri, Rev. Aragonesa Entomol. 4: 49–60.

Vives E. and González C.F. 1998. Coleópteros Crisomélidos de la colección Español Auqué (2ª parte) (Coleoptera: Chrysomelidae). Zapateri, Rev. Aragonesa Entomol. 8: 171–178.

Biodiversity and Conservation (2006) 15:217–230
DOI 10.1007/s10531-004-7065-7 

# Topographic heterogeneity plays a crucial role for grasshopper diversity in a southern African megabiodiversity hotspot

SOLOMON GEBEYEHU[1] and MICHAEL J. SAMWAYS[2,*]
[1]*Forestry and Agricultural Biotechnology Institute, University of Pretoria, Pretoria, 0002, South Africa;* [2]*Department of Entomology and Centre for Agricultural Biodiversity, University of Stellenbosch, Matieland 7602, South Africa; *Author for correspondence (e-mail: Samways@sun.ac.za)*

Received 28 May 2004; accepted in revised form 18 November 2004

**Key words:** Diversity, Grasshoppers, Southern Africa, Topography

**Abstract.** Topographic heterogeneity as a determinant of insect diversity pattern has been little studied. Responses of grasshopper assemblages to three hill sizes were assessed in the arid Succulent Karoo, South Africa. This area is one of the world's 25 hotspots for conservation priorities. Small hills overall were more speciose than medium or large hills. There were also significantly higher densities of small-sized grasshoppers on small hills than on medium or large ones. The slopes of the three hill sizes did not differ significantly either in their species richness or abundance, and there was no significant difference in species richness between summits only of the three hill sizes. Patterns of grasshopper species dominance were markedly variable among sites, but with clear differences between small and larger hills, associated with vegetation characteristics. Vegetation cover and grass cover was less on the small hills. Grasshopper taxonomic groups varied among the three hill sizes, with small hills being taxonomically more diverse, supporting species from four families and nine subfamilies, while medium and large hills only supported Acrididae. It is concluded that topography has a remarkably strong effect on various aspects of grasshopper spatial heterogeneity and that small hills in particular are a major factor to consider in spatial conservation planning.

## Introduction

An understanding of pattern, its causes and consequences is central to understanding evolutionary processes such as speciation as well as ecological processes of succession, community development and the spread and persistence of a species (Wiens 1976). It is also central in conservation planning. The ability to translate fine-scale information across scales could facilitate the search for mechanistic explanations of broad-scale patterns, thus promoting a unification of pattern and process in ecology (With and Crist 1996). Nevertheless each organism observes the environment on its own unique suite of scales of time and space (Levin 1989). Indeed, regional scaling is rarely explicit in population models so as to increase the generality of a given model (Collins and Glenn 1997).

Variation in species assemblages is not an absolute but only has meaning relative to a particular scale of observation. This means that it is vital to recognize that change is taking place on many scales, and that there are interactions among phenomena at different scales. Factors influencing the proximate and physiological or behavioural state, or the ultimate fitness of individuals, exhibit discontinuities on many scales in time and space. Furthermore, spatial patterning is expressed in both the vertical as well as the horizontal dimension.

Populations of small organisms with short generations are sensitive to short-term environmental fluctuations and have greater potential for close tracking of environmental variation than do populations of larger organisms (Belovsky and Slade 1995). They may thus be more responsive to physical features of their environment or patch. The patch structure of an environment is that which is recognized or relevant to the organism under consideration. Patchiness is thus organism-defined, and must be considered in terms of the perceptions of the organism rather than those of the investigator (Wood and Samways 1991).

The Succulent Karoo is one of the world's 25 biodiversity hotspots for conservation (Myers et al. 2000). It makes up one quarter of the land area in South Africa, is characterized by a flatland matrix punctuated by hills of various sizes. This begs the question: are these hills the same as the flatlands in their biodiversity or do they create spatial heterogeneity resulting in unique assemblages of fauna and flora? Spatial heterogeneity of landscape elements such as hills and flatlands will have important consequences for population dynamics depending on the dispersal rates of individuals and the spatial scales at which the process is considered (Turchin 1991).

While some workers have reported that topographic elements are important in grasshopper distribution pattern (Stabaev 1972; Claridge and Singhrao 1978; Samways 1990; Wachter et al. 1998), others have reported that topography as well as aspect were not significant (Narisu et al. 2000). However, none of these studies has explicitly addressed the significance of hills of different sizes as relative determinants of spatial heterogeneity, which is the aim of this study.

## Methods

### *Study area*

The study area was in the Succulent Karoo, the central arid plateau of South Africa, with many mesas and inselbergs of various sizes. The major land use system is sheep grazing and some cattle grazing. Temperatures are low during winter months (June–August) with a mean monthly minimum of 3 °C and high during summer months (December–February) with a mean monthly maximum of 36 °C, and rainfall of 125–375 mm per year.

The study sites were near Middelberg, Eastern Cape Province on four farms totaling 360 km$^2$, at a flatland elevation of 1200 m a.s.l. Each farm had hills ranging from gentle undulations to large mesas. Three size categories of hills were selected on each farm: 1251–1350 m a.s.l. were small (SH), 1351–1450 m a.s.l. were medium (MH), and 1451–1550 m a.s.l. were large (LH). Each hill size category was replicated four times. Flatlands surrounding each hill were sampled, as well as the slope, and the summit. A transect 50 m × 15 m made a single replicate. Six replicates constituted one site. Flatlands, slopes and summits of each hill size category had one sampling site. The distance between transects was > 10 m to avoid double counts of grasshoppers. Each transect on the slope ran along the contour to maintain uniform elevation. Data from the transects across the sides on the slope of a hill were pooled together to yield a single value for the slope.

Sampling was on sunny days with < 15% cloud cover and when wind movement was minimal (< 30 km $h^{-1}$) from January to May 2000, during good rains and vegetation growth, and high grasshopper abundance, until the first frosts killed most of the grasshoppers. A voucher collection is housed in the School of Botany and Zoology, University of KwaZulu-Natal.

### *Sampling methods*

Sampling of grasshoppers was by visual counts. Taxonomic verification was done with close-focus binoculars or after capturing individuals. Only adults were counted and assigned to one of three size categories: ≤ 20 mm long were small, 21–35 mm were medium, and > 35 mm were large. These size categories were similar to With and Crist (1996).

The vegetation was estimated using percent basal cover of grasses and shrubs in 15 randomly placed 4 $m^2$ quadrats within grasshopper transects. Each quadrat was subdivided into 20 units of 20 $cm^2$, and the relative proportions of grasses, shrubs, rock and bare ground were estimated. Percent grass greenness was estimated: 0% = not green; 25% = slightly green; 50% = moderately green; 75% = green, and 100% = very green. Cragginess was estimated: 0 = smooth; 1 = slightly craggy; 2 = moderately craggy; 3 = craggy, and 4 = very craggy, while vegetation density was estimated: 0 = bare; 1 = very sparse vegetation; 2 = sparse; 3 = moderate; 4 = dense, and 5 = very dense. Average height of grasses was estimated at 40 random points in a transect and the tallest inflorescence was the maximum grass height. Surface soil temperature at a depth of 15 cm was measured in each quadrat using a soil thermometer. A slender iron peg marked at 15 cm was inserted at four points in each quadrat to make a hole in the soil for the thermometer probe, which was inserted into the hole for 5 min prior to recording temperature. The average of the four readings was taken. The depth of 15 cm was selected as arid and semi-arid grasshopper species generally lay their eggs between 11–15 cm (Braker 1989).

Both univariate and multivariate statistics were used in data analyses. Detrended Correspondence Analysis (DCA) was used in preference over Correspondence Analysis (CA) to avoid the arch effect, as recommended by (Ter Braak 1986). Canonical Correspondence Analysis (CCA) was used to relate species and site scores to underlying environmental variables. The length of an arrow representing an environmental variable is equal to the rate of change in the weighted average as inferred from the biplot, and is a measure of how much species distributions differ along that environmental variable. All grasshopper abundance data were square-root transformed and all vegetation proportion data were log-transformed to maintain normality and to satisfy the requirements of ANOVA and the multivariate analyses. CANOCO version 4 was used for multivariate analyses, and PRIMER was used for Hill's diversity indices.

## Results

A total of 19 grasshopper species belonging to 3 families and 9 sub-families were recorded (Table 1). Hill sizes differed significantly in grasshopper species and individuals. Small hills had the highest number of species ($F = 6.40$; $p < 0.01$), followed by medium and large hills, which were not significantly different. Small hills also had the highest grasshopper abundance ($F = 3.63$; $p < 0.05$), followed by medium and large hills which were not significantly different (Figure 1a, b).

Some of the measured environmental variables were significantly different on the three hill sizes, whereas others were not (Table 2). Grass cover was significantly lower on small hills than on medium and large hills. Percent greenness was highest on medium hills, and lowest on small hills. Vegetation density

*Table 1.* Grasshopper species recorded during this study.

| Family | Subfamily | Species | Code |
|---|---|---|---|
| Pamphagidae | Porthetinae | *Hoplolopha horrida* (Burmiester 1838) | Hoho |
| Lentulidae | Lentulinae | *Lentula callani* (Dirsh 1956) | Leca |
| Acrididae | Calliptaminae | *Sphodromerus gilli* (Uvarov 1929) | Spgi |
| | | *Acorypha pallidicornis* (Stål 1876) | Acpa |
| | Euryphyminae | *Rhachitopis crassus* (Walker 1870) | Rhcr |
| | | *Calliptamulus hyalinus* (Uvarov 1922) | Cahy |
| | | *Calliptamicus* sp. | Casp |
| | Eyprepocnemidinae | *Heteracris* sp. | Hesp |
| | Cyrtacanthacridinae | *Cyrtacanthacris tatarica* (Linnaeus 1758) | Cyta |
| | Acridinae | *Acrida turrita* (Linnaeus 1758) | Actu |
| | Oedipodinae | *Locustana pardalina* (Walker 1870) | Lopa |
| | | *Oedaleus nigrofasciatus* (De Geer 1773) | Oeni |
| | | *Acrotylus insubricus* (Scopoli 1786) | Acin |
| | | *Sphingonotus scabriculcus* (Stål 1876) | Spsc |
| | | *Picnodictya flavipes* (Miller 1932) | Pifl |
| | Gomphocerinae | *Pseudogmothela* sp. | Psgm |
| | | *Dnopherula crassipes* (Uvarov 1921) | Dncr |
| | | *Pnorisa squalus* (Stål 1861) | Pnsq |

Species codes are those used in the multivariate analyses.

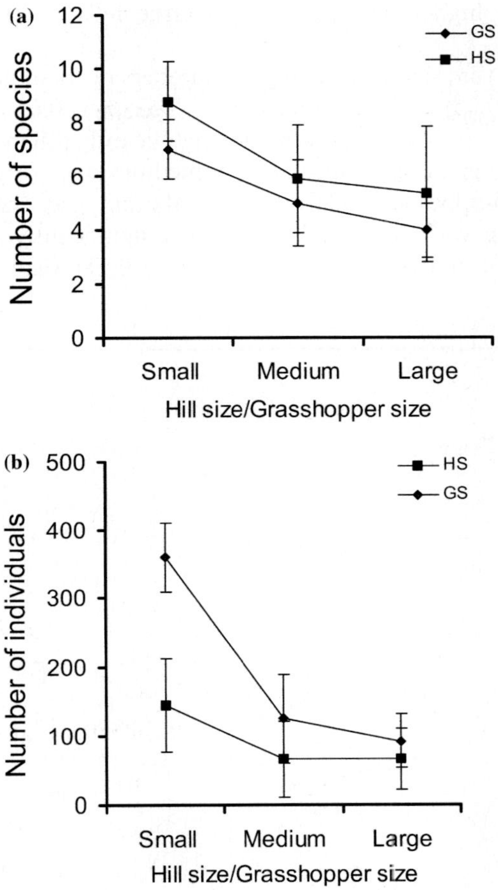


*Figure 1.* Mean number of grasshopper species (a) and individuals (b) on three hill and grasshopper sizes.

*Table 2.* ANOVA results of measured environmental variables (±SE) across the three hill sizes.

| Variable | SH | MH | LH | *F* | *p* |
|---|---|---|---|---|---|
| Grass (%) | 19.2 (11.6)b | 42.5 (10.8)a | 40.0 (5.8)a | 1.7 | 0.02* |
| Shrub (%) | 18.3 (10.9) | 9.2 (2.2) | 14.2 (3.0) | 0.5 | 0.60 |
| Rock (%) | 41.7 (23.2) | 40.0 (20.8) | 25.8 (12.9) | 0.2 | 0.82 |
| Bare (%) | 5.8 (3.8) | 8.3 (4.2) | 20.0 (11.5) | 0.7 | 0.50 |
| Avgrht | 40.0 (7.6) | 48.0 (13.4) | 43.0 (11.7) | 0.2 | 0.80 |
| Maxgrht | 69.2 (7.4) | 70.8 (19.2) | 58.3 (12.0) | 0.5 | 0.71 |
| Greenness of grass (%) | 25.0 (0.9) | 50.0 (7.2) | 33.3 (8.3) | 4.0 | 0.07 |
| Cragginess (scale) | 1.8 (0.6) | 2.3 (1.2) | 2.7 (1.3) | 0.2 | 0.80 |
| Veg. density (scale) | 2.2 (0.4)b | 3.3 (0.2)a | 3.5 (0.3)a | 5.2 | 0.04* |
| Soil temp. (EC) | 33.3 (2.6) | 33.2 (2.5) | 31.3 (4.8) | 0.1 | 0.90 |

*Significant at 5% level of probability.
SH = small hills; MH = medium hills; LH = large hills.

was significantly higher on medium and large hills but low on small hills (Figure 2).

Overall, there were significantly higher numbers of small grasshopper species ($F = 2.82$; $p < 0.05$) and individuals ($F = 3.52$; $p < 0.05$) than medium and large grasshopper species, which were not significantly different, although there were higher number of large species than medium ones (Figure 1a, b). When comparing grasshopper size relative to specific hill size, the only statistically significant finding was that small hills had a significantly higher number of small grasshopper individuals ($F = 5.36$; $p < 0.05$) than medium or large hills.

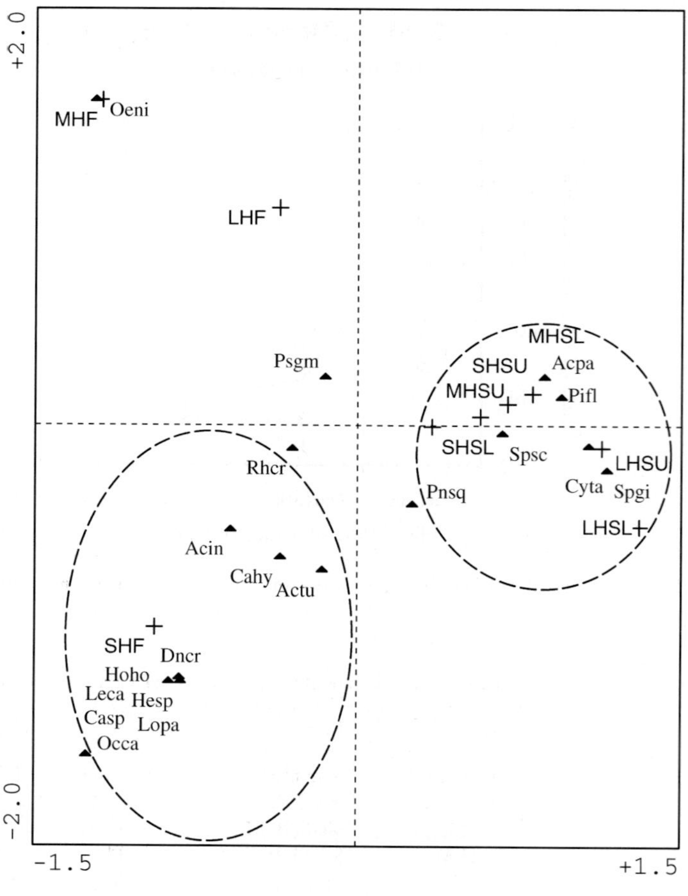


*Figure 2.* CCA triplot of grasshopper species (▲), sampling sites (+) and environmental variables (arrows) of three hill sizes. Species codes are as in Table 1. Site abbreviations: SHF = small hill flatlands; MHF = medium hill flatlands; LHF = large hill flatlands; SHSL = small hill slopes; MHSL = medium hill slopes; LHSL = large hill slopes; SHSU = small hill summit; MHSU = Medium hill summit; LHSU = large hill summit. Environmental variables abbreviations: Maxgrht = maximum grass height; Avgrht = average grass height; Soil temp = soil temperature; Veg density = vegetation density.

Hills's diversity and evenness indices varied among sites (Table 3). Flatlands of small hills had the highest values of *N*1 and *N*2. In contrast, slopes of large hills had the lowest *N*1, followed by flatlands of the medium hills, which had the lowest *N*2 followed by slopes of large hills. Slopes of medium hills had the highest E followed by flatlands of small hills and large hills.

DCA revealed that flatlands surrounding small hills had a separate grasshopper assemblage from their slopes and summits (Figure 2), while those surrounding medium hills were characterized by only one species, *Oedaleus nigrofasciatus*. Flatlands surrounding large hills had no detectable association with any one of the other sites, but showed a weak association with only *Pseudogmothela* sp. The eigenvalues of the first two axes of DCA (Figure 3) and CCA, and the intraset correlations between the CCA axes and the environmental variables are given in Table 4.

*Table 3.* Hill's diversity indices and measures of evenness for all sites sampled on the three hill sizes.

| Site | *N*1 | *N*2 | *E* |
|---|---|---|---|
| SHF | 11.50 | 10.50 | 0.88 |
| SHSl | 7.30 | 6.81 | 0.86 |
| SHSu | 6.47 | 5.42 | 0.77 |
| MHF | 2.61 | 2.13 | 0.69 |
| MHSl | 7.18 | 6.23 | 0.85 |
| MHSu | 5.77 | 5.72 | 0.90 |
| LHF | 4.17 | 4.03 | 0.88 |
| LHSl | 2.29 | 2.17 | 0.75 |
| LHSu | 4.12 | 3.61 | 0.79 |

Site abbreviations: SHF = small hill flatlands; MHF = medium hill flatlands; LHF = large hill flatlands; SHSL = small hill slopes; MHSL = medium hill slopes; LHSL = large hill slopes; SHSU = small hill summit; MHSU = medium hill summit; LHSU = large hill summit.

*Table 4.* Eigenvalues of the first two axes of DCA and CCA, and the intraset correlations between CCA axes and the measured environmental variables.

| | DCA | | CCA | |
|---|---|---|---|---|
| | Axis 1 | Axis 2 | Axis 1 | Axis 2 |
| Eigenvalue | 0.55 | 0.36 | 0.56 | 0.38 |
| Grass | | | −0.59 | 0.31 |
| Shrub | | | −0.71 | −0.57 |
| Rock | | | 0.84 | −0.26 |
| Bare | | | −0.59 | 0.34 |
| Avgrht | | | 0.92 | −0.38 |
| Maxgrht | | | 0.52 | −0.66 |
| Greenness of grass | | | −0.10 | 0.66 |
| Cragginess | | | 0.90 | −0.19 |
| Vegetation density | | | −0.32 | 0.33 |
| Soil temperature | | | 0.58 | −0.18 |

Avgrht = average grass height; Maxgrht = maximum grass height.

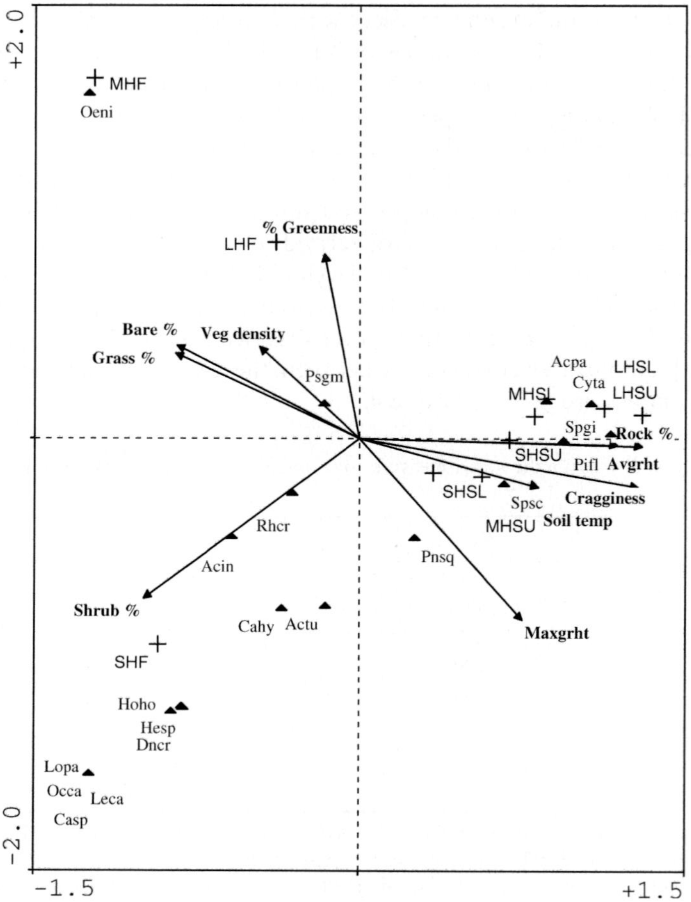


*Figure 3.* DCA of grasshopper species (▲) and sampling sites (+). Species codes are as in Table 3. Site abbreviations: SHF = small hill flatlands; MHF = medium hill flatlands; LHF = large hill flatlands; SHSL = small hill slopes; MHSL = medium hill slopes; LHSL = large hill slopes; SHSU = small hill summit; MHSU = medium hill summit; LHSU = large hill summit.

Grasshopper dominance varied markedly depending on hill size and on elevation within a hill. Small hills were more speciose, but with less dominance than for medium and large hills (Figure 4a). *Acorypha pallidicornis* (35%), and *Picnodictya flavipes* (20%) were the most dominant small hill species. Slopes showed less dominance, with *A. pallidicornis* (25%) the dominant species. Flatlands, in contrast, were not dominated by any particular species, with *Acrotylus insubricus* (17%) and *Locustana pardalina* (13%) being the common species.

Flatlands around medium hills had fewer species and greater dominance than did slopes and summits (Figure 4b). Dominant flatland species were *O. nigrofasciatus* (>65%), and *Pseudogmothela* sp.(>20%). Slopes and summits were dominated by *A. pallidicornis* (about 40 and 30% respectively).

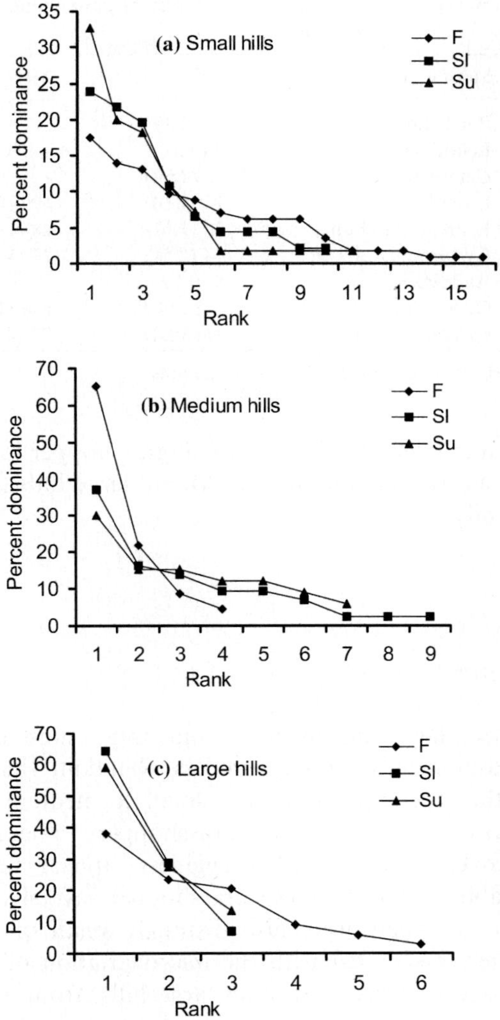


*Figure 4.* Rank-abundance curves of grasshopper species from the surrounding flatlands (F), slopes (Sl) and summits (Su) of hills of different sizes.

Slopes and summits of large hills had only three species, and were dominated by *Cyrtacanthacris tatarica* (about 60%) and *A. pallidicornis* (about 65%) respectively (Figure 4c). Flatlands of large hills had fairly evenly ranked species compared to slopes and summits, with *O. nigrofasciatus* contributing 40%.

In general, there were variations among the three hill sizes with respect to the occurrence and distribution of higher taxonomic levels of grasshoppers (Table 5). Small hills had representatives from three families and nine sub-families, but medium and large hills had only members of Acrididae, in five sub-families. Acrididae, Eyprepocnemidinae and Cyrtacanthacridinae were not

*Table 5.* Relative proportion of higher taxonomic classes of grasshoppers (percentage of total number of individuals) across the three hill sizes.

| Family | Sub-family | SH | MH | LH |
|---|---|---|---|---|
| Pamphagidae | Porthetinae | 37 (2.1) | – | – |
| Lentulidae | Lentulinae | 8 (0.5) | – | – |
| Acrididae | Calliptaminae | 139 (8.0) | – | – |
| | Euryphyminae | 156 (9.0) | 109 (6.30) | 143 (8.3) |
| | Eyprepocnemidinae | 20 (1.6) | 28 (1.6) | 69 (4) |
| | Cyrtacanthacridinae | 44 (2.5) | 29 (1.7) | 102 (5.9) |
| | Acridinae | 8 (0.5) | – | – |
| | Oedipodinae | 251 (14.5) | 224 (13) | 83 (4.8) |
| | Gomphocerinae | 163 (9.4) | 77 (4.45) | 4 (0.2) |

SH = small hills; MH = medium hills; LH = large hills.

on medium and large hills. Nearly 50% of all grasshopper individuals were on small hills, followed by medium hills (27%) and large hills (23%). Lentulidae was the rarest family.

## Discussion

### *Topography and grasshoppers*

Grasshopper assemblages on medium and large hills overall were less speciose, less taxonomically varied and less abundant than those of small hills, indicating that the transition from small to medium hills is a critical threshold for most Succulent Karoo grasshoppers. The sparse vegetation and low grass cover, coupled with cragginess, appear to provide a wide range of microhabitats and promote grasshopper diversity and abundance. Most of these grasshoppers were also distinctly small or large, with small grasshoppers largely associated with the microvariation of habitat on small hills, while large species were mostly on large hills, from which they readily dispersed.

The slopes of all hills had similar but low species richness, with most species being geophilous and adapted to rocky and craggy patches (Uvarov 1977). On the flatlands below, there was high shrubbiness as a result of sheep and cattle removing palatable grasses and leaving unpalatable shrubs. Summits in particular are rarely grazed, resulting in tall grasses being mostly on slopes and summits of medium and large hills, although such a habitat did not necessarily encourage high grasshopper diversity.

Hills are the only habitat for some species. *A. pallidicornis* dominated small hill slopes, and with *P. flavipes*, also dominated small-hill summits with high rockiness and cragginess. Neither occurred on the flatlands. *A. pallidicornis* also dominated summits and slopes of medium hills, although the slopes of large hills were dominated by *C. tatarica*.

*Significance of variable hill sizes for Orthoptera conservation*

Most grasshopper species in the Succulent Karoo are rare, including the brown locust, *Locustana pardalina*, which is a pest when it swarms. Moderate livestock grazing on small hills appears to encourage grasshopper species richness and abundance. This is due to more heterogeneity of patches on small hills created by short and long grasses when the palatable grasses are reduced in height and the unpalatable ones are left behind. Such patch heterogeneity is not common on larger hills because sheep and cattle rarely climb these hills as it takes them far away from water points which are on the flatlands. This trend also occurs on the highly grazed and trampled sites that are closer to water points inside and outside the Mountain Zebra National Park (MZNP) (Gebeyehu and Samways 2001), which is located about 100 km from the current study area. This implies that if grazing pressure on small hills continues, as with overstocking, many grasshopper species, except possibly *L. pardalina*, could undergo local extinction. In contrast, large hills being less accessible to sheep and cattle, are dominated by high grass cover and high greenness (Table 2), serving as refugia for certain grasshopper species that are highly sensitive to grazing disturbance. In general, heterogeneity in the landscape created by hills of various sizes is a desirable attribute for Orthoptera conservation.

Any region may be scaled along at least two axes, distance and organisms. Distance scaling is simply measured as the aerial extent of the region in question (Collins and Glenn 1997). Given that heterogeneity increases as distance between two habitat patches increases, the extent of the area sampled also increases. This means that patterns of species' regional distribution will shift in such a way that the number of regionally rare species increases and the number of regionally abundant species decreases (Brown 1984; Hanski and Gyllenberg 1993). These patterns were observed in this study. There was less dominance on the flatlands, where most species occurred as rare and the number of regionally abundant species decreased.

Since the occurrence of the critical threshold is a function of whether or not a particular species perceives the landscape as connected, it is unlikely that a single threshold value can adequately describe the response of all species in a community to changes in landscape pattern (Wood and Samways 1991; With and Crist 1995), particularly as each organism observes the environment in its own particular way (Levin 1989). Also, the critical threshold is not an inherent property of a landscape, but emerges from the interplay of species interactions with landscape structure (Plotnick and Gardner 1993; Ingham and Samways 1996; With and Crist 1996). Indeed, the patterns of occurrence and distribution shown by the grasshopper species in this study concur with this notion. For example, what *C. tatarica* (large grasshopper) views as a connected landscape is viewed by *A. insubricus* or *O. nigrofasciatus* (which are small grasshoppers) as patchy flatlands and hills. Similarly, while *O. nigrofasciatus* views a patch of bare ground and grasses as connected, *Or. dasycnemis* views bare ground as a critical threshold because its preferred habitat is one made of homogenous stands of tall, green grasses. At a regional scale, flatlands and small hills are generally considered as connected. But at the landscape scale, the various topographies result in clumped rather than random distributions. From a conservation perspective, this means that planning is more than just about area size and shape, but must also include maximum topographic heterogeneity.

### *The significance of grasshopper size*

Most of the grasshopper species in this study were small in size, and were mostly associated with small hills, whereas large species were generally on large hills. Large *C. tatarica* has sustained flight and covers many tens of metres, unlike most of the small species. Since rate of movement is an allometric consequence of hind-leg length in grasshoppers (Gabriel 1985; Bennett-Clark 1990), this species is highly mobile and interacts with landscape at a coarser scale than do small species. While large hills are the most significant topographic features, small hills are important in their own right as landscape patches of high grasshopper abundance and richness.

## Conclusion

Since animal movement is constrained by processes operating at different scales, extrapolation of information on fine-scale movement across broad scales may provide poor quantitative prediction of the spatial dynamics of populations. Even the most detailed spatially explicit population models cannot predict the exact location of individuals across a landscape, or reproduce precise statistical properties of population distributions (Dunning et al. 1995). The best application of this type of modelling approach is in making comparative and qualitative statements about likely population responses to a set of potential or real landscape scenarios. Indeed, such is the case with the interaction between grasshopper assemblage patterns in the Succulent Karoo, as shown in this study. While small scale studies (Gebeyehu and Samways 2001) serve to show a fine-scale picture of patch preference by individual grasshoppers, landscape-level studies serve to make qualitative as well as quantitative comparisons between major physical elements of the landscape and their influence on assemblage patterns of small herbivores such as these grasshoppers.

## Acknowledgements

We thank Stuart Green and Margaret Kieser for assistance with identification of species, and the European Union (FP5) for financial support.

## References


Belovsky G.E. and Slade J.B. 1995. Dynamics of two Montana grasshopper populations: relationships among weather, food abundance and interspecific competition. Oecologia 101: 383–396.

Bennett-Clark H.C. 1990. Jumping in Orthoptera. In: Chapman R.F. and Joern A. (eds), Biology of Grasshoppers. Wiley, New York, pp. 389–406.

Braker H.E. 1989. Evolution and ecology of oviposition on host plants by acridid grasshoppers. Biol. J. Linn. Soc. 38: 389–406.

Brown J.H. 1984. On the relationship between abundance and distribution of species. Am. Nat. 124: 253–279.

Claridge M.F. and Singhrao J.S. 1978. Diversity and altitudinal distribution of grasshoppers (Acridoidea) on a Mediterranean mountain. J. Biogeogr. 5: 239–250.

Collins S.L. and Glenn S.M. 1997. Effects of organismal and distance scaling on analysis of species distribution and abundance. Ecol. Appl. 7: 543–551.

Dunning J.B., Stewart D.J., Danielson B.J., Noon B.R., Root T.L., Lamberson R.H. and Stevens E.E. 1995. Spatially explicit population models: current forms and future uses. Ecol. Appl. 5: 3–11.

Gabriel J.M. 1985. The development of the locust jump mechanism. I. Allometric growth and its effect on jumping performance. J. Exp. Biol. 118: 313–326.

Gebeyehu S. and Samways M.J. 2001. Grasshopper assemblage response to a restored national park (Mountain Zebra National Park) South Africa. Biodivers. Conserv. 11: 283–304.

Hanski I. and Gyllenberg M. 1993. Two general metapopulation models and the core-satellite species hypothesis. Am. Nat. 142: 17–41.

Ingham D. and Samways M.J. 1996. Application of fragmentation and variegation models to epigaeic invertebrates in South Africa. Conserv. Biol. 10: 1353–1358.

Levin S.A. 1989. Challenges in the development of a theory of community and ecosystem structure and function. In: Roughgarden J., May R.M. and Levin S.A. (eds), in Perspectives in Ecological Theory. Princeton University Press, Princeton, pp. 242–255.

Myers N., Mittermeier R.A., Mittermeier C.G., da Fonseca G.A.B. and Kent J. 2000. Biodiversity hotspots for conservation priorities. Nature 403: 853–858.

Narisu J.A., Lockwood A. and Schell S. 2000. Rangeland grasshopper movement as a function of wind and topography: implications for pest management. J. Orthoptera Res. 9: 111–120.

Plotnick R.E. and Gardner R.H. 1993. Lattices and landscapes. In: Gardner R.H. (ed.), in Biology: Predicting Spatial Effects in Ecological Systems. American Mathematical Society, Providence, RI, USA, pp. 129–157.

Samways M.J. 1990. Land forms and winter habitat refugia in the conservation of montane grasshoppers in South Africa. Conserv. Biol. 4: 375–382.

Stabaev I.V. 1972. Periodic changes in the ecological distribution of grasshoppers in the temperate and extreme continental Steppe regions, and their importance for the local ecosystems. In: Hemming C.F. and Taylor T.H.C. (eds), Proceedings of the International Study Conference on Current and Future Problems of Acridology. London pp. 207–213.

Ter Braak C.J.F. 1986. Cannonical Correspondence Analysis: A new eigenvector technique for multivariate direct gradient analysis. Ecology 67: 1167–1179

Turchin P. 1991. Translating foraging movements in heterogenous environments into the spatial distribution of foragers. Ecology 72: 1253–1266.

Uvarov B.P. 1977. Grasshoppers and Locusts. Centre for Overseas Pest Research, London.

Wachter D.H., O'Neill K.M. and Kemp W.P. 1998. Grasshopper (Orthoptera: Acrididae) communities on an elevational gradient in southwestern Montana. J. Kansas Entomol. Soc. 71: 35–43.

Wiens J.A. 1976. Population responses to patchy environments. Annu. Rev. Ecol. Systemat. 7: 81–120.

With K.A. and Crist T.O. 1995. Critical thresholds in species' responses to landscape structure. Ecology 76: 2446–2459.

With K.A. and Crist T.O. 1996. Translating across scales: simulating species distributions as the aggregate response of individuals to heterogeneity. Ecol. Model. 93: 125–137.

Wood P.A. and Samways M.J. 1991. Landscape element pattern and continuity of butterfly flight paths in an ecologically landscaped botanic garden, Natal, South Africa. Biol. Conserv. 58: 149–166.

Biodiversity and Conservation (2006) 15:231–245
DOI 10.1007/s10531-004-7182-3


# Landscape effects on the genetic structure of the ground beetle *Poecilus versicolor* STURM 1824

ANNA-CHRISTINE SANDER*, TOBIAS PURTAUF, STEPHANIE I. J. HOLZHAUER and VOLKMAR WOLTERS

*Department of Animal Ecology, Justus-Liebig-University, Heinrich-Buff Ring 26-32, D-35392 Giessen, Germany; *Author for correspondence (e-mail: Anna-Christine.Sander@bio.uni-giessen.de; phone: +49-641-99-35717; fax: +49-641-99-35709)*

Received 8 April 2004; accepted in revised form 22 November 2004

**Key words:** Carabidae, Landscape structure, Land-use change, *Poecilus versicolor*, Population dynamics

**Abstract.** The impact of landscape structure on the genetic structure of an abundant and widespread carabid beetle *Poecilus versicolor* was analysed in a low mountain range (Lahn-Dill-Bergland, Germany) by means of RAPD-PCR. Habitat patch and landscape characteristics were included as independent variables into a GIS oriented correlative approach. Results indicated a high overall genetic diversity of the beetle population and suggest that the mobility of *P. versicolor* is much higher than previously thought. An equilibrium among migration and genetic drift has not been reached yet, therefore it is very likely that revealed differences in allele frequencies reflect current pattern of genetic diversity. Landscape characteristics at a scale of 1000 × 1000 m surrounding each study site significantly affected the genetic population structure of the carabid beetle, while it is only indirect affected by patch conditions. Opposite effects of grassland and arable land on genetic diversity demonstrated that grassland in the surrounding landscape facilitates dispersal, while arable land apparently decreases successful dispersal, but both factors increase population density on the study site. We conclude that the local population genetic structure of a widespread and highly mobile species as *P. versicolor* is strongly affected by the amount of suitable habitat in a landscape of large habitat proportion. The inclusion of landscape characteristics offers a powerful way for analysing effects on genetic diversity. Further studies on conservation genetics should incorporate a landscape perspective in order to assess the loss of local genetic diversity.

## Introduction

Landscape structure critically affects the long-term persistence of populations by modulating a variety of demographic and genetic factors (Hedrick and Gilpin 1997; Hanski 1998; Manel et al. 2003). One important aspect in human dominated landscapes is the impact of management induced habitat fragmentation on population genetic structure. Despite the generally presumed negative effect of this process on genetic diversity (Gibbs 2001), however, fragmentation can also increase genetic diversity by enhancing adaptive evolution in spatially segregated populations (Grüm 1994; Fahrig 2003). Thus, the loss of spatial patterning associated with the continuous increase of area covered by grassland in regions of Europe where conditions are unfavourable for arable farming (Stanners and Bourdeau 1995) may even lead to a loss of genetic diversity.

We have tested this hypothesis by investigating the genetic structure of the eurytopic grassland beetle *Poecilus versicolor* STRUM 1824 (Coleoptera: Carabidae) in the Lahn-Dill-Bergland, a low mountain range located in Central Hesse (Germany). *P. versicolor* is common in the study region where it occurs in higher density on grasslands (suitable habitat) than on e.g. arable land (unsuitable habitat; Purtauf et al. 2004). Flight activity is rarely observed although the beetle is macropterous (Den Boer 1990; Ribera et al. 2001). According to its dispersal capacity, *P. versicolor* is expected to form spatially structured subpopulations (Den Boer 1990; Lindroth 1992). Decreasing agricultural activity has markedly increased the proportion of grassland in the Lahn-Dill-Bergland (Hietel et al. 2004). We expect increasing amounts of grassland to reduce genetic differentiation of *P. versicolor* by causing a shift from locally distinct subpopulations to a more continuous population (Hiebeler 2000; Vandermeer and Carvajal 2001; Whitlock 2001).

Quantitative information on landscape characteristics are provided by a large spectrum of landscape metrics and indices (e.g. McGarigal and Marks 1995). We selected the amount of suitable habitat as measure of landscape fragmentation because the isolation of a habitat patch and the fragmentation of the surrounding landscape can be accurately viewed as a measure of increase or decrease of suitable habitat (Bender et al. 2003). Widespread and abundant species with a rather low mobility are particularly suited for investigating the genetic response to changes of landscape structure because they are very sensitive to the spatial configuration of their environment (Holsinger and Vitt 1997; Roslin 2001). By focussing on a widespread species in a landscape with increasing habitat proportion we *inter alia* aimed at complementing the many studies on the response of rare species to habitat loss in regions of agricultural intensification (Van Dongen et al. 1998; Ramirez and Haakonsen 1999; Stow et al. 2001; Vucetich et al. 2001). More specifically, we included patch and landscape parameters as independent variables into a GIS oriented correlative approach to address the following questions:

(i) Is the population of *P. versicolor* structured into local subpopulations?
(ii) Can genetic diversity be explained by the amount of suitable and unsuitable habitat in the surrounding landscape?
(iii) Does local density affect genetic diversity?

The genetic structure of *P. versicolor* was analysed by means of RAPD-PCR. Despite some methodological shortcomings as sensitivity to reaction conditions and expression of polymorphisms of nongenetic origin (e.g. Avise 1994; Bielawski et al. 1995; Grosberg et al. 1996), this approach has proven to provide very useful information on population genetic structure and taxonomy (e.g. Nebauer et al. 2000; Osakabe et al. 2000; Bandeira and Nilsson 2001; Vucetich et al. 2001; Cotrim et al. 2003; Rout et al. 2003; Kjølner et al. 2004).

## Material and methods

### *Study area and study sites*

The study was carried out along a gradient of grassland proportion in 17 interspersed landscape sectors located in the Lahn-Dill-Bergland, Central Hesse, Germany (Figure 1). Grassland proportion in the landscape sectors increased continuously along the gradient. This low mountain range comprises some 900 km$^2$. Due to poor soil quality, agricultural use is decreasing for economical reasons. Most of the study area consists of moderately managed grassland, but a small fraction is still in agricultural use. As study sites we selected 17 grassland patches that were moderately grazed by sheep ($n = 13$; about 20 ind. ha$^{-1}$ for a period of max. two weeks year$^{-1}$), horse, or cattle ($n = 4$).

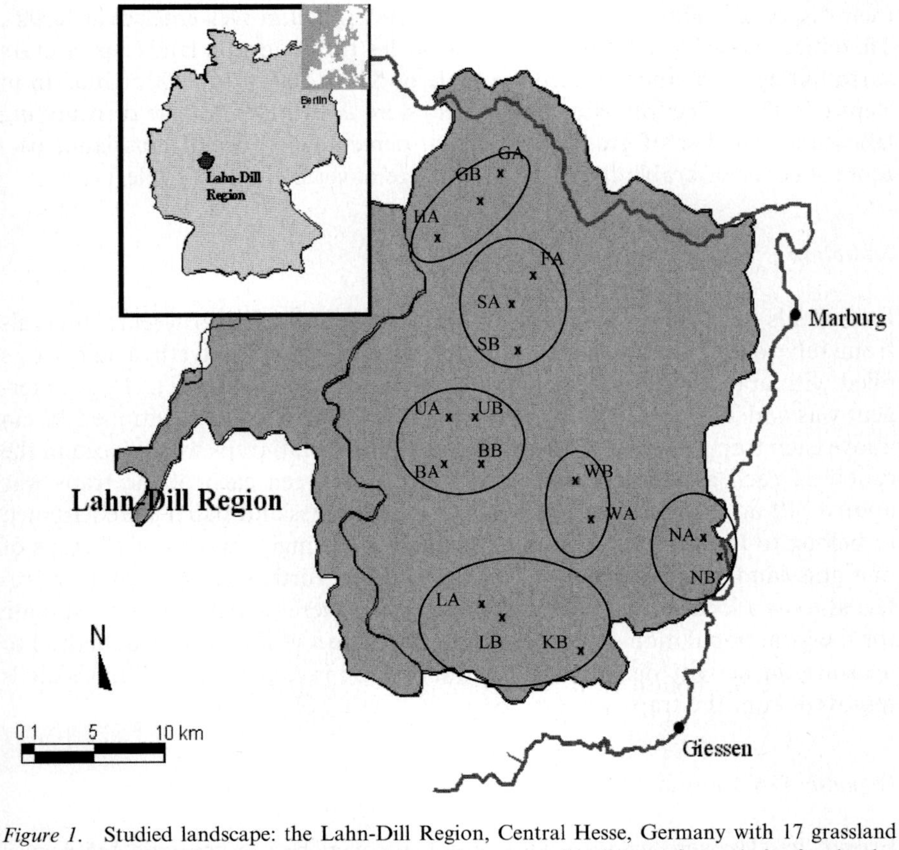


*Figure 1.* Studied landscape: the Lahn-Dill Region, Central Hesse, Germany with 17 grassland patches and groups of grassland patches with smallest geographic distance for analysis of molecular variance (AMOVA).

*Patch and landscape characteristics*

Landscape characteristics were derived from a digital map based on the official German land use database ATKIS (Amtliches Topographisch-Kartographisches Informations system, Hessische Verwaltung für Regionalentwicklung, Kataster und Flurneuordnung, Wiesbaden). ATKIS-data were reclassified into the following land-use types: arable land, grassland, forest, settlement, water, and others, respectively. Patch characteristics (slope, insolation, elevation above sea level) were derived from the Digital Elevation Model (Hessische Verwaltung für Regionalentwicklung, Kataster und Flurneuordnung, Wiesbaden). Insolation was calculated using the formula given by Shary et al. (2002). All maps were transformed into a 20 m grid. Digital terrain and GIS analysis was done using ArcView 3.2 (ESRI). The characteristics of the landscape surrounding each study site were analysed using a Computer Aided Sampling and Landscape Characterization Tool: CAS (CAS: ENVIRO, http://www.geo-extend.com).

Species perceive landscape structure at different spatial scales, according to their dispersal ability (e.g. With and King 2001; Steffan-Dewenter et al. 2002). Therefore, we analysed different spatial scales for quadratic landscape sectors surrounding each study site from a scale of 500 × 500 m to 2000 × 2000 m in steps of 500 m. The following parameters were determined for the surrounding landscape: number of grassland patches, percentage cover of grassland, percentage cover of arable land and percentage cover of forest (Table 1).

*Sampling*

Individuals of *P. versicolor* were sampled with pitfall traps at weekly intervals from July to August 2002. The traps consisted of 500 ml polyethylene beakers filled with approx. 170 ml of a 75% ethanol/glycerine solution (2:1). A detergent was added to reduce surface tension. Acrylic glass cover positioned 10 cm above each trap prevented flooding by rain. Five pitfall traps were placed in the centre of each grassland patch. The distance between each of the traps was approx. 10 m. Individuals trapped on the same grassland patch were assumed to belong to the same local subpopulation. Accordingly, beetles of all traps of one grassland patch were clustered together for further analysis and are referred to as a local subpopulation. These counts were used as a rough estimate for the local population density as pitfall traps are a well-established method to measure the activity density of the ground-living invertebrates. All individuals removed from the traps were stored in 99% ethanol at −20 °C.

*Genomic DNA extraction*

Frozen beetles were grinded with liquid nitrogen before genomic DNA was extracted ($n = 77$ beetles). The procedure followed the manufacturers protocol

*Table 1.* Landscape characteristics (in a quadratic window of 1000 × 1000 m) and patch characteristics at the 17 study sites (Lahn-Dill Region, Germany) with $D$ = local density; and $n$ = number of samples used for genetic analysis.

| Site | Landscape characteristics | | | Patch characteristics | | | | Population characteristics | |
|---|---|---|---|---|---|---|---|---|---|
| | Grassland [%] | Arable land [%] | Forest [%] | Number of grassland patches | Slope | Insolation [J/m$^2$] | Elevation above sea level [m] | $D$ | $n$ |
| NB | 15.1 | 7.6 | 76.7 | 2 | 7.1 | 77.5 | 261 | 4 | 4 |
| NA | 28.0 | 22.7 | 36.9 | 2 | 6.2 | 88.4 | 218 | 4 | 4 |
| WB | 30.7 | 27.2 | 42.0 | 2 | 4.4 | 84.6 | 307 | 10 | 6 |
| WA | 25.2 | 31.7 | 39.8 | 4 | 5.1 | 85.3 | 334 | 80 | 6 |
| KB | 10.9 | 16.3 | 72.3 | 3 | 7.7 | 86.1 | 289 | 6 | 6 |
| GB | 42.6 | 33.8 | 7.7 | 5 | 9.0 | 84.1 | 364 | 62 | 5 |
| GA | 15.3 | 30.6 | 39.3 | 7 | 9.1 | 83.8 | 429 | 9 | 4 |
| SA | 20.9 | 16.3 | 61.0 | 8 | 6.2 | 85.5 | 524 | 2 | 2 |
| FA | 43.3 | 39.9 | 12.7 | 3 | 3.6 | 85.6 | 493 | 26 | 8 |
| SB | 23.1 | 26.6 | 44.2 | 5 | 9.6 | 82.1 | 420 | 5 | 3 |
| BB | 18.1 | 21.2 | 13.2 | 7 | 8.7 | 80.3 | 354 | 4 | 2 |
| UB | 21.5 | 16.2 | 54.7 | 10 | 9.6 | 80.5 | 348 | 5 | 2 |
| UA | 27.6 | 15.6 | 39.9 | 6 | 10.5 | 76.8 | 311 | 24 | 6 |
| LA | 17.8 | 19.4 | 60.7 | 7 | 5.2 | 84.8 | 313 | 3 | 3 |
| LB | 19.9 | 40.4 | 28.8 | 12 | 5.0 | 87.1 | 335 | 27 | 7 |
| HA | 19.1 | 44.2 | 22.8 | 3 | 8.0 | 86.3 | 515 | 43 | 5 |
| BA | 23.0 | 32.0 | 36.2 | 5 | 8.4 | 87.5 | 360 | 11 | 5 |

of the Qiagen DNeasy Tissue Kit. Amounts of 40–80 ng/μl genomic DNA were used for subsequent PCR procedures.

*RAPD-PCR conditions*

A total of 40 decamer oligonucleotide primers ($C = 10$ μM) were tested for polymorphic banding pattern on a subset of samples from different grassland patches. Ten random primers were chosen for the analyses (Table 2): (i) -I3, -I4, -I12, -I18, -I19, -J10, -K19 (Operon Technologies, Inc.), and (ii) -J10, -E14 and -E19 (Carl Roth GmbH + Co). PCR was carried out in 25 μl reactions consisting of 1 μl template DNA, 1 $\rho$M Primer, 2 mM $MgCl_2$, 0.1 mM dNTP's and 0.5 U Taq Polymerase. Temperature and time profile was as follows: (i) denaturation at 94 °C for 10 min, (ii) 45 cycles of 94 °C for 20 s, 34 °C for 40 s and 72 °C for 40 s, and (iii) elongation at 72 °C for 10 min. Samples were electrophoresed on 1.4% agarose gels at 70 mA for approx. 2 h. Gels were stained with 1% ethidiumbromide and photographed under UV light. Precautions were taken to ensure PCR-reproducibility and PCR-conditions were optimised following Bielawski et al. (1995).

*Data analysis*

Molecular data were derived from the RAPD-PCR. Bands with the same mobility were considered as homologous (Grosberg et al. 1996); faint bands were not scored. Although agarose gels stained with ethidiumbromide do not always reveal all bands present, faint bands were considered as absent and well-distinguished bands as present. Images were digitised and presence and absence of bands was coded in a binary matrix (1 and 0, respectively). Detected markers were analysed using the RFLP-Scan Software (Scanalytics 1994).

*Table 2.* Oligonucleotide sequences used for RAPD-PCR, with minimum basepair length (bp min) and maximum basepair length (bp max) of amplified fragments and total number of bands.

| Primer | Sequence | bp min | bp max | Total bands |
|---|---|---|---|---|
| Operon I3 | 5′ > CTGGGGCTG < 3′ | 173 | 2224 | 36 |
| Operon I4 | 5′ > CCGCCTAGTC < 3′ | 189 | 1899 | 21 |
| Operon I12 | 5′ > AGAGGGCACA < 3′ | 159 | 1093 | 25 |
| Operon I18 | 5′ > TGCCCAGCCT < 3′ | 171 | 2970 | 37 |
| Operon I19 | 5′ > AATGCGGGAG < 3′ | 210 | 1463 | 19 |
| Operon J10 | 5′ > AAGCCCGAGG < 3′ | 144 | 1012 | 21 |
| Operon K19 | 5′ > CACAGGCGGA < 3′ | 192 | 1708 | 30 |
| Roth E14 | 5′ > TGCGGCTGAG < 3′ | 195 | 1599 | 28 |
| Roth E19 | 5′ > ACGGCGTATG < 3′ | 80 | 867 | 31 |
| Roth J10 | 5′ > AAGCCCGAGG < 3′ | 125 | 1076 | 41 |

Similarity indices were estimated using the Dice coefficient of similarity (Nei and Li 1979 in NTSys PC 2.01; Rohlf 1998) within and between local subpopulations.

To test for significant differences in allele frequencies we performed an 'Exact test for population differentiation' (Raymond and Rousset 1995), using the program TFPGA. A Markov Chain Monte Carlo approach was employed that gives a good approximation of the exact probability of the observed differences in allele frequencies. For dominant marker sets, this analysis is rather based on markers than on allele frequencies. We run the analysis with 20 batches, 2000 permutations per batch and 1000 dememorization steps to estimate the *p*-value.

We applied a hierarchical analysis of molecular variance (AMOVA) and the corresponding *F*-statistics to investigate the population genetic heterogeneity (Arlequin ver. 2.000; Schneider et al. 2000). The significance of the partitioning of variance components was tested using 1000 permutations. Partitioning is based on the geographical location of the grassland patches, with pairs of grassland patches grouped by smallest geographic distance. Six groups were formed: one group containing four local subpopulations, two groups containing two local subpopulations and three groups with three local subpopulations (see Figure 1). Finally, the randomness in the spatial distribution of individual genotypes was analysed by means of spatial autocorrelation statistics (Degen 2000). Therefore, the spatial genetic structure at RAPD's was calculated as mean genetic distance (Tanimoto distance) between all pairs of individuals belonging to ten spatial distance classes that where measured as Euclidean ground distance and defined according to the spatial autocorrelation statistics software, SGS by Degen et al. (2001). Level of significance was tested using the Monte Carlo permutation procedure with 2000 permutations (Degen et al. 2001).

To test for relationships between landscape and patch parameters on genetic similarity and population density of *P. versicolor*, respectively, general regression models (GRM) were conducted using a forward stepwise multiple regression approach and a multivariate analysis of variances (General Regression Model: GRM; STATISTICA 6.0, StatSoft, Inc., Tulsa, USA). First, a GRM was used to test the effect of differences in landscape pattern on genetic similarity. Independent variables were local density of *P. versicolor*, landscape and patch characteristics (continuous variables) as well as management intensity (categorical variable: grazed by horse and cattle vs. grazed by sheep and mulched once a year). The dependent variable was the genetic similarity of local subpopulation. This parameter was chosen to hinder a potential negative correlation with geographic distance, as expected by isolation by distance (Slatkin 1993). All data were log-transformed to account for potential monotonous non-linear relationships. In a second GRM approach, we tested whether local density was correlated to landscape and patch characteristics (percentage cover of land use types, number of habitat patches, patch characteristics). GRMs were calculated separately for each spatial scale (i.e., 500 × 500 m to 2000 × 2000 m in steps of 500 m). Here we only show results for the scale of 1000 × 1000 m, since this provided the highest variance explanation.

## Results

A total of 325 *P. versicolor* individuals were trapped and a subset of 77 beetles was used for genetic analysis (see Table 1). Two hundred and eighty-nine polymorphic markers were detected but no subpopulation specific markers were revealed by RAPD-PCR due to abundant polymorphism. An overview of scored markers is given in Table 2.

There is significant differentiation in allele frequencies, as revealed by the exact test of population differentiation ($\chi^2 = 558.5$; $p = 0.0$).

The Dice index of genetic similarity of beetles within local subpopulation ranges from approx. 50% (NB, NA, SA, UA) to 67% in the study site of LA and displays a rather high diversity. Genetic similarity within subpopulations is not higher than among subpopulations. The average genetic similarity among all subpopulations amounts to 52%. Further more, the pair-wise mean genetic similarity among local subpopulations shows that most of the study sites display again an average similarity around 50%.

According to the results of the AMOVA, 92% of the overall genetic diversity was found within the local subpopulations of *P. versicolor* and still 9% are found between local subpopulations of neighbouring grassland patches. Very low population differentiation within and among local populations is indicated by the accompanied *F*-statistics (Table 3).

Results of the spatial autocorrelation analysis show a higher genetic distance than expected for smaller distance classes (class I) and lower than expected for higher distance classes (class IV), as displayed in Figure 2, so that an evidence for isolation by distance cannot be found.

The effects of the relationship of genetic similarity and local density, respectively, and landscape and patch characteristics are presented in Table 4. Only four of the nine independent variables included into the first GRM model

*Table 3.* Analysis of molecular variances (AMOVA) and *F*-statistics of the genetic diversity of *Poecilus versicolor*. Local subpopulations are grouped hierarchically according to their geographic location. Note that variance can be slightly negative in absence of genetic differentiation (Schneider et al. 2000).

| Source of variation | df | SS | Var | % | *F*-statistics |
|---|---|---|---|---|---|
| Among groups | 5 | 105.352 | −0.186 | −1.10 | −0.011 |
| *Among populations* | | | | | |
| Within groups | 11 | 242.963 | 1.524 | 9.01 | 0.089* |
| Within populations | 61 | 950.889 | 15.588 | 92.09 | 0.079* |
| Total | 78 | 1299.205 | 16.926 | | |

*$p < 0.01$.

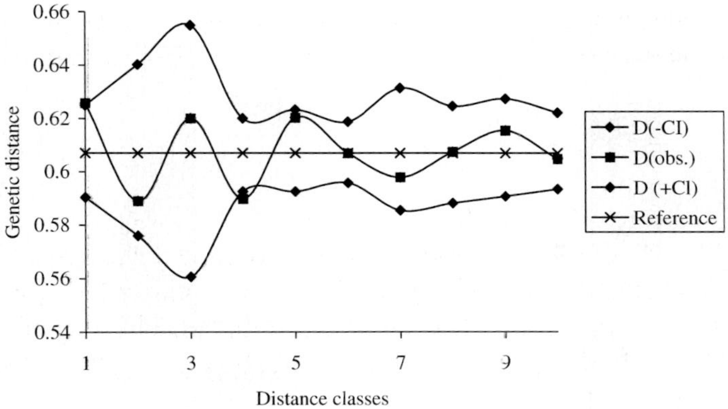


*Figure 2.* Distogram of the mean genetic distance of *P. versicolor* plotted by spatial distance classes. A significant observed correlation between genetic and spatial distance is indicated if points are plotted outside of the limits of the confidence interval, $p < 0.05$.

*Table 4.* Parameter estimates of the GRM (forward stepwise selection) on the effect of patch, landscape and population characteristics, respectively, on (i) pair wise genetic similarity within population units and (ii) local population density ($r^2 = 0.26$; $r^2 = 0.91$) ($p$ significant at $< 0.05$).

| Dependent variable | Genetic similarity | | | Local population density | | |
|---|---|---|---|---|---|---|
| | Level of effect | Variances [%] | *p* | Level of effect | Variance [%] | *p* |
| *Landscape characteristics* | | | | | | |
| Grassland [%] | −0.15 | 11.69 | < 0.001 | 0.52 | 1.08 | 0.010 |
| Arable land [%] | 0.14 | 9.12 | < 0.001 | 0.65 | 1.96 | < 0.001 |
| Forest [%] | n.s. | n.s. | n.s. | −0.66 | 4.42 | < 0.010 |
| Number of grassland patches | n.s. | n.s. | n.s. | 0.71 | 9.89 | < 0.001 |
| *Patch characteristics* | | | | | | |
| Slope | n.s. | n.s. | n.s. | −0.80 | 20.27 | < 0.001 |
| Insolation | n.s. | n.s. | n.s. | −9.26 | 7.33 | < 0.001 |
| Elevation above sea level [m] | 0.15 | 2.76 | 0.03 | −1.96 | 14.91 | < 0.001 |
| Management intensity | n.s. | n.s. | n.s. | −0.40 | 11.95 | < 0.001 |
| *Population characteristics* | | | | | | |
| Local population density | −0.05 | 3.57 | 0.01 | | | |
| Intercept | | 5.52 | < 0.001 | | 11.67 | < 0.001 |
| Error | | 67.32 | | | 16.51 | |

had a significant impact on genetic similarity among individuals of *P. versicolor*. The variance explained by the regression model was approx. 27%, with the landscape characteristics grassland and arable field having the strongest impact (Figure 3, Table 4). Genetic similarity within local subpopulations increased with decreasing percentage cover of grassland and with increasing

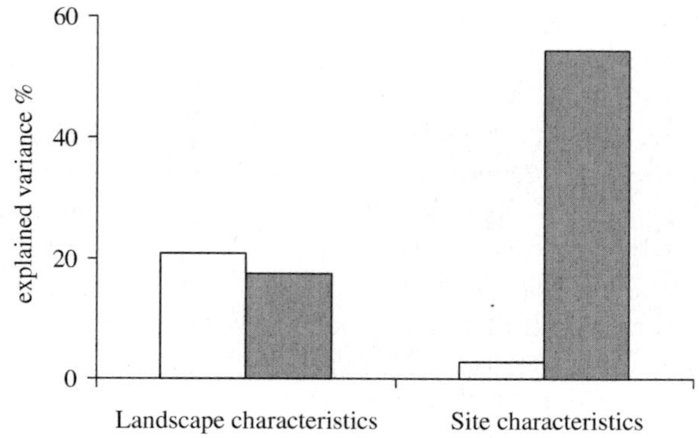


*Figure 3.* Impact of landscape and patch characteristics on the variance explanation of genetic similarity (white bars) and the local density (grey bars) of *P. versicolor* (results of the General Regression Model; see Table 4 for further explanations).

percentage cover of arable land in the surroundings. Moreover, genetic similarity was positively correlated to elevation above sea level. The negative correlation to density indicates that genetic similarity declined with the number of *P. versicolor* individuals on the grassland patch.

Additionally, the GRM on effects of patch and landscape characteristics on local population density of *P. versicolor* showed a significant effect of patch conditions and landscape characteristics (Table 4). The GRM explained about 91% of total variance, with the category patch conditions having the strongest impact (Figure 3). All parameters of patch conditions chosen for our study were negatively correlated to local population density, with the strength of the effect decreasing in the following order: slope > elevation > management intensity > insolation. In contrast, landscape parameters had a positive effect on the density of *P. versicolor*. It increased with the number of grassland patches as well as with the percentage cover of grassland and arable land in the surrounding. Percentage cover of forests in the neighbouring landscape, in contrast, had a negative influence.

## Discussion

The main objective of this study was to highlight the effect of landscape structure on the population genetic structure of the abundant carabid species *P. versicolor*. Our findings emphasise that the decrease of habitat fragmentation increases dispersal and decreases genetic similarity of the local subpopulations, which leads to an overall high genetic diversity of the studied population. The main results are: (i) the population of *P. versicolor* shows significant differences in allele frequencies, so Hardy–Weinberg-equilibrium is not assumed, (ii) the genetic diversity within local subpopulations is very high and spatial segregation among subpopulations is low, (iii) the genetic similarity within the local subpopulations is mainly influenced by the amount of suitable and unsuitable habitat in the surrounding landscape, while local differences in population density are mainly driven by patch conditions, (iv) local genetic similarity declines with increasing population density and increasing amount of habitat in the surrounding landscape. Thus, referring to the questions raised in the introductory section, our data showed that (i) the population of *P. versicolor* is not structured into spatially segregated subpopulations, (ii) the increasing amount of grassland in the landscape leads to a decrease of genetic similarity of local subpopulations and an increase of beetles on the habitat patch, so that spatial segregation of local subpopulations are dissolved and (iii) that local population density decreases genetic similarity within the subpopulation.

*P. versicolor* reveals a high migration rate, which reduces the genetic differentiation across all local subpopulations (Nichols and Freeman 2004). However, an equilibrium among migration and genetic drift has not been reached yet. Therefore, it is very likely that the revealed differences in allele frequencies are due to recent changes in the landscape structure which influence the successful dispersal between local subpopulations of *P. versicolor*.

Several authors have assumed that populations of *P. versicolor* are structured into local subpopulations due to the limited dispersal capacity of this species (Den Boer 1990; Lindroth 1992). But the absence of a clear spatial structure, indicated by the slightly negative variance components of the AMOVA (Schneider at al. 2000) and the results of the autocorrelation analysis, contradicts this contention. Several studies failed to detect isolation-by-distance patterns of different insect populations in either continuous or fragmented landscapes (Van Dongen et al. 1998; Knutsen et al. 2000; Wood and Pullin 2002). But only few studies take measures of landscape structure into account (e.g. Keller et al. 2004; Krauss et al. 2004; Castric et al. 2001). The absence of correlation between spatial distance and genetic diversity, as well as the lack of spatially differentiated populations and high levels of genetic diversity within subpopulations indicates high mobility and dispersal rate of the species. Thus, genetic diversity does not increase at higher spatial scale, as the high gene flow counteracts the effect of genetic drift at larger spatial scale (e.g. Keyghobadi et al. 1999; Vandewoestijne et al. 1999).

Our data show a negative correlation between population density and genetic similarity. Migrants have a strong impact on both the effective local population size and the structure of the local gene pool (Hedrick and Gilpin 1997; Holt et al. 2002). While differences in local population densities are mainly driven by migration (Hedrick and Gilpin 1997; Gibbs 2001; Holt et al. 2002), gene flow may be responsible for genetic differences between pairs of populations (Bossart and Prowell 1998).

Many ecological processes affecting populations and communities operate at a local spatial scale (Rosenzweig 1995). Here, the attractiveness of grasslands for individuals of *P. versicolor* declines with increasing insolation, elevation, slope, and management intensity. Thus, habitats with higher altitude, a steeper topography, and more insolation are unfavourable and are likely to have lower population densities (Brouat et al. 2004; Keller et al. 2004). This is supported by the negative effect of increased altitude on the genetic diversity of local subpopulations, as the stochastic loss of genetic diversity is more probably in smaller populations.

Since *P. versicolor* colonises several types of open land (incl. arable fields; Den Boer 1990), we found that population density increases with higher amount of arable land, grassland and number of patches of grassland. In this context of a landscape with large habitat proportion, the positive effect of the number of grassland patches indicates a higher dispersal of the beetles with a higher habitat fragmentation. This is likely for organisms where migration is an important determinant of population density (Fahrig 2003). As higher population density positively affects genetic diversity of the local subpopulations, the factors mentioned above might have an indirect effect on the genetic structure of the beetle population if they influence the rate of successful dispersal. Considering the increased genetic similarity of *P. versicolor* in landscapes with a high amount of unsuitable habitat (i.e., arable land) this suggests a direct effect of habitat fragmentation. Studies on carabid communities in the study region (Purtauf et al. 2004) have shown that abundance of *P. versicolor* is only half on arable land compared to grassland. Increasing amount of arable land in the surrounding landscape thus seems to decrease successful dispersal and reproduction. This leads to a spatially more explicit genetic differentiation of local subpopulations of *P. versicolor*. The opposite effect is indicated by increasing amount of suitable habitat (i.e., grassland): genetic similarity of local subpopulations decreases and individual numbers increased. Larger habitat proportions thus favours successful dispersal and counteract population subdivision.

By analysing the recent landscape structure we were able to explain approx. 25% of the observed population genetic structure. The increase of grassland in some areas of the study region (Hietel et al. 2004) facilitates exchange among formerly segregated populations. However, we cannot fully exclude that there might be a time lag in our study between changes of landscape structure and genetic diversity as has been reported for other carabid and plant species (Petit and Burel 1998; Lindborg and Eriksson 2004). This assumptions need further analysis.

To conclude, genetic diversity of a widespread and highly mobile species as *P. versicolor* is strongly affected by the amount of suitable habitat in the surrounding landscape. Our results further more support the conclusion that common and widespread species with large populations and low extinction rates provide a good opportunity for studying the impact of land-use change on genetic diversity (Vucetich et al. 2001). This implies that the focus in maintaining genetic diversity should incorporate a landscape perspective to prevent the loss of local genetic diversity. Further studies have to prove whether our findings also apply to landscapes with a higher degree of fragmentation. This will be particularly important considering agricultural landscapes in many parts of Europe (Suarez-Seoane and Baudry 2002).

## Acknowledgements

We thank Thorsten Behrens for providing analysis tools for GIS. Jens Dauber, Thomas Wilke and two anonymous reviewers are acknowledged for critical comments on early versions of the manuscript. We are grateful to Klemens Ekschmitt for assistance in statistical analyses. This study was supported by the German Research Foundation (DFG) in the context of the SFB 299.

## References

Avise J.C. 1994. Molecular Markers, Natural History and Evolution. Chapman and Hall, New York.

Bandeira S.O. and Nilsson P.G. 2001. Genetic population structure of the seagrass *Thalassodendron ciliatum* in sandy and rocky habitats in southern Mozambique. Marine Biol. 139: 1007–1012.

Bender D.J., Tischendorf L. and Fahrig L. 2003. Using patch isolation metrics to predict animal movement in binary landscapes. Landscape Ecol. 18: 17–39.

Bielawski J.P., Noack K. and Pumo D.E. 1995. Reproducible amplification of RAPD markers from vertebrate DNA. BioTechniques 18: 257–260.

Bossart J.L. and Prowell D.P. 1998. Genetic estimates of population structure and gene flow: limitations, lessons and new directions. Trends Ecol. Evol. 13: 202–206.

Brouat C., Chevallier H., Meusnier S., Noblecourt T. and Rasplus J.-Y. 2004. Specialization and habitat: spatial environmental effects on abundance and genetic diversity of forest generalist and specialist *Carabus* species. Mol. Ecol. 13: 1815–1826.

Castric V. Bonney F. and Bernatchez L. 2001. Landscape structure and hierarchical genetic diversity in the brook charr *Salvelinus fontinalis* Mitchill. Evolution 55:1016–1028.

Cotrim H.C., Chase M.W. and Pais M.S. 2003. *Silene rothmaleri* P. Silva (Caryophyllaceae), a rare, fragmented but genetically diverse species. Biodiv. Conserv. 12: 1083–1098.

Degen B. 2000. SGS: Spatial Genetic Software. Computer program and user's manual. http://kourou.cirad.fr/genetique/software.html.

Degen B., Petit R. and Kremer A. 2001. SGS-spatial genetic software: a computer program for analysis of spatial genetic and phenotypic structures of individuals and populations. J. Heredity 92: 447–449.

Den Boer P.J. 1990. The survival value of dispersal in terrestrial arthropods. Biol. Conserv. 54: 175–192.

Fahrig L. 2003. Effects of habitat fragmentation on biodiversity. Ann. Rev. Ecol., Evol. Syst. 34: 487–515.

Gibbs J.P. 2001. Demography versus habitat fragmentation as determinants of genetic variation in wild populations. Biol. Conserv. 100: 15–20.

Grosberg R.K., Levitan D.R. and Cameron B.B. 1996. Characterization of genetic structure and genealogies using RAPD-PCR markers: a random primer for the novice and nervous. In: Ferraris J.D. and Palumbi S.R. (eds), Molecular Zoology. John Wiley and Sons, Inc., New York, pp. 65–100.

Grüm L. 1994. Habitat preference and competition among *Carabus*. In: Desender K., Dufrêne M., Loreau M., Luff M.L. and Maelfait J.-P. (eds), Carabid Beetles: Ecology and Evolution. Kluwer Academic Publishers, pp.295–298.

Hanski I. 1998. Metapopulation dynamics. Nature 396: 41–49.

Hedrick P.W. and Gilpin M.E. 1997. Genetic effective size of a metapopulation. In: Hanski I. and Gilpin M.E. (eds), Metapopulation Biology, Ecology, Genetics and Evolution. Academic Press, London, New York, Tokyo, pp. 165–181.
Hiebeler D. 2000. Populations on fragmented landscapes with spatially structured heterogeneities: landscape generation and local dispersal. Ecology 81: 1629–1641.
Hietel E., Waldhardt R. and Otte A. 2004. Analysing land-cover changes in relation to environmental variables in Hesse, Germany. Landscape Ecol. 19: 473–489.
Holsinger K.E. and Vitt P. 1997. The future of conservation biology: What's a geneticist to do? In: Pickett S.T.A., Ostfeld R.S., Shachak M. and Liken G.E. (eds), The Ecological Basis of Conservation. Chapman and Hall, New York, pp. 202–216.
Holt A.R., Warren P.H. and Gaston K.J. 2002. The importance of biotic interactions in abundance – occupancy relationships. J. Anim. Ecol. 71: 846–854.
Keller I., Nentwig W. and Largiadèr C.R. 2004. Recent habitat fragmentation due to roads can lead to significant genetic differentiation in an abundant flightless ground beetle. Mol. Ecol. 13: 2983–2994.
Keyghobadi N., Roland J. and Strobeck C. 1999. Influence of landscape on the population genetic structure of the alpine butterfly *Parnassius smintheus* (Papilionidae). Mol. Ecol. 8: 1481–1495.
Kjølner S., Såstad M., Taberlet P. and Brochmann C. 2004. Amplified fragment length polymorphism versus random amplified polymorphic DNA markers: clonal diversity in *Saxifraga cernua*. Mol. Ecol. 13: 81–86.
Knutsen H., Rukke B.A., Jorde P. and Ims R.A. 2000. Genetic differentiation among populations of the beetle *Bolitophagus reticulatus* (Coleoptera: Tenebrionidae) in a fragmented and a continuous landscape. Heredity 84: 667–676.
Krauss J., Schmitt T., Seitz A., Steffan-Dewenter I. and Tscharntke T. 2004. Effects of habitat fragmentation on the genetic structure of the monophagous butterfly *Polymmatus coridon* along its northern range margin. Mol. Ecol. 13: 311–320.
Lindborg R. and Eriksson O. 2004. Historical landscape connectivity affects present species diversity. Ecology 85(7): 1840–1845.
Lindroth C.H. 1992. Ground Beetles (Carabidae) of Fennoscandia. A zoogeographic study. In: Part I Specific Knowledge Regarding the Species. Intercept, Andover, Hampshire.
Manel S., Schwartz M.K., Luikart G. and Taberlet P. 2003. Landscape genetics: combining landscape ecology and population genetics. Trends Ecol. Evol. 18: 189–197.
McGarigal K. and Marks B. 1995. FRAGSTATS: Spatial Pattern Analysis Program for Quantifying Landscape Structure. USDA Forest Service, Pacific Northwest Research Station, Portland, Oregon, General Technical Report PNW-GTR-351.
Nebauer S.G., Castillo-Agudo L. and Segura J. 2000. An assessment of genetic relationships within the genus *Digitalis* based on PCR-generated RAPD markers. Theor. Appl. Genet. 100: 1209–1216.
Nichols R.A. and Freeman L.M. 2004. Using molecular markers with high mutation rates to obtain estimates of relative population size and to distinguish the effects of gene flow and mutation: a demonstration using data from endemic mauritian skinks. Mol. Ecol. 13: 775–787.
Osakabe M.H., Hinomoto N., Toda S. and Goka K. 2000. Molecular cloning and characterization of a microsatellite locus found in an RAPD marker of a spider mite, *Panonychus citri* (Acari: Tetranychidae). Exp. Appl. Acarol. 24: 385–395.
Petit S. and Burel F. 1998. Effects of landscape dynamics on the metapopulation of a ground beetle (Coleoptera, Carabidae) in a hedgerow network. Agri., Ecosyst. Environ. 69: 243–252.
Purtauf T., Dauber J. and Wolters V. 2004. Carabid communities in the spatio-temporal mosaic of a rural landscape. Landscape Urban Plann. 67: 185–193.
Ramirez M.G. and Haakonsen K.E. 1999. Gene flow among habitat patches on a fragmented landscape in the spider *Argiope trifasciata* (Araneae: Araneidae). Heredity 83: 580–585.
Raymond M.L. and Rousset F. 1995. An exact test for population differentiation. Evolution 49: 1280–1283.
Ribera J., Dolédec S., Downie J.S. and Foster G.N. 2001. Effect of land disturbance and stress on species traits of ground beetle assemblages. Ecology 82: 1112–1129.

Rohlf F.J. 1998. NTSYS-pc: Numerical Taxonomy and Multivariate Analysis System, Version 2.01. Setauket, Exeter Software, New York.
Rosenzweig M.L. 1995. Species Diversity in Space and Time. Cambridge University Press, Cambridge.
Roslin T. 2001. Spatial population structure in a patchily distributed beetle. Mol. Ecol. 10: 823–837.
Rout G.R., Bhatacharya D., Nanda R.M., Nayak S. and Das P. 2003. Evaluation of genetic relationships in *Dalbergia* species using RAPD markers. Biodiv. Conserv. 12: 197–206.
Schneider S., Roessli D. and Excoiffier L. 2000. Arlequin ver. 2.000: A Software for Population Genetic Data Analysis. Genetics and Biometry Laboratory, University of Geneva, Switzerland.
Shary P.A., Sharaya L.S. and Mitusov A.V. 2002. Fundamental quantitative methods of land surface analysis. Geoderma 107: 1–35.
Slatkin M. 1993. Isolation by distance in equilibrium and non-equilibrium populations. Evolution 47: 264–279.
Stanners D. and Bordeau P. 1995. Europe's Environment. The Dobris Assessment. European Environmental Agency, Copenhagen.
Steffan-Dewenter I., Münzenberg U., Bürger C., Thies C. and Tscharntke T. 2002. Scale-dependent effects of landscape context on three pollinator guilds. Ecology 83(5): 1421–1432.
Stow A.J., Sunnucks P., Briscoe D.A. and Gardner M.G. 2001. The impact of habitat fragmentation on dispersal of Cunningham's skink (*Egernia cunninghami*): evidence from allelic and genotypic analyses of microsatellites. Mol. Ecol. 10: 867–878.
Suarez-Seoane S. and Baudry J. 2002. Scale dependence of spatial patterns and cartography on the detection of landscape change: relationships with species' perception. Ecography 25: 499–511.
Vandermeer J. and Carvajal R. 2001. Metapopulation dynamics and the quality of the matrix. Am. Natur. 158(3): 211–220.
Vandewoestijne S., Nève G. and Baguette M. 1999. Spatial and temporal population genetic structure of the butterfly *Aglais urticae* L. (Lepidoptera, Nymphalidae). Mol. Ecol. 8: 1539–1543.
Van Dongen S., Backeljau T., Matthysen E. and Dhondts A.A. 1998. Genetic population structure of the winter moth (*Operophtera brumata* L.) (Lepidoptera, Geometridae) in a fragmented landscape. Heredity 80: 92–100.
Vucetich L.M., Vucetich J.A., Joshi C.P., Waite T.A. and Peterson R.O. 2001. Genetic (RAPD) diversity in *Peromyscus maniculatus* populations in a naturally fragmented landscape. Mol. Ecol. 10: 35–40.
Whitlock M.C. 2001. Dispersal and the genetic properties of metapopulations. In: Clobert J., Nichols J.D., Danchin E. and Dhondt A.A. (eds), Dispersal. Oxford University Press Inc, New York, pp. 273–282.
With K.A. and King A.W. 2001. Analysis of landscape sources and sinks: the effect of spatial pattern on avian demography. Biol. Conserv. 100: 75–88.
Wood B.C. and Pullin A.S. 2002. Persistence of species in a fragmented urban landscape: the importance of dispersal ability and habitat availability for grassland butterflies. Biodiv. Conserv. 11: 1451–1468.

Biodiversity and Conservation (2006) 15:247–260
DOI 10.1007/s10531-004-6899-3


# The response of ground beetles (Coleoptera: Carabidae) to selection cutting in a South Carolina bottomland hardwood forest

MICHAEL D. ULYSHEN[1,*], JAMES L. HANULA[1], SCOTT HORN[1], JOHN C. KILGO[2] and CHRISTOPHER E. MOORMAN[3]
*[1]USDA Forest Service, Southern Research Station, 320 Green St., Athens, GA 30602-2044, USA; [2]USDA Forest Service, Southern Research Station, Savannah River Site, P.O. Box 700, New Ellenton, SC 29809, USA; [3]NC State University, Department of Forestry, Box 8003, 3028E Biltmore Hall, Raleigh, NC 27695-8003, USA; *Author for correspondence (e-mail: mulyshen@hotmail.com)*

Received 20 April 2004; accepted in revised form 29 November 2004

**Key words:** Canopy gaps, Carabidae, Disturbance, Edge effect, Ground beetles, Group selection harvesting

**Abstract.** We compared the response of ground beetles (Coleoptera: Carabidae) to the creation of canopy gaps of different size (0.13, 0.26, and 0.50 ha) and age (1 and 7 years) in a bottomland hardwood forest (South Carolina, USA). Samples were collected four times in 2001 by malaise and pitfall traps placed at the center and edge of each gap, and 50 m into the surrounding forest. Species richness was higher at the center of young gaps than in old gaps or in the forest, but there was no statistical difference in species richness between old gaps and the forests surrounding them. Carabid abundance followed the same trend, but only with the exclusion of *Semiardistomis viridis* (Say), a very abundant species that differed in its response to gap age compared to most other species. The carabid assemblage at the gap edge was very similar to that of the forest, and there appeared to be no distinct edge community. Species known to occur in open or disturbed habitats were more abundant at the center of young gaps than at any other location. Generalist species were relatively unaffected by the disturbance, but one species (*Dicaelus dilatatus* Say) was significantly less abundant at the centers of young gaps. Forest inhabiting species were less abundant at the centers of old gaps than in the forest, but not in the centers of young gaps. Comparison of community similarity at various trapping locations showed that communities at the centers of old and young gaps had the lowest similarity (46.5%). The community similarity between young gap centers and nearby forest (49.1%) and old gap centers and nearby forest (50.0%) was similarly low. These results show that while the abundance and richness of carabids in old gaps was similar to that of the surrounding forest, the species composition between the two sites differed greatly.

## Introduction

Southeastern bottomland hardwood forests are important for water quality and control, nutrient cycling, wildlife habitat, and they support among the most diverse plant and animal communities in North America (Kellison and Young 1997). To protect this unique ecosystem, and to satisfy increasing demand for forest products, the remaining stands must be maintained and managed properly. According to Guldin (1996), proper forest management attempts to imitate natural rates of succession and disturbance in order to minimize the environmental impacts of timber removal. One promising method for use in bottomland hardwood forests is group selection cutting, an uneven age forest management practice that emulates small-scale natural disturbances (i.e. tree falls, insect outbreaks, wind damage, etc.) to create small openings throughout the forest (Hunter 1990; Guldin 1996; Meadows and Stanturf 1997).

Ground beetles (Carabidae) are taxonomically well known, easily and inexpensively surveyed, and respond quickly to environmental change (Rainio and Niemelä 2003). These attributes have made them useful bioindicators in numerous studies involving disturbance (Allegro and Sciaky 2003; Rainio and Niemelä 2003).

While the response of ground beetles to clearcuts in the conifer forests of Europe and northeastern North America has been well studied Niemelä et al. 1993; Altegrim et al. 1997; Beaudry et al. 1997; Niemelä 1997; Duchesne et al. 1999; Heliola et al. 2001; Koivula 2002a; Koivula et al. 2002; Magura et al. 2003; Pearce et al. 2003), little work has been done on alternative harvesting methods (Altegrim et al. 1997; Werner and Raffa 2000; Koivula 2002b; Koivula and Niemelä 2003; Vance and Nol 2003; Moore et al. 2004), or in hardwood forests (Lenski 1982a; Warriner et al. 2002; Vance and Nol 2003; Moore et al. 2004).

Here we report the results of the first study to examine the response of carabids to group selection cutting in a bottomland hardwood forest in the southeastern United States. We compare the abundance and species richness of carabids in canopy gaps of different size (0.13, 0.26, and 0.50 ha) and age (1 or 7 years) to those at gap edge and in the surrounding forest.

## Materials and methods

### *Study site*

This study was conducted from May to November 2001 on the Savannah River Site (SRS), an 80,269-ha nuclear production facility near Aiken, South Carolina. The SRS is owned and operated by the United States Department of Energy (DOE) as a National Environmental Research Park. Our study site was an approximately 120-ha stand of 75–100 year-old bottomland hardwoods. Common forest trees included numerous oak species (*Quercus* spp.), bald cypress (*Taxodium distichum* (L.) Richard), sweetgum (*Liquidambar styraciflua* L.), red maple (*Acer rubrum* L.), and loblolly pine (*Pinus taeda* L.). The mid-story consisted predominantly of red mulberry (*Morus rubra* L.), ironwood (*Carpinus caroliniana* Walter) and American holly (*Ilex opaca* Aiton). The understory was dominated by dwarf palmetto (*Sabal minor* (Jacquin) Persoon) and switchcane (*Arundinaria gigantean* (Walter) Muhl.). Pre-harvest basal area of the stands was 33 $m^2$/ha (Pauley et al. 1996). The study site often experiences seasonal flooding (January–April) with some low-lying areas remaining under water much of the year. Total rainfall in 2001 was 104 cm with the wettest month being June (23.4 cm) and the driest being December (1.2 cm).

### *Gaps*

Of the 24 gaps used in this study, 12 were created in December 1994 ('old gaps') and 12 in August 2000 ('young gaps'). There were four replicates of three different sizes (0.13, 0.26, and 0.50 ha) for each gap age. The gap area was defined as the area surrounded by the boles of the peripheral dominant forest trees. The gaps were located throughout the 120 ha bottomland hardwood forest, and were spaced at least 200 m apart. Vegetation in old gaps was 1–8 m in height and consisted of pioneer species such as sweetgum, sycamore (*Platanus occidentalis* L.), green ash (*Fraxinus pennsylvanica* Marshall), black willow (*Salix nigra* Marshall), tulip poplar (*Liriodendron tulipifera* L.), oaks, switchcane, and dwarf palmetto. Young gaps contained small stump sprouts or seedling of these species as well as fireweed (*Erechtites hieracifolia* (L.) Raf.), blackberries (*Rubus* spp.), and plumegrass (*Erianthus giganteus* (Walter) Muhl.), other native grasses, and various sedge species (*Cyperus* spp.).

### *Beetle sampling and identification*

Ground beetles were sampled at the center and edge of each gap and in the surrounding forest 50 m from gap edges during four 7-day trapping periods (17–23 May, 10–16 July, 7–13 September, and 3–9 November). Each sample location had a malaise and two pitfall traps to capture flying and crawling beetles, respectively. Malaise traps ('Canopy Traps', Sante Traps, Lexington, KY) differed from the traditional design in that they contained collecting jars at the top and bottom so insects that fall when encountering a barrier were also collected. The traps were suspended from 3 m tall metal hangers.

Pitfall traps consisted of a 480 ml plastic cup buried to ground level. A small funnel (8.4 cm diameter) inserted into the cup directed captured beetles into a smaller 120 ml specimen cup below. The pitfall was positioned at the intersection of four 0.5 m long drift fences. Two pitfall traps were placed 5 m apart at each sample station, and the samples from each were combined for each location (center, edge, and forest). The collecting jars for both pitfall and malaise traps were filled with NaCl–2% formaldehyde solution to preserve specimens and a drop of detergent to reduce surface tension (New and Hanula 1998). Once collected, beetles were brought back to the lab and immediately stored in 70% alcohol. Specimens were sorted to morphospecies and later identified using a reference collection and a key to South Carolina Carabidae (Ciegler 2000). In the interest of accuracy, we were unable to assign species-level names to several morphospecies.

We assigned the most abundant species to categories (open-habitat species (fields, meadows, and disturbed areas), generalist species (open or forested areas), and forest species (forested areas)) based on known habitat data (Larochelle and Lariviere 2003). Not all species were classified into these categories due to inadequate information or to the genus-level identification of several morphospecies (Table 1).

*Statistical analysis*

We combined malaise and pitfall trap captures at each location before analyzing the results. A 3-way analysis of variance with gap age, trap location, and gap size as the main effects showed a significant interaction between gap age and trap location so we analyzed the data for each gap age separately. The General Linear Model procedure of SAS (SAS Institute 1985) was used for all analyses and the Ryan–Einot–Gabriel–Welsch Multiple Range Test was used to determine differences ($\alpha < 0.05$ unless otherwise stated) in relative abundance of insects between trap locations or gap sizes for each gap age (Day and Quinn 1989). We used Raabe's percent similarity (Southwood 1966) to compare similarity among trap locations and trap types.

**Results**

In total, 5498 ground beetles were collected representing 26 tribes, 60 genera, and 87 species. Species richness was higher at the center of young gaps than in old gaps or in the forest, but there was no statistical difference in species richness between old gaps and the forests surrounding them (Figure 1). Carabid abundance followed the same trend (Figure 2), but only with the exclusion of *Semiardistomis viridis* (Say), a very abundant species (23% of the total number) that differed in its response to gap age compared to most other species (Table 1). There was no statistical difference in abundance or species richness among gaps of differing size (Figure 3).

We were able to classify 19 of the 31 most abundant (>25 individuals) species as open-habitat species, generalists, or forest dwellers (Table 1). In general, carabids associated with open-habitat responded positively to canopy gap creation, and were more abundant at the centers of young gaps than at other young or old gap locations (center, edge, or forest) (Figure 4). The number of open-habitat species at the centers of old gaps was comparable to that of the surrounding forest. Likewise, the abundance of generalist carabids was similar among both young and old gap locations (Figure 4). Carabids that prefer forest habitats were less abundant at the centers of young and old gaps than in their respective forest locations, but this was only significant ($p < 0.1$) for old gaps (Figure 4).

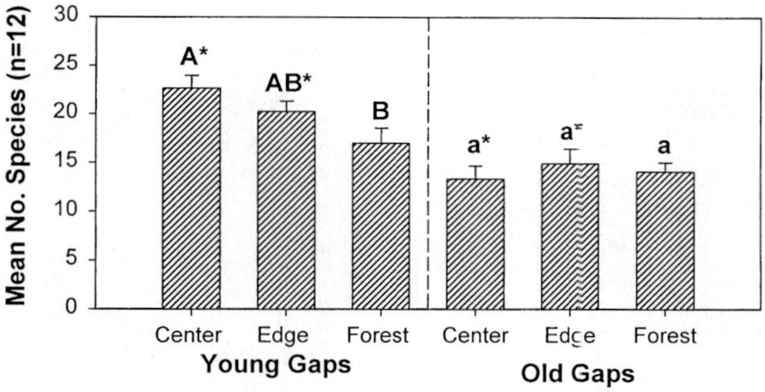


*Figure 1.* Mean (±SE) richness of carabids collected in malaise and pitfall traps in a bottomland hardwood forest, South Carolina, USA in 2001. The traps were placed at the center, edge, and in the forest surrounding 'young' (created in 2000) and 'old' (created in 1994) canopy gaps. Within graphs (for each gap age), bars with the same letter above them are not significantly different (Ryan–Einot–Gabriel–Welsch Multiple Range Test, $p < 0.05$). Asterisks denote significant differences ($p < 0.05$) between the same trap locations (e.g. center vs. center) in old and young gaps.

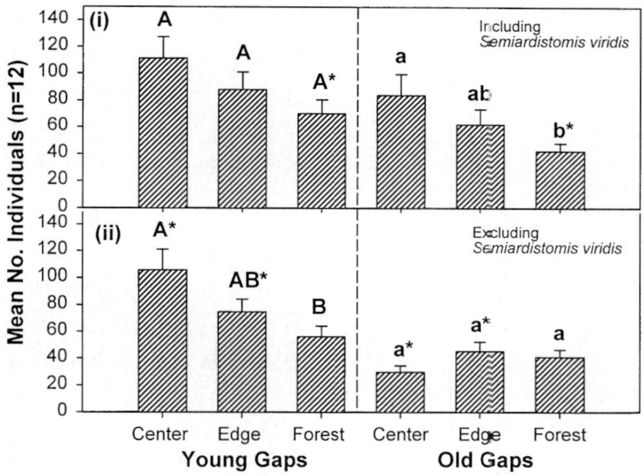


*Figure 2.* Mean (±SE) number of carabids collected in malaise and pitfall traps in a bottomland hardwood forest, South Carolina, USA in 2001. The traps were placed at the center, edge, and in the forest surrounding 'young' (created in 2000) and 'old' (created in 1994) canopy gaps. (b) depicts total beetle abundances excluding *Semiardistomis viridis*. Within graphs (for each gap age), bars with the same letter above them are not significantly different (Ryan–Einot–Gabriel–Welsch Multiple Range Test, $p < 0.05$). Asterisks denote significant differences ($p < 0.05$) between the same trap locations (e.g. center vs. center) in old and young gaps.

Of the 31 most abundant species, ten species exhibited a significant difference among young gap locations and five differed significantly among old gap locations (Table 1). Eight of the ten species that differed among young gap locations were more abundant in the centers of young gaps than in the surrounding forest. Conversely, only two (*Acupalpus* sp. 2 and *S. viridis*) were more abundant at the center of old gaps than in the surrounding forest

*Table 1.* Mean (±SE) number of the most abundant (>25 specimens collected) carabid species collected in malaise and pitfall traps in 2001 at different locations in bottomland hardwood forest gaps ($n$ = 12) created in 1994 (old) and 2000 (new).

| Species | Young gaps | | | Old gaps | | |
|---|---|---|---|---|---|---|
| | Center | Edge | Forest | Center | Edge | Forest |
| Open-habitat species | | | | | | |
| *Acupalpus testaceous* Dejean | A 1.50 ± 1.04 | A 1.17 ± .63 | A .42 ± .23 | a 1.17 ± .56 | a 2.25 ± 1.92 | a 0 |
| *Clivina bipustulata* (Fabricius) | A* 5.50 ± 1.64 | A 2.33 ± .67 | A 3.67 ± .96 | a* .92 ± .36 | a 2.25 ± .72 | a 2.00 ± .28 |
| *Harpalus pennsylvanicus* (De Geer) | A* 2.00 ± .77 | B .25 ± .18 | B .08 ± .08 | a* .17 ± .11 | a .08 ± .08 | a .17 ± .17 |
| *Notiobia terminata* (Say) | A* 1.92 ± .74 | B* .42 ± .19 | B 0 | a* 0 | a* 0 | a 0 |
| *Poecilus chalcites* (Say) | A .33 ± .19 | A .33 ± .19 | A .83 ± .39 | a .17 ± .11 | a .42 ± .23 | a .50 ± .29 |
| *Scarites* spp. Fabricius | A* 3 ± .70 | B* 1.17 ± .37 | B .92 ± .29 | a* .17 ± .11 | a* .17 ± .17 | a .58 ± .26 |
| Generalist species | | | | | | |
| *Brachinus alternans* Dejean | A* 21.83 ± 4.93 | A* 16.25 ± 5.05 | A 9.50 ± 3.48 | a* 5.33 ± 2.25 | a* 4.08 ± 1.74 | a* 3.67 ± 1.97 |
| *Carabus sylvosus* Say | A 1 ± .66 | A 1.5 ± .87 | A .58 ± .23 | a .25 ± .13 | a .17 ± .11 | a .42 ± .19 |
| *Dicaelus dilatatus* Say | A .08 ± .08 | AB .58 ± .19 | B 1.08 ± .36 | a .17 ± .11 | a .83 ± .34 | a .92 ± .36 |
| *Dicaelus elongatus* Bonelli | A* .08 ± .08 | A .75 ± .33 | A 1.17 ± .56 | a* 1.00 ± .35 | a .5 ± .26 | a .33 ± .14 |
| *Galerita* spp. Fabricius | A .58 ± .34 | A .75 ± .33 | A .42 ± .26 | a 1.08 ± .36 | a .42 ± .26 | a .67 ± .43 |
| *Stenolophus ochropezus* (Say) | A 1.75 ± .62 | A* 1.08 ± .42 | A .33 ± .26 | a 1.42 ± .74 | a .08 ± .08 | a 0 |
| Forest species | | | | | | |
| *Chlaenius aestivus* Say | A 5.08 ± 1.39 | A 7.58 ± 2.92 | A 5.67 ± 2.60 | a 5.42 ± 1.78 | a 3.92 ± 1.49 | a 3.50 ± 1.98 |
| *Chlaenius erythropus* Germar | A 1.42 ± .71 | A 1.25 ± .65 | A .83 ± .21 | a .58 ± .29 | a 1.58 ± .99 | a .42 ± .19 |

| | | | | | | |
|---|---|---|---|---|---|---|
| *Cyclotrachelus brevoorti* (LeConte) | A .25 ± .13 | A .75 ± .28 | A 1.08 ± .80 | a .08 ± .08 | a 1.00 ± .83 | a .5 ± .34 |
| *Diplocheila assimilis* (LeConte) | A .83 ± .58 | A .50 ± .42 | A .75 ± .35 | a .25 ± .13 | ab 1.33 ± .61 | b 2.25 ± .72 |
| *Lophoglossus gravis* LeConte | A 2.50 ± .77 | A 6.58 ± 2.28 | A 5.17 ± 1.36 | a 2.42 ± .82 | a 4.92 ± 1.32 | a 4.25 ± 1.49 |
| *Olisthopus* spp. Dejean | A .33 ± .19 | A .92 ± .67 | B 4.83 ± 1.10 | a 0.58 ± .19 | b 1.67 ± .64 | b 4.08 ± .88 |
| *Piesmus submarginatus* (Say) | A .75 ± .75 | A 1.83 ± .76 | A .67 ± .36 | a .08 ± .08 | a 1.5 ± .60 | a 3.67 ± 1.77 |
| Unknown habitat requirements | | | | | | |
| *Acupalpus* sp. 2 | A* 8.67 ± 2.81 | B* 3.00 ± 1.04 | B 1.33 ± 0.80 | a* 1.17 ± 0.32 | b* 0.50 ± 0.15 | b 0.25 ± 0.13 |
| *Agonum aeruginosum* Dejean | A .67 ± .28 | A .25 ± .18 | A .83 ± .21 | a .08 ± .08 | a .08 ± .08 | a .33 ± .19 |
| *Agonum decorum* (Say) | A* 3.17 ± .89 | AB* 2.33 ± .69 | B .42 ± .26 | a* .08 ± .08 | a* .58 ± .36 | a .33 ± .19 |
| *Chlaenius* sp. 3 | A .92 ± .42 | A 4.17 ± 1.73 | A 3.75 ± 1.41 | a .75 ± .45 | a 1.42 ± .68 | a 1.58 ± .63 |
| *Clivina rubicunda* LeConte | A* 2.0 ± .62 | B .42 ± .15 | B 0 | a* 0 | a .17 ± .11 | a .08 ± .08 |
| *Cymindis* spp. Latreille | A .33 ± .14 | A 0.75 ± .33 | A .92 ± .36 | a .33 ± .19 | a 1.17 ± .34 | a 1.5 ± .65 |
| *Loxandrus* sp. LeConte | A* 4.83 ± 1.09 | A 2.58 ± 0.50 | A 2.42 ± 0.70 | a* 0.67 ± 0.22 | a 2.08 ± 0.82 | a 1.75 ± 0.73 |
| *Micratopus aenescens* (LeConte) | A* 14.83 ± 5.96 | A 7.17 ± 1.67 | A* 2.67 ± .53 | b* 1.08 ± .34 | a 5.33 ± 1.80 | ba* 4.25 ± .84 |
| *Oodes amaroides* Dejean | A* 3.42 ± .92 | B* 1.50 ± .54 | B .25 ± .18 | a* .42 ± .19 | a* .25 ± .13 | a .17 ± .11 |
| *Oodes* sp. 2 | A .83 ± .59 | A .83 ± .59 | A .83 ± .44 | a .25 ± .18 | a .17 ± .11 | a .17 ± .11 |
| *Paratachys* spp. Casey | A* 4.33 ± 1.51 | B .58 ± .19 | AB 2.08 ± 1.02 | a* .92 ± .51 | a 1.67 ± .90 | a .83 ± .30 |
| *Semiardistomis viridis* (Say) | A* 5.42 ± 1.90 | A 13.25 ± 5.54 | A* 14.08 ± 3.40 | a* 54.33 ± 12.29 | b 16.67 ± 6.27 | b* 1.58 ± .43 |

For each species and gap age, values with the same letter are not significantly different (Ryan–Einot–Gabriel–Welsch Multiple Range Test, $p < 0.05$).
*denotes significant differences ($p < 0.05$) between the same trap locations (e.g. center vs. center) in young and old gaps.

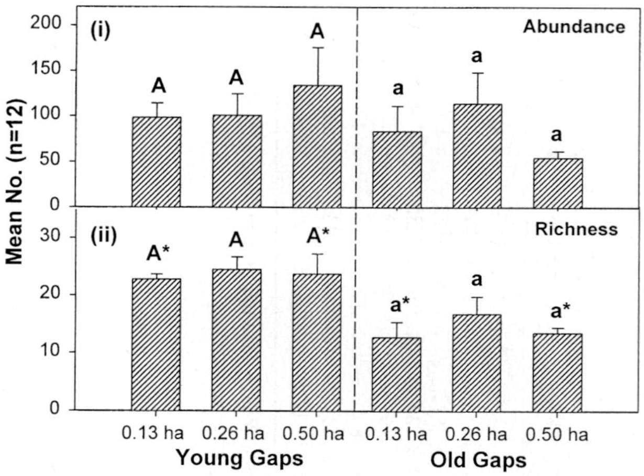


*Figure 3.* Mean (± SE) abundance (a) and richness (b) of carabids collected in malaise and pitfall traps in 2001 in bottomland hardwood forest gaps of different size (0.13, 0.26, and 0.50 ha) created in 1994 and 2000 in South Carolina, USA. Within graphs (for each gap age), bars with the same letter above them are not significantly different (Ryan–Einot–Gabriel–Welsch Multiple Range Test, $p < 0.05$). Asterisks denote significant differences ($p < 0.05$) between the same trap locations (e.g. center vs. center) in old and young gaps.

(Table 1). While *Acupalpus* sp. 2 was more abundant at the centers of both young and old gaps, *S. viridis* was much more numerous in old gap centers than at the edge or in the forest. It was also significantly more abundant in old gap centers than in young gap centers. Several abundant species appeared to respond positively to recent disturbance, but only *Notiobia terminata* (Say) was found exclusively in young gaps (Table 1).

The most similar carabid assemblages were those at the edges of gaps and the forests surrounding them (Table 2). The edges of old and young gaps also had a high degree of similarity (72%). The least similar carabid communities were those at the centers of young and old gaps (Table 2), but carabid assemblages in gap centers and surrounding forests also had relatively low similarity.

## Discussion

Many studies have shown an overall increase in the species richness and/or abundance of carabids following disturbance (Eryschov and Trophimova 1984; Niemelä et al. 1993, 1994; Thompson and Allen 1993; Beaudry et al. 1997; Heliola et al. 2001; Warriner et al. 2002; Koivula et al. 2002). While some studies have found no overall change in carabid abundance or species richness, they have identified significant affects at the species level (Atlegrim et al. 1997) as well as differences in species composition between disturbed and undisturbed sites (Greenburg and Thomas 1995; Butterfield 1997; Werner and Raffa 2000).

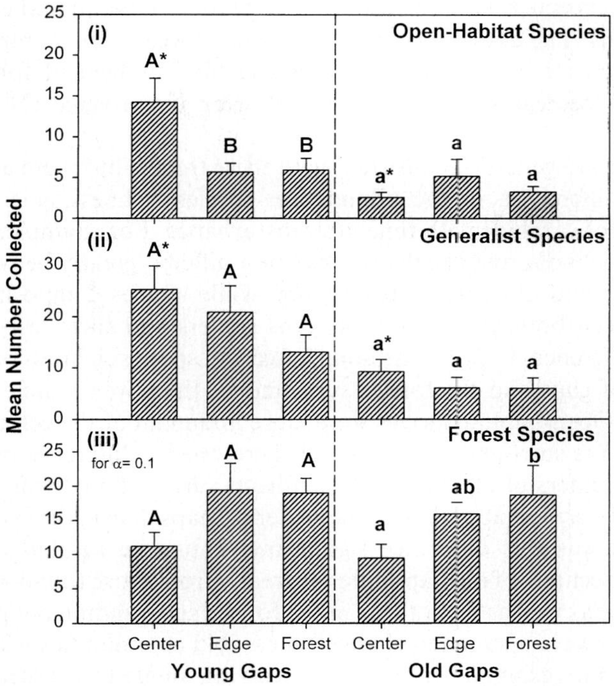


*Figure 4.* Mean (±SE) number of carabids collected in malaise and pitfall traps in a bottomland hardwood forest, South Carolina, USA in 2001. The traps were placed at the center, edge, and in the forest surrounding 'young' (created in 2000) and 'old' (created in 1994) canopy gaps. The species (Table 2) were categorized as preferring open-habitat (a), being generalists (b), or preferring intact forests (c) based on information in Larochelle and Lariviere (2003). Within graphs (for each gap age), bars with the same letter above them are not significantly different (Ryan–Einot–Gabriel–Welsch Multiple Range Test, $p < 0.05$). Asterisks denote significant differences ($p < 0.05$) between the same trap locations (e.g. center vs. center) in old and young gaps.

*Table 2.* Raabe's percent similarity of carabids in new (1 year) vs. old (7 years) canopy gaps by location (center, edge, or 50 m into surrounding forest) in a South Carolina bottomland hardwood forest, 2001.

| Comparison | Percent similarity |
|---|---|
| New Edge vs. New Forest | 76.21 |
| Old Edge vs. Old Forest | 75.30 |
| New Edge vs. Old Edge | 72.33 |
| New Forest vs. Old Forest | 67.66 |
| New Center vs. New Edge | 60.64 |
| Old Center vs. Old Edge | 58.47 |
| Old Center vs. Old Forest | 50.01 |
| New Center vs. New Forest | 49.11 |
| New Center vs. Old Center | 46.49 |

As might be expected, habitat specificity appears to determine the response of many carabids. The abundance of open-habitat species, for example, has been shown to increase in disturbed areas, while the numbers of forest-dwelling species often decreases or disappears following disturbance (Niemelä et al. 1993).

Our results are generally consistent with these trends, but there appears to be substantial differences in the abundance, species richness, and community composition of carabids with time after disturbance. For example, the carabid abundance, richness, and species composition differed greatly between the centers of young and old gaps. Furthermore, while species composition differed greatly between both young and old gaps centers and their respective forest locations, differences between the abundance and species richness of carabids at the centers of gaps and the forests surrounding them was significant only for young gaps. Open-habitat species were more abundant at the centers of young gaps than in the surrounding forest, but there was no difference in abundance between the centers of old gaps and the forests surrounding them. Conversely, forest species were less abundant at the centers of gaps than in the forest, but only for old gaps was this difference significant. Thus, the carabid communities present at the centers of old gaps differed greatly from those found at the centers of young gaps as well as from those in the forests surrounding old gaps.

Past studies have also noted changes in carabid communities with time after disturbance. For example, in a study involving single-tree selection cutting, Vance and Nol (2003) found reduced activity densities in recently (0.5–3 years) cut stands compared to reference stands, while the activity densities for certain species was higher in older (15–20 years) cut stands. The authors attribute these differences to significant reductions in leaf litter in the recently cut stands, and to differences in the vegetation in older stands. The importance of factors such as vegetation structure, temperature, humidity, light intensity, and soil moisture to ground beetles is well supported by past research (Lenski 1982a; Cardenas and Bach 1989; Thompson and Allen 1993; Magura et al. 1997; Antvogel and Bonn 2001; Warriner et al. 2002).

Reduced competitive exclusion may have played a role in the higher abundance and species richness of carabids observed in young gaps (Allen and Thompson 1977; Lenski 1982a, b), but we suspect that it had relatively little effect in this study. The increase in habitat heterogeneity following disturbance was probably much more important. For example, timber removal created large amounts of coarse woody debris and greatly increased the complexity of the gap floor. While young gaps contained an abundance of CWD, little remained in the old gaps. Differences in vegetation between young and old gaps were similarly dramatic. In contrast to young gaps in which there were scattered clumps of grasses, tree sprouts, and herbaceous growth, old gaps were covered in a dense growth of young trees competing for sunlight. Because young and old gaps were so different in habitat structure, it is not surprising that carabid abundance, species richness, and composition differed greatly between the two locations.

Because the carabid communities at the edges of young and old gaps were so similar to those in the surrounding forest, we have little indication of a distinct edge community. Although researchers in Hungary reported unique edge communities as well as several species unique to edge habitats (Magura and Tothmeresz 1997; Magura et al. 2001; Magura 2002), the results from other studies are similar to our own (Spence et al. 1996; Heliola et al. 2001; Kotze and Samways 2001).

The carabid community in seven-year old gaps is far from recovered, despite comparable abundance and species richness between old gaps and the surrounding forest. This is indicated by the low degree of similarity between the two sites. In fact, carabids at the centers of old gaps are only slightly more similar to those in the forest than are the carabids at young gap centers (50.0 and 49.1% similar, respectively). These results emphasize the fact that abundance should not be used alone (Moore et al. 2004) to determine the recovery time of carabid assemblages.

Although Vance and Nol (2003) found an increase in both open-habitat species and forest generalists 0.5–3 and 15–20 years after single-tree selection harvests, we could identify no common trend among carabids between young and old canopy gaps. The response of carabids to young and old gaps differed greatly, even among species with similar habitat preferences. For example, of the six common open-habitat species in this study, three were significantly more abundant at young gap centers than at the edges or in the forest surrounding young gaps, but there were no differences among old gap locations. Furthermore, of the 31 most abundant morphospecies collected, 13 exhibited a significant difference among either young or old gap locations. Of these, only five differed significantly among old gap locations and just two responded similarly to young and old gaps.

These differences in abundance between young and old gaps are probably due to the specific habitat requirements of each species. Given this, it is interesting to note that just two species (*S. viridis* and an *Acupalpus* species) were more abundant at the centers of old gaps than in the nearby forest. While the *Acupalpus* species was more abundant at the center of young gaps than old gaps, *S. viridis* was more abundant at the centers of old gaps than at any other young or old gap location. This result further emphasizes the importance of time considerations when studying the effects of disturbance on ground beetles, as well as the species-specific response of carabids to disturbance.

While many species tend to be more abundant in disturbed habitats, several have been shown to exist there exclusively (Niemelä et al. 1993; Thompson and Allen 1993; Beaudry et al. 1997; Warriner et al. 2002). For example, in this study, *N. terminata* was collected only in the center or at the edge of young gaps. Similarly, a number of forest species were found in much greater numbers in the forest than elsewhere. While we found no substantial evidence for the presence of strict forest specialists, such species may have been collected in low numbers (and could not be analyzed statistically) or not at all. Since total carabid abundance in the forest near young gaps was different from that near old gaps, gap creation had a definite effect on the carabid community at least 50 m into the surrounding forest. Because of this, obligate forest species, if present, may have found the forests surrounding the gaps to be unsuitable. Although many carabids will eventually recolonize an area after disturbance (Koivula et al. 2002) some forest specialists are unable to reestablish populations in regenerating clear-cut stands (Beaudry et al. 1997) and old growth species with poor dispersal ability may face local extinction if stands of mature forest are not preserved (Halme and Niemelä 1993; Spence et al. 1996; Beaudry et al. 1997; Heliola et al. 2001; Koivula et al. 2002). Because group selection harvesting disturbs smaller patches of bottomland hardwood forest at any one time and is more similar to natural levels of disturbance, it may lessen the detrimental effects of disturbance on these sensitive forest species.

How group selection cutting compares to other forestry practices, remains unclear. Recent work in Finland has found small (0.16 ha) openings to be less disruptive of community structure than larger clear-cut stands (Koivula 2002b; Koivula and Niemelä 2003) but much more comparative work is needed to ascertain the advantages of various harvesting techniques with respect to environmental health. Different forests have different natural rates of disturbance (Hunter 1990; Guldin 1996) so the effects of a particular management technique may depend upon the forest type under consideration. Because few carabids were negatively affected by gap creation, and none seemed to be completely eliminated by the disturbance, we feel that group selection cutting may be particularly well suited to bottomland hardwood forests and deserves further consideration.

## Acknowledgements

We thank H. Lee, Jr. and C. Smith for assisting with the identification of specimens, D. Dyer and J. Campbell for helping with field work, and J. McHugh for editing an early draft of the manuscript. We also thank the U.S. Department of Energy-Savannah River and the USDA Forest Service-Savannah River for access and logistical support. This research was funded by the National Research Initiative Competitive Grants Program of the USDA Cooperative State Research Education and Extension Service (CSREES Grant No. 00-35101-9307).

## References

Allegro G. and Sciaky R. 2003. Assessing the potential role of ground beetles (Coleoptera, Carabidae) as bioindicators in poplar stands, with a newly proposed ecological index (FAI). Forest Ecol. Manag. 175: 275–284.

Allen R.T. and Thompson R.G. 1977. Faunal composition and seasonal activity of Carabidae (Insecta: Coleoptera) in three different woodland communities in Arkansas. Ann. Entomol. Soc. Am. 70: 31–34.

Altegrim O., Sjoeberg K. and Ball J.P. 1997. Forestry effects on a boreal ground beetle community in spring: selective logging and clear-cutting compared. Entomol. Fennica 8: 19–26.

Antvogel H. and Bonn A. 2001. Environmental parameters and microspatial distribution of insects: a case study of carabids in an alluvial forest. Ecography 24: 470–482.

Beaudry S., Duchesne L.C. and Cote B. 1997. Short-term effects of three forestry practices on carabid assemblages in a jack pine forest. Can. J. Forest Res. 27: 2065–2071.

Butterfield J. 1997. Carabid community succession during the forestry cycle in conifer plantations. Ecography 20: 614–625.

Cardenas A.M. and Bach C. 1989. The effect of some abiotic factors on the distribution and selection of habitat by the carabid beetles in the central Sierra Morena Mountains (SW Cordoba, Spain). Vie et milieu 39: 93–103.

Ciegler J.C. 2000. In: Wheeler A.G. (ed.), Ground Beetles and Wrinkled Bark Beetles of South Carolina (Coleoptera: Geadephaga: Carabidae and Rhysodidae). Clemson University, 149 pp.

Day R.W. and Quinn G.P. 1989. Comparisons of treatments after an analysis of variance in ecology. Ecol. Monogr. 59: 433–463.

Duchesne L.C., Lautenschlager R.A. and Bell F.W. 1999. Effects of clear-cutting and plant competition control methods on carabid (Coleoptera: Carabidae) assemblages in northwestern Ontario. Environ. Monitor. Assess. 56: 87–96.

Eryschov V.I. and Trophimova O.L. 1984. Changes in the population of ground beetles (Coleoptera, Carabidae) in a clearing in the mountain taiga of the Kuznetsky Alatau foothills. Zool. zhurnal Moscow 63: 848–852.

Greenburg C.G. and Thomas M.C. 1995. Effects of forest management on coleopteran assemblages in sand pine scrub. Florida Entomol. 78: 271–285.

Guldin J.M. 1996. The role of uneven-aged silviculture in the context of ecosytem management. Western J. Appl. Forest. 11: 4–13.

Halme E. and Niemelä J. 1993. Carabids beetles in fragments of coniferous forest. Ann. Zool. Fenn. 30: 17–30.

Heliola J., Koivula M. and Niemelä J. 2001. Distribution of carabid beetles (Coleoptera, Carabidae) across a boreal forest-clearcut ecotone. Conserv. Biol. 15: 370–377.

Hunter M.L. 1990. Wildlife, Forests, and Forestry. Prentice-Hall, Englewood Cliffs, New Jersey, USA.

Kellison R.C. and Young M.J. 1997. The bottomland hardwood forest of the southern United States. Forest Ecol. Manag. 90: 101–115.

Koivula M. 2002a. Boreal carabid-beetle (Coleoptera: Carabidae) assemblages in thinned uneven-aged and clear-cut spruce stands. Ann. Zool. Fenn. 39: 131–149.

Koivula M. 2002b. Alternative harvesting methods and boreal carabid beetles (Coleoptera: Carabidae). Forest Ecol. Manag. 167: 103–121.

Koivula M., Kukkonen J. and Niemelä J. 2002. Boreal carabid-beetle (Coleoptera: Carabidae) assemblages along the clear-cut originated succession gradient. Biodivers. Conserv. 11: 1269–1288.

Koivula M. and Niemelä J. 2003. Gap felling as a forest harvesting method in boreal forests: responses of carabid beetles (Coleoptera: Carabidae). Ecography 26: 179–187.

Kotze D.J. and Samways M.J. 2001. No general edge effects for invertebrates at Afromontane forest/grassland ecotones. Biodivers. Conserv. 10: 443–466.

Larochelle A. and Lariviere M.-C. 2003. A Natural History of the Ground Beetles (Coleoptera:Carabidae) of America North of Mexico. Pensoft Series Faunistica No. 7. Pensoft Publishers, 583 pp.

Lenski R.E. 1982a. The impact of forest cutting on the diversity of ground beetles (Coleoptera: Carabidae) in the southern Appalachians. Ecol. Entomol. 7: 385–390.

Lenski R.E. 1982b. Effects of forest cutting on two *Carabus* species: evidence for competition for food. Ecology 63: 1211–1217.

Magura T., Thothmeresz B. and Bordan Z. 1997. Comparison of the carabid communities of a zonal oak–hornbeam forest and pine plantations. Acta Zool. Acad. Sci. Hungaricae 43: 173–182.

Magura T. and Tothmeresz B. 1997. Testing edge effect on carabid assemblages in an oak-hornbeam forest. Acta Zool. Acad. Sci. Hungaricae 43: 303–312.

Magura T., Tothmeresz B. and Molnar T. 2001. Forest edge and diversity: carabids along forest-grassland transects. Biodivers. Conserv. 10: 287–300.

Magura T. 2002. Carabids and forest edge: spatial pattern and edge effect. Forest Ecol. Manag. 157: 23–37.

Magura T., Tothmeresz B. and Elek Z. 2003. Diversity and composition of carabids during a forestry cycle. Biodivers. Conserv. 12: 73–85.

Meadows J.S. and Stanturf J.A. 1997. Silvicultural systems for southern bottomland hardwood forests. Forest Ecol. Manag. 90: 127–140.

Moore J.-D., Ouimet R., Houle D. and Camire C. 2004. Effects of two silvicultural practices on ground beetles (Coleoptera: Carabidae) in a northern hardwood forest, Quebec, Canada. Can. J. Forest Res. 34: 959–968.

New K.C. and Hanula J.L. 1998. Effect of time elapsed after prescribed burning in longleaf pine stands on potential prey of the Red-Cockaded Woodpecker. Southern J. Appl. Forest. 22(3): 175–183.

Niemelä J. 1997. Invertebrates and boreal forest management. Conserv. Biol. 11: 601–610.

Niemelä J., Langor D. and Spence J.R. 1993. Effects of clear-cut harvesting on boreal ground-beetle assemblages (Coleoptera:Carabidae) in western Canada. Conserv. Biol. 7: 551–561.

Niemelä J., Spence J.R., Langor D., Haila Y. and Tukia H. 1994. Logging and boreal ground-beetle assemblages on two continents: implications for conservation. In: Gaston K., Samways M. and New T. (eds), Perspectives Insect in Conservation. Intercept, Andover, pp. 29–50.

Pauley E.F., Collins B.S. and Smith W.P. 1996. Early establishment of cherrybark oak in bottomland hardwood gaps: effects of seed predation, gap size, herbivory, and competition. In: Flynn K.M. (ed.), Proceedings of the Southern Forested Wetlands Ecology and Management Conference. Clemson, S.C, pp.132–136.

Pearce J.L., Venier L.A., McKee J., Pedlar J. and McKenney D. 2003. Influence of habitat and microhabitat on carabid (Coleoptera: Carabidae) assemblages in four stand types. Can. Entomol. 135: 337–357.

Rainio J. and Niemelä J. 2003. Ground beetles (Coleoptera: Carabidae) as bioindicators. Biodivers. Conserv. 12: 487–506.

SAS Institute 1985. SAS Guide for Personal Computers, 6th ed. SAS Institute, Cary, NC, 378 pp.

Southwood T.R.E. 1966. Ecological Methods with Particular References to the Study of Insect Populations. Methuen & Co. Ltd, London, 391 pp.

Spence J.R., Langor D.W., Niemelä J., Carcamo H.A. and Currie C.R. 1996. Northern forestry and carabids: the case for concern about old-growth species. Ann. Zool. Fenn. 33: 173–184.

Thompson L.C. and Allen R.T. 1993. Site preparation affects ground beetles in a clearcut bottomland hardwood forest in southeastern Arkansas. In: Brissette J.C. (ed.), Proceedings of the Seventh Biennial Southern Silvicultural Research Conference, 1992 November 17–19, Mobile, AL. Gen. Tech. Rep. SO-93. U.S. Department of Agriculture, Forests Service, Southern Forest Experiment Station, New Orleans, LA, pp. 57–64.

Vance C.C. and Nol E. 2003. Temporal effects of selection logging on ground beetle communities in northern hardwood forests of eastern Canada. Ecoscience 10: 49–56.

Warriner M.D., Nebeker T.E., Leininger T.D. and Meadows J.S. 2002. The effects of thinning on beetles (Coleoptera: Carabidae, Cerambycidae) in bottomland hardwood forests. In: Outcalt K.W. (ed.), Proceedings of the Eleventh Biennial Southern Silvicultural Research Conference, Gen. Tech. Rep. SRS-48. U.S. Department of Agriculture, Forest Service, Southern Research Station, Asheville, NC, pp. 569–573.

Werner S.M. and Raffa K.F. 2000. Effects of forest management practices on the diversity of ground-occurring beetles in mixed northern hardwood forests of the great lakes region. Forest Ecol. Manag. 139: 135–155.

Biodiversity and Conservation (2006) 15:261–280
DOI 10.1007/s10531-004-8231-7 

# Factors influencing bug diversity (Insecta: Heteroptera) in semi-natural habitats

CORINNE ZURBRÜGG[1] and THOMAS FRANK[1,2,*]
[1]*Zoological Institute, University of Bern, Baltzerstrasse 6, CH-3012 Bern, Switzerland;* [2]*Present address: Institute of Zoology, Department of Integrative Biology, University of Natural Resources and Applied Life Sciences, Gregor Mendelstrasse 33, A-1180 Vienna, Austria; *Author for correspondence (e-mail: thomas.frank@boku.ac.at; phone: +43-1-47654-3201; fax: +43-1-47654-3203)*

Received 9 July 2004; accepted in revised form 7 December 2004

**Key words:** Agroecosystems, Bug community, Canonical correspondence analysis, Ecological compensation areas, Flower abundance, Functional groups, Insect diversity, Species richness, Vegetation structure

**Abstract.** We investigated the abundance and species richness of heteropteran bugs and explored environmental factors which influence bug diversity in three types of semi-natural habitats (wildflower areas, extensively used meadows, extensively grazed pastures). To cover this topic, it is essential to know how much the relatively young wildflower areas contribute to biodiversity compared with well-established extensive meadows and pastures. Total bug species richness and phytophagous bug species richness were significantly higher in wildflower areas and meadows than in pastures. In wildflower areas, we found the highest number of zoophagous bug species and species overwintering in the egg-stage. Species overwintering as adults were most abundant in meadows. Total number of bug species as well as species richness in either trophic groups and overwintering strategies were significantly positively correlated with vegetation structure. Except for overwintering strategies, the same was true to bug abundance. The bug community based on the number of individuals per species was significantly explained by flower abundance and vegetation structure, accounting for 18.4 and 16.8% of the variance, respectively. Our results indicate that vegetation structure and flower abundance are key factors for bug species richness, abundance and bug species composition. Since wildflower areas and meadows clearly increased bug species richness and contained several specialised bug species that did not occur in pastures, we recommend the promotion of wildflower areas and extensively used meadows in order to restore both high heteropteran diversity and overall insect biodiversity in agricultural landscapes.

## Introduction

In recent decades a dramatic decrease of biological diversity in European agricultural landscapes can be observed (Edwards et al. 1999; Marshall and Moonen 2002). The rapid decline in plant and animal species diversity in modern agricultural landscapes can be explained by two main causes. Firstly, most species disappear from agroecosystems due to habitat destruction by increased farming intensity, a deterministic cause of extinction (Tscharntke and Kruess 1999). Secondly, reductions in population size caused by habitat fragmentation lead to further stochastic species losses (Baur and Erhardt 1995; Steffan-Dewenter and Tscharntke 2000). European countries are dominated by agricultural landscapes (Jedicke 1994). Therefore, habitat management schemes in these areas have a particularly high potential to restore overall biodiversity and enable the survival of many species in modern cultivated landscapes. Invertebrates play an important role as major contributors to total biodiversity on farmland and as food for vertebrates, such as farmland birds (Fuller et al. 1995; McCracken and Bignal 1998). In response to an increasing awareness of the rapid decline of biological diversity, several approaches are currently being introduced to reduce this process. Since 1993, Swiss farmers have been financially supported to maintain ecological compensation areas such as wildflower areas, extensively managed grasslands, hedges or orchards (Ullrich and Edwards 1999). Such non-cropped areas provide food resources, shelter and hibernation sites for insects and spiders which make different demands on their habitat (Lagerlöf and Wallin 1993; Steffan-Dewenter and Tscharntke 1997; Frank 1999; Keller and Häni 2000). Considering the differential habitat use of arthropods means that only a mosaic of different habitats can guarantee high species diversity in the agricultural landscape (Greiler 1994; Duelli and Obrist 2003). The purpose of ecological compensation areas is not only to enhance biodiversity in intensively used arable land, but also to increase the numbers of natural enemies of herbivores that feed on arable crops and their potential for natural pest control (Thomas et al. 2001; Barone and Frank 2003). Because every habitat type has specific structural characteristics we assume that various habitat types contain differential insect communities. Consequently, the creation of different kinds of compensation areas is desirable to achieve high insect biodiversity. In this study three types of semi-natural habitats are explored, namely wildflower areas, extensively used meadows and extensively grazed pastures. Wildflower areas (a term synonymous with wildflower or weed strips) were developed in the late 1980s, thereby being a relatively young type of ecological compensation area on Swiss farmland (Nentwig 1988). The number of wildflower areas is continuously growing and this type of semi-natural area was previously shown to enhance species richness and abundance of arthropods remarkably (e.g. Lys and Nentwig 1992; Frank 1998). While wildflower areas turned out to be important habitats for encouraging arthropod diversity, nothing is known about how much they contribute to biodiversity compared with well-established compensation areas such as extensively used meadows and extensively grazed pastures. Such meadows and pastures are well-tried types of ecological compensation areas being more common than wildflower areas. Extensive use of meadows and pastures is known to increase overall diversity of insects, including heteropteran bugs (Morris 2000; Di Giulio et al. 2001; Kruess and Tscharntke 2002).

Tscharntke and Greiler (1995) showed that invertebrate diversity in grassland ecosystems could be predicted by using botanical parameters such as floral diversity or other characteristics of vegetation structure. Considering the influence of environmental parameters (plant species richness, vegetation structure, flower abundance, field size, surrounding landscape structure) on insect diversity in different semi-natural habitats appears to be useful to quantify the effect of measures enhancing biodiversity in the agricultural landscape. True bugs (Heteroptera) were chosen as an indicator group for insect diversity because they are an ecologically very diverse group, including phytophagous and zoophagous species as well as generalists and specialists (Dolling 1991). Furthermore both larvae and adults live in the same habitat and react sensitively to environmental changes (Morris 1969, 1979; Otto 1996). Additionally, bug species richness was found to correlate strongly with total arthropod richness in cultivated landscapes, making bugs an excellent group for biodiversity evaluation (Duelli and Obrist 1998).

The objectives of our investigation were (i) to show whether bug species richness, abundance and assemblages differ significantly among wildflower areas, extensively used meadows and pastures, and (ii) to determine common key factors significantly influencing bug species richness, abundance and the bug community based on the number of individuals per species. It is essential to know how much each type of semi-natural habitat contributes to heteropteran diversity in agroecosystems, particularly in terms of the comparison between the newly established wildflower areas with the well-tried extensive meadows and pastures. This knowledge can be used as a tool to make recommendations about which types of ecological compensation areas should particularly be promoted.

## Material and methods

### *Research area and study sites*

The study was carried out from the end of May to the end of September 2002 in the western part of Bern, an intensively used arable region in Switzerland. The area containing the study sites measured about 9 $km^2$. Three types of semi-natural habitats were studied (wildflower areas, extensively used meadows, extensively grazed pastures), using five replicates for each habitat type. Wildflower areas have a minimum width of 3 m and are sown with a standard wildflower mixture of indigenous arable weeds, meadow and ruderal plant species (Günter 2000). They are maintained for at least 2 years and a maximum of 6 years and the use of pesticides and fertilisers is not allowed. From the second year on, one half of the area may be mown in a yearly rotation after the flowering period. Extensively used meadows are sown with a standard mixture consisting of 95% grass and 5% herb seeds. No fertilisers and pesticides are allowed. Extensively used meadows have to be mown at least once a year but not before 15 June. The five meadows surveyed were mown twice in the sampling period. The five wildflower areas and the five meadows were sown in spring 1999. The five pastures studied were extensively managed since spring 1999, the time before they were managed intensively. Mean grazing intensity in the extensively used pastures surveyed was 2.7 ± 0.2 cattle $ha^{-1}$. Except for cow-pats, usage of additional fertilisers and pesticides is prohibited (Charollais et al. 1999). Wildflower areas were dominated by *Achillea millefolium*, *Hypericum perforatum*, *Leucanthemum vulgare*, *Origanum vulgare*, *Pastinaca sativa* and *Tanacetum vulgare*. Plants with a high coverage in meadows were *Centaurea jacea*, *Leucanthemum vulgare*, *Trifolium pratense* and the grasses *Dactylis glomerata* and *Trisetum flavescens*. In pastures, the same grasses as in the meadows were dominant. Pastures were further dominated by *Trifolium repens* and *Taraxacum officinale*. The 15 study sites were selected to lie in the same climate zone providing similar site conditions in terms of mean annual rainfall, temperature and altitude, which was about 600 m a.s.l. The size of the study sites ranged from 0.05 to 0.4 ha.

### *Sampling methods and bug parameters*

Between the end of May and the end of September 2002, six samples were taken from each study site every 2 or 3 weeks. Sampling was only carried out when the weather conditions were favourable for bug activity, i.e. air temperature of minimum 17 °C, sunshine, dry vegetation and moderate air conditions. Sampling was restricted to the period between 9.30 a.m. and 17.00 p.m., and the sampling order of the study sites varied between sampling dates. The heteropteran bugs were collected using a standardised sweep-net method (Otto 1996). The sweep-net had a diameter of 40 cm and was fitted with a heavy cloth suitable for use in dense vegetation. For each sample, 100 sweeps were made at a constant pace over a transect of about 80 m. The net was emptied after every 25th sweep, resulting in four subsamples per site at each sampling date. Afterwards, the four subsamples were pooled and insects were killed immediately with ethyl acetate ($C_4H_8O_2$). For data analysis, bug abundance (total number of adults and larvae per site) and bug species richness (total number of adult bug species per site) were used. Moreover, two functional groups were analysed considering the trophic level and the overwintering strategy of bugs, by separating into zoophagous (including zoophytophagous species) and phytophagous species, as well as species overwintering as eggs and species overwintering as adults. The adult bugs were determined with the help of entomological handbooks and publications (Wagner 1952, 1966, 1967, 1970–1975; Péricart 1983, 1984, 1987, 1998) and the nomenclature followed Günther and Schuster (2000). Larvae were only counted, but not determined to species level.

### *Environmental factors*

To examine the influence of vegetation factors on bug species richness, bug abundance and bug communities, plant species richness, flower abundance and vegetation structure were analysed. Plant species richness was surveyed once in June 2002 based on five 1 $m^2$ plots randomly chosen in every site. Vegetation structure and flower abundance were analysed six times during the bug sampling period. The sampling locations were ordered every 2 m along a transect of 50 m, resulting in 26 replicates per sampling date. These 50 m transects were located in the transects where the heteropteran bugs were collected. Flower abundance was estimated in a 30 cm × 30 cm square using the following scale: 0 = 0 flowers, 1 = 1–25, 2 = 26–50, 3 = 51–75, 4 = 76–100,5 = > 100 flowers/900 $cm^2$. To investigate the vegetation structure, a simplified version of the point quadrat method was used (Künzle 2002): (a) the sampling was carried out along a transect, (b) instead of a needle a 150 cm long iron rod measuring 8 mm in diameter was used and (c) individual plant species were not taken into account. The iron rod was marked at the heights of 15 cm (soil level), 55 cm (40 cm mark), 95 cm (80 cm mark) and 135 cm (120 cm mark). It was put 15 cm vertically into the soil and every part of a plant, which was in contact with the rod, was counted for each height level separately. For data analysis, however, the arithmetic mean (mean number of plant parts touching the iron rod up to the height of 120 cm) of six subsamples with 26 replicates each was used. Field size and the surrounding landscape structure were calculated using a 1:5000 map. The surrounding landscape structure was surveyed within a square by measuring the area of natural landscape in the surrounding of 300 m width of each site, thereby considering the dispersal range of heteropteran bugs inhabiting open land (Ullrich and Edwards 1999). In this area the environment was separated into two habitat types: natural landscape (extensively managed meadows and pastures, wildflower areas, ruderal sites, orchards, hedges, woodlands) and others (intensively managed arable land, roads, buildings). None of the surroundings of different study sites did overlap. Field size and percentage of natural landscape structure were used for statistical analyses.

*Statistical analyses*

For the analysis of the data, all samples were pooled over time, resulting in one sample per site. Flower abundance and vegetation structure were logarithmic and percentage of natural landscape structure was square root transformed to achieve normal distribution and homogeneity of residuals (Zar 1996). Percentage data were also arcsine transformed. However, for analysis we used square root transformed data, because percentage data were better normally distributed when square root transformed. Bug data were transformed as necessary (for details see Section Result). One-way-ANOVA was performed using the program Systat 10.0 to ascertain differences in bug species richness and abundance as well as differences in environmental factors between the three habitat types. The Tukey-test was carried out for multiple comparisons. To examine the influence of environmental factors on bug species richness and abundance, multiple stepwise linear regression models (backward option) were calculated using the program Systat 10.0. Curve estimations were used to test for best fitting curves of the most explanatory factors using the program SPSS 11.0. In addition to ANOVA and regression models, where only one single value like species richness or abundance is analysed, canonical correspondence analysis (CCA) and correspondence analysis (CA) were calculated using the programme Canoco 4.5 (Ter Braak and Smilauer 2002). Using CCA and CA, the whole bug community was characterised based on the number of individuals per species and site. Species represented by less than five individuals and occurring in only one site were omitted to reduce noise (Voigt et al. 2003). CCA and CA were performed using log $(x + 0.1)$-transformed species data in order to prevent high values from unduly influencing the ordination and to consider zero values. Ordination by CCA was calculated to analyse the influence of environmental factors on the bug community. The significance for each factor in the CCA was obtained by a Monte Carlo test run with 499 permutations. Ordination by CA was used to compare the similarity of the bug species assemblages among the three habitat types.

## Results

### *Bug abundance and species richness in the three habitat types*

Altogether we recorded 5029 individuals consisting of 1554 adults and 3476 larvae of 75 bug species. In wildflower areas, we observed 69 species and 1820 individuals (639 adults, 1181 larvae). In meadows, there were 53 species and 1523 individuals (468 adults, 1055 larvae) and 31 species and 1686 individuals (447 adults, 1239 larvae) in pastures. Seventy per cent of all adult individuals and 28 species belonged to the Miridae, which were the dominating family in all habitat types (Table 1). Seventy-two per cent of all recorded species and 85.5% of the adult individuals were phytophagous, whereas the minority were zoophagous or zoophytophagous, mainly Nabidae and Anthocoridae. Percentages of phytophagous individuals were rather equal in all three habitat types, but percentages of phytophagous species were lower in wildflower areas (69.4%) than in meadows and pastures (77%). Considering the overwintering strategy, 72% of the observed species and 66% of all individuals belonged to bugs overwintering as adults. 12.9% of the species and 20.4% of the individuals collected in pastures overwinter in the egg-stage. In wildflower areas 30.6% and in meadows 20.7% of the species belonged to bugs overwintering as eggs, and percentages of individuals overwintering as eggs were also higher in wildflower areas (42.9%) than in meadows (34.1 %).

The number of bug individuals overwintering in the egg-stage was significantly higher in wildflower areas than in pastures (Figure 1a). By contrast, no differences were found in individuals overwintering as adults, in either trophic group, and in total bug abundance. The total number of bug species in wildflower areas and meadows was significantly higher than in pastures (Figure 1b). The same was true to phytophagous bug species (Figure 1c). For

*Table 1.* Bug families in terms of total numbers of species (sp) and adult individuals (ind) occurring in all sites and in three types of semi-natural habitats, and percentage of individuals per sum.

| Family | All sites ($n$ = 15) | | | Wildflower areas ($n$ = 5) | | | Meadows ($n$ = 5) | | | Pastures ($n$ = 5) | | |
|---|---|---|---|---|---|---|---|---|---|---|---|---|
| | sp | ind | % | sp | ind | % | sp | ind | % | sp | ind | % |
| Alydidae | 1 | 2 | 0.1 | 0 | 0 | 0.0 | 1 | 2 | 0.4 | 0 | 0 | 0.0 |
| Anthocoridae | 4 | 17 | 1.1 | 4 | 14 | 2.2 | 2 | 2 | 0.4 | 1 | 1 | 0.2 |
| Berytidae | 1 | 6 | 0.4 | 0 | 0 | 0.0 | 1 | 4 | 0.9 | 2 | 2 | 0.5 |
| Coreidae | 2 | 7 | 0.5 | 2 | 5 | 0.8 | 1 | 2 | 0.4 | 0 | 0 | 0.0 |
| Lygaeidae | 9 | 27 | 1.7 | 6 | 11 | 1.7 | 5 | 9 | 1.9 | 7 | 7 | 1.6 |
| Miridae | 28 | 1087 | 69.9 | 19 | 404 | 63.2 | 20 | 318 | 68.0 | 13 | 365 | 81.6 |
| Nabidae | 8 | 191 | 12.3 | 8 | 81 | 12.7 | 6 | 64 | 13.7 | 5 | 46 | 10.3 |
| Pentatomidae | 9 | 66 | 4.3 | 6 | 17 | 2.6 | 6 | 39 | 8.3 | 4 | 10 | 2.2 |
| Piesmatidae | 1 | 8 | 0.5 | 0 | 0 | 0.0 | 1 | 7 | 1.5 | 1 | 1 | 0.2 |
| Pyrrhocoridae | 1 | 8 | 0.5 | 0 | 0 | 0.0 | 0 | 0 | 0.0 | 1 | 8 | 1.8 |
| Rhopalidae | 5 | 121 | 7.8 | 23 | 106 | 16.6 | 5 | 13 | 2.8 | 1 | 2 | 0.5 |
| Saldidae | 1 | 2 | 0.1 | 1 | 1 | 0.2 | 1 | 1 | 0.2 | 0 | 0 | 0.0 |
| Scutelleridae | 2 | 5 | 0.3 | 0 | 0 | 0.0 | 2 | 5 | 1.1 | 0 | 0 | 0.0 |
| Tingidae | 3 | 7 | 0.5 | 0 | 0 | 0.0 | 2 | 2 | 0.4 | 1 | 5 | 1.1 |
| Sum | 75 | 1554 | 100 | 69 | 639 | 100 | 53 | 468 | 100 | 31 | 447 | 100 |

Numbers refer to six samples per site. $n$ = number of sites per habitat type.

zoophagous species and species overwintering as eggs, significantly larger numbers were observed in wildflower areas compared to pastures (Figures 1d, e). Meadows contained significantly more species overwintering as adults than pastures (Figure 1f).

### *Factors explaining bug abundance and species richness*

We tested the influence of five environmental factors on bug species richness and abundance using stepwise multiple regression. The variability of adult individuals (Figure 2a), total bug abundance ($y = 509.001 \log x^{1.97}$, $F = 5.53$, $R^2 = 0.299$, $p = 0.035$, $n = 15$), and the number of phytophagous individuals ($\sqrt{y} = 12.465 \log x^{1.28}$, $F = 10.91$, $R^2 = 0.456$, $p = 0.006$, $n = 15$) and zoophagous individuals ($\sqrt{y} = 5.432 \log x^{2.02}$, $F = 7.64$, $R^2 = 0.37$, $p = 0.016$, $n = 15$) were best explained by vegetation structure. There was a significant negative relation between individuals overwintering as adults and surrounding landscape structure in a perimeter of 300 m, which accounted for 55.5% of the variance (Figure 2b). The distribution of bug individuals that are overwintering as eggs was best explained by a negative relationship with field size accounting for 69.3% of the variance (Figure 2c). Total bug species richness and phytophagous bug species increased significantly with vegetation structure, which explained 40.2% (Figure 2d) and 54.5% (Figure 2e) of the variance, respectively. Also the variability of zoophagous bug species

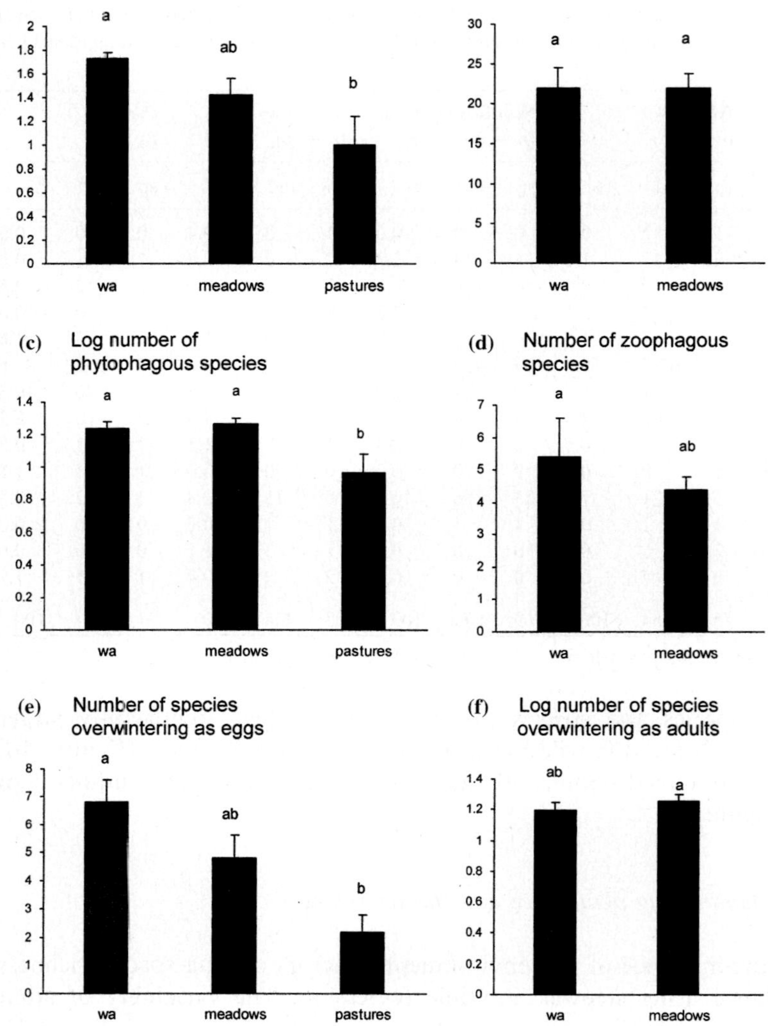


*Figure 1.* Distribution of bug individuals and bug species (mean ± SE) in wildflower areas (wa), meadows and pastures. Different letters above bars indicate significant differences between habitat types (Tukey test, $p < 0.05$). (a) Number of individuals overwintering in the egg-stage (logarithmic transformed): ANOVA, $F = 5.137$, $p = 0.024$, $n = 15$. (b) Number of bug species: $F = 6.704$, $p = 0.011$, $n = 15$. (c) Number of phytophagous bug species (logarithmic transformed): $F = 5.496$, $p = 0.020$, $n = 15$. (d) Number of zoophagous bug species: $F = 3.948$, $p = 0.048$, $n = 15$. (e) Number of bug species overwintering as eggs: $F = 9.852$, $p = 0.003$, $n = 15$. (f) Number of bug species overwintering as adults (logarithmic transformed): $F = 4.536$, $p = 0.034$, $n = 15$.

($y = -5.299 + 12.178 \log x$, $F = 13.618$, $R^2 = 0.512$, $p = 0.003$, $n = 15$), species overwintering as adults ($y = -2.133 + 8.037 \log x - 4.766 \log x^2$, $F = 6.92$, $R^2 = 0.535$, $p = 0.01$, $n = 15$) and species overwintering as eggs

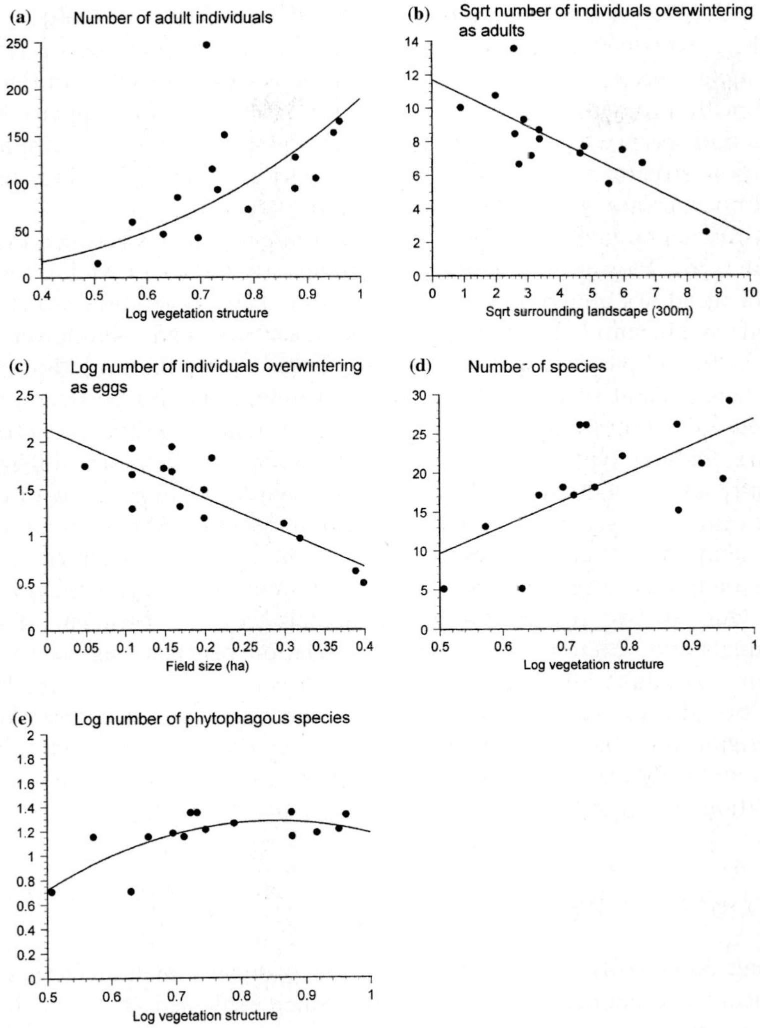


*Figure 2.* Effects of environmental factors on bug abundance and bug species richness. (a) Relation between number of adult individuals and vegetation structure: $y = 188.12 \log x^{2.67}$, $F = 13.52$, $R^2 = 0.51, p = 0.003, n = 15$. Vegetation structure is logarithmic transformed. (b) Negative relation between number of individuals overwintering as adults and surrounding landscape: $\sqrt{y} = 11.696 - 0.965\sqrt{x}, F = 18.451, R^2 = 0.591, p = 0.001, n = 15$. Surrounding landscape and individuals overwintering as adults are square root transformed. (c) Negative relation between number of individuals overwintering as eggs and field size (ha): $\log y = 2.132 - 0.037x, F = 32.582$, $R^2 = 0.715, p < 0.001, n = 15$. Individuals overwintering as eggs are logarithmic transformed. (d) Relation between total number of bug species and vegetation structure: $y = -7.667 + 34.47 \log x$, $F = 10.396, R^2 = 0.444, p = 0.007, n = 15$. Vegetation structure is logarithmic transformed. (e) Relation between number of phytophagous bug species and vegetation structure: $\log y = -2.0132 + 7.7209 \log x - 4.5299 \log x^2, F = 7.19, R^2 = 0.545, p = 0.009, n = 15$. Vegetation structure and phytophagous bug species richness are logarithmic transformed.

($y = -5.777 + 13.687 \log x$, $F = 18.017$, $R^2 = 0.581$, $p = 0.001$, $n = 15$) were best explained by vegetation structure. Whereas total species richness, zoophagous species and species overwintering as eggs showed a continous increase with progressing vegetation structure (linear model), phytophagous species and species overwintering as adults were saturated at a certain level of vegetation structure (quadratic model), suggesting that these bug features would not increase with additional vegetation structure.

Environmental factors surveyed showed different patterns between the three habitat types. Flower abundance was significantly higher in wildflower areas than in meadows and pastures (Tukey, $p < 0.005$). Vegetation structure increased significantly from pastures to meadows and wildflower areas ($p < 0.05$), and pastures were significantly larger in size than wildflower areas ($p < 0.005$). Plant species richness and surrounding landscape structure in a perimeter of 300 m to the study sites showed no significant differences between the three habitat types. Since vegetation structure and field size differed significantly among habitat types and revealed significant relations with certain bug features, we tested whether these relations were caused by habitat effect rather than by an environmental factor. For that, we calculated multiple regression models where habitat types were dummy coded using meadows as a basis. Only in two of nine regression models (relation between adult bug individuals and vegetation structure, and relation between bug species overwintering as adults and vegetation structure) was there a significant habitat effect, but always explaining less variance than vegetation structure. Accordingly, significant relations found were predominantly due to the environmental factor, and only two relations were caused by both the environmental factor (vegetation structure) plus habitat effect.

### *Factors influencing bug communities*

The bug community was examined relative to environmental factors using canonical correspondence analysis (CCA), which explained 44.2% of the total variance. Of the five environmental factors considered, only flower abundance contributed significantly to the distribution of heteropteran bugs accounting for 18.4% of the variance. The remaining variance was explained by the other factors (Table 2). A further CCA model excluding flower abundance, which was highly correlated with vegetation structure (Spearman's $r_s = 0.861$, $p < 0.001$, $n = 15$), explained 36.7% of the total variance of the bug community. In this model vegetation structure was the only significant factor accounting for 16.8% of the variance. The same results were obtained when CCA models were analysed including also the rare species with less than five individuals. A group of species (*Aelia acuminata*, *Dicyphus globulifer*, *Sticto-pleurus punctatonervosus*, *Capsus ater*, *Megalocoleus molliculus*, *Plagiognathus arbustorum*) was preferentially or exclusively found in wildflower areas (Table 3) and strongly correlated with flower abundance and vegetation

immediately removing the vegetation cover was not as big as the more continuous disturbance by grazing. Morris (1979) showed for *N. elongata* that the timing of life cycle in relation to the timing of management is important. *N. elongata* has two generations per year and was therefore less susceptible to cutting than univoltine grassland mirids. Furthermore, *N. elongata* usually was not significantly more common in untreated grassland compared with cut plots. *N. elongata* was the most abundant bug species in our study sites. It reached much higher densities in pastures and meadows than in wildflower areas, which shows that its phenology well fitted to the management regimes and further supports the observations described above.

There were significantly more zoophagous bug species in wildflower areas than in pastures, whereas numbers of zoophagous individuals showed no differences between the habitat types. The reason why we only found differences in species number is due to the fact that several rare species occurred in wildflower areas, but were absent from pastures. In the regression models the variability of both zoophagous bug species and zoophagous individuals were best explained by vegetation structure, which confirms previous findings revealing that zoophagous bug species were positively correlated with vegetation structure (Künzle 2002). Highly structured vegetation supports large insect populations by providing a greater potential surface for colonisation and more resources, such as oviposition, resting and overwintering sites (May 1973; Price et al. 1980; Lawton and Strong 1981; Lawton 1983). These advantages could explain the positive response of zoophagous bug species and individuals to vegetation structure. Due to denser and higher vegetation, predators find more hiding places and cover from their own enemies (Lagerlöf and Wallin 1993; Morris 2000). Furthermore, richly structured habitats are colonised faster by prey populations (White and Hassal 1994), which may favour the nutritional conditions of zoophagous bugs. Although pollen seems of poor value for development and reproduction of bugs, it perhaps represents an indirect cue for finding developing prey populations (Fauvel 1999). Prey populations such as thrips, aphids and psyllids were shown to be attracted by flower abundance (Russel 1989; Fauvel 1999), which was highest in wildflower areas in our study. We found most individuals of the zoophagous bugs *O. niger*, *O. majusculus*, *H. mirmicoides* and *N. pseudoferus* in wildflower areas.

The abundance of bugs, which are overwintering as eggs, was negatively associated with field size. Our results agree with previous research on Hemiptera (Sanderson 1992) and are supported by a study reflecting high butterfly densities in small habitats as an accumulation of individuals from the surrounding landscape, as these fragments provide the only attractive habitat patches (Steffan-Dewenter and Tscharntke 2000). In our study sites, field size was strongly negatively correlated with vegetation structure (Spearman's $r_s = -0.856$, $p < 0.001$, $n = 15$). This means that the smaller wildflower areas showed a complex vegetation structure in contrast to the larger pastures that featured a low vegetation complexity, indicating that wildflower areas offer better possibilities for laying eggs. One may assume that the negative relationship between bug individuals overwintering as eggs and field size was caused by the different vegetation structures in habitat types, rather than by field size. However, partial correlation using vegetation structure as control variable revealed that there was a real relationship between bug individuals overwintering as eggs and field size (partial correlation coefficient = −0.625, $p = 0.017$). Species overwintering as eggs depend on food plants in their overwintering site for development in spring, and thus cannot use annual plants (Wagner 1966; Ullrich 2001). This agrees with our study where many host plants of bug species overwintering in the egg-stage, such as *Achillea*, *Tanacetum*, other Asteraceae and *Urtica*, were most abundant in wildflower areas. The amount of natural landscape surrounding a site had a negative effect on bug individuals overwintering as adults. Such individuals can disperse directly after overwintering, thereby being independent of host plants at hibernation sites. Therefore, we assume that bugs overwintering as adults in fairly monotonous areas colonise semi-natural areas where they find more favourable conditions for nutrition and reproduction. However, bugs overwintering in more heterogenous areas may stay there and not colonise our study sites, which may have caused the negative relation between adult overwinterers and surrounding landscape structure.

In the CCA, frequencies of certain bug species were strongly related to vegetation structure and flower abundance, which can be explained by the fact that these two environmental factors were positively correlated. Many species showing a correlation with flower abundance and vegetation structure belonged to the Miridae and were only abundant in wildflower areas. *P. arbustorum* feeds on Urticaceae, *M. molliculus* on *Tanacetum* and *Achillea*, and *D. globulifer* on *Melandryum*. These findings are supported by the fact that they depend on perennial host plants, which almost exclusively occurred in wildflower areas. *Leptopterna dolobrata* is feeding on grasses (*Alopecurus*, *Dactylis*, *Phleum*). Thus this species would be expected to be more frequent in meadows and pastures where these grasses are common. In former surveys *L. dolobrata* was adversely affected in its abundance by intensive management, thus was reduced by both frequency of cutting and early cuts (Di Giulio et al. 2001). Since the females lay their eggs on the bottom part of grass stems (Kullenberg 1944), damage to the eggs does probably not occur, but larval development may be the critical phase because it takes place in June, when the extensively used meadows were cut. This may have been the reason for very low numbers of *L. dolobrata* in meadows and pastures.

In general, our results indicate that vegetation structure and flower abundance are of high predictive value for bug species richness, abundance and bug species composition. Wildflower areas and extensively used meadows had higher vegetation structure than extensively grazed pastures, and wildflower areas also comprised higher flower abundance compared with pastures. Wildflower areas and extensively used meadows contained a number of specialised bug species, while the pastures were characterised by common and widespread species, which also occurred in meadows or wildflower areas. Since wildflower areas and meadows clearly increased total bug species richness and the majority of species occurring in extensively grazed pastures could also be found in the other two habitat types surveyed, we recommend the promotion of wildflower areas and extensively used meadows in order to restore high heteropteran diversity in modern cultivated landscapes. This recommendation is of practical relevance because the semi-natural habitats studied are part of agri-environment schemes supported by the Swiss government and management prescriptions can be modified easily.

## Acknowledgements

We are grateful to J.-P. Airoldi and P. Kehrli for statistical advice. Special thanks to I. Künzle for the help in determining the nabid species and advice in field work, and R. Egli and B. Tschanz for assistance with field work. R. Heckmann checked the determination of critical bug species. Special thanks to P. Borges and W. Rabitsch, and two anonymous referees for critically commenting on the draft version of the manuscript.

## References

Barone M. and Frank T. 2003. Habitat age increases reproduction and nutritional condition in a generalist arthropod predator. Oecologia 135: 78–83.

Baur B. and Erhardt A. 1995. Habitat fragmentation and habitat alterations: principal threats to most animal and plant species. Gaia 4: 221–226.

Borges A.V. and Brown V.K. 2001. Phytophagous insects and web-building spiders in relation to pasture vegetation complexity. Ecography 24: 68–82.

Brown V.K., Gibson C.W.D. and Kathirithamby J. 1992. Community organisation in leaf hoppers. Oikos 65: 87–106.

Charollais M., Kuchen S., Mulhauser G., Schiess-Bühler C. and Schüpbach H. 1999. Wegleitung für den ökologischen Ausgleich auf dem Landwirtschaftsbetrieb. Landwirtschaftliche Beratungszentrale LBL, Lindau.

Curry J.P. 1994. Grassland Invertebrate-ecology, Influence on Soil Fertility and Effects on Plant Growth. Chapman and Hall, London.

Di Giulio M., Edwards P.J. and Meister E. 2001. Enhancing insect diversity in agricultural grasslands: the roles of management and landscape structure. J. Appl. Ecol. 38: 310–319.

Di Giulio M. and Edwards P.J. 2003. The influence of host plant diversity and food quality on larval survival of plant feeding heteropteran bugs. Ecol. Entomol. 28: 51–57.

Dolling W.R. 1991. The Hemiptera. Oxford University Press, Oxford.

Dramstad W. and Fry G. 1995. Foraging activity of bumblebees (*Bombus*) in relation to flower resources on arable land. Agric. Ecosyst. Environ. 53: 123–135.

Duelli P. and Obrist K.M. 1998. In search for the best correlates for local organismal biodiversity in cultivated areas. Biodivers. Conserv. 7: 297–309.

Duelli P. and Obrist K.M. 2003. Regional biodiversity in an agricultural landscape: the contribution of seminatural habitat islands. Basic Appl. Ecol. 4: 129–138.

Edwards P.J., Kollmann J. and Wood D. 1999. Determinants of agrobiodiversity in the agricultural landscape. In: Wood D. and Lenné J.M. (eds), Agrobiodiversity: Characterization, Utilization and Management. CAB International, CABI Publishing, Wallingford Oxon, pp. 183–210.

Fauvel G. 1999. Diversity of Heteroptera in agroecosystems: role of sustainability and bioindication. Agric. Ecosyst. Environ. 74: 275–303.
Frank T. 1998. Attractiveness of sown weed strips on hoverflies (Syrphidae, Diptera), butterflies (Rhopalocera, Ledidoptera), wild bees (Apoidea, Hymenoptera) and thread-waisted wasps (Sphecidae, Hymenoptera). Mitt. Schweiz. Entomol. Ges. 71: 11–20.
Frank T. 1999. Density of adult hoverflies (Dipt., Syrphidae) in sown weed strips and adjacent fields. J. Appl. Entomol. 123: 351–355.
Fuller R.J., Gregory R.D., Gibbons D.W., Marchant J.H., Wilson J.D., Baillie S.R. and Carter N. 1995. Population declines and range contractions among farmland birds in Britain. Conserv. Biol. 9: 1425–1441.
Gibson C.W.D., Hambler C. and Brown V.K. 1992. Changes in spider (Araneae) assemblages in relation to succession and grazing management. J. Appl. Ecol. 29: 133–142.
Greiler H.J. 1994. Insektengesellschaften auf selbstbegrünten und eingesäten Ackerbrachen. Agrarökologie 11: 1–136.
Günter M. 2000. Anlage und Pflege von mehrjährigen Buntbrachen unter den Rahmenbedingungen des schweizerischen Ackerbaugebietes. Agrarokölogie 37: 1–154.
Günther H. and Schuster G. 2000. Verzeichnis der Wanzen Mitteleuropas (Insecta: Heteroptera). Mitt. Int. Entomol. Vereins 2: 1–69.
Huusela-Veistola E. and Vasarainen A. 2000. Plant succession in perennial grass strips and effects on the diversity of leafhoppers (Homoptera, Auchenorrhyncha). Agric. Ecosyst. Environ. 80: 101–112.
Jedicke E. 1994. Biotopverbund. Ulmer, Stuttgart.
Keller S. and Häni F. 2000. Ansprüche von Nützlingen und Schädlingen an den Lebensraum. In: Nentwig W. (ed), Streifenförmige ökologische Ausgleichsflächen in der Kulturlandschaft: Ackerkrautstreifen, Buntbrache, Feldränder. Verlag Agrarökologie, Bern, pp. 199–217.
Kruess A. and Tscharntke T. 2002. Contrasting responses of plant and insect diversity to variation in grazing intensity. Biol. Conserv. 106: 293–302.
Kullenberg B. 1944. Studien über die Biologie der Capsiden. Ph.D. thesis, University of Uppsala.
Künzle I. 2002. Early succession of bug communities (Insecta: Heteroptera) on wildflower areas. Diploma thesis, University of Bern.
Lagerlöf J. and Wallin H. 1993. The abundance of arthropods along two field margins with different types of vegetation composition: an experimental study. Agric. Ecosyst. Environ. 43: 141–154.
Lawton J.H. 1983. Plant architecture and the diversity of phytophagous insects. Ann. Rev. Entomol. 28: 23–39.
Lawton J.H. and Strong D.R.J. 1981. Community patterns and competition in folivorous insects. Am. Nat. 118: 317–338.
Lys J.-A. and Nentwig W. 1992. Augmentation of beneficial arthropods by strip-management. 4. Surface activity, movements and activity density of abundant carabid beetles. Oecologia 92: 373–382.
Marshall E.J.P. and Moonen A.C. 2002. Field margins in Northern Europe: their functions and interactions with agriculture. Agric. Ecosyst. Environ. 89: 5–21.
May R.M. 1973. Stability and Complexity in Model Ecosystems. Princeton University Press, Princeton.
McCracken D.I. and Bignal E.M. 1998. Applying the results of ecological studies to land-use policies and practices. J. Appl. Ecol. 35: 961–967.
Morris M.G. 1967. Differences between the invertebrate faunas of grazed and ungrazed chalk grasslands. I. Responses of some phytophage insects to cessation of grazing. J. Anim. Ecol. 36: 459–474.
Morris M.G. 1969. Differences between the invertebrata faunas of grazed and ungrazed chalk grasslands. III. The heteropterous fauna. J. Appl. Ecol. 6: 475–487.
Morris M.G. 1979. Responses of grassland invertebrates to management of cutting. II. Heteroptera. J. Appl. Ecol. 16: 417–432.

Morris M.G. 2000. The effects of structure and its dynamics on the ecology and conservation of arthropods in British grasslands. Biol. Conserv. 95: 129–142.

Murdoch W.W., Evans F.C. and Peterson C.H. 1972. Diversity and pattern in plants and insects. Ecology 53: 819–829.

Nentwig W. 1988. Augmentation of beneficial arthropods by strip-management. 1. Succession of predacious arthropods and long-term change in the ratio of phytophagous and predacious arthropods in a meadow. Oecologia 76: 597–606.

Otto A. 1996. Die Wanzenfauna montaner Magerwiesen und Grünbrachen im Kanton Tessin (Insecta, Heteroptera). Ph.D. thesis, ETH, Zürich.

Péricart J. 1983. Hémiptères Tingidae Euro-méditerranéens. Faune de France 69. Fédération Francaise des Sociérés de Sciences Naturelles, Paris.

Péricart J. 1984. Hémiptères Berytidae Euro-méditerranéens. Faune de France 70. Fédération Francaise des Sociérés de Sciences Naturelles, Paris.

Péricart J. 1987. Hémiptéres Nabidae d' Europe Occidental et du Maghreb. Faune de France 71. Fédération Francaise des Sociérés de Sciences Naturelles, Paris.

Péricart J. 1998. Hémiptères Lygaeidae d' Europe Occidental et du Maghreb. Faune de France 84 A, B, C. Fédération Francaise des Sociérés de Sciences Naturelles, Paris.

Price P.W., Bouton C.E., Gross P., McPheron B.A., Thompson J.N. and Weis A.E. 1980. Interactions among three trophic levels, the influence of plants on interactions between insect herbivores and natural enemies. Ann. Rev. Ecol. Syst. 11: 41–65.

Rieger C. 1978. Zur Verbreitung von *Trigonotylus coelestialium* (Kirklady), 1902. Mitt. Bayer. Entomol. 27: 83–90.

Russel E.P. 1989. Enemies hypothesis: a review of the effect of vegetational diversity on predatory insects and parasitoids. Environ. Entomol. 18: 590–599.

Sanderson R.A. 1992. Diversity and eveness of Hemiptera communities on naturally vegetated derelict land in NW-England. Ecography 15: 154–160.

Steffan-Dewenter I. and Tscharntke T. 1997. Early succession of butterfly and plant communities on set-aside fields. Oecologia 109: 294–302.

Steffan-Dewenter I. and Tschantke T. 2000. Butterfly community structure in fragmented habitats. Ecol. Lett. 3: 449–456.

Ter Braak C.J.F. and Smilauer P. 2002. Canoco Reference Manual and CanoDraw for Windows User's Guide. Software for Canonical Community Ordination (version 4.5). Biometris, Wageningen and Cescé Budejovice.

Thomas C.F.G., Parkinson L., Griffiths A., Fernandez Garcia A. and Marshall E.J.P. 2001. Aggregation and temporal stability of carabid beetle distributions in field and hedgerow habitats. J. Appl. Ecol. 38: 100–116.

Tscharntke T. and Greiler H.J. 1995. Insect communities, grasses and grasslands. Ann. Rev. Entomol. 40: 535–558.

Tscharntke T. and Kruess A. 1999. Habitat fragmentation and biological control. In: Hawkins B.A. and Cornell H.V. (eds), Theoretical Approaches to Biological Control. Cambridge University Press, Cambridge, pp. 190–205.

Ullrich K.S. 2001. The influence of wildflower-strips on plant and insect (Heteroptera) diversity in an arable landscape. Ph.D. thesis, ETH, Zürich.

Ullrich K.S. and Edwards P.J. 1999. The colonization of wildflower-strips by insects (Heteroptera). In: Maudsley M.J. and Marshall E.J.P. (eds), Heterogeneity in Landscape Ecology. Proceedings of International Association for Landscape Ecology, UK, pp. 131–138.

Voigt W., Perner J., Davis A.J., Eggers T., Schumacher J., Bährmann R., Fabian B., Heinrich W., Köhler G., Lichter D., Marstaller R. and Sander F. 2003. Trophic levels are differentially sensitive to climate. Ecology 84: 2444–2453.

Wagner E. 1952. Blindwanzen oder Miriden. Fischer, Jena.

Wagner E. 1966. I. Pentatomorpha. Fischer, Jena.

Wagner E. 1967. II. Cimicomorpha. Fischer, Jena.

Wagner E. 1970–1975. Die Miridae des Mittelmeerraumes und der Makronischen Inseln (Hemiptera, Heteroptera) Teil 1-3. Entomologische Abhandlungen 37. Akademische Verlagsgesellschaft Geest & Portig, Leipzig.
White P.C.L. and Hassal M. 1994. Effects of management on spider communities of headlands in cereal fields. Pedobiologia 38: 169–184.
Zar J.H. 1996. Biostatistical Analysis. Prentice-Hall International, London.

Biodiversity and Conservation (2006) 15:281–295
DOI 10.1007/s10531-005-1536-3 

# Insect colonisation of fruiting bodies of the wood-decaying fungus *Fomitopsis pinicola* at different distances from an old-growth forest

MATTIAS JONSSON* and GÖRAN NORDLANDER
*Department of Entomology, Swedish University of Agricultural Sciences, Uppsala, Sweden; *Author for correspondence (e-mail: Mattias.Jonsson@entom.slu.se; phone: +46-18-672338; fax: +46-18-672890)*

Received 6 July 2004; accepted in revised form 21 January 2005

**Key words:** Boreal forest, Colonisation, Conservation, Dispersal, Forest management, Fungi, Insect

**Abstract.** We studied the colonisation rate of insects inhabiting fruiting bodies of the wood-decaying fungus *Fomitopsis pinicola* both within and at different distances (up to 1610 m) from an old-growth forest reserve. The colonisation rate of most species was not affected by the distance from the reserve, and none of the species were affected by the size of local sources of fruiting bodies in the managed forest. We suggest that many insect species inhabiting fruiting bodies of wood-decaying fungi can colonise fruiting bodies at a high enough rate to persist in managed forests of Fennoscandia. However, the colonisation rates of the fungivorous beetle *Cis quadridens* and the predatory fly *Medetera apicalis* were negatively affected by distance from the reserve. *Cis quadridens* is rare in many managed forests, but often quite common at sites with high substrate densities. The rarity of this species may therefore be due to weak ability to colonize distant patches. The same may also be true for *M. apicalis*, but less is known about the biology of this species. *Medetera apicalis* was the most common insect predator in the old-growth forest, but it was rare at the largest distances from it in the managed forest. Therefore, it seems likely that the overall pressure from natural enemies significantly declined with distance from the reserve.

## Introduction

To survive in a patchy landscape a species must be able to colonise new habitat patches at a high enough rate to compensate for local extinctions (Hanski 1997). Knowledge about colonisation rates is, therefore, important for understanding the population structure, and for designing appropriate management strategies to protect threatened species exploiting patchy environments. Colonisation is difficult to quantify, so colonisation rates are commonly inferred from presence-absence data (Ims and Yoccoz 1997; Thomas and Hanski 1997), but relying solely on them to estimate colonisation rates may lead to erroneous conclusions (Ims and Yoccoz 1997; Lewis et al. 1997). Local aggregation patterns may occur in highly mobile species, giving the false impression of local populations (Harrison and Taylor 1997), and causing colonisation rates to be underestimated (Jonsell et al. 2003; Lewis et al. 1997). On the other hand, aggregations of individuals may represent entirely separate (relictual) populations with little or no movement between them (Harrison and Taylor 1997), in which case colonisation rates may be overestimated (Lewis et al. 1997). To avoid these potential pitfalls, direct studies of dispersal and colonisation are essential.

Many insects are specialised on well-defined substrate patches that are more or less ephemeral, e.g. dead wood, dung and certain types of living plant tissue. Direct studies of such insects' colonisation rates are feasible, since the turnover rates of their populations are high (Hanski 1987). In addition, it is often possible to experimentally manipulate the distribution of their substrates. Such experiments can be used not only to investigate colonisation rates of single species (Whitlock 1992), but also to analyse differences in colonisation patterns between guilds of species and to pinpoint potential effects of landscape structure on ecosystem functions such as predation or pollination (Kruess and Tscharntke 1994; Steffan-Dewenter 2003).

Fruiting bodies of wood-decaying fungi constitute a species-rich and well-defined type of dead wood substrate that can be manipulated easily (Jonsell et al. 2001; Komonen 2003). Basic knowledge about the substrate requirements and spatial distribution patterns of species dwelling in wood-deacying fungi is good compared with other saproxylic species, i.e. species dependent on dead wood or other species that depend on this resource (Speight 1989; Komonen 2003). In Fennoscandia, the amount of dead wood has declined to just 2–30% of the amounts found in old-growth forests, due to forest management (Fridman and Walheim 2000; Siitonen 2001). Therefore, many saproxylic species have declined and are now red-listed (Gärdenfors 2000; Rassi et al. 2000).

Insect species associated with fruiting bodies of wood-decaying fungi differ in their sensitivity to forest management (Komonen et al. 2000; Jonsell and Nordlander 2002). For instance, Komonen et al. (2000) found that the parasitoid fly *Elfia cingulata* (Robineau-Desvoidy) was more sensitive to forest fragmentation than its tineid moth host *Agnathosia mendicella* (Denis & Schiffermüller), and the fungal host *Fomitopsis rosea* (Alb. & Schwein.: Fr). Similarly, in inventories of fruiting bodies of *Fomes fomentarius* (L.:Fr) and *Fomitopsis pinicola* (Schwartz:Fr.) Jonsell and Nordlander (2002) found three species to be significantly more common in less intensively managed forests: the tenebrionid beetle *Oplocephala haemorrhoidalis* (F.), the ciid beetle *Cis quadridens* Mellie and the tineid moth *Scardia boletella* (F.). One reason why these species are sensitive to fragmentation and management may be that they have weak abilities to colonise isolated patches (Komonen et al. 2000; Jonsson et al. 2001; Jonsell and Nordlander 2002). For *O. haemorrhoidalis* support for this hypothesis has been found recently in comparative studies with its common relative *B. reticulatus* (Jonsson 2003; Jonsson et al. 2003a).

We here present results from a colonisation study of insects in fruiting bodies of *F. pinicola*, in which colonisation rates were analysed inside an old-growth forest reserve and at 14 different distances up to 1610 m away from the border of this forest into an intensively managed forest. Colonisation was measured as successful reproduction, a measure that represents the outcome of several processes of which none must fail: the coloniser must fly to and locate the patch as well as produce offspring there (Ims and Yoccoz 1997). The reserve contained much higher densities of fungal fruiting bodies than the managed forest, so we expected to find a general decline in colonisation frequency with distance from the reserve. Our main objective was to see if species known to be sensitive to forest management show a steeper decline in colonisation rate with distance from the reserve than species considered less sensitive. Such a pattern would give further support for the hypothesis that they have weak abilities to colonise distant patches. We also wanted to see if the distance from reserve affected the pressure from natural enemies.

## Methods

Fruiting bodies of *Fomitopsis pinicola* used in the experiment were collected in May and June 1997 at a clear-cutting about 40 km N of Uppsala, central Sweden. All sampled fruiting bodies were alive and most of them had grown on large-diameter spruce stumps, about 0.5 m high. Shortly after collection all fruiting bodies were frozen at −60 °C for at least 24 h to kill potential insect inhabitants (Gehrken et al. 1991). However, most of the fruiting bodies were probably already devoid of insects as most associated insect species prefer dead fruiting bodies (Jonsell et al. 2001). Fruiting bodies treated in this way have a similar colonising fauna to naturally occurring, newly dead fruiting bodies (Jonsell et al. 1999). Some of the largest fruiting bodies were divided, so that each piece of fruiting body had a breadth and width between 5 and 15 cm. Before the experiment, fruiting bodies were divided into size classes and distributed into 15 different groups, each with 16 fruiting bodies, so that each group of fruiting bodies had a similar size distribution.

The experiment was carried out in 1997 inside and in the vicinity of the nature reserve 'Fiby urskog', a 70 ha old-growth forest located about 20 km west of Uppsala. The old-growth forest, which contains high amounts of dead wood and fruiting bodies of *F. pinicola*, is known to host a high diversity of saproxylic insects (Lundblad 1950). 'Fiby urskog' constitutes a well-delimited old-growth island in a landscape with a long history of management where dead wood is scarce. The surroundings of the old-growth forest include both young and mature managed forests, agricultural land and a small lake. To obtain structural and microclimatic conditions as similar to the old-growth forest as possible, we located the colonisation study in a mature managed spruce-dominated forest bordering the old-growth forest to the south–west (Figure 1).

Fruiting bodies were exposed to insects in groups of 16 inside the old-growth forest (80 m from the nearest border), at the edge of the old-growth forest, and at distances of 90, 160, 240, 320, 380, 510, 580, 680, 840, 1030, 1160, 1360 and 1610 m from the old-growth forest in the managed forest (Figure 1). The sites

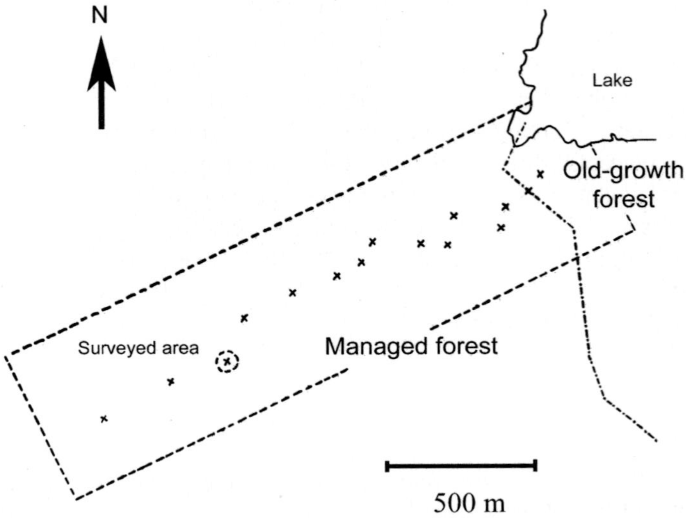


*Figure 1.* Map of the study area at Fiby urskog 20 km west of Uppsala. The locations of experimental groups of fruiting-bodies are marked with 'x'. The circle around one of the groups shows the size of the area where naturally occurring fruiting bodies were counted and included in the analyses as local sources. The large rectangle shows the whole area that was surveyed for naturally occurring fruiting bodies.

were chosen so that extremes in degree of sun-exposure were avoided. We also avoided organising the experimental groups on an absolute straight line to decrease the risk of dependence among them. All fruiting bodies were nailed on separate wooden poles (Norway spruce) about 1 m above the ground. The 16 fruiting bodies in each group were placed in a 4×4 grid, spaced about 2 m apart, each pointing in a random direction.

The experiment was divided into two periods (23 May–25 June and 26 June–4 August), between which the fruiting bodies were exchanged and the results were analysed separately for each time period. In this way most of the flight periods of associated insects were included in the study (M. Jonsell, pers. comm.). To avoid fruiting bodies drying out during times of low precipitation, they were sprayed with water once during each time period. At the end of each time period all fruiting bodies were marked and put separately into 1-l containers of waxed paper to which a glass vial was attached. As the photopositive insects emerged, they tended to accumulate in the vial and were then removed for determination. Initially, insects were reared at room temperature, but since many of them need to be exposed to a period of winter-cold to break diapause, all rearing containers were kept outdoors from October until February. Thereafter, rearing continued at room temperature during spring and early summer 1998. In July 1998, all fruiting bodies were frozen until they were dissected to detect any insects that had not entered the glass vials. The insects reared were determined as far as possible and categorised as monophagous taxa specialised on fruiting bodies of *F. pinicola*, oligophagous taxa that use different species of wood-decaying fungi, polyphagous taxa that also feed on substrates other than fruiting bodies of wood-decaying fungi, parasitoids and predators (Jonsell et al. 1999; Jonsell et al. 2001).

During autumn 1997 and spring 1998, a complete survey of dead fruiting bodies of *F. pinicola* and *Piptoporus betulinus* (Bull. ex Fr.) was carried out in 14 ha of the old-growth forest surrounding the experimental group located in it, and in 102 ha of the managed forest bordering the old-growth stand (Figure 1). In this way the whole area within a 200 m radius of each group of fruiting bodies was surveyed. We surveyed *P. betulinus* since it was one of the most common wood-decaying fungi in the area and some of the fauna it hosts are also hosted by *F. pinicola*.

### *Statistical analysis*

Multiple linear regression was used to analyse the effects of distance from the reserve and local sources of fungal fruiting bodies on the average number of colonising taxa per fruiting body per group of 16 fruiting bodies and the proportion of fruiting bodies colonised by certain species. For parasitoids, the response variable was the proportion of parasitized hosts. The group of fruiting bodies placed 80 m inside the border of the reserve was not included in these analyses. All proportions were arcsine transformed and count data were log transformed before analysis. The size of local sources was determined as the number of dead fruiting bodies occurring within a 35 m radius of each group of experimental fruiting bodies (this distance was chosen because it was the largest radius that could be used without any circles overlapping) (Figure 1). For species considered monophagous on *F. pinicola* (Table 1), only fruiting bodies of this species were considered as local sources, whereas fruiting bodies of *P. betulinus* were also considered local sources for oligophagous and polyphagous species (Table 1). Colonisations in June and July were analysed separately.

All taxa that were determined to species or genus rank, and that colonised at least 11 of the 14 distances were analysed. For flies of the genus *Medetera*, only males could be determined to species level. As 95% of all males belonged to the species *Medetera apicalis* (Zetterstedt), we assume that the majority of female *Medetera* sp. belonged to this species as well. We therefore analysed all females of *Medetera* sp. as being *M. apicalis*.

## Results

The density of dead fruiting bodies of *F. pinicola* was 16 times higher in the surveyed part of the old-growth nature reserve than in the adjacent managed forest, and the density of dead fruiting bodies of *P. betulinus* was 18 times

*Table 1.* Percentage of exposed fruiting bodies of *Fomitopsis pinicola* (240 in total each month) colonised by different insect taxa during June and July 1997.

| Taxon | Feeding mode | % fruiting bodies colonised in June | % fruiting bodies colonised in July |
|---|---|---|---|
| Coleoptera | | | |
| *Cis quadridens* Mellie | Mon | 7 | 26 |
| *Cis glabratus* Mellie | Mon | 27 | 83 |
| *Cis alter* Silfverb. | Oligo | < 1 | < 1 |
| *Cis bidentatus* (Ol.) | Oligo | | 2 |
| *Ennearthron cornutum* (Gyll.) | Oligo | < 1 | 1 |
| *Dorcatoma punctulata* Muls. & Rey | Mon | 1 | 17 |
| *Bolitophagus reticulatus* (L.) | Mon other sp. | | < 1 |
| *Aridius nodifer* (Westwood) | Pol | | < 1 |
| *Orthoperus* sp. | Pol | | < 1 |
| *Corticaria* sp. | Pol | < 1 | |
| Staphylinidae | Pred | | 4 |
| Diptera | | | |
| *Leucophenga quinquemaculata* Strobl | Oligo/Pol? | 18 | 1 |
| *Medetera abstrusa* Thuneberg | Pred | < 1 | < 1 |
| *Medetera apicalis* (Zetterstedt) | Pred | 2 | 17 |
| *Medetera* sp. | Pred | 5 | 26 |
| *Tephrochlamys* sp. | Pol? | 2 | < 1 |
| Cecidomyidae | Pol | 13 | 35 |
| Sciaridae | Pol | 90 | 85 |
| Drosophilidae | Pol? | 2 | 6 |
| Sphaeroceridae | Pol? | < 1 | |
| Hymenoptera | | | |
| *Choeras parasitellae* (Bouché)[a] | Par | | 3 |
| *Lissonota* sp.[a] | Par | | 10 |
| Lepidoptera | | | |
| *Archinemapogon yildizae* Kocak | Pol | 3 | 10 |

[a]Hosts are tineid moths such as *A. yildizae*. Feeding modes: Mon = monophagous species that strongly prefers fruiting bodies of *F. pinicola*, Oligo = oligophagous species that strongly prefers fruiting bodies, of more than one species of wood-decaying fungi, Pol = polyphagous species that also utilise substrates other than fruiting bodies of wood-decaying fungi, Pred = predators, and Par = parasitoids. *Bolitophagus reticulatus* prefers fruiting bodies of *Fomes fomentarius*.

higher in the nature reserve. Throughout the managed area the number of dead fruiting bodies remained at a low level (Figure 2).

In total, 22 different taxa colonised the fruiting bodies exposed to insect colonisation (Table 1). There was no significant effect of either distance from the old-growth forest or of local sources of fruiting bodies on the average number of colonising taxa per fruiting body (Table 2; Figure 3). However, in July the effect of distance from the old-growth forest was close to being statistically significant (Table 2; Figure 3).

We analysed the effects of distance from the old-growth forest and local sources on the colonisation rate of six species. There was a significant negative effect of distance from the forest reserve on the proportion of fruiting bodies

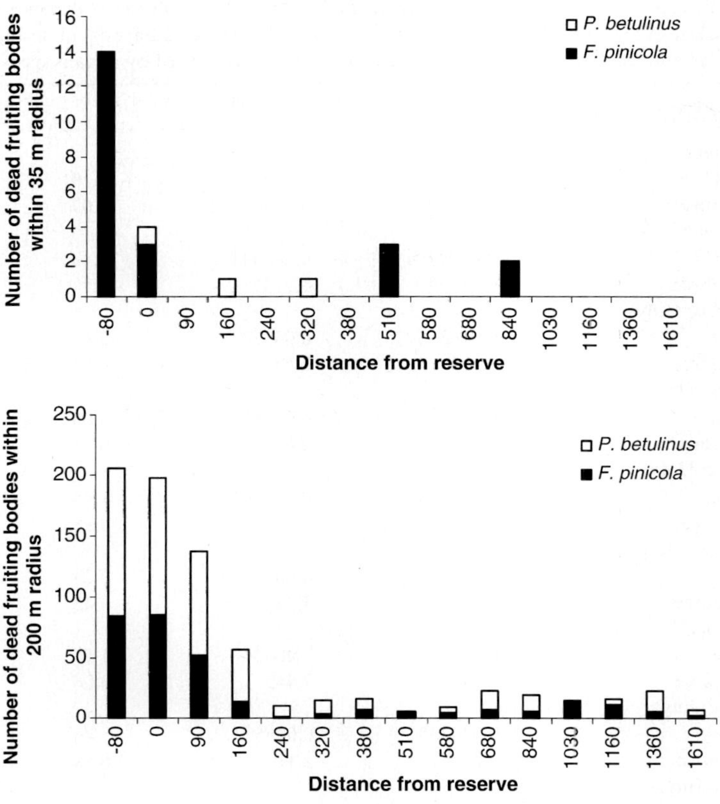


*Figure 2.* Number of dead fruiting bodies naturally occurring within circles of different radii around the experimental groups of fruiting bodies.

colonised by the fungivorous beetle *Cis quadridens* and the predatory fly *Medetera apicalis* (Table 2; Figure 4). For other species analysed, no significant effect of distance from the reserve was detected (Table 2; Figure 4). For none of the species, did the size of local sources of dead fruiting bodies have any significant effect on the proportion of colonised fruiting bodies (Table 2). For other taxa, which were either too rare for analysis or were only determined to family level, we only explored the data graphically. For these species no effect of distance from the reserve seemed to be present. The proportion of parasitized *Archinemapogon yildizae* Kocak was also unaffected by the distance from the reserve and local sources of fruiting bodies (Table 2; Figure 5).

## Discussion

The ability to colonise fruiting bodies far from the nature reserve differed between insect taxa. For several taxa there was no evidence of any decline in colonisation rate with distance from the reserve, and all taxa were unaffected

*Table 2.* Results of multiple linear regression models testing the effects of distance from the reserve and size of local source of fruiting bodies on the (log) average number of colonising taxa per fruiting body and the (arcsine) proportions of fruiting bodies colonised by certain species.

| Taxa (month), variable | Coeff | SE Coeff | *t* | *p* |
|---|---|---|---|---|
| *Average number of taxa* (June) | | | | |
| Distance | 0.00001117 | 0.0000582 | 0.19 | 0.851 |
| Local source | −0.02369 | 0.02208 | −1.07 | 0.306 |
| *Average number of taxa* (July) | | | | |
| Distance | −0.0009478 | 0.00004351 | −2.18 | 0.052 |
| Local source | −0.01150 | 0.0165 | −0.7 | 0.501 |
| *C. quadridens* (July) | | | | |
| Distance | −0.016356 | 0.005767 | −2.84 | 0.016 |
| Local source | 2.327 | 2.478 | 0.94 | 0.368 |
| *C. glabratus* (June) | | | | |
| Distance | 0.006185 | 0.006160 | 1.00 | 0.337 |
| Local source | −1.426 | 2.646 | −0.54 | 0.601 |
| *C. glabratus* (July) | | | | |
| Distance | −0.009538 | 0.006463 | −1.48 | 0.168 |
| Local source | 2.707 | 2.776 | 0.98 | 0.350 |
| *D. punctulata* (July) | | | | |
| Distance | −0.001817 | 0.007874 | −0.23 | 0.822 |
| Local source | 1.515 | 3.383 | 0.45 | 0.663 |
| *L. quinquemaculata* (June) | | | | |
| Distance | 0.010098 | 0.006447 | 1.57 | 0.146 |
| Local source | −0.143 | 2.445 | −0.06 | 0.954 |
| *M. apicalis* (July) | | | | |
| Distance | −0.030405 | 0.005573 | −5.46 | <0.001 |
| Local source | −1.216 | 2.114 | −0.58 | 0.577 |
| *A. yildizae* (July) | | | | |
| Distance | 0.007503 | 0.004402 | 1.70 | 0.116 |
| Local source | 2.406 | 1.670 | 1.44 | 0.178 |
| *Prop. parasitism* (July) | | | | |
| Distance | −0.01790 | 0.01484 | −1.25 | 0.239 |
| Local source | −2.695 | 5.452 | −0.49 | 0.631 |

For the parasitoids (*C. parasitellae* and *Lissonota* sp. combined) the response variable is the (arcsine) proportions of parasitized hosts (*A. yildizae*). Colonisations in June and July were analysed separately. Overall regression statistics for significant tests: *Cis quadridens* (July) $F = 5.50$, $p = 0.022$, $r^2 = 0.409$, *Medetera apicalis* (July) $F = 16.36$, $p = 0.001$, $r^2 = 0.702$. $r^2$ values are adjusted for df.

by the size of local sources in the managed forest. There was also no significant effect of distance or size of local sources on the average number of colonising taxa per fruiting body (although distance had a near significant effect in July). These findings fit well with previous results from inventories showing that many species inhabiting fruiting bodies of wood-decaying fungi occur at similar frequencies (per fruiting body) in managed and unmanaged forests (Jonsell and Nordlander 2002). They are also consistent with results from a previous colonisation study (Jonsell et al. 1999). These studies, taken together, suggest that many species inhabiting fruiting bodies of wood-decaying fungi in

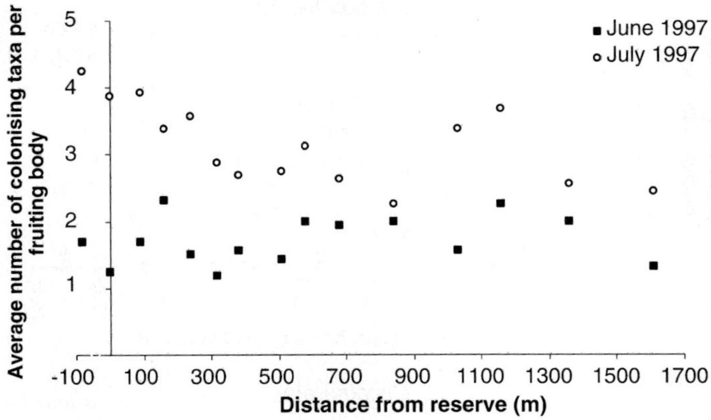


*Figure 3.* Mean number of colonising taxa per fruiting body inside (−80 m) and at different distances from the border of the old-growth forest.

Fennoscandia can colonise the substrate available in managed forests at a high enough rate to maintain populations in them.

However, this is probably not true for all species. For the specialist fungivore *Cis quadridens* and the predatory fly *Medetera apicalis* the colonisation rate was negatively affected by distance from the reserve. The colonisation rate of both of these species declined steeply within a few 100 m of the reserve, and at the largest distances both species were very rare or absent. In Sweden, inventories have shown that *C. quadridens* is generally rare in managed forests, but quite common in forests with high amounts of dead wood and fruiting bodies of wood-decaying fungi (Jonsell and Nordlander 2002). *Cis quadridens* was the only nationally red-listed species found in this study (categorised as Near Threatened in Sweden; Gärdenfors 2000). Our results indicate that the poor performance of *C. quadridens* in managed forests with low quantities of suitable substrate is due to a weak ability to colonize distant resources. This is in agreement with the small-scale colonisation patterns found in a previous study (Jonsell et al. 1999).

The predatory fly *M. apicalis* has been confirmed to be the same species as the *Medetera impigra* Collin recorded in some previous studies of insects in fruiting bodies of *F. pinicola* (Jonsell et al. 1999; Jonsell and Nordlander 2002). These two closely related species belong to the 'apicalis group' and are difficult to separate even by taxonomic specialists (Igor Grichanov, pers. comm.), so we will refer to the individuals determined as *M. impigra* in the above studies as *M. apicalis*.

A negative effect of distance on the colonisation rate of *M. apicalis* was also found by Jonsell et al. (1999), although *M. apicalis* was generally less common in that study. Similarly, in inventories of forest stands with different densities of dead wood and forestry histories, *M. apicalis* tended to be less common in managed forests than in old growth stands (Jonsell and Nordlander 2002). As

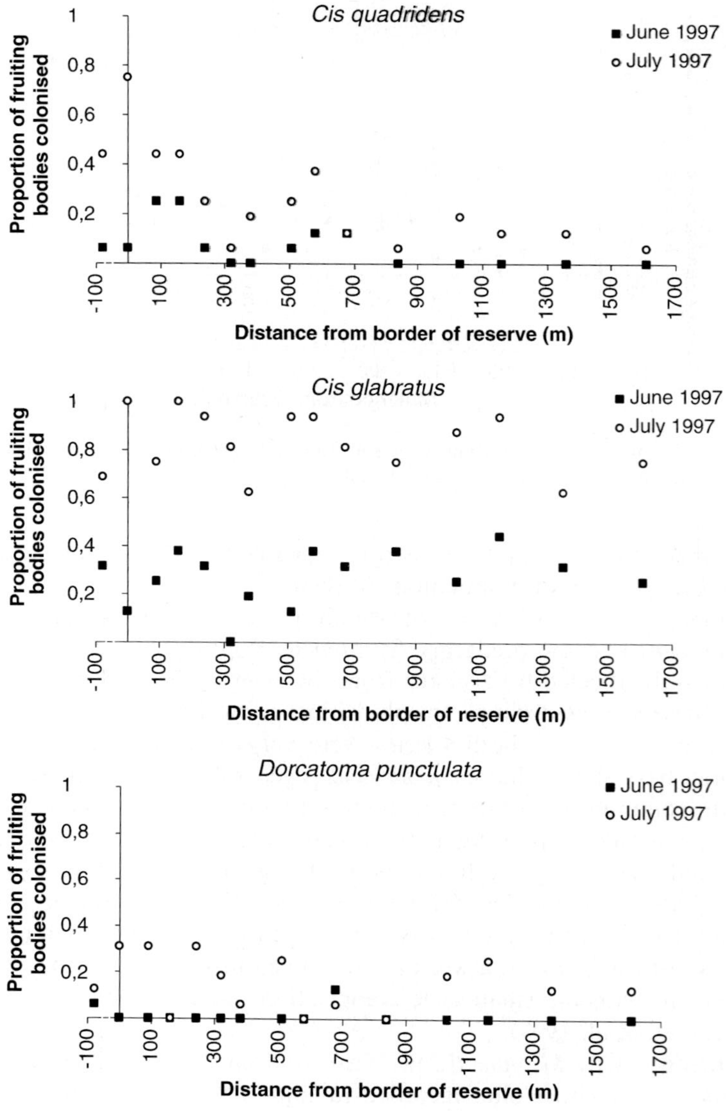


*Figure 4.* The proportion of fruiting bodies colonised by different species inside (−80 m) and at different distances from the border of the old-growth forest.

for *C. quadridens*, our results suggest that *M. apicalis* has a weak ability to colonize distant resources. However, as the biology of *M. apicalis* is poorly known this conclusion must be considered preliminary. *Medetera apicalis* has been recorded from fruiting bodies of several species of wood-decaying fungi (Bickel 1985; Jonsell et al. 1999; Jonsell and Nordlander, 2002), and also from under the bark of a number of tree species (Bickel 1985). It is not entirely clear,

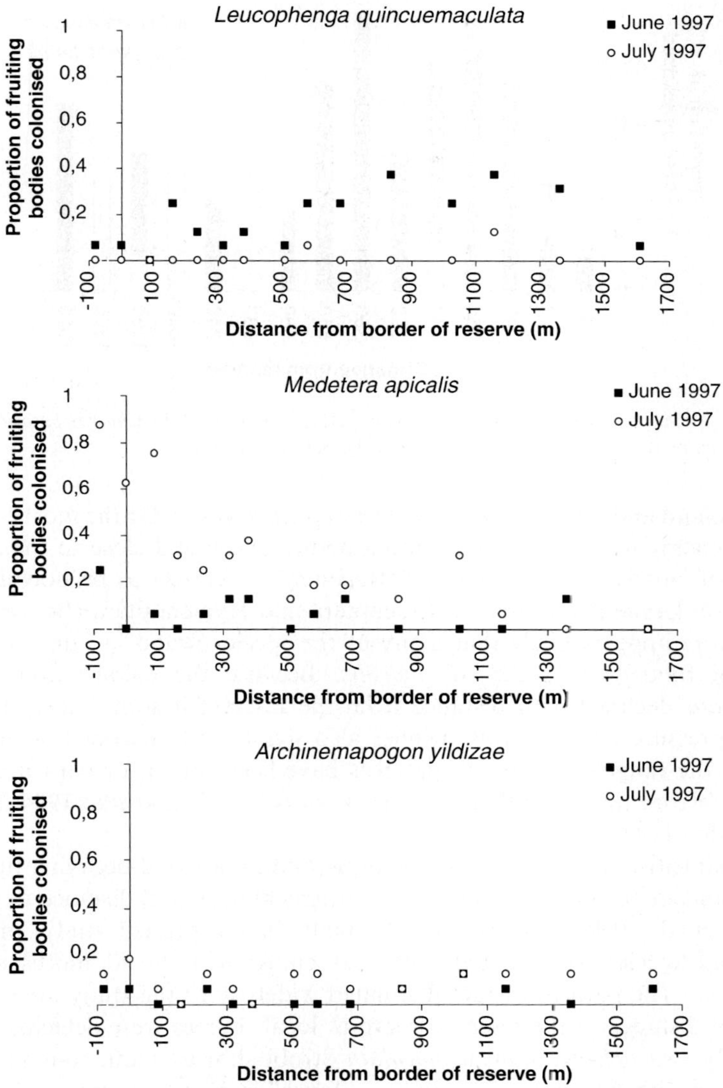


*Figure 4.* Continued.

however, whether all records really concern the same species. In any case, as we found only low amounts of dead wood in the managed forest, the amount of substrate for *M. apicalis* is likely to be much lower in it than in the nature reserve.

Organisms of higher trophic levels may be particularly vulnerable to habitat loss and fragmentation (Holt 1996). This may have an effect not only on natural enemies themselves, but also on the dynamics of their prey, which would be released from predation and parasitism (Kruess and Tscharntke

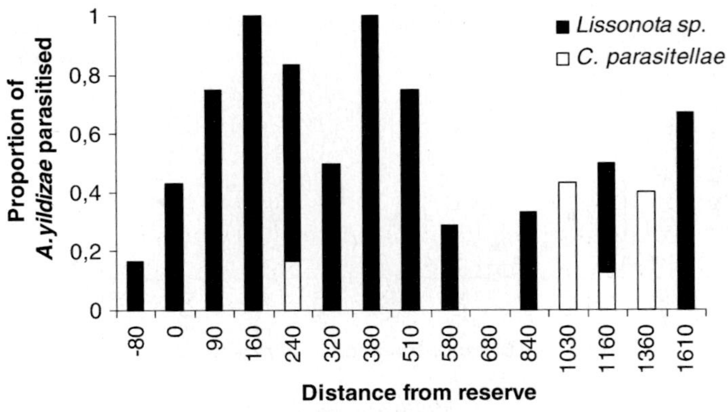


*Figure 5.* Mean proportion of *Archinemapogon yildizae* parasitized by *Choeras parasitellae* and *Lissonota* sp. in July at different distances from the old-growth forest.

1994; Roland and Taylor 1997). *Medetera apicalis* was by far the most common natural enemy reared from the fruiting bodies inside and close to the reserve. Larvae of closely related species of *Medetera* are known to be polyphagous, feeding on larvae from Diptera, Coleoptera and Hymenoptera (Beaver 1966). It therefore appears likely that many of the species found in our study were potential food sources for *M. apicalis*. Because the colonisation rate of *M. apicalis* declined with distance from the reserve, it seems likely that the overall pressure from natural enemies also significantly declined, at least for some of the fungivores. Similar patterns have been found for parasitoids and their herbivore hosts in other systems (Kruess and Tscharntke 1994; Roland and Taylor 1997).

The parasitism rate of species in fruiting bodies of wood-decaying fungi has been found to be negatively affected by fragmentation and distance to sources (Jonsell et al. 1999; Komonen et al. 2000). In the present study, only two parasitoid species were reared, both associated with tineid moths such as *A. yildizae*. The parasitism rate fluctuated widely over the study area, but no effects of distance to the reserve or size of local sources were detected.

The fly *Leucophenga quinquemaculata* Strobl showed reductions in colonisation rate with distance in the study by Jonsell et al. (1999), but we found no negative effect of distance from the reserve on the colonisation rate of this species. Instead, the highest colonisation rate of *L. quinquemaculata* was found at a distance of about 600–1400 m. This pattern was probably not due to a release from predation, as *L. quinquemaculata* was most common in June while the common predator *M. apicalis* primarily colonised in July. The presence of some unknown substrate in the managed forest used by *L. quinquemaculata* may explain the observed colonisation pattern, since the habitat requirements of this species is inadequately known. In Sweden, *L. quincuemaculata* has only been found in fruiting bodies of *F. pinicola* (Jonsell et al. 2001), but in Norway it has also been reared from *P. betulinus* (Bächli and Thunes 1992).

Specialist species with small population sizes may be particularly sensitive to habitat loss and fragmentation (Holt 1996). The anobiid beetle *Dorcatoma punctulata* Muls. & Rey is a specialist on fruiting bodies of *F. pinicola* and usually occurs at low population densities (Jonsell et al. 2001; Jonsell and Nordlander 2002). However, both our study and that by Jonsell et al. (1999) showed that *D. punctulata* is one of the most efficient long distance colonisers of the inhabitants of *F. pinicola*. It has been suggested that this species invests substantially in long distance dispersal to avoid competition (Jonsell et al. 1999) or natural enemies (Jonsson et al. 2003b). An important reason why species differ in their responses to habitat fragmentation is probably that they differ in dispersal behaviour (Kareiva 1987). One aspect of the dispersal behaviour of *D. punctulata* that may contribute to its good colonisation ability is its strategy for finding a mate during dispersal. Modelling has shown that the mate finding strategy of *Dorcatoma* is particularly efficient at low densities of substrate and conspecific individuals (Jonsson et al. 2003b).

To conclude, we have shown that several species inhabiting fruiting bodies of wood-decaying fungi seem to be able to colonise substrates at a high enough rate to persist also in managed forests. However, both *Cis quadridens* and *Medetera apicalis* showed steeply decreasing colonisation rates with distance from the reserve, suggesting that the majority of successfully colonising individuals had only flown a couple of hundred metres. Weak ability to colonize distant resources may therefore be an important reason for the rarity of these and other saproxylic species in managed forests (cf. Jonsson 2003). The distribution of natural enemies may also influence the colonisation of other species. *Medetera apicalis* was the dominating predator inside and close to the nature reserve, but its incidence declined steeply with distance away from the old-growth stand in the managed forest. In addition, other studies have found parasitoids to be more sensitive to forest management than their fungivorous hosts (Jonsell et al. 1999; Komonen et al. 2000). Thus, forest management may lead to a decline in the pressures from natural enemies.

## Acknowledgements

We thank Henrik Nordenhem, Lisa Nordenhem, Fredrik Söderman and Olle Terenius for help with the fieldwork. Igor Grichanov and Jevgeni Jakovlev helped determining the flies. Mats Jonsell, Thomas Ranius and two anonymous reviewers provided valuable comments on previous versions of the manuscript. The study was financed by the Oscar and Lili Lamm Memorial Foundation.

## References

Bächli G. and Thunes K.H. 1992. *Leucophenga quinquemaculata* Strobl (Diptera, Drosophilidae) from Norway. Fauna Norvegica. Serie B 39: 81–84.

Beaver R.A. 1966. The biology and immature stages of two species of *Medetera* (Diptera: Dolichopodidae) associated with the bark beetle *Scolytus scolytus* (F.). Proc. R. entomol. Soc. London (A) 41: 145–154.

Bickel D.J. 1985. A revision of the nearctic *Medetera* (Diptera: Dolichopodidae). Tech. Bull. 1692: 1–109.

Fridman J. and Walheim M. 2000. Amount, structure and dynamics of dead wood on managed forestland in Sweden. Forest Ecol. Manag. 131: 23–26.

Gärdenfors U. (ed.) 2000. The 2000 Red-List of Swedish Species. Swedish Species Information Centre, Swedish University of Agricultural Sciences, Uppsala, Sweden.

Gehrken U., Strømme A., Lundheim R. and Zachariassen K.E. 1991. Inoculative freezing in overwintering tenebrionid beetle, *Bolitophagus reticulatus* Panz. J. Insect Physiol. 37: 683–687.

Hanski I. 1987. Colonization of ephemeral habitats. In: Gray A.J., Crawley M.J. and Edwards P.J. (eds), Colonization, Succession and Stability. Blackwell Scientific Publications, Oxford, pp. 155–185.

Hanski I. 1997. Metapopulation dynamics: From concepts and observations to predictive models. In: Hanski I.A. and Gilpin M.E. (eds), Metapopulation Biology: Ecology, Genetics and Evolution. Academic Press, San Diego, pp. 69–91.

Harrison S. and Taylor A.D. 1997. Empirical evidence for metapopulation dynamics. In: Hanski I.A. and Gilpin M.E. (eds), Metapopulation Biology: Ecology, Genetics and Evolution. Academic Press, San Diego, pp. 27–42.

Holt R.D. 1996. Food webs in space: An island-biogeographical perspective. In: Polis G.A. and Winemiller K.O. (eds), Food Webs. Chapman Hall, New York, pp. 313–323.

Ims R.A. and Yoccoz N.G. 1997. Studying transfer processes in metapopulations: Emigration, migration and colonization. In: Hanski I.A. and Gilpin M.E. (eds), Metapopulation Biology: Ecology, Genetics and Evolution. Academic Press, San Diego, pp. 247–265.

Jonsell M. and Nordlander G. 2002. Insects in polypore fungi as indicator species: a comparison between forest sites differing in amounts and continuity of dead wood. Forest Ecol. Manage. 157: 101–118.

Jonsell M., Nordlander G. and Jonsson M. 1999. Colonization patterns of insects breeding in wood-decaying fungi. J. Insect Conserv. 3: 145–161.

Jonsell M., Nordlander G. and Ehnström B. 2001. Substrate associations of insects breeding in fruiting bodies of wood-decaying fungi. Ecol. Bull. 49: 173–194.

Jonsell M., Schroeder M. and Larsson T. 2003. The saproxylic beetle *Bolitophagus reticulatus*: its frequency in managed forests, attraction to volatiles and flight period. Ecography 26: 421–428.

Jonsson M. 2003. Colonisation ability of the threatened tenebrionid beetle *Oplocephala haemorrhoidalis* and its common relative *Bolitophagus reticulatus*. Ecol. Entomol. 28: 159–167.

Jonsson M., Jonsell M. and Nordlander G. 2001. Priorities in conservation biology: a comparison between two polypore-inhabiting beetles. Ecol. Bull. 49: 195–204.

Jonsson M., Johannesen J. and Seitz A. 2003a. Comparative genetic structure of the threatened tenebrionid beetle *Oplocephala haemorrhoidalis* and its common relative *Bolitophagus reticulatus*. J. Insect Conserv. 7: 111–124.

Jonsson M., Kindvall O., Jonsell M. and Nordlander G. 2003b. Modelling mating success of saproxylic beetles in relation to search behaviour, population density and substrate abundance. Animal Behav. 65: 1069–1076.

Kareiva P. 1987. Habitat fragmentation and the stability of predator-prey interactions. Nature 326: 388–390.

Komonen A. 2003. Hotspots of insect diversity in boreal forests. Conserv. Biol. 17: 976–981.

Komonen A., Penttilä R., Lindgren M. and Hanski I. 2000. Forest fragmentation truncates a food chain based on an old-growth forest bracket fungus. Oikos 90: 119–126.

Kruess A. and Tscharntke T. 1994. Habitat fragmentation, species loss, and biological control. Science 264: 1581–1584.

Lewis O.T., Thomas C.D., Hill J.K., Brookes M.I., Crane T.P.R., Graneu Y.A., Mallet J.L.B. and Rose O.T. 1997. Three ways of assessing metapopulation structure in the butterfly *Plebejus argus*. Ecol. Entomol. 22: 283–293.

Lundblad O. 1950. Studier över insektsfaunan i Fiby urskog. Kungliga svenska vetenskapsakademiens avhandlingar i naturskyddsärenden No: 6. Stockholm, (In Swedish).

Rassi P., Mannerkowski I., Peltonen S.-L. and Alanen A. 2000. 2nd Red Data Book of Finland. Ministry of Environment, Helsinki.

Roland J. and Taylor P.D. 1997. Insect parasitoid species respond to forest structure at different spatial scales. Nature 386: 710–713.

Siitonen J. 2001. Forest management, coarse woody debris and saproxylic organisms: Fennoscandian boreal forests as an example. Ecol. Bull. 49: 11–41.

Speight M.C.D. 1989. Saproxylic Invertebrates and their Conservation. Council of Europe, Strasbourg.

Steffan-Dewenter I. 2003. Landscape context affects trap-nesting bees, wasps, and their natural enemies. Ecol. Entomol. 27: 631–637.

Thomas C.D. and Hanski I. 1997. Butterfly metapopulations. In: Hanski I.A. and Gilpin M.E. (eds), Metapopulation Biology: Ecology, Genetics and Evolution. Academic Press, San Diego, pp. 247–265.

Whitlock M.C. 1992. Nonequilibrium population structure in forked fungus beetles: extinction, colonization, and the genetic variance among populations. Am. Nat. 139: 952–970.

Biodiversity and Conservation (2006) 15:297–318
DOI 10.1007/s10531-005-2007-6 

# Dragonfly assemblages in arid tropical environments: a case study from western Namibia

FRANK SUHLING[1,*], GÖRAN SAHLÉN[2,3], ANDREAS MARTENS[4], EUGENE MARAIS[5] and CARSTEN SCHÜTTE[1]

*[1]Institut für Geoökologie, Technische Universität Braunschweig, Langer Kamp 19c, D-38106 Braunschweig, Germany; [2]Ecology and Environmental Science, Halmstad University, P.O. Box 823, SE-30118 Halmstad, Sweden; [3]Systematic Zoology, Evolutionary Biology Centre, Uppsala University, Norbyvägen 18D, SE-75236 Uppsala, Sweden; [4]Biology, Karlsruhe University of Education, Bismarckstrasse 10, D-76133 Karlsruhe, Germany; [5]National Museum of Namibia, Windhoek, P.O. Box 1203, Windhoek, Namibia; *Author for correspondence (e-mail: f.suhling@tu-bs.de)*

Received 10 August 2004; accepted in revised form 31 January 2005

**Key words:** Freshwater wetland conservation, Indicators, Odonata, Species composition

**Abstract.** Dragonflies have been proposed as indicators for the ecosystem health of freshwater wetlands. For their useful functioning as indicators it is, however, necessary to identify species compositions in specific habitats and species-habitat associations, particularly in the tropics, where such knowledge is still weak. We examined the dragonfly species composition of 133 localities in the arid environment of western Namibia. An analysis of nestedness indicated that distinct, and predictable patterns of species associations can be expected. Discriminant analyses revealed that most of the nine habitat types separated by structural and hydrological parameters are well discriminated by their dragonfly assemblages. Spring brooks in particular host a specific assemblage, which is threatened due to the habitat restriction of several species, as well as by recent habitat loss and degradation. Using a hierarchical method of several criteria we demonstrated the selection of a set of potential indicator species from the species set, most of these being useful indicators for spring brook assemblages. The conservation status of certain habitats and species is discussed. We propose that dragonflies will have a high indicator potential for threatened freshwater wetlands in such areas and may also serve as an indication of the sustainable use of water resources including evaluating measures to rehabilitate environments.

## Introduction

One of the challenges for the future lies in protecting the ecological integrity and biodiversity of freshwater aquatic systems, particularly in the tropics. Anthropogenic habitat alterations may cause significant changes in freshwater biodiversity (Ward 1998). These environments are essential resources for development (Ward 1998; Crisman et al. 2003) and face ever-increasing pressure especially in arid countries where freshwater is disproportionately important to humans and other species (Barnard and Shikongo 2000; Day 2003). Gaining knowledge about tropical freshwater communities and of potential indicators of freshwater ecosystem health is therefore crucial.

When compared to the attributes desired for indicators (cf. McGeoch 1998; McGeoch and Chown 1998; Simberloff 1998) dragonflies are among the most promising animals to serve as an indicator group, e.g. for species richness and ecosystem health of freshwater wetlands (Brown 1991; Sahlén and Ekestubbe 2001; Clausnitzer and Jödicke 2004). However, in order to use dragonflies as indicators, basic knowledge of assemblages and habitat preferences of species is required (Corbet 1993). There are several recent approaches to assess and compare dragonfly communities and species richness in relation to habitat in the tropics (Cleary et al. 2004) and particularly in Africa (Samways and Steyler 1996; Clausnitzer 2003; Dijkstra and Lempert 2003). However, the general knowledge of habitat associations of African Odonata is still scarce and requires further research action (e.g., Suhling et al. 2003, 2004a; Clausnitzer 2004a,b; Dijkstra and Vick 2004).

Our first aim was to assess dragonfly communities on a large scale in Namibia. If the dragonfly species composition differed significantly between various types freshwater habitats, i.e. the species composition is nested, then it would indicate that the dragonfly community is not randomly organised. Distinct and predictable patterns of occurrence might be expected with a high level of nestedness. Distinct types of habitats may likewise have a nested species composition, hosting certain species assemblages. Our second goal was to identify species that might be indicative of different assemblages. We suggest a set of five criteria, including frequency, habitat specificity and criteria derived from the statistical analyses we applied to select such species. In our context the presence of particular species in distinct habitats would indicate the completeness of the typical community expected at such sites (cf. Sahlén and Ekestubbe 2001). Such indicator species might then be used to identify threatened environments and monitor the impact of human activities on the aquatic biodiversity of Namibia.

## Materials and methods

### *Study area*

Namibia is the most arid country of the Afrotropical region, i.e. south of the Sahara. The only perennial rivers occur along the northern and southern borders of the country. Natural permanent surface water in the interior parts of Namibia only occurs at widely separated springs around mountains and in ephemeral river courses (Breen 1991). Water is therefore one of the most relevant, and limited, resources in Namibia (Heyns et al. 1998; Christelis and Struckmeyer 2001). Development and changes in human lifestyle during the 20th century have affected the way in which water is managed (Stern and Lau 1990; Seely 1998). Large impoundments have been built to ensure reliable water supply for industrial development, urban centres and irrigated agriculture, which altered flood regimes and destroyed perennial wetlands in ephemeral rivers. Large-scale extraction of groundwater to provide water for smaller urban centres, mining and intensive livestock agriculture has caused to a fall in water tables, a loss of spring habitats (Jacobsen et al. 1995; Seely 1998), and changes in vegetation structure through die-back of large trees and other vegetation tapping the aquifers. Some large-scale water transfer schemes have been established, channelling or piping water over long distances to major urban and industrial centres to meet the increasing demand for water. These schemes are likely to become more extensive due to the planned development of dams on the perennial rivers along both the northern and southern borders in order to meet projected requirements of the 21st century.

Our study was conducted in western Namibia approximately between 17 ° and 25 ° S and 13 ° and 18 ° E (Figure 1), an area characterised by arid climate and therefore mainly by savannah, karoo and desert biomes (Mendelsohn et al. 2002). We restricted our study mainly to the western ephemeral river catchments that originate in the central Namibian highlands and flow into the Atlantic Ocean (cf. Jacobsen et al. 1995). Additionally, we selected sites in areas adjacent to the watershed to consider habitats that were otherwise underrepresented in the study, i.e. large impoundments and spring brooks.

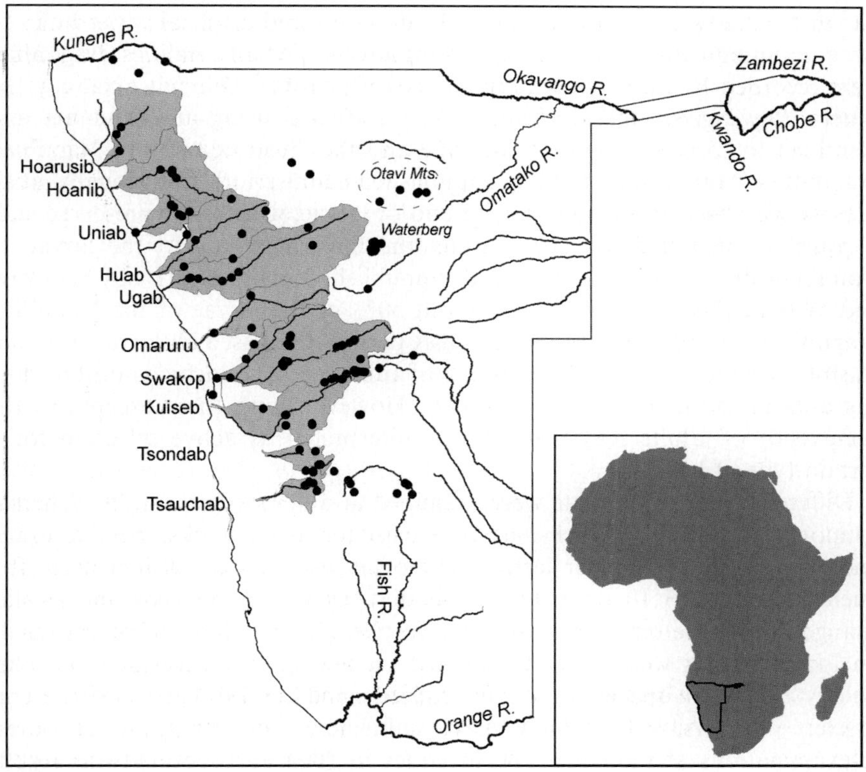


*Figure 1.* Map of Namibia showing the distribution of the sample sites and the western ephemeral river catchments and the Fish River catchment. Note that some few sites do not belong to the western catchments.

These sites were mainly in the Kunene River catchment, in the Otavi Mountains, at the Waterberg and in the upper Fish River catchment.

In terms of altitude the study area is broadly declining from about 2000 m a.s.l. in central Namibia down to sea level, with the study sites ranging from 29 to 1786 m a.s.l. The average annual rainfall in the area is varying from about 550 mm in the Otavi Mountains to less than 50 mm at the coast. The average maximum temperatures during the hottest month increase from 20 °C at the coast to about 30–34 °C in central Namibia and are higher than 36 °C in the southern part of the study area. The average minimum of the coldest month ranges between 10 °C at the coast and 4 °C inland, with about 1–10 days of frost per year. All climate and geographical data are from the Atlas of Namibia (Mendelsohn et al. 2002).

*Recording of dragonflies and habitats*

Between January 2001 and May 2004 we recorded dragonflies at 133 localities (Figure 1 and Appendix) that represented all major types of freshwater habitats in the study area, from ephemeral rain pools and artificial water holes to large impoundments and perennial spring brooks. At all localities dragonflies were recorded by identifying adults *in vivo* if possible. Difficult taxa, e.g. the genera *Pseudagrion* and *Orthetrum*, were collected using an insect net and identified to species using a microscope and the illustrated key to Namibian dragonflies (Martens and Suhling unpublished manuscript). Additionally, at all sites we searched for odonate exuviae and/or larvae, which were preserved and, if possible, identified to species using the illustrated key to the larvae of Namibian dragonflies (Suhling et al. unpublished manuscript) and Samways and Wilmot (2003). Because only about 50% of the larvae of the Namibian dragonflies are described, full species lists per locality based on larvae were not feasible. We are aware that species lists of adults will also include non-breeding vagrants in the data set (Sahlén 1999). However, with these exceptions the occurrence of adults may generally be interpreted as active selection for a certain type of habitat.

Different types of habitats were examined at different frequencies. Whereas temporary rain pools, which may only exist for a few weeks, were recorded once, perennial and longer lasting temporary waters were visited more frequently, i.e. up to 10 times (cf. Appendix), as the species assemblage may change during the ongoing season due to phenological differences. Especially on localities that were only investigated on one or two occasions, the phenology of certain species may have caused an under-estimation of the true species numbers, which may have influenced our analysis. However, re-examinations at more than 40 localities in 2004 corroborated our results concerning the species assemblages (Suhling unpublished data). From all records we produced a presence/absence matrix for the species at all localities. The entry for a species at a given locality was 1 if either an adult or a larva/

exuvia was recorded at least once, and it was 0 for a given locality if a species was never encountered. For analysis (below) the localities were sorted into nine functional types of habitats. These are:

(1) *Wetlands below dams* (Number of localities $n = 9$): Leakage at most of the large impoundments (see 9) results in small perennial running waters and wetlands immediately below the dams. Some of these are similar to spring brooks with respect to their wetland habitat structures. All sites were well vegetated by various submerged plants and reeds.

(2) *Spring brooks* ($n = 12$): Small perennial running waters fed by strong springs, which run for stretches of up to 2 km before they vanish into the ground or evaporate. Many streams consisted of linked pools and included small waterfalls and rapids. All spring brooks sustain trees in their surroundings. However, some sites were completely shaded, e.g. at Zebra River, whereas others were widely exposed to the sun, e.g. Ongongo Fall. All sites contained semiaquatic vegetation, i.e. reeds (*Phragmites* sp. *Typha latifolia* and Cyperaceae), submerged vegetation (*Chara* sp., *Potamogeton* spp.) and/or mosses.

(3) *Degraded spring brooks* ($n = 10$): Here we grouped former spring brooks that have been extensively altered by humans. Most sites have been changed into perennial spring-fed ponds by the construction of dams just below the spring, e.g. in the Otavi Mountains and at Klein Barmen. Others have become degraded due to water extraction reducing the length of perennial section (Otjisongombe) and subjected to heavy cattle grazing. All have in common that the normal structure of a spring brook has been lost.

(4) *Spring pools* ($n = 10$): Perennial springs with weak discharge or in depressions may form pools or small chains of pools. Most of these are densely vegetated by rushes; even those that are subjected to heavy grazing by cattle or game. All of them are widely exposed to the sun.

(5) *Ephemeral river sections* ($n = 34$): Wetlands – some being perennial – along the courses of the large ephemeral rivers result from the resurgence of underground water due to geology or topography (Jacobsen et al. 1995). Unlike the other habitat types, these wetlands may be subjected to extensive disturbance or even complete alteration due to strong floods. Consequently, vegetation and bottom substrate, which mainly consists of sand, may be washed away. We were able to register a rapid succession after such an event in the Ugab River, with the vegetation recovering within a few months. The vegetation consisted mainly of reeds and the shorelines were covered with fast growing shrubs. The water had appreciable levels of dissolved salts resulting in high conductivity of 2–8 mS/cm, with a peak value of 42 mS/cm (Swakop River near Swakopmund).

(6) *Temporary waters* ($n = 23$): This category includes ponds as well as small springs that contain water only after heavy rains and may persist for some months during and shortly after the rainy season. Most sites contain no vegetation or only some scattered terrestrial plants. One site, however, contained some rushes.

(7) *Artificial waters* ($n$ = 14): Man-made waters including water holes for cattle and game, fish ponds and even some swimming pools. A common feature of most artificial sites are that they are concrete constructions and extremely poor in wetland structures. However, in large parts of the study area they represent the only permanent freshwater habitats during the dry season.

(8) *Farm dams* ($n$ = 12): Small earth dams in drainage gullies and smaller ephemeral rivers may form larger ponds that – after being filled – may persist for several months. Because the ponds are in riverbeds they are affected by mechanical stress (see type 4) so that aquatic vegetation is very sparse, if not absent altogether. Also the shores are often free of vegetation due to grazing cattle and the varying water levels.

(9) *Lakes* ($n$ = 8): Large impoundments forming perennial lakes. All these impoundments were created in larger ephemeral rivers for water supply to towns. Due to the variability of annual rainfall, evaporation and the use of water the water level may vary by several meters between years and even seasons. Most lakes therefore contained very little aquatic vegetation, of which the most common were *Potamogeton* spp.

One locality, an artificial canal with high current velocity, did not fit to any of these categories and was therefore omitted in the analyses of assemblage patterns and of habitat specificity of the species, but was used in the nestedness analysis of all localities (see below).

## *Analysis of nestedness*

The use of nestedness as a tool for analysing species composition in fragmented habitats is controversial (e.g., Simberloff and Martin 1991; Wright and Reeves 1992; Atmar and Patterson 1993; Lomolino 1996; Worthen 1996). Several different methods are in use, among them the Nestedness Temperature Calculator, NTC (Atmar and Patterson 1995), which is available on the World Wide Web. Fischer and Lindenmayer (2002) noted that this method has been used indiscriminately. They showed that even randomly generated data sets may indicate significant nesting if all species is treated as equally common. Bearing this in mind we decided to use two methods to corroborate whether our species assemblages were nested, *viz.* the NTC and the Standardised Nestedness Score ($C$) described by Wright and Reeves (1992).

First we included all species and localities in a presence–absence matrix and analysed the nestedness of species in the study area. Second we analysed the nestedness of each habitat type (see above) separately. Contrary to Sahlén and Ekestubbe (2001), we included all species in the matrix, also the obligate migrants. Migrating species have a more random occurrence and will elevate the temperature in the NTC (Atmar and Patterson 1993) and hence lower the $C$-score. But since the ecology of all species in the area is not known, we cannot exclude known migrants while other unknown migrants may be hiding in the rest of the species pool. The size, shape and fill of the matrix will also affect the temperature in the NTC (Atmar and Patterson 1995). A rectangular matrix as well as an empty one (more zeros than ones) will result in a lower temperature than a square one or one, which contains more ones than zeroes. This may cause a non-nested composition to be classified as a nested one, hence our use of the *C*-score. In order to be able to compare the methods we used the same packing of the matrix for *C* as when calculating matrix temperature in NTC. As we use the *C*-score to verify the statistics of the NTC, the *z*-score statistics (Wright and Reeves 1992) or *Q*-value (cf. also McCulloch 1985) was not calculated. Considering that the *C*-score varies between 0 and 1 and there is no consensus on how low score a nested community may have still being nested, we decided to compare our *C*-scores with those presented in other analyses of odonate communities.

*Analysis of assemblage patterns*

We performed a discriminant function analysis using SPSS 11.0 to determine if the nine types of habitats we distinguished (see above) were, indeed, separate with regard to the odonate assemblages of the localities. Given a set of independent variables, discriminant analysis attempts to find linear combinations of those variables (discriminant functions) that best separate the groups of cases (here types of habitats). A matrix of presence/absence data as used in the nestedness analysis of species served as independent variables in the analysis. In addition, the procedure produces Eigenvalues, which provide information about the relative efficacy of each discriminant function, and Wilks' lambda values as measures of how well each function separates cases into groups. Wilks' lambda is equal to the proportion of the total variance in the discriminant scores not explained by differences among the groups, i.e. smaller values indicate greater discriminatory ability of the function. By associated chi-square statistics we tested the hypothesis that the means of the functions listed are equal across groups. Canonical correlations indicate which variables (species) correlate best with the respective functions. Finally, the analysis provides classification results, i.e. how well the distinguished types of habitats are predicted by the assemblage structure.

*Selection of indicative species*

For the selection of indicative species we used a set of five criteria in a stepwise order. We decided that species had to match all five criteria to serve as potential indicators for the health of dragonfly assemblages of certain habitats. (1) We analysed the habitat specificity of the species by comparing the number of sites to the habitat (see above) at each site. We assumed that generalist species and migrants were found in most, if not all, of the nine types of habitat, while habitat specialist species were expected in maximally one-third (i.e. 1–3) of the habitat types. (2) We selected species from the group of 'moderately common' species to compare to general species richness according to Sahlén and Ekestubbe (2001). As moderately common we counted those that were recorded at $<20\%$ and $\geq 3\%$ of all localities surveyed. (3) As a second criterion of habitat specificity we used univariate ANOVAs analyses on the equality of the distribution of each species, which is used as test of the potential of each independent variable in the discriminant analysis. A species was only selected when its distribution was significantly different from random. (4) We assumed that a potential indicator should not be too rare in its specific habitat type. We therefore accepted only species that occurred at least at 25% of the localities of a particular habitat. (5) We selected species, of which the distribution were correlated with one of the significant discriminant functions according to the canonical correlation analyses derived from the discriminant analysis.

## Results

### *Nestedness*

The matrix (59 species, 133 localities; fill 13.3%) was nested in the NTC giving a temperature of 4.38 °, which was significantly different from the temperature generated by 1000 Monte-Carlo simulations of random distributions ($49.6 \pm 1.6$°sd; $p < 0.001$. The *C*-score for the matrix was 0.351. All of the nine separate habitats were also nested using both methods, with *C*-scores ranging from 0.388 and 0.389 in spring brooks and artificial waters to 0.676 and 0.895 in wetlands/lakes and spring pools respectively (Table 1). All of the individual habitats had higher *C*-scores than the total species pool in the region. All temperatures but one derived from the NTC were higher than that of the total species pool, the lowest temperature (3.01 °) in temporary waters and the highest (28.38 °) in degraded spring brooks (Table 1).

### *Assemblage composition and diversity*

Discriminant analyses used eight discriminant functions of which the first four functions were significant. The first function (Eigenvalue = 45.44, Wilks' lambda = 0.00004, $\chi^2$ = 1009.709, df = 416, $p < 0.001$) explained 78.4% of the variance. Function 2 (Eigenvalue = 5.17, Wilks' lambda = 0.002, $\chi^2$ = 623.97, df = 357, $p < 0.001$) explained 8.9%, function 3 (Eigenvalue = 2.73, Wilks' lambda = 0.012, $\chi^2$ = 441.08, df = 300, $p < 0.001$) explained 4.7%, and function 4 (Eigenvalue = 1.99, Wilks' lambda = 0.046, $\chi^2$ = 300.87, df = 300, $p$ = 0.004) explained 3.44%. The first two functions together explained in total 87.3% of the variance.

In total, 81.1% of all habitats were correctly classified according to their odonate assemblage structures. But, the classification results varied between the habitats (Table 2), from about 50% (spring pools) to 100% (spring

*Table 1.* Comparison of the nestedness values (temperatures and *C*-scores) and species diversity (total number species and median, minimum, and maximum number of species per type of habitat) of the nine habitat types.

| | Wetlands /lakes | Spring brooks | Degraded spring brooks | Spring pools | Ephem. R. sections | Temp. waters | Artificial waters | Farm dams | Lakes |
|---|---|---|---|---|---|---|---|---|---|
| Replicates | 9 | 12 | 10 | 10 | 34 | 23 | 14 | 12 | 8 |
| Nestedness temperatures (°) | 9.21*** | 23.85*** | 28.38** | 17.93** | 11.82*** | 3.01*** | 19.22** | 25.56*** | 18.44*** |
| *C*-scores | 0.676 | 0.388 | 0.638 | 0.895 | 0.605 | 0.430 | 0.389 | 0.469 | 0.500 |
| Total species | 40 | 41 | 26 | 14 | 24 | 27 | 19 | 21 | 24 |
| Median | 13 | 13 | 9 | 6 | 8 | 2 | 2 | 7 | 12 |
| Min. species # | 6 | 4 | 4 | 1 | 1 | 1 | 1 | 3 | 3 |
| Max. species # | 35 | 26 | 17 | 9 | 15 | 23 | 10 | 12 | 17 |

** $p < 0.01$, *** $p < 0.001$.

*Table 2.* Classification results of a discriminant analysis showing groups predicted from the dragonfly assemblage pattern (presence/absence data of species) in relation to the original habitat type groups according to the classification given in the methods ($n$ = original numbers of habitats included).

| Original group | $n$ | Group predicted | | | | | | | | | % Correctly classified |
|---|---|---|---|---|---|---|---|---|---|---|---|
| | | 1 | 2 | 3 | 4 | 5 | 6 | 7 | 8 | 9 | |
| (1) Wetlands below lakes | 9 | **8** | – | – | – | – | 1 | – | – | – | 88.9 |
| (2) Spring brooks | 12 | – | **12** | – | – | – | – | – | – | – | 100.0 |
| (3) Degraded spring brooks | 10 | – | – | **8** | – | 1 | – | – | 1 | – | 80.0 |
| (4) Spring pools | 10 | – | – | – | **5** | 3 | 2 | – | – | – | 50.0 |
| (5) Ephemeral river sections | 34 | – | – | – | 1 | **25** | 7 | – | 1 | – | 73.5 |
| (6) Temporary waters | 23 | – | – | – | – | 2 | **21** | 2 | 1 | – | 91.3 |
| (7) Artificial waters | 14 | – | – | – | – | 2 | – | **11** | 1 | – | 78.6 |
| (8) Farm dams | 12 | – | – | – | – | 2 | 1 | – | **9** | – | 75.0 |
| (9) Lakes | 8 | – | – | – | – | – | – | – | – | **8** | 100.0 |

Bold are the numbers of localities correctly classified.

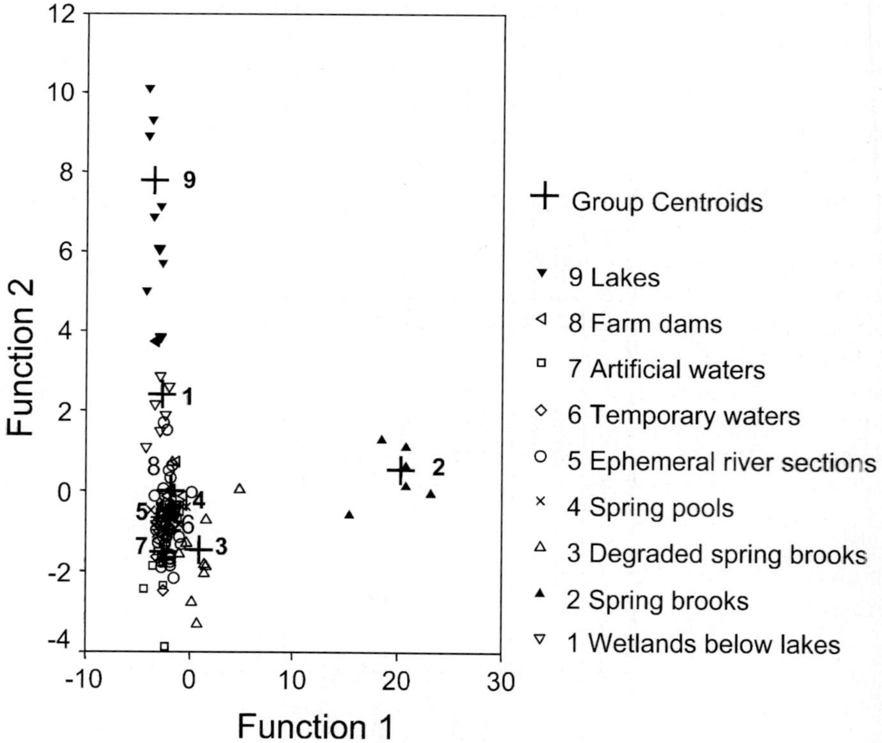


*Figure 2.* Diagram showing the sorting of examined localities according to the first two Canonical Discriminance Functions. The symbols depict the different habitat types and the group centroids.

brooks, lakes). The habitat types do, however, seem to be sufficiently discriminated. Particularly the habitat types lakes and spring brooks were clearly separated (Figure 2). Species correlated (canonical correlation) with the first discriminant function mainly occurred in perennial spring brooks, while species correlated to the second function were mainly recorded from lakes. The nine habitat types differed widely in species richness (Table 1). The highest total species numbers were noted from wetlands below lakes and spring brooks, whereas the lowest numbers were noted from spring pools (Table 1). The median number of species per habitat followed a similar pattern. Degraded spring brooks had much lower species numbers than natural spring brooks. The variation between minimum and maximum species numbers was high in all kinds of habitats, suggesting a relative high heterogeneity. This is also indicated by the high, although significant, nestedness temperatures and low *C*-scores (Table 1).

*Selected indicator species and their habitats*

Twenty-five species were present in more than four (> 50%) of all types of habitats and most species of this group were also present in more than 20% of all localities examined. Although *F*-tests (Table 3) indicated significant differences in the distribution of a number of these species, all appeared not to be very specific at least in the selection of adult habitats. Additionally, 22 species were recorded at less than four localities so that no useful information about their habitats and assemblage associations could be derived.

Of the remaining 12 species two showed no significant difference in distribution (Table 3), which suggests low habitat specificity. Hence, according to our criteria, 10 species remain, which may be useful indicators for the health of their assemblages. Canonical correlations indicate that the proportions of *Crocothemis sanguinolenta* (correlation 0.20), *Anax speratus* (0.17), *Orthetrum julia* (0.12), *Pseudagrion kersteni* (0.12), *Trithemis stictica* (0.12), all showing particularly high specificity to spring brooks, correlated best with discriminant function 1. Species that correlate best with function 2 were *Ceratogomphus pictus* (0.51) and *Ictinogomphus ferox* (0.35), both mainly recorded in lakes. Whereas function 3 was represented by *Azuragrion nigridorsum* (0.42) and *Agriocnemis exilis* (0.20), *Palpopleura jucunda* (0.10) represented function 4.

## Discussion

The species assemblages in our study are nested, which means that distinct, and predictable, patterns of species associations can be expected. Discriminant analyses revealed that most out of nine habitat types separated by structural and hydrological parameters are well discriminated by their dragonfly assemblages, particularly spring brooks and large impoundments (lakes).

*Table 3.* Distribution of species recorded.

| Species | Total | | Habitat type | | | | | | | | | | Function |
|---|---|---|---|---|---|---|---|---|---|---|---|---|---|
| | *n* | M | 1 | 2 | 3 | 4 | 5 | 6 | 7 | 8 | 9 | | |
| *Trithemis kirbyi ardens* (Gerstäcker, 1891) | 102 | 9 | 0.89 | 1.00 | 0.80 | 0.60 | 0.79 | 0.52 | 0.93 | 0.75 | 0.75 | * | |
| *Crocothemis erythraea*(Brullé, 1832) | 86 | 9 | 1.00 | 0.92 | 0.90 | 0.80 | 0.65 | 0.30 | 0.36 | 0.67 | 0.88 | * | |
| *Orthetrum chrysostigma* (Burmeister, 1839) | 85 | 9 | 1.00 | 0.92 | 0.90 | 0.80 | 0.76 | 0.22 | 0.43 | 0.50 | 0.50 | * | |
| *Pantala flavescens* (Fabricius, 1798) | 85 | 9 | 0.78 | 0.58 | 0.50 | 0.80 | 0.62 | 0.61 | 0.57 | 0.92 | 0.38 | | |
| *Anax imperator* Leach, 1815 | 78 | 9 | 1.00 | 0.67 | 0.70 | 0.60 | 0.59 | 0.30 | 0.36 | 0.83 | 0.75 | * | |
| *Sympetrum fonscolombii* (Selys, 1840) | 63 | 9 | 1.00 | 0.25 | 0.20 | 0.10 | 0.53 | 0.43 | 0.36 | 0.83 | 0.63 | * | |
| *Ischnura senegalensis* (Rambur, 1842) | 58 | 9 | 0.78 | 0.50 | 0.60 | 0.50 | 0.56 | 0.13 | 0.07 | 0.25 | 0.88 | * | |
| *Paragomphus genei* (Selys, 1841) | 55 | 9 | 0.56 | 0.58 | 0.40 | 0.20 | 0.68 | 0.17 | 0.29 | 0.17 | 0.38 | * | |
| *Trithemis annulata* (Palisot de Beauvois, 1807) | 40 | 9 | 0.56 | 0.25 | 0.50 | 0.10 | 0.21 | 0.17 | 0.21 | 0.42 | 0.75 | * | 2 |
| *Orthetrum trinacria* (Selys, 1841) | 40 | 8 | 0.56 | 0.08 | 0.60 | – | 0.24 | 0.13 | 0.14 | 0.58 | 0.88 | * | |
| *Pseudagrion massaicum* Sjöstedt, 1909 | 35 | 9 | 0.44 | 0.50 | 0.30 | 0.10 | 0.32 | 0.04 | 0.14 | 0.17 | 0.63 | * | |
| *Diplacodes lefebvrii* (Rambur, 1842) | 33 | 9 | 0.56 | 0.42 | 0.30 | 0.20 | 0.32 | 0.09 | 0.07 | 0.25 | 0.13 | | 4 |
| *Trithemis arteriosa* (Burmeister, 1839) | 30 | 8 | 0.56 | 0.83 | 0.40 | 0.20 | – | 0.04 | 0.29 | 0.08 | 0.25 | * | |
| *Lestes pallidus* Rambur, 1842 | 27 | 8 | 0.22 | 0.17 | 0.20 | – | 0.15 | 0.39 | 0.07 | 0.33 | 0.25 | | |
| *Africallagma glaucum* (Burmeister, 1839) | 17 | 5 | 0.67 | 0.33 | 0.00 | – | 0.12 | 0.09 | – | 0.08 | – | * | 4 |
| *Brachythemis leucosticta* (Burmeister, 1839) | 14 | 6 | 0.22 | – | 0.10 | – | 0.06 | 0.04 | – | 0.08 | 0.88 | * | 2 |
| *Zygonyx torridus* (Kirby, 1899) | 14 | 4 | 0.22 | 0.50 | – | – | 0.15 | 0.04 | – | – | – | * | |
| *Ceriagrion glabrum* (Burmeister, 1839) | 12 | 5 | 0.33 | 0.17 | 0.50 | 0.10 | 0.03 | – | – | – | – | * | 3 |
| *Diplacodes luminans* (Karsch, 1893) | 10 | 7 | 0.22 | 0.17 | 0.20 | – | 0.03 | 0.04 | 0.07 | 0.08 | – | | |
| *Tramea basilaris* (Palisot de Beauvois, 1807) | 10 | 6 | 0.11 | 0.08 | 0.30 | – | – | 0.04 | 0.14 | 0.08 | – | | |
| *Pseudagrion salisburyense* Ris, 1921 | 10 | 4 | 0.11 | 0.50 | – | – | 0.03 | – | – | – | 0.25 | * | 2 |
| *Orthetrum julia falsum* Longfield, 1955 | 10 | 2 | – | 0.58 | 0.20 | – | – | – | – | 0.08 | – | * | 1 |
| *Anax ephippiger* (Burmeister, 1839) | 9 | 6 | 0.11 | 0.17 | – | – | 0.06 | 0.04 | 0.14 | – | 0.13 | | |
| *Ictinogomphus ferox* (Rambur, 1842) | 8 | 3 | 0.22 | – | 0.10 | – | – | – | – | – | 0.63 | * | 2 |
| *Ceratogomphus pictus* Selys, 1854 | 8 | 2 | 0.22 | – | – | – | – | – | – | – | 0.75 | * | 2 |
| *Crocothemis sanguinolenta* (Burmeister, 1839) | 8 | 1 | – | 0.67 | – | – | – | – | – | – | | * | 1 |
| *Orthetrum brachiale* (Palisot de Beauvois, 1817) | 7 | 4 | 0.22 | – | – | – | 0.03 | – | – | – | – | * | 4 |
| *Pseudagrion sublacteum* (Karsch, 1893) | 7 | 4 | 0.22 | 0.17 | – | – | 0.06 | 0.04 | – | – | – | | 4 |

| | | | | | | | | | | | | | |
|---|---|---|---|---|---|---|---|---|---|---|---|---|---|
| *Agriocnemis exilis* Selys, 1872 | 7 | 3 | 0.22 | 0.25 | 0.20 | – | – | – | – | – | – | * | 4 |
| *Azuragrion nigridorsum* (Selys, 1876) | 7 | 2 | – | 0.17 | 0.50 | – | – | – | – | – | – | * | 3 |
| *Anax speratus* Hagen, 1867 | 7 | 1 | – | 0.58 | – | – | – | – | – | – | – | * | 1 |
| *Palpopleura jucunda* Rambur, 1842 | 6 | 3 | 0.11 | 0.33 | – | – | – | 0.09 | – | – | – | * | 4 |
| *Pseudagrion kersteni* (Gerstäcker, 1869) | 5 | 1 | – | 0.42 | – | – | – | – | – | – | – | * | 1 |
| *Trithemis stictica* (Burmeister, 1839) | 5 | 1 | – | 0.42 | – | – | – | – | – | – | – | * | 1 |
| *Palpopleura lucia* (Drury, 1773) | 4 | 4 | – | – | 0.10 | 0.10 | 0.03 | 0.04 | – | – | – | | |
| *Rhyothemis semihyalina* (Desjardins, 1832) | 4 | 3 | 0.11 | 0.08 | – | – | 0.03 | – | – | – | 0.13 | | 2 |
| *Tholymis tillarga* (Fabricius, 1798) | 4 | 3 | 0.11 | 0.17 | – | – | – | 0.04 | – | – | – | | |
| *Pseudagrion nubicum* Selys, 1871 | 3 | 3 | 0.11 | 0.08 | – | – | – | – | – | – | 0.13 | | 2 |
| *Orthetrum a. abbotti* Calvert, 1892 | 3 | 2 | 0.11 | 0.17 | – | – | – | – | – | – | – | | 4 |
| *Trithemis donaldsoni* (Calvert, 1899) | 3 | 2 | 0.11 | – | – | – | – | – | – | – | 0.25 | * | 2 |
| *Trithemis furva* Karsch, 1899 | 3 | 2 | 0.11 | 0.17 | – | – | – | – | – | – | – | * | 4 |
| *Nesciothemis farinosa* (Förster, 1898) | 3 | 1 | – | 0.25 | – | – | – | – | – | – | – | * | 1 |
| *Tramea limbata* (Desjardins, 1832) | 2 | 2 | 0.11 | – | – | – | – | – | – | 0.08 | – | | 4 |
| *Trithemis hecate*Ris, 1912 | 2 | 2 | – | – | 0.10 | – | – | – | 0.07 | – | – | | |
| *Urothemis assignata* (Selys in Lucas, 1872) | 2 | 2 | 0.11 | 0.08 | – | – | – | – | – | – | – | | 4 |
| *Urothemis edwardsii* (Selys, 1849) | 2 | 2 | 0.11 | – | – | – | – | – | – | – | 0.13 | | 2 |
| *Anax tristis* Hagen, 1867 | 2 | 1 | – | – | – | – | – | – | 0.07 | – | – | | |
| *Palpopleura deceptor* (Calvert, 1899) | 2 | 1 | 0.11 | 0.33 | – | – | – | – | – | 0.08 | – | | |
| *Trithemis monardi* Ris, 1931 | 2 | 1 | – | – | 0.20 | – | – | – | – | – | – | * | 3 |
| *Acisoma panorpoides ascalaphoides* Rambur, 1842 | 1 | 1 | 0.11 | – | – | – | – | – | – | – | – | | |
| *Aeshna minuscula* McLachlan, 1896 | 1 | 1 | – | 0.08 | – | – | – | – | – | – | – | | |
| *Bradinopyga cornuta* Ris, 1911 | 1 | 1 | – | – | – | – | – | 0.04 | – | – | – | | |
| *Crocothemis divisa* Karsch, 1898 | 1 | 1 | – | – | – | – | – | 0.04 | – | – | – | | |
| *Hemistigma albipunctum* (Rambur, 1842) | 1 | 1 | 0.11 | – | – | – | – | – | – | – | – | | |
| *Olpogastra lugubris* Karsch, 1895 | 1 | 1 | – | – | – | – | – | 0.04 | – | – | – | | |
| *Orthetrum machadoi* Longfield, 1955 | 1 | 1 | 0.11 | – | – | – | – | – | – | – | – | | |
| *Phyllomacromia picta* (Selys, 1871) | 1 | 1 | – | – | – | – | – | – | – | – | 0.13 | * | |
| *Pseudagrion glaucescens* Selys, 1876 | 1 | 1 | – | 0.08 | – | – | – | – | – | – | – | | |
| *Pseudagrion hamoni* Fraser, 1955 | 1 | 1 | – | 0.08 | – | – | – | – | – | – | – | | |

*n* is the total no. of localities colonised, M the no. of habitats. The distribution per habitat is given as relative frequency of localities colonised per habitat; the asterisks indicate $p < 0.05$ from *F*-test (df = 8, 123). Function indicates the discriminant function to which the species is correlated (see Methods and Results). Species matching all indicator criteria (see Materials and methods) are marked in grey.

*Community nestedness*

Wright and Reeves (1992) considered an average score of 0.58 might be regarded as typical for terrestrial habitat systems, while for freshwater systems no such value is available. The value in this study is low (0.351), thus indicating a more loosely ordered species composition. This is comparable to other heterogeneous habitats, e.g. the least nested habitat in Sahlén and Ekestubbe (2001; data re-analysed by GS) has a *C*-score of 0.358 but a *C*-score from a North American river surveyed by Worthen (2003) was only 0.250. Although this river had the 'least nested' species assemblage, many sites within this locality were pristine (Worthen 2003) and the odonate species belonged to several ecological groups. Thus, a low *C*-score does not necessarily indicate a degraded species assembly, but rather a more varied one. The species composition in our study area is probably varied, including species with many different ecological preferences.

*Dragonfly assemblages and diversity*

Although the different habitat types were generally well discriminated by their dragonfly assemblages, ephemeral river sections, temporary waters, and farm dams, all suffering high degrees of abiotic disturbance, i.e. drying out and/or flash floods, displayed very similar assemblages. This fit well with the general theory that communities of habitats subjected to harsh conditions are mostly affected by abiotic factors and are mainly colonised by generalist species (cf. Menge and Sutherland 1976; Peckarsky 1983). In fact the great majority of the recorded species are widespread in the study area and colonise all habitat types (cf. Table 3). Due to rapid development (cf. Johansson and Suhling 2004; Suhling et al. 2005b) and high dispersal ability these species are able to cope with such adverse habitat conditions. Hosting only generalist species, farm dams in Namibia do therefore not play an important role as potential refugia for dragonflies, unlike in South Africa (Samways 1989). Artificial waters like water holes and spring pools belong to this group, the former probably due to their poor habitat structures (see Materials and methods). All spring pools we examined were highly disturbed by grazing cattle or game.

Very well defined by their dragonfly assemblages, by contrast, are lakes and spring brooks, and, to a minor extent also wetlands below lakes and degraded spring brooks. Lakes and particularly wetlands below lakes have high species diversity and contribute highly to the regional $\gamma$-diversity. For instance, they add species like *Ceratogomphus pictus*, *Ictinogomphus ferox* and *Trithemis donaldsoni* to the fauna of our study area, which otherwise only occur along the large rivers (Martens et al. 2003). The wetlands below lakes also provide suitable habitats for a number of species that depend on well-vegetated perennial wetlands, such as *Urothemis edwardsi* and *Hemistigma albipunctum*. Although these habitats are in many ways similar to spring brooks, obviously no species specialised to such spring brooks was able to use them as replacement habitats. Despite contributing to $\gamma$-diversity and hosting a specific assemblage lakes and wetlands below lakes are artificial and are not in need of special conservation measures.

Spring brooks, on the other hand, hold a very diverse and unique assemblage containing a number of species that were exclusively or almost exclusively recorded here in the entire Namibia (cf. Martens et al. 2003), including *Pseudagrion kersteni*, *Aeshna minuscula*, *Anax speratus*, *Crocothemis sanguinolenta*, *Orthetrum julia* and *Trithemis stictica*. Except for *A. minuscula*, which is mainly restricted to South Africa (Samways 1999), most are widespread in tropical Africa, some of them being common in certain regions, e.g. in Kenya (V. Clausnitzer personal communication). However, in Namibia all these species are rare, most probably due to habitat restrictions (cf. Suhling et al. 2003, 2005).

*Threats and conservation of spring assemblages*

Most perennial springs in the interior of Namibia occur in the mountain ranges of Damaraland and Kaokofeld, the Otavi Mts., the Waterberg and the Naukluft and Tsaris Mts. To our knowledge natural undisturbed spring brooks currently only remain in remote or protected areas of the Kaokofeld/Damaraland and the Naukluft and Tsaris regions. Historical distribution data (up to the 1990s, cf. Martens et al. 2003) demonstrated that *P. kersteni*, *C. sanguinolenta*, and *O. julia* occurred until fairly recently at springs in the Waterberg and the Otavi Mts. During our own fieldwork we only recorded *O. julia*. Overextraction of the aquifer in the region is a probable major cause as we have personally observed that some of the spring brooks that were still flowing strongly during the drought years of the early 1980s and 1990s have completely dried up. Today, even large springs, such as the one in Grootfontein in the Otavi Mts. (meaning 'big spring'), which contained *P. kersteni*, are completely dry in most years. In other cases the springs were capped so that only the strongly shaded spring itself remained, where *O. julia* as a shade tolerant species can exist. Sadly, many of these springs occurred in proclaimed conservation areas, indicating that the conservation ethic often does not extend to include natural water bodies. All still existing spring brooks in the Otavi Mts. were degraded due to the construction of dams at the spring outflow and, consequently, typical spring brook species were absent while the generalists were still present.

*Potential indicators*

Indicators should demonstrate the health of communities in selected habitats (McGeoch 1998; Simberloff 1998). In our study we aimed to identify species that would reflect the completeness or conservation status of their respective communities. We identified species (Table 4), which are mainly associated with two distinct habitat types (cf. Figure 2). Two lake species, *C. pictus* and *I. ferox*, appear to be good indicators for healthy lake assemblages. However, these lakes are impoundments created during the 20th century and hence are not original to the region (see above). Indicators are therefore not necessarily needed. *Azuragrion nigridorsum* occurs in natural and particularly degraded spring brooks and may mainly need perennial waters with vegetation. It is probably less sensitive to destruction of the original spring structure. The same may apply to *Agriocnemis exilis* and *Palpopleura jucunda*.

The majority of potential indicator species occurs exclusively in spring brooks (cf. Table 3), which are natural habitats in Namibia and host a number of unique assemblages and endemic species in various taxonomic groups. The potential value of reliable indicators to determine the conservation status of freshwater habitat refugia in arid areas is therefore highly relevant. Indicators should be sensitive enough to indicate the particular vulnerability and early recognition of habitat decline. Thus, the most sensitive indicators will be the first to show the effects of habitat deterioration. Indicators serve as a kind of trip-wire to show that habitat destruction has occurred, or are in the process of occurring. As we demonstrated above at least some of the selected species, *viz. P. kersteni*, *C. sanguinolenta*, *O. julia*, severely suffered from habitat destruction. Hence, at least these species are proven indicators according to the sensitivity standard. *Anax speratus* and *Trithemis stictica* should also meet the requirements due to their habitat restriction, although historical records to confirm local extinctions do not exist. These five species appear to be good indicators for the health of spring brook ecosystems. We suggest, however, that at least two species should be recorded at a given locality to assume good health of the assemblage, as the presence of one species may be accidental. If at least two species are around, the probability of good conditions is indeed favourable.

Hence, we suggest indicators for at least perennial spring brooks, which are under severe pressure in Namibia. We cannot comment on potential indicators for the health of the large perennial river systems, as we did not deal with the more humid northern parts of the country, where a bigger species pool occurs.

*Table 4.* Overview of the species finally proposed as indicators of the health of freshwater assemblages in central Namibia.

| Indicator species | Type of habitat/assemblage |
|---|---|
| *Anax speratus* | Spring brooks |
| *Crocothemis sanguinolenta* | Spring brooks |
| *Orthetrum julia falsum* | Spring brooks |
| *Pseudagrion kersteni* | Spring brooks |
| *Trithemis stictica* | Spring brooks |
| *Ictinogomphus ferox* | Lakes |
| *Ceratogomphus pictus* | Lakes |

The types of assemblages to which the species belong are indicated.

However, several rare species of dragonflies are present in that region, which depend on certain habitat conditions, such as undisturbed riverine forests and swamps (Suhling et al. 2004a). We therefore assume that dragonflies will have a high indicator potential for most kinds of freshwater wetlands in the entire area and may also serve as an indication of the sustainable use of water resources including evaluating measures to rehabilitate environments.

Namibia generally recognises that wetlands provide essential ecological services and cannot be allowed to degrade to such an extent that costly measures have to be taken to rehabilitate or even re-establish wetland processes. Though Namibia has recognised a decline in some species associated with freshwater ecosystems (cf. Bethune 1998; Curtis et al. 1998) the decline of such habitats has largely passed unnoticed. The establishment of an Index of Biological Integrity for wetlands and the implementation of regulations for the protection of wetlands have been identified as critical issues for management. If such an Index is to be introduced to evaluate the status, and possible vulnerability, of wetlands, then the identification of likely indicators is essential. We believe that our results indicate that odonates are at least sensitive indicators for natural spring brooks assemblages and because monitoring is repeatable and simple (i.e. without complex apparatus or training) they form a valuable tool for evaluating ecosystem integrity.

## Acknowledgements

GS would like to thank Bertil Nilsson, Halmstad University for invaluable help sorting out the mathematics of nestedness. The study was granted by German Ministry of Research and Education, BIOLOG Programme (BMBF 01LC0024) and was part of Biodiversity Transect Analysis in Africa (BIOTA). We gratefully acknowledge permission from the Ministry of Environment and Tourism to conduct biological research in all parts of Namibia.

## Appendix

Table listing the localities surveyed. Presented are the geographical coordinates, the altitude, the river catchment to which they belong, their habitat type and the total number of surveys per locality. Table 1.

| Locality | Degree S | Degree E | Altitude (m a.s.l.) | River catchment | Habitat type | No. of surveys |
|---|---|---|---|---|---|---|
| Fish R. Hardap | 24.498 | 17.863 | 1132 | Fish | Wetland below lake | 3 |
| Oanob River | 23.297 | 17.054 | 1520 | Fish | Wetland below lake | 4 |
| Kamanjab R. | 19.624 | 14.838 | 1250 | Huab | Wetland below lake | 4 |
| River at C28/Farm | 22.692 | 16.548 | 1662 | Kuiseb | Wetland below lake | 3 |
| Omdel River | 21.900 | 14.544 | 150 | Omaruru | Wetland below lake | 2 |

*Appendix*. Continued.

| Locality | Degree S | Degree E | Altitude (m a.s.l.) | River catchment | Habitat type | No. of surveys |
|---|---|---|---|---|---|---|
| Augeigas River | 22.539 | 16.956 | 1626 | Swakop | Wetland below lake | 6 |
| Gross Barmen 3 | 22.116 | 16.735 | 1238 | Swakop | Wetland below lake | 4 |
| Koch River | 22.549 | 16.938 | 1711 | Swakop | Wetland below lake | 4 |
| S. Von Bach River | 22.016 | 16.953 | 1333 | Swakop | Wetland below lake | 10 |
| Ongongo Fall | 19.140 | 13.820 | 730 | Hoanib | Spring brook | 4 |
| Baynes River | 17.231 | 12.805 | 969 | Kunene | Spring brook | 1 |
| Spring Swartbooisdrif | 17.263 | 13.700 | 778 | Kunene | Spring brook | 2 |
| Tsams Ost | 24.254 | 16.110 | 1439 | Tsams R. | Spring brook | 2 |
| Gororosib River | 24.280 | 16.237 | 1452 | Tsauchab | Spring brook | 4 |
| Naukluft River | 24.263 | 16.230 | 1482 | Tsauchab | Spring brook | 8 |
| Tsauchab River 2 | 24.503 | 16.115 | 1100 | Tsauchab | Spring brook | 3 |
| Zebra River Spring | 24.598 | 16.300 | 1510 | Tsauchab | Spring brook | 3 |
| Köcherbaumschlucht | 24.153 | 16.327 | 1215 | Tsondab | Spring brook | 2 |
| Noab Fountain | 23.922 | 16.275 | 1396 | Tsondab | Spring brook | 2 |
| Aub Gorge | 19.727 | 13.800 | 992 | Uniab | Spring brook | 3 |
| Palmwag | 19.887 | 13.937 | 925 | Uniab | Spring brook | 8 |
| Sesfontein | 19.123 | 13.620 | 614 | Hoanib | Degraded spring brook | 1 |
| Warmquelle | 19.185 | 13.817 | 648 | Hoanib | Degraded spring brook | 2 |
| Fransfontein | 20.209 | 15.018 | 1137 | Huab | Degraded spring brook | 2 |
| Didimala | 19.527 | 18.019 | 1533 | Omatako | Degraded spring brook | 1 |
| Lone Star | 19.509 | 18.175 | 1423 | Omatako | Degraded spring brook | 2 |
| Otjosongombe Spring | 20.474 | 17.276 | 1533 | Omatako | Degraded spring brook | 3 |
| Waterberg Spring | 20.510 | 17.243 | 1524 | Omatako | Degraded spring brook | 2 |
| Klein Barmen | 22.141 | 16.641 | 1209 | Swakop | Degraded spring brook | 2 |
| Neuras | 24.463 | 16.238 | 1198 | Tsauchab | Degraded spring brook | 3 |
| Otavifontein Dam | 19.670 | 17.378 | 1462 | Ugab | Degraded spring brook | 2 |
| Okondeka Etosha | 18.995 | 15.868 | 1110 | Cuvelai | Spring pool | 1 |
| Ongongo | 19.131 | 13.821 | 760 | Hoanib | Spring pool | 2 |
| Ongongo | 19.131 | 13.819 | 732 | Hoanib | Spring pool | 2 |
| Ongongo | 19.133 | 13.817 | 729 | Hoanib | Spring pool | 2 |
| Gai-As | 20.767 | 14.020 | 581 | Huab | Spring pool | 2 |
| Weener Farm Spring | 23.446 | 16.218 | 1042 | Kuiseb | Spring pool | 1 |
| Otjisandjima Spring | 17.462 | 13.246 | 1130 | Omatako | Spring pool | 2 |
| Mariabronn | 19.503 | 18.047 | 1486 | Omatako | Spring pool | 3 |
| Awaxas Spring | 19.761 | 13.849 | 964 | Uniab | Spring pool | 3 |
| Bergsig | 20.221 | 14.068 | 1026 | Uniab | Spring pool | 6 |
| Fish R. Gamis | 24.264 | 16.598 | 1329 | Fish | Ephemeral R. section | 2 |
| Fish R. Kabib | 24.617 | 16.943 | 1417 | Fish | Ephemeral R. section | 2 |
| Fish R. Trib. Farm Lever | 24.641 | 17.676 | 1400 | Fish | Ephemeral R. section | 2 |
| Fish R. Usib | 24.475 | 16.879 | 1351 | Fish | Ephemeral R. section | 2 |
| Fish R. Mariental | 24.656 | 17.935 | 1100 | Fish | Ephemeral R. section | 3 |
| Khowarib Gorge | 19.267 | 13.891 | 732 | Hoanib | Ephemeral R. section | 3 |
| Hoarusib River | 18.801 | 12.922 | 278 | Hoarusib | Ephemeral R. section | 1 |
| Hoarusib River 2 | 18.516 | 12.866 | 279 | Hoarusib | Ephemeral R. section | 1 |
| Hoarusib River 3 | 18.395 | 12.945 | 541 | Hoarusib | Ephemeral R. section | 1 |
| Huab River | 20.316 | 14.217 | 474 | Huab | Ephemeral R. section | 4 |
| Gaub R., Weener Farm | 23.470 | 16.218 | 1000 | Kuiseb | Ephemeral R. section | 3 |
| Gaub River, Pass | 23.483 | 15.767 | 752 | Kuiseb | Ephemeral R. section | 5 |

*Appendix*. Continued.

| Locality | Degree S | Degree E | Altitude (m a.s.l.) | River catchment | Habitat type | No. of surveys |
|---|---|---|---|---|---|---|
| Kuiseb R. Friedenau | 22.697 | 16.735 | 1621 | Kuiseb | Ephemeral R. section | 3 |
| Kuiseb R. side canyon | 23.305 | 15.758 | 750 | Kuiseb | Ephemeral R. section | 2 |
| Kuiseb River Bridge | 23.300 | 15.774 | 740 | Kuiseb | Ephemeral R. section | 3 |
| River at C28 | 22.670 | 16.617 | 1680 | Kuiseb | Ephemeral R. section | 2 |
| Pool at Neudam | 22.504 | 17.373 | 1783 | Nossob | Ephemeral R. section | 2 |
| Augeigas River | 22.585 | 16.972 | 1630 | Swakop | Ephemeral R. section | 3 |
| Kloake | 22.564 | 17.023 | 1642 | Swakop | Ephemeral R. section | 4 |
| River at B 1 | 22.347 | 17.051 | 1471 | Swakop | Ephemeral R. section | 2 |
| Stengel River | 22.546 | 16.938 | 1700 | Swakop | Ephemeral R. section | 5 |
| Swakop Groß Barmen | 22.122 | 16.711 | 1232 | Swakop | Ephemeral R. section | 2 |
| Swakop on B1 | 22.034 | 16.936 | 1331 | Swakop | Ephemeral R. section | 3 |
| Swakop R. Mouth | 22.679 | 14.589 | 48 | Swakop | Ephemeral R. section | 7 |
| Tsauchab River | 24.504 | 16.093 | 1085 | Tsauchab | Ephemeral R. section | 5 |
| River at C 35 | 20.629 | 14.865 | 911 | Ugab | Ephemeral R. section | 2 |
| Ugab at Bridge C 35 | 20.862 | 14.959 | 618 | Ugab | Ephemeral R. section | 8 |
| Ugab at Sorris Sorris | 20.956 | 14.838 | 551 | Ugab | Ephemeral R. section | 8 |
| Ugab Brandberg West | 20.970 | 14.108 | 227 | Ugab | Ephemeral R. section | 8 |
| Ugab Rest Camp | 21.016 | 14.685 | 471 | Ugab | Ephemeral R. section | 8 |
| Ugab Rhino Camp | 20.961 | 14.135 | 254 | Ugab | Ephemeral R. section | 8 |
| Uniab Delta | 20.190 | 13.197 | 57 | Ugab | Ephemeral R. section | 8 |
| Aub River | 19.723 | 13.801 | 998 | Uniab | Ephemeral R. section | 3 |
| Uniab Spring | 19.915 | 13.988 | 1002 | Uniab | Ephemeral R. section | 8 |
| Rainpools in Etosha | 19.193 | 16.182 | 1130 | Cuvelai | Temporary pond | 1 |
| Fish R.: Pond at C 4 | 24.418 | 16.841 | 1351 | Fish | Temporary pond | 2 |
| Pond at D1998 | 23.194 | 15.383 | 900 | Kuiseb | Temporary pond | 1 |
| Pond at C28 | 22.747 | 16.431 | 1786 | Kuiseb | Temporary pond | 2 |
| Rainpool at C 28 | 22.604 | 16.809 | 1688 | Kuiseb | Temporary pond | 1 |
| Rainpool at C36 | 21.266 | 15.173 | 986 | Omaruru | Temporary pond | 1 |
| Spitzkoppe | 21.815 | 15.184 | 1122 | Omaruru | Temporary pond | 2 |
| Waterberg Pool 2 | 20.483 | 17.235 | 1686 | Omatako | Temporary pond | 2 |
| Waterberg Pool 3 | 20.462 | 17.231 | 1613 | Omatako | Temporary pond | 2 |
| Waterberg Pool 4 | 20.353 | 17.262 | 1687 | Omatako | Temporary pond | 2 |
| Waterberg Pool 5 | 20.344 | 17.294 | 1686 | Omatako | Temporary pond | 2 |
| Waterberg Pool 6 | 20.375 | 17.406 | 1701 | Omatako | Temporary pond | 2 |
| Gross Barmen 2 | 22.069 | 16.865 | 1238 | Swakop | Temporary pond | 3 |
| Karibib Pond | 21.942 | 15.849 | 1199 | Swakop | Temporary pond | 3 |
| Leopard Quelle | 22.398 | 15.734 | 781 | Swakop | Temporary pond | 2 |
| Pool near Stengel Dam | 22.540 | 16.939 | 1691 | Swakop | Temporary pond | 1 |
| Python Valley | 22.436 | 15.728 | 780 | Swakop | Temporary pond | 2 |
| Rainpool at B2 | 22.095 | 15.227 | 1077 | Swakop | Temporary pond | 1 |
| Sand Pit at B1 | 22.028 | 16.931 | 1329 | Swakop | Temporary pond | 5 |
| Tsaobis Kudu Ponds | 22.379 | 15.749 | 740 | Swakop | Temporary pond | * |
| Bergplaas | 20.401 | 16.229 | 1325 | Ugab | Temporary pond | 1 |
| Rainpool on C40 | 19.823 | 15.423 | 1291 | Ugab | Temporary pond | 1 |
| Pool E Brandberg West | 20.974 | 14.267 | 320 | Uniab | Temporary pond | 1 |
| Birds Paradise | 22.962 | 14.520 | 29 | Kuiseb | Artificial waters | 3 |
| Ghaub Farm | 19.466 | 17.726 | 1550 | Omatako | Artificial waters | 1 |
| Goanikontes Oasis | 22.669 | 14.820 | 176 | Swakop | Artificial waters | 7 |
| Pool de la Bat Camp | 20.509 | 17.243 | 1524 | Swakop | Artificial waters | 4 |
| Puccinis | 22.569 | 17.077 | 1668 | Swakop | Artificial waters | 4 |

*Appendix.* Continued.

| Locality | Degree S | Degree E | Altitude (m a.s.l.) | River catchment | Habitat type | No. of surveys |
|---|---|---|---|---|---|---|
| Tsaobis waterhole 1 | 22.426 | 15.709 | 740 | Swakop | Artificial waters | 4 |
| Tsaobis waterhole 2 | 22.470 | 15.728 | 714 | Swakop | Artificial waters | 4 |
| Tsaobis waterhole 3 | 22.381 | 15.763 | 749 | Swakop | Artificial waters | 6 |
| Tsaobis waterhole 4 | 22.393 | 15.809 | 732 | Swakop | Artificial waters | 6 |
| Zoopark ponds | 22.567 | 17.086 | 1700 | Swakop | Artificial waters | 5 |
| Buellsport | 24.149 | 16.362 | 1417 | Tsauchab | Artificial waters | 3 |
| Urikos | 24.444 | 16.171 | 1172 | Tsauchab | Artificial waters | 2 |
| Solitaire | 23.894 | 16.005 | 1105 | Tsondab | Artificial waters | 4 |
| Outjo Fountain | 20.105 | 16.148 | 1319 | Ugab | Artificial waters | 2 |
| Kamanjab Dam | 19.624 | 14.838 | 1250 | Huab | Farm dam | 4 |
| Friedenau Dam | 22.686 | 16.745 | 1670 | Kuiseb | Farm dam | 2 |
| Gaub River Dams | 23.479 | 15.769 | 749 | Kuiseb | Farm dam | 3 |
| Otjosongombe Dam | 20.478 | 17.308 | 1444 | Omatako | Farm dam | 2 |
| Arandis | 22.347 | 15.087 | 751 | Swakop | Farm dam | 6 |
| Dam at C 32 | 21.983 | 15.838 | 1222 | Swakop | Farm dam | 3 |
| Habib Dam | 22.118 | 15.824 | 1241 | Swakop | Farm dam | 3 |
| Habib Dam 2 | 22.088 | 15.796 | 1292 | Swakop | Farm dam | 3 |
| Habis Dam 3 | 22.121 | 15.843 | 1240 | Swakop | Farm dam | 2 |
| Koch Dam | 22.550 | 16.938 | 1714 | Swakop | Farm dam | 4 |
| Stengel Dam | 22.547 | 16.938 | 1715 | Swakop | Farm dam | 5 |
| Grootberg Pass | 19.840 | 14.115 | 1411 | Uniab | Farm dam | 4 |
| Hardap Dam | 24.465 | 17.813 | 1129 | Fish | Lake | 4 |
| Oanob Dam | 23.301 | 17.031 | 1550 | Fish | Lake | 3 |
| Otjivero Dam | 22.283 | 17.957 | 1700 | Nossob | Lake | 3 |
| Omatako Dam | 21.149 | 17.178 | 1368 | Omatako | Lake | 2 |
| Augeigas Dam | 22.538 | 16.953 | 1689 | Swakop | Lake | 3 |
| Avis Dam | 22.572 | 17.129 | 1738 | Swakop | Lake | 3 |
| Gross Barmen Dam | 22.109 | 16.745 | 1239 | Swakop | Lake | 3 |
| S. Von Bach Dam | 22.012 | 16.950 | 1377 | Swakop | Lake | 10 |
| Canal at Otjosongombe | 20.500 | 17.293 | 1433 | Omatako | Canal | 4 |

## References

Atmar W. and Patterson B.D. 1993. The measure of order and disorder in the distribution of species in fragmented habitat. Oecologia 96: 373–382.

Atmar W. and Patterson B.D. 1995. The Nestedness Temperature Calculator: A Visual Basic Program, including 294 Presence–Absence Matrices. AICS Research, Inc., University Park, NM and The Field Museum, Chicago, IL.

Barnard P. and Shikongo S.T. 2000. Namibias National report to the Fifth Conference of Parties on Implementation of the Convention on Biological Diversity. Nairobi, Kenya. URL: http//www.dea.met.gov.na/programmes/biodiversity.html.

Bethune S. 1998. Wetland habitats. In: Barnard and P. (ed.), Biological diversity in Namibia – A Country Study. Namibian National Biodiversity Task Force, Windhoek, pp. 60–66.

Breen C.M. 1991. Are intermittently flooded wetlands of arid environments important conservation sites? Madoqua 17: 61–65.

Brown K.S. Jr. 1991. Conservation of neotropical environments and their habitats: insects as indicators. In: Collins N.M. and Thomas J.A. (eds), The Conservation of Insects and their Habitats. Academic Press, London, pp. 350–404.

Christelis G. and Struckmeyer W. 2001. Groundwater in Namibia – an Explanation of the Hydrogeological Map. Ministry of Agriculture, Water and Rural Development, Windhoek.

Clausnitzer V. 2003. Dragonfly communities in coastal habitats of Kenya: indication of biotope quality and the need of conservation measures. Biodiv. Conserv. 12: 333–356.

Clausnitzer V. 2004a. Guardians of the watershed. In: Clausnitzer V. and Jödicke R. (eds), Guardians of the watershed. Global status of dragonflies: critical species, threat and conservation. Int. J. Odonatol. 7: 111.

Clausnitzer V. 2004b. Critical species of Eastern Africa. In: Clausnitzer V. and Jödicke R. (eds), Guardians of the watershed. Global status of dragonflies: critical species, threat and conservation. Int. J. Odonatol. 7: 189–206.

Cleary D.F.R., Mooers A.Ø., Eichhorn K.A.O., Van Tol J., de Jong R. and Menken S.B.J. 2004. Diversity and community composition of butterflies and odonates in an ENSO-induced fire affected habitat mosaic: a case study from East Kalimantan, Indonesia. Oikos 105: 426–446.

Corbet P.S. 1993. Are Odonata useful as bioindicators? Libellula 12: 91–102.

Corbet P.S. 1999. Dragonflies – Behaviour and Ecology of Odonata. Harley Books, Colchester.

Crisman T.L., Chapman L.J., Chapman C.A. and Kaufman L.S. 2003. Conservation, Ecology, and Management of African Freshwaters. University of Florida Press, Gainesville, 514 p.

Curtis B., Roberts K., Griffin M., Bethune S., Hay C.L. and Kolberg H. 1998. Species richness and conservation of Namibian freshwater macro-invertebrates, fish and amphibians. Biodiv. Conserv. 7: 447–466.

Day J. 2003. Management of water resources and freshwater ecosystems in Southern Africa. In: Crisman T.L., Chapman L.J., Chapman C.L. and Kaufman L.S. (eds), Conservation, Ecology and Management of African Fresh Waters. University of Florida Press, Gainesville, Florida, pp. 41–61.

Dijkstra K.-D.B. and Lempert J. 2003. Odonate assemblages of running waters in the Upper Guinean forest. Archiv für Hydrobiologie 157: 397–412.

Dijkstra K.-D.B. and Vick G.S. 2004. Critical species of Odonata in western Africa. In: Clausnitzer V. and Jödicke R. (eds), Guardians of the watershed. Global status of dragonflies: critical species, threat and conservation. Int. J. Odonatol. 7: 229–238.

Fischer J. and Lindenmayer D.B. 2002. Treating the nestedness temperature calculator as a "black box" can lead to false conclusions. Oikos 99: 193–199.

Heyns P., Montgomery S., Pallet J. and Seely M. 1998. Namibia's Water, a Decision Makers Guide. DRFN and DWA, Windhoek.

Jacobsen P.J., Jacobsen M. and Seely M.K. 1995. Ephemeral Rivers and their Catchments: sustaining People and Development in Western Namibia. DRFN, Windhoek.

Johansson F. and Suhling F. 2004. Behaviour and growth of dragonfly larvae along a permanent-temporary water habitat gradient. Ecol. Entomol. 29: 196–202.

Lande R. 1993. Risks of population extinction from demographic and environmental stochasticity and random catastrophes. Am. Nat. 142: 911–927.

Lomolino M.V. 1996. Investigating causality of nestedness of insular communities: Selective immigrations or extinctions? J. Biogeogr. 23: 699–703.

Martens A., Jödicke R. and Suhling F. 2003. Annotated checklist of the Odonata of Namibia. Cimbebasia 18: 139–162.

McCulloch C.E. 1985. Variance tests for species association. Ecology 66: 1676–1681.

McGeoch M.A. 1998. The selection, testing and application of terrestrial insects as bioindicators. Biol. Rev. 73: 181–201.

McGeoch M.A. and Chown S.L. 1998. Scaling up the value of bioindicators. Trends Ecol. Evol. 13: 46–47.

Mendelsohn J., Jarvis A., Roberts C. and Robertson T. 2002. A Portrait of the Land and its People. Atlas of Namibia. David Philip Publishers, Cape Town, 200pp.

Menge B.A. and Sutherland J.P. 1976. Species diversity gradients: synthesis of the roles of predation, competition, and temporal heterogeneity. Am. Nat. 110: 351–369.

Peckarsky B.L. 1983. Biotic interactions or abiotic limitations? A model of lotic community structure. In: Barnes J.R. (ed.), Stream Ecology: Application and Testing of General Ecological Theory. Plenum Press, New York, pp. 303–324.

Sahlén G. 1999. The impact of forestry on dragonfly diversity in Central Sweden. Int. J. Odonatol. 2: 177–186.

Sahlén G. and Ekestubbe K. 2001. Identification of dragonflies (Odonata) as indicators of general species richness in boreal forest lakes. Biodiv. Conserv. 10: 673–690.

Samways M.J. 1989. Farm dams as nature reserves for dragonflies (Odonata) at various altitudes in the Natal Drakensberg mountains, South Africa. Biol. Conserv. 48: 181–187.

Samways M.J. 1999. Diversity and conservation status of South African dragonflies (Odonata). Odonatologica 28: 13–62.

Samways M.J. and Steyler N.S. 1996. Dragonfly (Odonata) distribution patterns in urban and forest landscapes, and recommendations for riparian management. Biodiv. Conserv. 78: 279–288.

Samways M.J. and Wilmot B.C. 2003. Odonata. In: de Moor I.J., Day J.A. and de Moor F.C. (eds), Guides to the Freshwater Invertebrates of southern Africa. Vol. 7: Insecta I. Water Research Commission, Gezina, South Africa, pp. 160–212.

Seely M. 1998. Environmental change. In: Barnard P. (ed.), Biological diversity in Namibia – A Country Study. Namibian National Biodiversity Task Force, Windhoek, pp. 67–72.

Simberloff D. 1998. Flagships, umbrellas, and keystones: Is single-species management passé in the landscape era? Biol. Conserv. 83: 247–257.

Simberloff D. and Martin J.L. 1991. Nestedness of insular avifaunas: simple summary statistics masking complex species patterns. Ornis Fennica 68: 178–192.

Stern C. and Lau B. 1990. Namibian water resources and their management: a preliminary history. Archeia 15. National Archives of Namibia, Windhoek, 79 p.

Suhling F., Jödicke R. and Schneider W. 2003. Odonata of African arid regions – are there desert species? Cimbebasia 18: 207–224.

Suhling F., Martens A. and Marais E. 2004a. Critical species of Odonata in Southwest Africa. In: Clausnitzer V. and Jödicke E. (eds), Guardians of the watershed. Global status of dragonflies: critical species, threat and conservation. Int J. Odonatol. 7: 263–277.

Suhling F., Schenk K., Padeffke T. and Martens A. 2004b. A field study of larval development in a dragonfly assemblage in African desert ponds (Odonata). Hydrobiologia 528: 75–85.

Suhling F., Sahlén G., Kasperski J. and Gaedecke D. 2005. Behavioural and life history traits in temporary and perennial waters: comparisons among three pairs of sibling dragonfly species. Oikos 108: 609–617.

Ward J.V. 1998. Riverine landscapes: biodiversity patterns, disturbance regimes, and aquatic conservation. Biol. Conserv. 83: 269–278.

Worthen W.B. 1996. Community composition and nested-subset analyses: basic descriptors for community ecology. Oikos 76: 417–426.

Worthen W.B. 2003. Nested-subset structure of larval odonate assemblages in the Enoree River basin, USA. Int. J. Odonatol. 6: 79–89.

Wright D.H. and Reeves J.H. 1992. On the measurement of nestedness of species assemblages. Oecologia 92: 416–428.

Biodiversity and Conservation (2006) 15:319–345
DOI 10.1007/s10531-005-2574-6 

# Richness, abundance, and complementarity of fruit-feeding butterfly species in relict sacred forests and forest reserves of Ghana

J.L. BOSSART[1,*], E. OPUNI-FRIMPONG[2,3], S. KUUDAAR[2] and E. NKRUMAH[2]

*[1]Department of Applied Biology & Biomedical Engineering, Rose-Hulman Institute of Technology, Terre Haute, IN 47803, USA; [2]Forestry Research Institute of Ghana, Kumasi, Ghana; [3]Current address: School of Forest Resources and Environmental Science, Michigan, Technological University, Houghton, MI 49931, USA; *Author for correspondence (e-mail: bossart@rose-hulman.edu; phone: +1-812-872-6058; fax: +1-812-877-8025)*

Received 23 December 2003; accepted in revised form 8 February 2005

**Key words:** Afrotropics, Biodiversity hotspot, Community composition, Dominance–diversity curves, Evenness, Fragmentation, Richness estimators

**Abstract.** Sacred forest groves in Ghana are centuries old protected areas that were once part of continuous forest cover but now mostly exist as relict forest patches embedded in an agropastoral landscape. We conducted a year-long survey of the fruit-feeding butterfly fauna of four sacred groves and two forest reserves in the moist semi-deciduous forest zone of Ghana to characterize resident species diversity and complementarity among communities. Joint analysis of frugivorous butterfly diversity at these six forest fragments, which ranged in size from 6 to 5000 ha, was used to evaluate the conservation potential of these ancient indigenous reserves. A total of 6836 individuals were trapped across all sites, representing 79 species and five subfamilies. Community diversity was characterized in terms of, (a) number of species accumulated versus sampling effort, (b) rarefied species richness, (c) nonparametric richness estimates, (d) species evenness, (e) Simpson's Index of Diversity, and (f) complementarity of communities. Diversity of the fruit-feeding butterfly communities, quantified in terms of both species evenness and rarefied species richness, was higher at the larger forest reserves than at the small sacred forest groves. Additionally, although all sites had species trapped only at that site, the 5000-ha forest reserve harbored a resident community that was clearly distinctive from and more diverse than the other communities including the other forest reserve. Hence, our findings add to the burgeoning body of data that indicates large reserves are the foundation of successful conservation programs. Nonetheless, we found these small forest patches contribute to biodiversity conservation in at least three ways and these are identified and discussed. We also identify a number of species that appear more or less vulnerable to dynamics of forest fragmentation based on changes in their relative abundance across sites and we interpret these data in the context of potential indicator species and theoretical predictions of at-risk species.

## Introduction

The Upper Guinean forests of Ghana are recognized as among the most depleted and fragmented in the world and also among the most biologically unique (Hall and Swain 1981; Hawthorne 1988; Myers 1990; Whitmore 1997, 1998; FAO 1999; Goudie 2000; Myers et al. 2000). Estimates are that 80–90% of original high canopy forest in the country has been eliminated and nearly all that not set aside as forest reserve is gone (Hawthorne 1988; Hawthorne and Abu-Juam 1995). Pervasive external pressures in the form of residential development, bush fires, illegal logging, mining, and consumption of forest products, threaten remaining forest tracts and some formal reserves are virtually devoid of trees as a result (Hawthorne and Abu-Juam 1995). The Afrotropics have, unfortunately, generally not inspired the same concerted research attention as their Neotropical and Indo-Malayan counterparts (Owen 1971; Andre et al. 1992; Wagner and Cobbinah 1993; Laurance 1997) and empirical data from the fragmented forests of West Africa are exceedingly rare. This hinders development and implementation of the science-based conservation efforts necessary to counter these threats.

Only about 1% of forest cover in Ghana remains outside gazetted reserves and sacred forest groves account for most of this. Sacred groves are areas of land that were set aside by indigenous peoples hundreds of years ago because of their cultural significance and strictly protected via religious sanctions and taboos (Lebbie and Freudenberger 1996; Ntiamoa-Baidu 2001; UNESCO 2003). The potential conservation value of Ghana's indigenous reserves is high. Although originally embedded within continuous forest cover, they were transformed into isolated habitat "islands" as the landscape matrix surrounding them was converted from one of pristine forest to an agropastoral "sea". Their long history of protection coupled with the dramatic and rapid transformation of the surrounding landscape may have allowed for local persistence of rare or patchily distributed forest-endemic species. In some regions of the country, sacred groves also represent the only remaining examples of old-growth forest vegetation. In this context, they count as nuclei to guide reforestation or ecosystem recovery efforts. In a landscape increasingly devoid of forested areas outside the existing protected areas network, Ghana's indigenous reserves likely also serve as vital stepping stones that help link resident communities of discrete forest reserves.

Despite their high potential conservation value, Ghana's sacred forests remain largely undocumented, unexplored, and underappreciated as refugia for forest biodiversity. The present study is a preliminary assessment of the extent to which these indigenous reserves may contribute to the preservation of the country's forest-endemic species. To this end, we conducted a comparative analysis of community structure at four sacred groves and two much larger forest reserves to assess their relative similarities and dissimilarities. Our efforts focused on isolated forest fragments in the moist semi-deciduous forest zone of Ghana, the region of the country where remaining forests are most imminently imperiled.

Forest communities at the six sites were expected to differ as a consequence of the myriad factors that determine which, and how many, individuals and species occupy a habitat fragment (Bierregaard et al. 1992; Turner 1996). Species–area relationships (Connor and McCoy 1979), edge effects (Murcia 1995), nature of the matrix (Turner 1996), relative isolation (MacArthur and Wilson 1967) and environmental stochasticity, are among the more commonly cited determinants of community composition. Species richness has been the measure of local diversity used most often to compare communities of different habitat patches. Singly, small fragments generally support fewer resident, forest-endemic species (Turner 1996) and are therefore usually viewed as having lower conservation value than larger expanses of forest. But collectively, the number of species supported by multiple small fragments can surpass that found in larger fragments (Fischer and Lindenmeyer 2002; Tscharntke et al. 2002). General consequences of fragmentation on other aspects of community diversity, i.e., species abundance–dominance relationships, species identities, intraspecific variation, and trophic interactions, and what these consequences signify in terms of relative conservation value, are much less understood because comparatively few empirical data have been compiled.

We rely on a focal group, the fruit-feeding butterfly species, to quantify three aspects of community diversity, species richness, species evenness, and species composition, at the six forest fragments. Insects are superlative focal species for investigating how fragmentation impacts forest communities due to their abundance, diversity, endemism, ubiquity, and rapid response to environmental change (Brown 1997; Bossart and Carlton 2002). The frugivorous butterflies, in particular, have been beneficially exploited to study numerous aspects of tropical forest ecology in natural (DeVries 1988; DeVries et al. 1999; Hill et al. 2001), managed, (Fermon et al. 2000, 2003; Willott et al. 2000; Hamer et al. 2003, Stork et al. 2003) and degraded ecosystems (Kremen 1992; Daily and Ehrlich 1995; Shahabuddin and Terborgh 1999). This is largely because member species can readily be collected via fruit-baited traps, which facilitates systematic collection of data, they show a diversity of relative sensitivities to environmental change, and they are tightly intertwined with ecological systems as both primary consumers (herbivores) and as food items.

Taxon inventories are the cornerstone of viable *in situ* conservation programs because they identify the current state of the biological systems of interest and provide a guideline from which to operate. To the best of our knowledge, our biodiversity data are the first on insects from Ghana and only the second on this group from the greater western West Africa region. Our sampling scheme also incorporated a significant temporal component, which is an uncommon feature of biodiversity studies, where rapid, "snapshot" surveys are the norm.

## Methods

### *Study sites*

Ghana's old-growth forests represent the eastern most range extent of the Upper Guinean forests of western West Africa. These forests comprise four increasingly dry vegetational zones that lie diagonally from northwest to southeast, beginning from the wettest southwest corner of the country and extending outward as ever drier concentric bands towards the outer margin of the forest block (Figure 1). Virtually all forests remaining in Ghana now exist as isolated habitat islands entirely surrounded by a highly modified landscape matrix. The history of deforestation in Ghana indicates the country's forest islands were created within living memory, and some as recently as approximately 30 years ago (Hawthorne and Abu-Juam 1995 and references therein).

We conducted a joint analysis of the diversity of the fruit-feeding butterfly communities at six forest fragments (hereafter, sites); two large forest reserves and four remnant, sacred forest groves (Table 1; Figure 1), in order to evaluate the relative conservation value of these small forest fragments. Specific sites were identified in consultation with personnel of the Forestry Department of Ghana, Nature Conservation Research Center (NCRC), and Forestry Research Institute of Ghana (FORIG). Individual sacred groves fall under the jurisdiction of local tribal councils, and permission from village elders to enter and collect from each grove was required and secured. All six sites are representatives of moist semi-deciduous forest and all were formerly embedded within the millions of hectares of continuous forest cover that once was the Upper Guinean forests of West Africa. Survey sites were concentrated around Kumasi in the Ashanti administrative region to facilitate ongoing sampling of multiple sites (Figure 1). All sites derive from the same basic habitat type and same general region; hence, $\beta$-diversity is assumed to have been low relative to $\alpha$-diversity prior to deforestation and fragmentation.

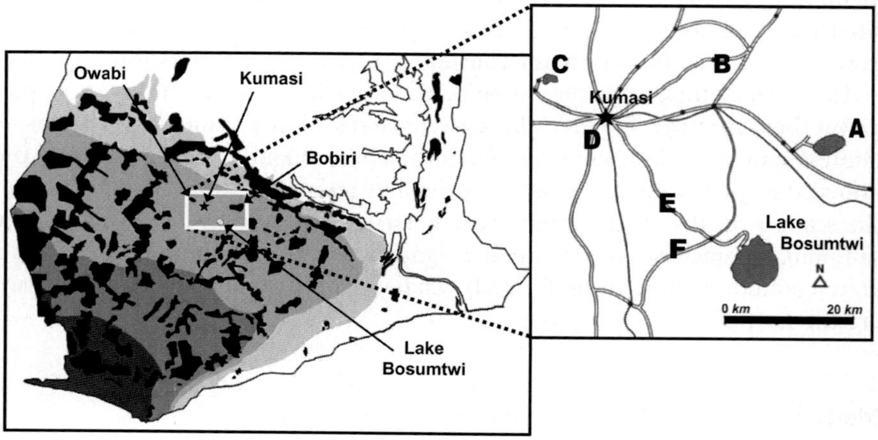


*Figure 1.* Map of survey sites in relation to Ghana's forest zones and gazetted forest reserves. Forest reserves are in black. Wet evergreen, moist evergreen, moist semi-deciduous, and dry semi-deciduous forest zones are depicted in decreasing intensities of gray, respectively, beginning in the lower southwest corner of the country. The inset shows the location of study sites: (A) Bobiri Forest Reserve, (B) Bonwire Sacred Grove, (C) Owabi Forest Reserve, (D) Kajease Sacred Grove, (E) Gyakye Sacred Grove, and (F) Asantemanso Sacred Grove. Kumasi, Lake Bosumtwi, and Owabi and Bobiri Forest Reserves are shown on both the map and the inset to aid in orientation.

*Table 1.* Site, trap, and capture specifics.

| Site | Size | Global coordinates | Total traps | Trap days[a] | Total captures | Captures/ trap day |
|---|---|---|---|---|---|---|
| Bobiri Forest Reserve | 5000 ha | 6°40′ N, 1°19′ W | 15 | 330 | 1581 | 4.79 |
| Owabi Wildlife Sanctuary | 1200 ha | 6°44′ N, 1°42′ W | 15 | 285 | 1843 | 6.47 |
| Asantemanso Sacred Grove | 259 ha | 6°28′ N, 1°33′ W | 10 | 180 | 988 | 5.54 |
| Gyakye Sacred Grove | 11.5 ha | 6°33′ N, 1°31′ W | 5 | 95 | 315 | 3.32 |
| Bonwire Sacred Grove | 8 ha | 6°46′ N, 1°28′ W | 8 | 144 | 670 | 4.65 |
| Kajease Sacred Grove | 6 ha | 6°38′ N, 1°39′ W | 8 | 160 | 1429 | 8.93 |

[a] Trap days is calculated as the number of traps at a site multiplied by the number of different days the site was sampled.

## *Trapping methods*

Trap sampling was initiated late June 2001 and continued through July 2002. Sampling was more frequent during the rainy season, when the senior author was in the country (approximately late May through early August), and less frequent during the rest of the year. In general, trapping occurred weekly from late June through mid August 2001, monthly between August 2001 and mid May 2002, then biweekly from mid May through late July 2002. Logistical constraints determined sampling schedule. Roads between sites are generally substandard and some are heavily traveled, and substantial time was required to move from site-to-site.

Additionally, significant hiking time was involved at half of the sites. For example, sampling at Bonwire required an approximate 90 min roundtrip hike just to reach the grove. A maximum of two sites could be inventoried per day. Sites in closest proximity were generally sampled on the same day. Additional sites were sampled on sequential days. All six sites were sampled within a 5-day-period during a given sampling bout.

Typical fruit-bait traps were hung in the forest understory, approximately 10 cm above the ground. An initial 4-day experiment indicated that fruit-baited traps hung close to the ground versus those hung at 1 m heights attracted a significantly greater number of captures and species. A total of 61 traps were installed across all sites. Because traps were assembled in the US and subsequently transported to Ghana, this number was set prior to identification of study sites, which constrained our subsequent experimental design, In general, more traps were hung at larger sites, fewer at smaller sites (Table 1). Traps were hung near forest foot paths where possible to facilitate access, but were always installed at least ≥3 m off the trail. In most cases, an access trail had to be cut into the forest to place traps because foot paths were nonexistent.

Conscious effort was taken to install all traps in similar micro habitats within areas of closed canopy forest. Five traps were hung in each of three discrete areas at Bobiri and Owabi, and in two at Asantemanso, in order to gain a more representative sample from these larger forests. Four traps were hung in each of two separate areas at Kajease and Bonwire, but the distance between areas was small and restricted by the size and irregular shape of these forests. Five traps total were hung at Gyakye. Individual traps within areas were separated from each other by at least 50 m and by no more than 200 m. Traps were baited with mashed, fermented banana, and butterflies were collected the following day approximately 24 h later. All specimens captured in the traps were collected for subsequent identification.

### *Species identifications*

Butterflies were identified using a variety of taxonomic treatises, including D'Abrera (1981, 1997, 2001), Hecq (1997, 1999, 2000, 2002), Henning (1988) and Larsen (In press). Difficult specimens were identified in consultation with T. B. Larsen, an internationally recognized authority on butterflies of West Africa and elsewhere. Spread specimens were assigned a unique number and then photographed six to a frame with a 300-mm macro lens. Both ventral and dorsal sides were photographed. One set of photographs was then retained by the senior author and a second set was forwarded to Larsen. This system allowed for interactive discussion via the internet on particular specimens when necessary to clarify species identifications.

### *Community diversity measures*

Community diversity was evaluated as: (a) observed richness, which was quantified both as number of species accumulated versus sampling effort and as rarefied species richness; (b) estimated total richness, calculated from a suite of nonparametric richness estimators; (c) species evenness, determined from species rank-abundance distributions; (d) Simpson's Index of Diversity, to capture overall heterogeneity; and (e) complementarity of communities, to quantify dissimilarity among communities with respect to species identities and abundances.

### *Statistical analyses*

Trap data at a given site on a specific date were pooled to generate a single sample for each site–date combination. Pooling of data minimizes variance associated with individual traps (DeVries et al. 1999), e.g., due to differential attractiveness of traps or sporadic destruction of samples by driver ants. An assessment of spatial and temporal variation within sites will be treated elsewhere.

Observed species richness was rarefied to standardize for sample size differences across sites. Rarefaction is a robust statistical technique that calculates estimates of species richness for sub samples of a specified size drawn at random from the total community (i.e., from the total collection) (Gotelli and Colwell 2001). Richness estimates for sites with larger overall samples are interpolated down to those for sites with smaller overall samples simply by specifying the number of individuals to be randomly drawn. Comparable sized sub samples can then be compared statistically across sites (Heck et al. 1975; Simberloff 1978). Rarefaction curves and 95% confidence intervals were estimated for all sites using EcoSim700 (Gotelli and Entsminger 2003) Rarefied estimates of richness were calculated after 1000 iterations.

Statistical estimates of total richness were calculated using EstimateS, Version 6.0bl (Colwell 2000). EstimateS uses curve fitting models to predict asymptotes of species accumulation curves and computes richness estimates based on a variety of nonparametric estimators and in many cases, their associated standard deviations. Different estimators differ with respect to how they deal with rare species (Chazdon et al. 1996). Input data were formatted as species (rows) by samples (columns) abundance matrices. Individuals within a species were randomly assigned to samples, which removes patchiness due to temporal differences in abundance when the patchiness parameter is set to zero. Sample order was randomized without replacement and mean richness estimates were calculated after 100 iterations of the random sampling algorithm.

Species rank-abundance distributions were graphed to evaluate species evenness among sites. Log10 $p_i$, where $p_i$ is the frequency of the $i$th species in the total sample, was calculated for each species then plotted against the relative rank for that species. Differences in species evenness among sites are apparent as differences in the comparative shape and steepness of the curves for different sites (Southwood and Henderson 2000; Magurran 2004). PROC GLM (GLM; Type III sum of squares, SAS 1990) was used to quantify heterogeneity-of-slopes across sites, where *Site* was designated as a class variable and *Relative Abundance* and *Rank* were designated as continuous variables, and where the interaction of *Site* × *Rank* tests heterogeneity of slopes. The slope of each linear regression curve was used to measure evenness of the community (Tokeshi 1993; Magurran 2004).

Indices of diversity integrate both species richness and evenness into a single measure in an attempt to capture overall heterogeneity at a site. We calculated Simpson's (inverse) Index of Diversity (1/D) using EstimatesS, Version 6.01b (Colwell 2000). This index is one of the most robust and most easily interpreted, although no diversity index is considered a perfectly unified measure (Magurran 2004). Values for the Simpson Index range from 1 to the number of species in the sample, with higher values indicating greater overall diversity (Krebs 1999).

Complementarity of community composition among sites was quantified using two quantitative indices of similarity, the *Morisita–Horn Index* (Magurran 2004) and *COMPAH's % Similarity* (Gallagher 1999). Quantitative, as opposed to qualitative, measures of similarity integrate both differences in species uniqueness between sites and differences in relative abundances of shared species. The *Morisita–Horn Index* is generally recognized as the best overall estimate of similarity because it is less sensitive than others to changes in species richness and sample size (Wolda 1981; Magurran 2004). The *Morisita–Horn Index* is defined as

$$\mathrm{MHS}_{ij} = \sum_{k=1}^{S} [x_{ik} x_{jk} / (d_i + d_j) N_i N_j],$$

where, $S$ = Number of species, $x_{ik}$ = Abundance of species $k$ in sample $i$, $N_i$ = Total individuals in sample $i$, and

$$d_i = \sum_{k=1}^{S} x_{ik}^2 / N_i^2.$$

However, no measure of similarity is free of limitations and the *Morisita–Horn* is sensitive to changes in the abundance of the most common species (Magurran 2004). We calculated a second similarity coefficient, *COMPAH's % Similarity*, for comparison and to corroborate results. The *Bray–Curtis Index* (which has multiple synonyms, e.g., *Pielou's % Similarity*, *Sorenson's Abundance Index*; Gallagher 1999) has also performed well when tested against a variety of similarity measures (Magurran 2004). *COMPAH's % Similarity* is analogous to the *Bray–Curtis Index* except that species abundances are first standardized by sample totals, which is appropriate when sample sizes differ between communities being compared *COMPAH's % Similarity* is defined as

$$\%\ \mathrm{SIM}_{ij} = \sum_{k=1}^{S} \min x_{ik}/N_i, x_{jk}/N_j],$$

where, $S$ = number of species. $x_{ik}$ = Abundance of species $k$ in sample $i$, and $N$ = total sample size.

## Results

Each site was sampled from 18 to 22 times throughout the course of the study (Table 1; trap days divided by total traps), resulting in a total of 6836 individuals captured across all sites combined. Trap productivity, in terms of number of individuals collected per trap day, was highest at Kajease and lowest at Gyakye. Seventy-nine species were collected in total from all sites combined, representing five subfamilies (Appendix 1). This collective total of 79 species accounts for ~33% of the fruit-feeding butterfly guild currently known for the country (Larsen 2001). All but five of the 79 species (*Bicyclus safitza*, *Melanitis leda*, *M. libya*, *Ypthimomorpha itonia*, and *Charaxes varanes)* are forest habitat endemics and of these five, only two specialize on savanna habitat (Appendix 1). Greater than 70% of those collected are considered either moist forest specialists or species found in all forest subtypes, which is as expected given the location of our study sites well within the moist semi-deciduous forest zone.

Considered solely with respect to this sub group of species, our trapping efforts netted approximately 42% of the fruit-feeding butterflies known from these two forest habitat subtypes (Appendix 1).

*Observed species richness*

The number of butterfly species trapped at each site ranged from a low of 27 at Gyakye to a high of 59 at Bobiri (Table 2; Appendix 1), However, species accumulation curves were still rising at all sites when the study was terminated, indicating that species saturation had not been reached (Figure 2). Only the Owabi sample was clearly approaching an asymptote. This was also the only sample in which the singleton curve was declining (Figure 2). which is as expected when sampling is nearing completion (Magurran 2004). The rate at which new species were being added to the collection at the other sites was still relatively high. In fact, this rate of increase was sufficiently high at Bonwire to suggest the associated accumulation curve might ultimately cross the curve for Owabi. Singleton curves at these other sites were either flat (e.g., Bobiri) or rising (e.g., Bonwire).

Observed species richness was higher at the forest reserves than at the sacred groves (Figure 2; Table 3), Of the two forest reserves, Bobiri was the more species rich. Rarefied values derived from random sub-samples of the total trap data from Bobiri were consistently higher than those derived from comparable sized random sub-samples of the total trap data from Owabi (Table 3, last two columns). Additionally, 17% more species were collected at Bobiri even though

*Table 2.* Species diversity data for each site.

| Site | Total species trapped | Singletons/ Doubletons[a] | Estimates of total richness[b] | | | Number additional species predicted[c] | Simpson's Index of Diversity[d] |
|---|---|---|---|---|---|---|---|
| | | | ACE / ICE | Chao1/ Chao2 | Jack1/ Jack2 | | |
| Bobiri | 59 | 10/3 | 66/66 | 72/69 | 70/75 | 6–15 | 10.07 |
| Owabi | 51 | 5/4 | 54/54 | 53/53 | 56/57 | 2–6 | 17.73 |
| Asantemanso | 41 | 10/6 | 51/51 | 48/48 | 50/54 | 7–13 | 8.64 |
| Gyakye | 27 | 6/3 | 32/33 | 31/33 | 34/38 | 4–11 | 8.03 |
| Bonwire | 38 | 13/1 | 53/54 | 79/79 | 50/61 | 12–41 | 13.72 |
| Kajease | 41 | 14/7 | 61/60 | 52/52 | 54/61 | 11–20 | 7.25 |

[a] Singletons and doubletons are the number of species represented by one or two individuals, respectively.
[b] Nonparametric richness estimators were used to estimate total species richness at a site. ACE and Chao1 are abundance-based richness estimators. All others are incidence-based estimators.
[c] The number of additional species estimated to be at each site is the difference between estimated total richness and the number of species actually collected from traps.
[d] Simpson's Index was calculated as1/D (the reciprocal), where larger numbers indicate greater diversity.

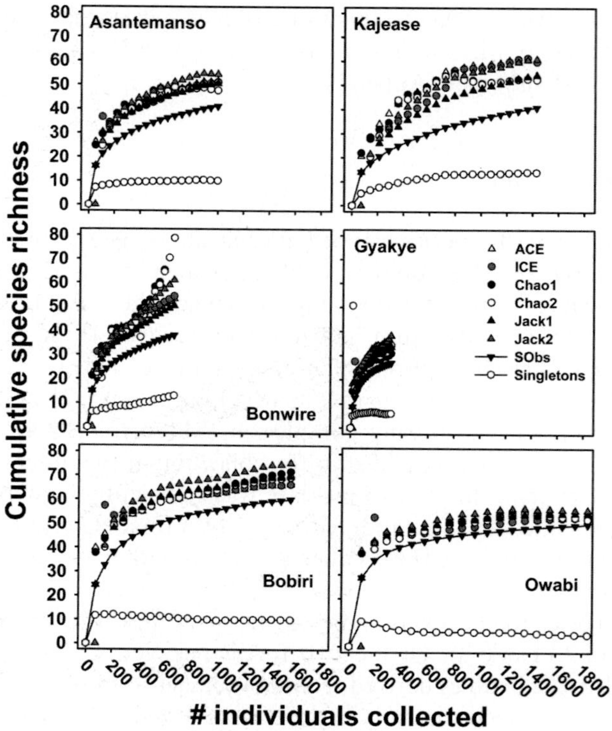


*Figure 2.* Randomized collectors curves. Number of observed and estimated species, and singletons accumulated at each site as a function of increasing numbers of individuals trapped through time. ACE, ICE, Chao1, Chao2, Jack1, and Jack2 = nonparametric estimates of species richness; Sobs = actual species observed in samples; Singletons = # of species represented by a single individual.

the total sample from this site was 15% smaller than that from Owabi. Of the sacred groves, sample species richness was lowest for Kajease. When richness was rarefied to sample sizes that were comparable to those collected from other sites, Kajease samples were always represented by fewer total species. Species richness at the remaining three sacred grove forests was generally similar when compared at the smallest sample size (Table 3). But the total number of 40 species represented by the 672 individuals collected at Bonwire exceeded the rarefied value of species richness at Asantemanso for a sample of similar size.

## *Estimated species richness*

In all cases but one (Bonwire), the Jackknife2 estimator generated the highest estimate of total richness at a site (Table 2; Figure 2). The rank order of all other estimators was inconsistent across sites. Estimates of total species

*Table 3.* Rarefied species richness.

| Comparison site | Actual trap data | Rarefied richness values[a] | | | | |
|---|---|---|---|---|---|---|
| | | Bonwire $N = 672$ | Asantemanso $N = 988$ | Kajease $N = 1429$ | Bobiri $N = 1581$ | Owabi $N = 1843$ |
| Gyayke | $S = 27, N = 315$ | 30.84 (26.84–34.83) [321] | 29.90 (25.84–33.96) [323] | 23.02 (13.43–27.6) [307] | 42.85 (38.03–47.67) [316] | 39.81 (35.76–43.86) [330] |
| Bonwire | $S = 40, N = 672$ | | 36.46 (33.36–39.55) [668] | 30.91 (26.38–35.44) [681] | 51.32 (47.08–55.56) [663] | 44.55 (41.08–48.01) [658] |
| Asantemanso | $S = 41, N = 988$ | | | 35.27 (31.56–38.98) [987] | 55.35 (51.75–59.95) [978] | 47.02 (44.31–49.73) [987] |
| Kajease | $S = 41, N = 1429$ | | | | 58.90 (56.91–60.88) [1419] | 48.90 (47.01–50.79) [1425] |
| Bobiri | $S = 60, N = 1581$ | | | | | 49.31 (47.79–50.83) [1571] |

Comparisons are presented in order of increasing total sample size collected from different sites.
Numbers in parentheses are 95% confidence intervals; numbers in brackets indicate sub-sample size associated with the rarefied value of species richness given.
$S$ = number of species collected at each site. $N$ = total sample size at each site.
[a]Each row compares observed species richness at a site that has a smaller overall sample to rarefied species richness at sites with larger overall samples Rarefied richness values are estimated for comparable sized, sub samples that represent random draws from the larger sample.

richness at Owabi, Asantemanso, and Gyakye were generally in good agreement regardless of the richness estimator used (*standard errors of the mean* of 0.67, 0.919, and 0.992 for Owabi, Asantemanso, and Gyakye, respectively). As few as a single species, and at most three species, separated the different estimates, except with respect to the Jackknife2 estimate, which was 6–12% higher than the others for Asantemanso, and 9–14% higher than the others for Gyakye. Estimates of total richness were more variable at the other three sites (*standard errors off the mean* of 1.43, 5.37, and 4.46 for Bobiri, Bonwire, and Kajease, respectively) and dependent upon the richness estimator used. At Bonwire, for example, the Chao1, Chao2, and Jacknife2 estimates ranged from 12 to 36% higher than the others. Incidence and abundance based estimates of total species richness were very similar (Table 2). ACE (abundance-based) and ICE (incidence-based) coverage estimators were identical or virtually identical in all cases. Similarly, Chao1 (abundance-based) and Chao2 (incidence-based) estimates were identical in all cases except at Bobiri, where the abundance-based estimate was slightly higher.

All estimates of total richness based on the Owabi sample were stable and converging on observed richness as sample size increased (Figure 2). The Chao2 estimates from the Kajease data stabilized at a sample size of approximately 1000, and the ICE estimates from the Asantemanso and Bobiri data, stabilized at a sample size of approximately 1000 and between 1300 and 1400, respectively. Richness estimates in all other cases had not stabilized and continued to increase with increasing sample size. This rate of increase was most extreme at Bonwire.

Estimates of total species richness at Bonwire and Kajease deviated the most from the number of species actually collected in traps (Table 2; Figure 2). At Bonwire, estimates of total richness spanned from 50 to 79, representing a 28–52% increase over the 38 actually collected. At Kajease, estimates of total richness represented a 22–33% increase over the 41 actually collected. Owabi showed the closest correspondence between number of species trapped and estimates of total species richness. Total species richness at Owabi Reserve was estimated to be only 4–6% higher than the number collected in traps.

Bobiri was estimated to have the most species rich community and Gyakye, the least. Among the four remaining sites, relative differences among forests depended upon whether estimates at the low end of the estimated range were compared, or estimates at the high end of the range. At the low end, all sites were estimated to have comparable species richness. At the high end, however, Bonwire and Kajease were predicted to be more speciose than Asantemanso or Owabi, with as many as 32 and 12% more species, respectively.

### *Species evenness and overall diversity*

The percentage of singletons (i.e., species represented by a single individual) trapped at each site ranged from a low of 10% of species at Owabi to a high of

34% of species at both Bonwire and Kajease (Table 2). The rank–abundance or dominance–diversity curve for Kajease declined at the most rapid rate (Figure 3a), indicating the community trapped at this site had the most uneven representation of different species. Indeed, three species, *Euphaedra ceres*, *Gnophodes betsimena*, and *Bicyclus vulgaris*, accounted for 56% of the 1429 individuals trapped at Kajease (Appendix 1). At the other extreme, 14 of the 41 total species collected at this site were represented by a single individual (Table 2).

Species evenness was higher and generally comparable at the other three sacred groves, except with respect to the most common species, where the curve for Bonwire initially declined more slowly. This initial slower drop off indicates this site had more relatively common species than Asantemanso or Gyakye, and their abundances were more equitable than those of commoner species at the other two sites (Figure 3a). For example, the single most common species at Asantemanso and Gyakye accounted for 1/4 of all individuals trapped, whereas the single most common species at Bonwire accounted for only 1/8 of all individuals trapped. In fact, Bonwire's common species were more evenly represented than those at any other site.

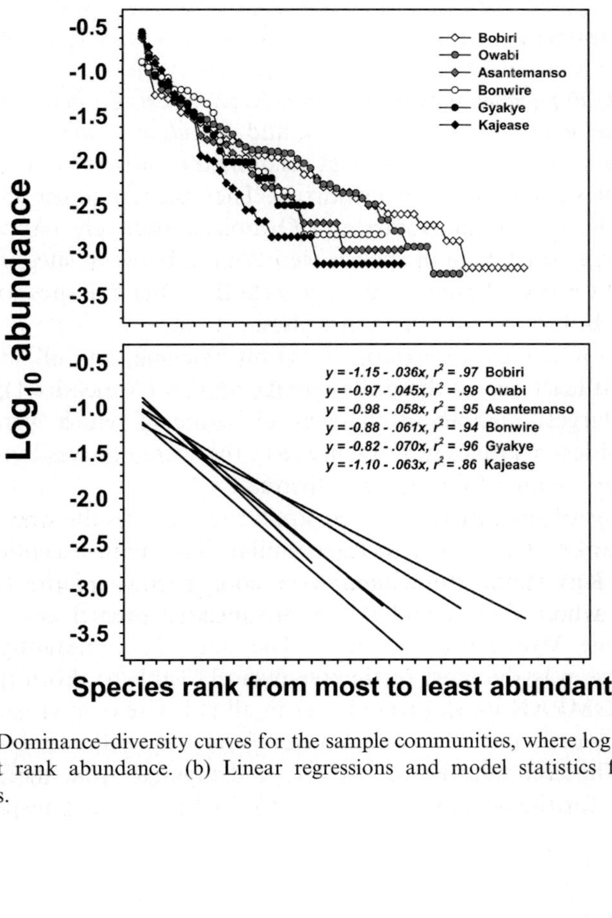


*Figure 3.* (a) Dominance–diversity curves for the sample communities, where log10 abundance is plotted against rank abundance. (b) Linear regressions and model statistics for dominance–diversity curves.

The dominance–diversity curves for the sample communities from the two forest reserves declined at the slowest relative rate, indicating these samples had the highest species evenness (Figure 3). Note the curves overlay each other until the least abundant species factor in, at which point they diverge because collectively more species at Bobiri are represented by each of the rarest abundance classes.

Regression slopes were highly heterogeneous among sites ($F = 41.67$, $p < 0.0001$, $df = 5$) and ranged from a low of $-0.036$ for the Bobiri sample to a high of $-0.070$ for the Gyakye sample (Figure 3b), The regression analysis indicated higher evenness at Kajease relative to Gyakye (slopes of $-0.063$ and $-0.070$, respectively). But the Kajease data were less adequately explained by a linear model ($r^2 = 0.86$ versus 0.96 for Kajease and Gyakye, respectively) and the slope of the regression was skewed to the right by the large number of singletons trapped from this site.

The sample collected from Owabi proved to be the most diverse when species richness and evenness were integrated into a single measure of overall heterogeneity (Simpson's Index; Table 2). The sample collected from Bonwire was next, followed by Bobiri, Asantemanso, Gyakye, and Kajease, respectively.

### *Site complementarily*

Eight species, *Euphaedra medon*, *E. ceres*, *E. phaetusa*, *E. themis*, *E. harpalyce*, *Gnophodes betsimena*, *Bicyclus funebris*, and *B. vulgaris*, were generally abundant at all six sites. These eight species accounted for 61–86% of all individuals trapped at the sacred groves (Appendix 1). They accounted for 35 and 53% of all individuals trapped at Bobiri and Owabi, respectively (Appendix 1). In general, a large percentage of species were shared between sites, ranging from the 80% that Gyakye shared on average with the other five sites to the 54% on average that Bobiri shared with other sites.

A number of species were trapped at only one site, and all sites were represented by at least one of these site-specific species (Appendix 1), Bobiri had, by far, the largest number of these at 13, some of which were fairly well represented. Eleven species were exclusive to the sacred groves but nearly all of these were represented by a single individual.

Both indices of similarity produced similar relative results when community pairs were ranked from most to least similar. The main exceptions were the relative rankings of the Bonwire/Gyakye comparison and the Bobiri/Owabi comparison, where the COMPAH index indicated greater community similarity than the Morisita–Horn index. The butterfly community trapped at Bobiri Reserve was the most distinctive overall (Table 4). Both the Morisita–Horn and COMPAH indices were lower in all pairwise comparisons involving Bobiri then when butterfly communities from any other two forests were compared. The communities at Kajease and Gyakye were the most similar (89 and 72% for the Morisita–Horn and COMPAH indices, respectively). On

*Table 4*. Complementarity of species assemblages among sites.

| Sites | Bobiri (60) | Owabi (51) | Asantemanso (41) | Bonwire (39) | Kajease (41) | Gyakye (27) |
|---|---|---|---|---|---|---|
| Bobiri (60) | – | 42 | 33 | 34 | 30 | 23 |
| Owabi (51) | 45/58 | – | 32 | 35 | 32 | 22 |
| Asantemanso (41) | 42/51 | 82/67 | – | 30 | 27 | 22 |
| Bonwire (39) | 41/49 | 74/61 | 74/60 | – | 27 | 21 |
| Kajease (41) | 36/31 | 65/52 | 86/71 | 68/58 | – | 20 |
| Gyakye (27) | 34/38 | 57/53 | 67/57 | 65/61 | 89/72 | – |

Number of shared species above the diagonal.
Morisita–Horn/COMPAH's quantitative indices of similarity (expressed as percentages) below the diagonal. The total number of species trapped at each site is in parentheses for ease of comparison.

average, communities trapped at the sacred groves were more similar to each other than they were to either forest reserve. There was an average similarity of 75 and 64% (Morisita–Horn and COMPAH indices, respectively) for all pairwise comparisons among the four sacred groves, an average of 70 and 58% between Owabi and the sacred groves, and an average of 38 and 42% between Bobiri and the sacred groves. Of the sacred groves, the community trapped at Asantemanso was the most similar to the communities trapped at the two forest reserves (Table 4). In fact, Asantemanso exhibited, on average, the greatest similarity of any site to all others.

## Discussion

### *Observed species diversity among sites*

The fruit-feeding butterfly communities trapped at the sacred groves were generally less speciose than those trapped at the forest reserves, which is consistent with theoretical expectations of species–area relationships, whereby smaller areas tend to support fewer species (Schoener 1976; Rosenzweig 1995; May and Stumpf 2000). The sacred grove communities were also generally less diverse with respect to the relative abundances of member species. In communities, such as Asantemanso, Kajease, and Gyakye, that are dominated by a few very abundant species, the vast majority of individuals in the community will be of these few predominant species. In communities where species are more equitably represented, randomly encountered individuals are more likely to derive from different species, which is a defining characteristic of a diverse community (Purvis and Hector 2000). Interestingly, Simpson's Index of overall diversity proved to be lower for the Bobiri sample than for the Owabi and Bonwire samples. But this diversity measure is heavily weighted by the relative abundance of the most abundant species, which at Bobiri accounted for 1/4 of all individuals trapped, i.e., *B. tentyris*. Hence, in a broad sense, our findings add to the already substantial body of data that indicate the primary success of biodiversity conservation hinges on protection of large habitat areas.

Trapping at the smallest grove, Kajease, yielded the overall most depauperate sample community. Relatively fewer species were collected at this site given the size of the sample, which was significantly larger than those collected from the other sacred forests, and which nearly rivaled the size of those collected from the two forest reserves. Additionally, over half of the total species collected from this site were represented by a single, or at most two, specimens. Indeed, the eight forest species common at all six sites accounted for nearly 90% of the entire sample collected at Kajease. Not surprisingly, Simpson's Index of Diversity was lower for Kajease than for any other site. This low observed richness and evenness was somewhat surprising, even considering the small size of this forest patch, because the forest retains much of its closed canopy and, judging by the size of the total sample we collected, apparently supports a large number of individuals. However, of the groves we inventoried, the landscape matrix surrounding Kajease is uniquely characterized by intensive residential development directly adjacent to, and in some cases within meters of, the grove. These activities have undoubtedly led to the degradation or disruption of key ecological processes in the grove, e.g., water drainage, and fostered the differential extinction of habitat sensitive species. This highly transformed landscape matrix has likely also served to hinder emigration from and immigration to this isolated forest patch by all but the most robust dispensers.

Bonwire was the most enigmatic and most diverse sacred grove we sampled. Sampling intensity at this site was very similar to that expended at Kajease in terms of patch size and trap days. Yet the number of individuals collected from Bonwire was less then half that collected at Kajease. Indeed, on multiple occasions our sampling efforts yielded only a few specimens and over time we had come to perceive this forest remnant as generally less diverse than the others. The relatively small sample collected from this 8-ha patch, however, ultimately proved to comprise a relatively diverse community, Observed richness at Bonwire exceeded the rarefied richness values at all other sacred grove sites, and member species of this community were more equitably represented than those at the other sacred grove sites. Indeed, when quantified as a single measure of diversity, Bonwire was second only to Owabi in terms of overall heterogeneity. That Bonwire is relatively distant from the village proper and located in a rural area has probably helped preserve the grove's ecological integrity and resident biodiversity. Bonwire, unfortunately, also proved to be the most imminently imperiled grove. Near the end of our study, strong winds associated with a rainy season storm uprooted virtually every old growth, emergent tree in the forest patch. This event stands as a clear testament to the enhanced vulnerability of small isolated forests to further degradation (Laurance et al. 1998).

Owabi was the only site at which most of the resident community of fruit-feeding butterflies attracted to our traps had been collected. The species accumulation curve was approaching an asymptote, the number of singletons in the sample was declining, and estimated and observed richness were converging, all signs that continued trapping would have resulted in the capture of few additional species. Accumulation curves were still climbing at most sites;

however, indicating that our trap data would have yielded additional species had our sampling continued. Bonwire appeared to be the most incompletely sampled site. The species accumulation and singleton curves were sharply increasing and estimated and observed richness showed no sign of convergence.

In any community, new species are initially accumulated quickly as common species are captured, but then are accumulated ever more slowly as rarer, infrequently captured species are added (Gotelli and Colwell 2001). Species missing from our trap collection do indeed include taxa considered to be rare (Larsen 2001; see Appendix 1). But our surveys also failed to trap a significant proportion of species (>1/3) thought to be generally to very common occupants of Ghana's moist forests. For example, a number of theoretically common Charaxinae were either entirely absent from our trap data or only rarely represented. Common species generally absent across all sites may have been relatively unattracted to our bait or vertically stratified in the forest, e.g., localized in the canopy DeVries 1988; DeVries et al. 1997, 1999; Fermon et al. 2003). Such resource or habitat affinities would have set boundaries on the species pool we were likely to sample, i.e., the pool of species attracted to banana and to traps hung near the forest floor. Species rare in samples that are generally abundant members of communities have been termed methodologically rare because their "rarity" is a function of the sampling method used not their actual abundance (Longino et al. 2002). But not all theoretically common species we trapped were equally rare across sites. Our survey of Bobiri, for example, resulted in 422 individuals of *Bebearia tentyris*, a species generally viewed as common. Indeed, this was by far the most abundant species at this site, accounting for more than 1/4 of all individuals captured (Appendix 1). Yet *B. tentyris* was hardly represented at the other forests, including the 1500-ha Owabi Wildlife Sanctuary. This wide dichotomy in relative abundance across sites suggests that *B. tentyris* may be particularly resource sensitive or vulnerable to changes in the landscape matrix. Although intrinsically rare species are generally perceived as more vulnerable to land conversion and fragmentation, numerically abundant species are not impervious to their effects (Vermeij 1993; Skinner 2000; Bossart and Carlton 2002).

### *Estimated species richness among sites*

Observed richness is a strongly biased measure of species diversity at a site because it is highly correlated with sample size and is inevitably an underestimate of true species richness unless sampling is nearly or wholly complete. Nonparametric richness estimators are a powerful and effective alternative for assessing total species richness because they reduce this bias (to varying degrees) and are independent of sample size above some minimum size (Colwell and Coddington 1994; Longino et al. 2002). The ICE and Chao2 estimators have been touted as particularly promising because they perform well at small sample sizes and are relatively insensitive to sample density and species patchiness (Chazdon et al. 1996; Longino et al. 2002; Magurran 2004).

Owabi is the only site where observed richness can be considered a viable measure of the actual number of fruit-feeding butterfly species in the community. At all other sites, observed richness clearly underestimated true species richness, particularly at Kajease and Bonwire where the increasing singleton curves and wide divergence between estimated and observed richness indicates sampling is far from complete, The ICE and Chao2-based estimates were the only ones that tended to level off with increasing sample size, which is consistent with earlier evaluations of estimator performance. Notably, the ICE and/or Chao2 estimates stabilized at approximately 1000 for three of the four samples, i.e., Asantemanso, Kajease, and Owabi, with Bobiri the one exception. This suggests sample sizes of at least 1000 are necessary for estimating true richness of these fragmented communities, and furthermore, may in many cases be the maximum size required.

Richness estimators, in general, are highly influenced by the number of rare species and observed and estimated richness will diverge considerably if the ratio of singletons to doubletons (or uniques to duplicates) is large. Additionally, the ranking of different sites based on relative richness will change unless this ratio remains approximately constant across sites. Of our samples, those from Owabi and Bonwire had the smallest and largest ratio of singletons to doubletons, respectively. Consequently, estimates of total species richness at Owabi showed the least divergence from the number actually collected, whereas estimates of total richness at Bonwire showed the greatest divergence. Because the ratio of singletons to doubletons was inconstant from site-to-site, relative richness of the different sites based on number of species actually collected also differed considerably from that based on estimated total richness. Bobiri and Owabi forest reserves are the most speciose communities and Kajease the least, when viewed in the context of species actually trapped. But when viewed in the context of estimated total richness, the Owabi fruit-feeding butterfly community appears to be no more speciose than that at Asantemanso, and perhaps even less speciose than that at Bonwire or Kajease, and the community at Gyakye sacred grove is clearly the least species rich.

### *Species composition and site complementarity*

Resident communities of small remnant forests and large forest blocks are expected to differ due to, for example, species–area relationships (Schoener 1976; Rosenzweig 1995; May and Stumpf 2000), population size and extinction risk dynamics (Lande 1988), extent of edge habitat and predominance of edge effects (Murcia 1995), and differential tolerance of individual species to relative habitat isolation and fragmentation (Rabinowitz 1981; Gaston 1998; Johnson 1998). Consequently, comparison of species assemblages among forest fragments of markedly different size should reveal clues about 'winners' and 'losers' with respect to species persistence and extinction in highly fragmented landscapes. Species similarly or more abundant in forest remnants should be those resistant to dynamics of forest fragmentation, whereas species that show notable decrease in abundance would presumably be those negatively impacted by fragmentation effects.

Eight forest butterfly species were commonly trapped at all six sites despite probable differences in the condition of individual forests and significant differences in forest size. This implies these eight species are relatively tolerant of, or perhaps even benefited by, extreme fragmentation of their resources and the variety of associated changes that small, isolated remnants of forest incur. In theory, effective dispersers that are undeterred from crossing a nonforest landscape matrix, for example, could move between patchily distributed forest fragments and repeatedly re-colonize such habitat islands (Hasting and Harrison 1994; Hanski and Gilpin 1996). Similarly, species characterized by traits that confer resistance to effects of forest deterioration, e.g., generalists versus specialists species (Kunin and Gaston 1997), might be able to maintain viable resident populations in small isolated forest remnants. The eight species abundant in our trap samples are known to be generally common, to have fairly wide ranges, to colonize both intact and disturbed forest, and to fly outside the forest proper (Larsen In press). These are all characteristics thought to facilitate persistence of forest species in highly transformed landscapes.

Species more abundant in the sacred groves versus the forest reserves included the grass-feeding Satyrine species whose increased presence is an apparent sign of forest disturbance (Larsen 1994). Grasses are generally uncommon on the Afrotropical forest floor (Owen 1971) but can colonize and spread in abundance when light penetration into the forest proper increases as a result of, for example, increased edge to interior habitat or degradation of the forest canopy. That some Satyrines might actually benefit from fragmentation and become more abundant in remnant patches is therefore not surprising. This coupled dynamic between grasses and Satyrines also signals the high potential of this group as biodiversity indicators of forest condition, an attribute that could be beneficially exploited to help direct limited conservation resources in economically disadvantaged countries, e.g., to identify priority sites for formal protection or to steer restoration efforts.

Fermon et al. (2000) found that certain Limenitinae were vulnerable to modification of forest habitat as a consequence of logging. Not surprisingly, this vulnerability apparently extends to modification of forest habitat at the broader landscape level as well. We found *Bebearia* and *Euriphene* species, as a group, to be those most sensitive to affects of forest fragmentation (Bossart et al. 2005). The sacred grove collections included 10 or fewer of the 19 total *Bebearia* and *Euriphene* species trapped and these captures were, in many cases, represented by a single individual. Four of the five *Bebearia* collected at Kajease, for example, were singletons (i.e., *B. absolon*, *B. cocalia*, *B. oxione* and *B. zonara*). Perhaps the most telling evidence of the apparent vulnerability of this group of fruit-feeders is that a number of the "absent" species are perceived as generally common residents of moist forests or all forest subtypes, e.g., *B. tentyris* (Appendix 1). Unlike rare species, whose absence could relate to their lower overall probability and rate of capture, the decreased numbers of these common species are difficult to explain except in the context of sensitivity to dynamics of forest fragmentation.

## Conclusions

Species diversity, measured as richness, evenness, and distinctiveness, was generally reduced in the sacred forest groves relative to the larger reserves, despite the long history of protection and restricted use of these indigenous conservation areas. Given these forest 'islands' were embedded within what once was the millions of hectares of continuous Upper Guinean forests of west Africa, our findings imply the geographic ranges of many of Ghana's forest-endemic fruit-feeding butterfly species have contracted as a consequence of the widespread destruction and fragmentation of the country's forest cover. The extrinsic and intrinsic drivers responsible for our observed patterns of community diversity are currently unknown as many factors disproportionately impact smaller versus larger habitat islands, e.g., edge effects, and influence which species occur where. Our studies in the region are ongoing, however, and ultimately we expect to uncover the key predictors, if they exist, that determine diversity of the fruit-feeding butterfly community of Ghana's relict forests.

Our preliminary assessment of frugivorous butterfly diversity in habitat patches adds to the burgeoning evidence that large reserves are the cornerstones of successful conservation strategies because they generally harbor greater forest biodiversity than smaller forest fragments. Nonetheless, small habitat fragments can contribute to the preservation of biodiversity in transformed landscapes (Fischer and Lindenmayer 2002; Tscharntke et al. 2002), and our survey data identify at least three ways that Ghana's indigenous reserves facilitate conservation of forest habitat specialists. First, a number of less common, forest-endemic species were collected from the sacred groves, three of which were collected exclusively from a sacred grove. *Bebearia barce*, which was trapped only at Asantemanso, is the most notable of these. That four individuals of this generally rare species of wet tropical forest were collected exclusively from Asantemanso implies this sacred grove alone supports a resident population. Second, although most aspects of species diversity were lower in the sacred groves versus the forest reserves, estimates of total species richness were comparable between Kajease, Bonwire, Asantemanso, and Owabi despite large differences in fragment size (6–1200 ha). This result was unexpected and at this point we have few insights. For many species local extinctions occur gradually and many of these forest endemic species may ultimately disappear from these fragments given sufficient time. But this explanation doesn't account for the fact that some species were only documented in these small fragments and others actually increased in abundance, Finally, the proportion of total species richness accounted for by Charaxinae species was approximately 2- to 3-fold higher at Asantemanso, Gyakye, and Kajease than at the forest reserves.

Thus, either these small forest fragments are able to support resident populations of these large, robust butterflies, or sacred groves facilitate dispersal of this guild of fruit-feeders among isolated forest patches. Either way, these small relicts of old growth forest are serving to foster persistence of forest species across a landscape matrix that is largely devoid of forest habitat.

The integration of small remnant fragments into mainstream conservation practice is a necessity to mitigate biodiversity loss in degraded landscapes, where large tracts of habitat no longer exist or where the economic cost of their protection is excessive. Sacred natural sites comprise an ancient and global system of small, indigenous reserves that largely remain outside mainstream conservation research and practice (UNESCO). Despite the cultural and biological significance of Ghana's sacred sites, few receive active protection. Many have been completely destroyed and many others are under imminent threat by encroaching farms and residential development, and degradation through consumption of forest products and bush fires, as traditional protective measures have broken down (Ntiamoa-Baidu 2001, GACON). Explicit integration of sacred forest groves into the protected areas network, either through community or governmental based conservation efforts is necessary to ensure these isolated examples of old growth forest continue to persist in a landscape where little unprotected forest habitat remains. Although their protection will never replace or offset the value of or need for formally protected, large contiguous tracts of forest, small islands of old-growth forest have clear and tangible conservation value and their protection is urgently needed to ensure this value is not lost.

## Acknowledgements

Dr. Torben Larsen is thanked profusely for steering the senior author in the direction of Ghana in the first place and for willingly offering and providing his expertise on butterfly identifications and on a multitude of other issues concerning research in the country, Dr. Joe Cobbinah, Director, Forestry Research Institute of Ghana, is thanked for bringing E. Opuni-Frimpong to the senior author's attention as an assistant on the project, and for generally helping with logistical support, especially concerning transport. His role in the successful completion of the project is gratefully acknowledged. Mr. Ntim Gyakari, Forestry Department of Ghana, offered invaluable assistance in identification of sacred groves for study. We are extremely indebted to the Chiefs and Elders of Asantemanso, Bonwire, Gyakye, and Kajease who granted access to the sacred sites. A number of different people provided invaluable help in the installation of traps at different sites, which on some occasions involved performance under less than optimal conditions! The senior author extends a very special thanks to Alfred, Emma, Boat, Atsu, and Karen. Funding was provided by The Philadelphia Zoological Society, The National Geographic Society, and internal grants from The College of New Jersey.

**Appendix 1.** Butterfly species trapped at each site.

| Species | Forest Zone–Abundance Categories | Forest Reserves | | Sacred Groves | | | |
|---|---|---|---|---|---|---|---|
| | | Bobiri | Owabi | Asante. | Bonwire | Gyakye | Kajease |
| Satyrinae | | | | | | | |
| *Bicyclus abnormis* | WF–NR | 1.27 | 2.49 | 0.10 | 0.15 | 0.00 | 0.07 |
| *B. dorothea* | ALF–VC | 0.38 | 1.25 | 0.50 | 0.15 | 0.00 | 0.49 |
| *B. funebris* | DRF–CO | 6.27 | 2.88 | 4.31 | 8.51 | 5.08 | 3.08 |
| *B. madetes* | MF–NR | 0.32 | 1.30 | 0.10 | 6.27 | 1.27 | 0.00 |
| *B. martius* | MF–CO | 0.70 | 2.22 | 0.50 | 11.19 | 4.76 | 0.07 |
| *B. procora* | WF–NR | 0.25 | 0.54 | 0.00 | 0.00 | 0.00 | 0.00 |
| *B. safitza* | GUI–NR | 0.25 | 0.22 | 0.10 | 0.15 | 0.00 | 0.28 |
| *B. sandace* | ALF–VC | 0.19 | 1.19 | 1.50 | 0.00 | 0.00 | 0.00 |
| *B. sangmelinae* | WF–NR | 1.14 | 0.60 | 0.00 | 0.00 | 0.00 | 1.05 |
| *B. taenias* | MF–CO | 0.13 | 1.19 | 0.50 | 3.43 | 2.22 | 0.00 |
| *B. vulgaris* | ALF–VC | 2.91 | 3.15 | 6.11 | 8.66 | 9.21 | 12.46 |
| *B. xeneas* | ALF–NR | 0.63 | 0.33 | 0.00 | 0.15 | 0.00 | 0.00 |
| *B. zinebi* | MF–NR | 0.51 | 1.95 | 0.00 | 6.12 | 0.00 | 0.00 |
| *Elymnias bammakoo* | MF–CO | 0.32 | 0.33 | 0.00 | 0.15 | 0.95 | 0.28 |
| *Gnophodes betsimena* | ALF–CO | 5.44 | 4.45 | 6.31 | 6.27 | 27.94 | 18.89 |
| *G. chelys* | MF–CO | 1.14 | 2.55 | 1.60 | 1.04 | 2.86 | 1.12 |
| *Hallelesis halyma* | MF–NR | 0.06 | 0.00 | 0.00 | 0.00 | 0.00 | 0.00 |
| *Melanitis leda* | UBQ–CO | 0.95 | 1.25 | 1.90 | 0.90 | 1.90 | 5.60 |
| *M. libya* | UBQ–NR | 0.25 | 0.22 | 0.60 | 0.15 | 0.00 | 0.56 |
| *Ypthimomorpha itonia* | SPE–NR | 0.00 | 0.05 | 0.00 | 0.00 | 0.00 | 0.00 |
| Limenitinae | | | | | | | |
| *Aterica galene* | ALF–CO | 4.93 | 4.40 | 4.31 | 1.64 | 0.95 | 0.98 |
| *Bebearia abesa* | MF–NR | 0.44 | 0.00 | 0.00 | 0.00 | 0.00 | 0.00 |
| *Bebearia absolon* | ALF–CO | 2.28 | 1.52 | 0.90 | 0.00 | 0.00 | 0.07 |
| *Bebearia cocalia* | ALF–CO | 1.58 | 0.92 | 0.00 | 0.00 | 0.00 | 0.07 |
| *Bebearia barce* | WF–RA | 0.00 | 0.00 | 0.40 | 0.00 | 0.00 | 0.00 |
| *Bebearia demetra* | MF–RA | 0.13 | 0.00 | 0.00 | 0.00 | 0.00 | 0.00 |
| *Bebearia mandinga* | ALF–CO | 0.89 | 0.43 | 0.20 | 0.15 | 0.00 | 0.00 |
| *Bebearia mardania* | ALF–CO | 1.64 | 1.30 | 0.20 | 0.60 | 0.00 | 0.00 |
| *Bebearia oxione* | MF–NR | 0.44 | 0.54 | 0.00 | 0.15 | 0.00 | 0.07 |
| *Bebearia phantasina* | ALF–CO | 1.20 | 0.00 | 0.00 | 0.00 | 0.00 | 0.00 |
| *Bebearia sophus* | ALF–CO | 1.71 | 6.62 | 1.20 | 0.30 | 0.63 | 0.77 |
| *Bebearia tentyris* | MF–CO | 26.69 | 1.30 | 0.20 | 0.75 | 0.63 | 0.00 |
| *Bebearia zonara* | MF–CO | 5.44 | 1.03 | 0.00 | 0.45 | 0.00 | 0.07 |
| *Euriphene amicia* | WF–NR | 0.06 | 0.00 | 0.00 | 0.00 | 0.00 | 0.00 |
| *Euriphene ampedusa* | ALF–NR | 0.19 | 0.81 | 0.20 | 1.04 | 0.32 | 0.00 |
| *Euriphene aridatha* | MF–NR | 0.51 | 0.43 | 0.20 | 0.75 | 0.00 | 0.07 |
| *Euriphene barombina* | ALF–VC | 2.34 | 2.39 | 3.31 | 1.04 | 0.32 | 0.00 |
| *Euriphene caerulea* | MF–CO | 0.06 | 0.00 | 0.00 | 0.00 | 0.00 | 0.00 |
| *Euriphene gambiae* | ALF–CO | 1.08 | 0.00 | 0.00 | 0.00 | 0.00 | 0.00 |
| *Euriphene simplex* | WF–NR | 0.89 | 0.49 | 0.70 | 0.15 | 0.00 | 0.00 |
| *Euphaedra ceres* | ALF–CO | 4.62 | 9.82 | 23.45 | 9.85 | 14.29 | 24.91 |
| *Euphaedra cyparissa* | DRF–NR | 0.00 | 0.00 | 0.00 | 0.00 | 0.00 | 0.07 |
| *Euphaedra diffusa* | DRF NR | 0.00 | 0.11 | 0.10 | 0.15 | 0.00 | 0.07 |
| *Euphaedra edwardsii* | MF–CO | 0.00 | 0.16 | 1.00 | 0.45 | 0.00 | 0.14 |
| *Euphaedra eupalus* | WF–RA | 0.25 | 3.15 | 0.00 | 0.00 | 0.00 | 0.00 |
| *Euphaedra gausape* | WF–NR | 0.06 | 0.00 | 0.00 | 0.00 | 0.00 | 0.00 |

*Appendix 1.* (Continued).

| Species | Forest Zone–Abundance Categories | Forest Reserves | | Sacred Groves | | | |
|---|---|---|---|---|---|---|---|
| | | Bobiri | Owabi | Asante. | Bonwire | Gyakye | Kajease |
| *Euphaedra harpalyce* | ALF–CO | 1.90 | 4.56 | 4.81 | 4.33 | 7.30 | 5.04 |
| *Euphaedra hebes* | WF–NR | 0.38 | 0.00 | 0.00 | 0.00. | 0.00 | 0.00 |
| *Euphaedra inanum* | MF–RA | 0.44 | 0.43 | 0.00 | 0.00 | 0.00 | 0.07 |
| *Euphaedra janetta* | ALF–CO | 1.08 | 1.41 | 1.60 | 1.04 | 0.95 | 0.42 |
| *Euphaedra medon* | ALF–CO | 10.31 | 8.90 | 16.03 | 12.84 | 4.13 | 10.43 |
| *Euphaedra phaetusa* | ALF–CO | 3.48 | 12.64 | 13.83 | 5.22 | 6.98 | 7.70 |
| *Euphaedra perseis* | WF–NR | 0.25 | 0.00 | 0.00 | 0.00 | 0.00 | 0.00 |
| *Euphaedra themis* | DRF–NR | 2.66 | 7.38 | 1.70 | 4.93 | 3.49 | 3.57 |
| *Euphaedra xypete* | MF–CO | 0.70 | 0.05 | 0.30 | 0.00 | 0.00 | 0.00 |
| *Eurytela dryope* | DRF–NR | 0.00 | 0.00 | 0.10 | 0.00 | 0.32 | 0.07 |
| *Catuna crithea* | ALF–CO | 0.13 | 0.11 | 0.00 | 0.00 | 0.00 | 0.00 |
| *Catuna* sp. B | ? | 0.00 | 0.16 | 0.00 | 0.00 | 0.00 | 0.00 |
| *Cymothoe caenis* | ALF–CO | 0.00 | 0.00 | 0.00 | 0.15 | 0.00 | 0.21 |
| *Cymothoe fumana* | MF–CO | 0.06 | 0.00 | 0.00 | 0.00 | 0.00 | 0.00 |
| *Harma theobene* | MF–CO | 0.00 | 0.11 | 0.00 | 0.15 | 0.00 | 0.07 |
| *Pseudacraea lucretia* | ALF–CO | 0.06 | 0.00 | 0.00 | 0.00 | 0.00 | 0.00 |
| Charaxinae | | | | | | | |
| *Charaxes bipunctatus* | WF–NR | 0.06 | 0.00 | 0.40 | 0.00 | 0.63 | 0.14 |
| *Charaxes cynthia* | ALF–CO | 0.00 | 0.05 | 0.00 | 0.00 | 0.32 | 0.00 |
| *Charaxes eupale* | ALF–CO | 0.00 | 0.00 | 0.00 | 0.00 | 0.32 | 0.00 |
| *Charaxes fulvescens* | ALF–CO | 0.19 | 0.00 | 0.10 | 0.00 | 0.00 | 0.00 |
| *Charaxes numenes* | ALF–NR | 0.06 | 0.11 | 0.10 | 0.00 | 0.00 | 0.14 |
| *Charaxes pleione* | ALF–CO | 0.00 | 0.00 | 0.10 | 0.00 | 0.00 | 0.00 |
| *Charaxes protoclea* | ALF–CO | 0.32 | 0.00 | 0.20 | 0.45 | 0.95 | 0.07 |
| *Charaxes tiridates* | ALF–CO | 0.00 | 0.05 | 0.00 | 0.00 | 0.00 | 0.21 |
| *Charaxes varanes* | GUI–CO | 0.06 | 0.00 | 0.00 | 0.00 | 0.32 | 0.14 |
| *Charaxes zingha* | MF–NR | 0.00 | 0.00 | 0.00 | 0.00 | 0.00 | 0.14 |
| *Palla decius* | MF–NR | 0.00 | 0.38 | 0.00 | 0.00 | 0.95 | 0.14 |
| *Pala violinitens* | MF–NR | 0.00 | 0.00 | 0.10 | 0.00 | 0.00 | 0.00 |
| Nymphalinae | | | | | | | |
| *Kallima rumia* | ALF–CO | 0.19 | 0.00 | 0.00 | 0.00 | 0.00 | 0.00 |
| *Hypolimnas anthedon* | ALF–CO | 0.00 | 0.00 | 0.10 | 0.00 | 0.00 | 0.14 |
| *Hypolimnas salmacis* | ALF–CO | 0.00 | 0.05 | 0.00 | 0.00 | 0.00 | 0.00 |
| *Junonia terea* | ALF–VC | 0.00 | 0.00 | 0.00 | 0.00 | 0.00 | 0.07 |
| Acraeinae | | | | | | | |
| *Alcrea alciope* | ALF–VC | 0.00 | 0.00 | 0.00 | 0.15 | 0.00 | 0.00 |
| Total collected | | 1581 | 1843 | 998 | 671 | 315 | 1429 |

Numbers are relative abundances based on the total sample size from that site.
Forest zone and abundance categories are from Emmel and Larsen (1997) and Larsen (2001). Forest zone designations: WF, Wet forest; MF, Moist forest; DRF, Dry forest; ALF, All forest subtypes; GUI, Guinea Savannah; UBQ, Ubiquitous; SPE, Special habitat. Abundance designations: VC, Very common; CO, Common; NR, Not rare; RA, Rare; VR, Very rare.

## References


Andre H.M., Lebrun P. and Noti M-I. 1992. Biodiversity in Africa: a plea for more data. Journal of African Zoology 106: 3–15.

Bierregaard R.O. Jr., Lovejoy T.E., Kapos V., dos Santos A.A and Hutchings R.W. 1992. The biological dynamics of tropical rainforest fragments. BioScience 42: 859–866.

Bossart J.L. and Carlton C.E. 2002. Insect conservation in America: status and perspectives. American Entomologist 48: 82–92.

Bossart J.L., Opuni-Frimpong E., Kuudaar S. and Nkrumah E. 2005. Fruit-feeding butterfly communities of Ghana: survey completeness and correlates of spatial diversity. In: Huber B.G., Sinclair B.J. and Lampe K-H. (eds), African Biodiversity: Molecules, Organisms, Ecosystems. Proceedings of the 5th International Symposium on Tropical Biology. Museum Koenig, Springer Verlag, Bonn, Germany, pp. 151–158.

Brown K.S. 1997. Diversity, disturbance, and sustainable use of neotropical forests: insects as indicators for conservation monitoring. Journal of Insect Conservation 1: 25–42.

Chazdon R.L., Colwell R.K., Denslow J.S. and Guariguata M.R. 1996. Statistical methods for estimating species richness of woody regeneration in primary and secondary rain forests of NE Costa Rica. In: Dalmeier F. and Comiskey J.A. (eds), Forest Biodiversity Research, Monitoring and Modeling: Conceptual Background and Old World Case Studies. Parthenon Publishing, Paris, pp. 289–313.

Colwell R.K. 2000. EstimateS: Statistical Estimation of Species Richness and Shared Species from Samples. Version 6.0b1. User's Guide and application published at: http://viceroy.eeb.uconn.edu/estimates.

Colwell R.K. and Coddington J.A. 1994. Estimating terrestrial biodiversity through extrapolation. Philosophical Transactions of the Royal Society of London: Series B 345: 101–118.

Connor E.F. and McCoy E.D. 1979. The statistics and biology of the species-area relationship. American Naturalist 113: 791–833.

Daily G.C. and Ehrlich P.R. 1995. Preservation of biodiversity in small rainforest patches, rapid evaluations using butterfly trapping. Biodiversity and Conservation 4: 35–55.

D'Abrera B. 1981. Butterflies of the Afrotropical Region. Landsdowne Press, Melbourne.

D'Abrera B. 1997. Butterflies of the Afrotropical Region, Part 1: Papilionidae, Pieridae, Acraeidae, Danaidae, Satyridae. Hill House Publishers, Melbourne.

D'Abrera B. 2001. The Concise Atlas of Butterflies of the World. Hill House Publishers, Melbourne.

DeVries P. 1988. Stratification of fruit-feeding nymphalid butterflies in a Cost Rican rainforest. Journal of Research on the Lepidoptera 26: 98–108.

DeVries P., Murray D. and Lande R. 1997. Species diversity in vertical, horizontal, and temporal dimensions of a fruit-feeding butterfly community in an Ecuadorian rainforest. Biological Journal of the Linnean Society 62: 343–364.

DeVries P., Walla T.R. and Greeney H.F. 1999. Species diversity in spatial and temporal dimensions of fruit-feeding butterflies from two Ecuadorian rainforests. Biological Journal of the Linnean Society 68: 333–353.

Emmel T.C. and Larsen T.B. 1997. Butterfly diversity in Ghana, West Africa. Tropical Lepidoptera 8(supplement 3): 1–13.

FAO 1999. Second Interim Report on the State of Tropical Forests by Forest Resources Assessment 1990 Project. Tenth World Forestry Congress, Paris.

Fermon H., Waltert M., Larsen T.B., Dall'Asta U. and Muhlenberg M. 2000. Effects of forest management on diversity and abundance of fruit-feeding Nymphalid butterflies in south-eastern Cote d'Ivoire. Journal of Insect Conservation 4: 173–189.

Fermon H., Waltert M. and Muhlenberg M. 2003. Movement and vertical stratification of fruit-feeding butterflies in a managed West African rainforest. Journal of Insect Conservation 7: 7–19.

Fischer J. and Lindenmayer D.B. 2002. Small patches can be valuable for biodiversity conservation: two case studies on birds in southeastern Australia. Biological Conservation 106: 129–136.

GACON. Jachie Sacred Grove. Ghana Association for the Conservation of Nature in cooperation with the Global Environmental Facility. Unknown publication date.

Gallagher E. 1999. COMPAH Documentation. User's Guide and application published at: http://www.es.umb.edu/edgwebp.htm.

Gaston K.J. 1998. Rarity as double jeopardy. Nature 394: 229–230.

Gotelli N.J. and Colwell R.K. 2001. Quantifying biodiversity: procedures and pitfalls in the measurement and comparison of species richness. Ecology Letters 4: 379–391.

Gotelli N.J. and Entsminger G.L. 2003. EcoSim: Null Models Software for Ecology. Version 7. Acquired Intelligence Inc. & Kesey-Bear, Burlington, VT. http://homepages.together.net/~gentsmin/ecosim.htm.

Goudie A. 2000. The Human Impact on the Natural Environment. 5th edn. MIT Press, Cambridge, Massachusetts.

Hall J.B. and Swain M.D. 1981. Distribution and Ecology of Vascular Plants in a Tropical Rain Forest: Forest Vegetation of Ghana. W. Junk Publishers, The Hague.

Hamer K.C., Hill J.K., Mustaffa B.N., Sherratt T.N., Maryati M. and Chey V.K. 2003. Ecology of butterflies in natural and selectively logged forests of northern Borneo: the importance of habitat heterogeneity. Journal of Applied Ecology 40: 150–162.

Hanski I. and Gilpin M. 1996. Metapopulation Biology: Ecology, Genetics and Evolution. Academic Press, London.

Hastings A. and Harrison S. 1994. Metapopulation dynamics and genetics. Annual Review of Ecology and Systematics 25: 167–188.

Hawthorne W.D. 1988. The flora and vegetation of Ghana's forests. In: Ghana Forest Inventory Proceedings. Overseas Dev. Agency/Ghana Forestry Dept, Accra, pp. 8–13.

Hawthorne W.D. and Abu-Juam M. 1995. Forest Protection in Ghana with Particular Reference to Vegetation and Plant Species. IUCN, Gland, Switzerland.

Heck K.L. Jr., Van Belle G. and Simberloff D. 1975. Explicit calculation of the rarefaction diversity measurement and the determination of sufficient sample size. Ecology 56: 1459–1461.

Hecq J. 1997. Euphaedra. Lambillionea, Tervuren.

Hecq J. 1999. Euphaedra. In: Bauer E. and Frankenbach T. (eds), Butterflies of the World, Part 4, Nymphalidae III. Goeke and Evers, Keltern, Germany.

Hecq J. 2000. Bebearia. In: Bauer E. and Frankenbach T. (eds), Butterflies of the World, Part 9, Nymphalidae IV. Goeke and Evers, Keltern, Germany.

Hecq J. 2002. Euriphene. In: Bauer E. and Frankenbach T. (eds), Butterflies of the World, Part 15, Nymphalidae VI. Goeke and Evers, Keltern, Germany.

Henning S.F. 1988. The Charaxinae Butterflies of Africa. Aloe Books, Capetown.

Hill J.K., Hamer K.C., Tangah J. and Dawood M. 2001. Ecology of tropical butterflies in rainforest gaps. Oecologia 128: 294–302.

Johnson C.N. 1998. Species extinction and the relationship between distribution and abundance. Nature 394: 272–274.

Krebs C.J. 1999. Ecological Methodology. Addison-Welsey, Menlo Park, CA.

Kremen C. 1992. Assessing the indicator properties of species assemblages for natural areas monitoring. Ecological Applications 2: 203–217.

Kunin W.E. and Gaston K. 1997. The Biology of Rarity: Causes and Consequences of Rare-common Differences. Chapman & Hall, New York.

Lande R. 1988. Genetics and demography in biological conservation. Science 241: 1455–1469.

Larsen T.B. 1994. The Butterflies of Ghana – Their Implications for Conservation and Sustainable Use. Report to IUCN and Dept. of Game & Wildlife, Ghana.

Larsen T.B. 2001. The Butterflies of Ankasa/Nini-Suhien and Bia Protected Area Systems in Western Ghana. Protected areas development programme. Western Region Ghana. ULG Northumbrian Ltd.

Larsen T.B. 2005. Butterflies of West Africa. Apollo Books, Svendborg, Denmark.

Laurance W.F. 1997. Introduction. In: Laurance W.F. and Bierregaard R.O. Jr. (eds), Tropical Forest Remnants: Ecology, Management, and Conservation of Fragmented Communities. University of Chicago Press, Chicago, Illinois, pp. 1–2.

Laurance W.F., Ferreira L.V., Rankin-de Merona J.M. and Laurance S.G. 1998. Rain forest fragmentation and the dynamics of Amazonian tree communities. Ecology 79: 7032–7040.

Lebbie A.R. and Freudenberger M.S. 1996. Sacred groves in Africa: forest patches in transition. In: Forest Patches in Tropical Landscapes. Island Press, Washington, D.C, pp. 300–326.

Longino J.T., Coddington J. and Colwell R.K. 2002. The ant fauna of a tropical rain forest: estimating species richness three different ways. Ecology 83: 689–702.

MacArthur R.H. and Wilson E.O. 1967. The Theory of Island Biogeography. Princeton Univ. Press, Princeton, New Jersey.

Magurran A. 2004. Measuring Biological Diversity. Blackwell Publishing, Oxford.

May R.M. and Stumpf M.P.H. 2000. Species-area relations in tropical forests. Science 290: 2084.

Murcia C. 1995. Edge effects in fragmented forests: implications for conservation. Trends in Ecology and Evolution 10: 58–62.

Myers N. 1990. The biodiversity challenge: expanded hot-spot analysis. The Environmentalist 10: 243–256.

Myers N., Mittermeier R.A., Mittermeier C.G., da Fonseca G.A.B. and Kent J. 2000. Biodiversity hotspots for conservation priorities. Nature 403: 853–858.

Ntiamoa-Baidu T. 2001. Indigenous versus introduced biodiversity conservation strategies: the case of protected area systems in Ghana. In: Weber W., White L.J.T., Vedder A. and Naughton-Treves L. (eds), in African Rain Forest Ecology & Conservation. Yale Univ. Press, New Haven, CT, pp. 385–394.

Owen D.F. 1971. Tropical Butterflies: The Ecology and Behaviour of Butterflies in the Tropics with Special Reference to African Species. Clarendon Press, London.

Purvis A. and Hector A. 2000. Getting the measure of biodiversity. Nature 405: 212–219.

Rabinowitz D. 1981. Seven forms of rarity. In: Synge H. (eds), The Biological Aspects of Rare Plant Conservation. Wiley, New York, pp. 205–217.

Rosenzweig M.L. 1995. Species Diversity in Space and Time. Cambridge University Press, New York.

SAS Institute Inc 1990. SAS/STAT user's guide. Version 6.

Schoener T.W. 1976. The species–area relationship within archipelagoes: models and evidence from island birds. Proceedings of the XVI International Ornithological Congress 6: 629–642.

Shahabuddin G. and Terborgh J.W. 1999. Frugivorous butterflies in Venezuelan forest fragments: abundance, diversity and the effects of isolation. Journal of Tropical Ecology 15: 703–722.

Simberloff D. 1978. Use of rarefaction and related methods in ecology. In: Dickson K.L., Cairns J. Jr. and Livingston R.J. (eds), Biological Data in Water Pollution Assessment: Quantitative and Statistical Analyses. Am. Soc. Testing Materials, Philadelphia, pp 150–165.

Skinner K.E. 2000. Past, present, and future of rangeland grasshopper management. Rangelands 22: 24–28.

Southwood T.R.E. and Henderson P.A. 2000. Ecological Methodology. Oxford, Blackwell Science Ltd, 464 pp.

Stork N.E., Srivastava D.S., Watt A.D. and Larsen T.B. 2003. Butterfly diversity and silvicultural practice in lowland rainforests of Cameroon. Biodiversity and Conservation 12: 387–410.

Tokeshi M. 1993. Species abundance patterns and community structure. Advances in Ecology Research 24: 112–186.

Tscharntke T., Steffan-Dewenter I., Kruess A. and Carsten T. 2002. Contribution of small habitat fragments to conservation of insect communities of grassland – cropland landscapes. Ecological Application 12: 354–363.

Turner I.M. 1996. Species loss in fragments of tropical rain forest: a review of the evidence. Journal of Applied Ecology 33: 200–209.

UNESCO 2003. The Importance of Sacred Natural Sites for Biodiversity Conservation. Paris.

Vermeij G.J. 1993. Biogeography of recently extinct marine species: implications for conservation. Conservation Biology 7: 391–397.

Wagner M.R. and Cobbinah J.R. 1993. Deforestation and sustainability in Ghana: the role of tropical forests. Journal of Forestry 6: 35–39.

Whitmore T.C. 1997. Tropical forest disturbance, disappearance, and species loss. In: Laurance W.F. and Bierregaaard R.O. Jr. (eds), Tropical Forest Remnants: Ecology, Management, and Conservation of Fragmented Communities. Univ. of Chicago Press, Chicago, Illinois, pp. 3–12.

Whitmore T.C. 1998. An Introduction to Tropical Rain Forests. 2nd ed.. Oxford Univ. Press, New York.

Willot S.J., Lim D.C., Compton S.G. and Sutton S.L. 2000. Effects of selective logging on the butterflies of a Bornean rainforest. Conservation Biology 14: 1055–1065.

Wolda H. 1981. Similarity indices, sample size, and diversity. Oecologia 50: 296–302.

Biodiversity and Conservation (2006) 15:347–369
DOI 10.1007/s10531-005-2785-x 

# Ecological impact assessment of the Aznalcóllar mine toxic spill on edaphic coleopteran communities in the Guadiamar River basin (Southern Iberian Peninsula)☆

ANA MARÍA CÁRDENAS* and JUAN M. HIDALGO
*Departamento de Zoología, Universidad de Córdoba, Campus Universitario Rabanales, Edif. Darwin, E-14071 Córdoba (Spain); *Author for correspondence (e-mail: ba1cataa@uco.es; phone: +34-57-218604; fax: +34-57-212002)*

Received 20 April 2004; accepted in revised form 14 February 2005

**Key words:** Coleoptera, Edaphic, Environmental impact, Guadiamar River basin, Protected areas

**Abstract.** This study is part of the Restoration Program of the animal populations from the Guadiamar River basin. The main objective is to evaluate the current state of populations of the most representative groups of edaphic coleoptera in the Guadiamar River basin. In 1998, a breach in the Aznalcóllar mines retaining wall (Southern Iberian Peninsula) led to toxic sludge pouring out into this area. Systematic samples were taken from monitoring plots, scattered throughout the affected area. In order to provide an overall assessment of the fauna collected from this area in relation to adjacent areas, ecological indices were applied, and temporal course and biogeographical comparisons were also made.The results indicate that, overall, the edaphic fauna has not been significantly affected by the spillage; however, a faunistic poverty was also detected, which becomes increasingly evident closer to the accident site. Moreover, temporal evolution analysis suggests that the most affected areas are undergoing a re-colonization process, although this varies widely between species and higher taxa.

## Introduction

In spring 1998, a breach formed in the tailing pond retaining wall of the Aznalcóllar mine, releasing an estimated 6 $hm^3$ of sludge containing pyrite and acidic waters into the Guadiamar River basin. A deluge of toxic flowed onto the floodplain, covering 4600 ha along a 62 km stretch of land with a depth of approximately 1.5 m. The toxic sludge was diverted by a dyke just before reaching the boundaries of Doñana National Park, declared a Biosphere Reserve by the UNESCO in 1980 (Anguas et al. 2002).

After the disaster, a series of emergency measures were taken in order to clean up and remove the sludge and a Plan was instigated for the integral management of the river basin: *Proyecto de Restauración del Corredor Verde y*

☆This research has been financed by the Project 'Caracterización, restauración y conservación de las comunidades animales de la cuenca del río Guadiamar' (Characterization, Re-establishment and Conservation of the animal communities from the Guadiamar River basin). Consejería de Medio Ambiente. Junta de Andalucía (Spain).

*del Guadiamar* (Green Corridor and Guadiamar Restoration Project). This Plan included the remediation of damaged fluvial and terrestrial ecosystems and the rehabilitation of the affected area into an ecological corridor between two singular ecosystems: the Sierra Morena and Doñana National Park (PICOVER 1999). Aguilar et al. (2003) and Toja et al. (2003) include chemical and ecotoxicological data that provide additional information about the extent of the environmental impact.

PICOVER (2003) describes the main sludge removal operations and action plans implemented in the restoration in the Guadiamar basin. These initiatives were carried out from the perspective of the environmental connectivity of protected areas in the Mediterranean basin (Castro, 2003; García and Montes 2003).

Given this context, the establishment of ecological corridors based on the restoration of fluvial sites has become an increasingly important conservation strategy (Naiman et al. 1993; Naiman and Roger 1997; Montes 1999).

Because soil macrofauna tends to be highly affected by environmental disturbances, this study – which forms part of the aforementioned Plan – examines the populations of edaphic coleoptera. Thus, the main aim of our study is to evaluate the current state of certain edaphic coleoptera populations in the Guadiamar River basin and other adjacent ecosystems, which can be considered source areas for future colonization of the proposed Green Corridor.

The term 'edaphic coleoptera' covers an extremely broad range of species, thus a comprehensive study is unfeasible. Therefore, for our results to be indicative, we have selected three taxa: Scarabaeoidea, Carabidae and Tenebrionidae which are highly sensitive to the environmental changes, thus they are considered effective biological indicators (Scholtz and Holm 1995; Boscaini et al. 2000).

## Study area

The Guadiamar River is located in the Southern Iberian Peninsula and is one of the main tributaries of the right bank of the Guadalquivir River. The basin is approximately 65 km in long and it constitutes a functional system of two interconnected natural areas (Montes 1999): Doñana National Park and the Sierra Morena, both of which are of great ecological value. Detailed information on the abiotic features referred to Doñana and S. Morena can be found in García-Canseco et al. (2002) and Cárdenas and Bach (1988a), respectively.

In the Guadiamar area, the geological horizon of the upper reach of the basin contains sulphide-rich mineral deposits which support the Aznalcóllar mining center. The predominant vegetation in the area comprises open oak meadowlands interspersed with Pine forests and Eucalyptus Groves, as well as shrubs (strawberry and mastic trees) and scrubland (rock rose and oleander). As a consequence of intense human managing, there are few of these vegetal formations left in the middle section of the basin, having been replaced by cereal, oleaginous, cotton and fruit farming. Rice fields and unmanaged marshlands predominate in the lower reach.

The Guadiamar basin has sub-humid Mediterranean climate, with an annual rainfall of approximately 600 mm. The location of the study area is shown in Map 1.

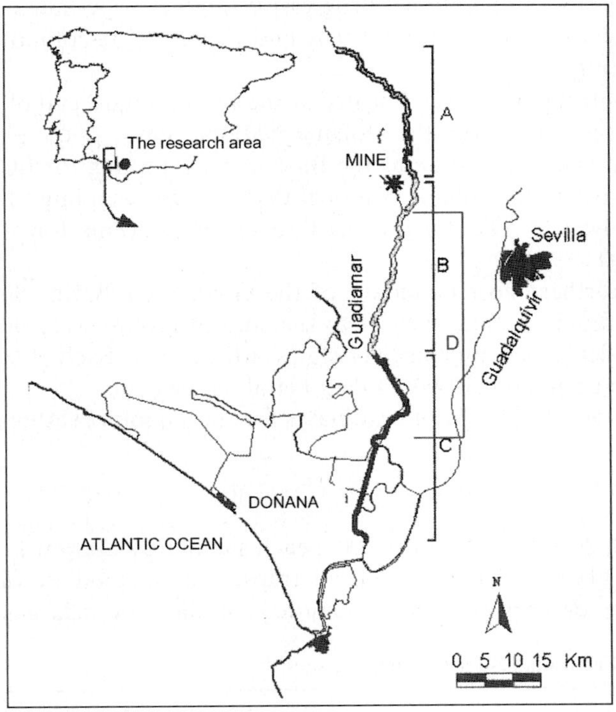


*Map 1.* The research area. A: Upper reach; B: Medium reach and C: Lower reach; D: Damaged section.

## Materials and methods

### *Sampling sites*

Since this research work was carried out by a multidisciplinary group, several common reference areas – Monitoring Plots – were established by the Research Program of the Green Corridor. These plots were considered representative of the different morphological and eco-dynamic conditions of the Guadiamar fluvial system. Comprehensive information about the area and the study sites can be found in Montes et al. (2003).

Four of these research sites (PS), distributed throughout the damaged area, were selected for sampling. Additionally, samples were taken from four plots called Biocenters (B) which were located near to the PS, but which were unaffected by the spillage and could, therefore, be considered a potential source of species moving towards the disturbed area. These plots were BDV (Oak Meadows), BOA (Olive Grove), BPA (Pine forest) and BEA (Eucalyptus Grove).

Moreover, two types of control plots were selected for comparison:

Type 1 control plot (PT): sampling site where toxic sludge was not removed during the clean-up operations. Progressive samples were taken to determine both species' tolerance to toxic levels and degree of re-colonization, if and when occurring.

Type 2 control plot (PCD): Located at the southernmost end of the Corridor and in forested areas from the Doñana National Park. These plots show the same predominant vegetation as the Biocenters, excepting for BOA (there are no Olive Grove in the Doñana National Park). These sampling sites are: PCD1 (Oak Meadows), PCD2 (Eucalyptus Grove), PCD3 (Pine forest) and PCD4 (River forest).

For the northernmost boundary of the Guadiamar Basin (Sierra Morena mountains) faunistic data referred to the studied groups were taken from literature: Cárdenas and Bujalance (1985); Cárdenas and Bach (1988a, b, 1989); Cárdenas and Hidalgo (1998); Hidalgo et al. (1998a, b).

Table 1 shows the UTM. co-ordinates for the sampling stations.

## *Sample design*

From March 2000 to September 2001, each sampling station was visited twice monthly to place and remove pit-fall traps. This method is widely used in ecology for descriptive and functional studies of edaphic arthropod

*Table 1.* UTM co-ordinates for the sampling sites.

| Sampling Sites | UTM H. 30 Co-ordinates | |
|---|---|---|
| | *X* | *Y* |
| Monitoring plots (PS) | | |
| PS1 | 216.150 | 4.151.781 |
| PS2-3 | 214.361 | 4.143.812 |
| PS4 | 211.264 | 4.134.032 |
| PS6-7 | 212.933 | 4.124.275 |
| Biocenter stations (B) | | |
| BDV (Oak Meadow) | 209.531 | 4.126.293 |
| BOA (Olive Grove) | 212.391 | 4.132.747 |
| BPA (Pine forest) | 214.323 | 4.129.313 |
| BEA (Eucalyptus Grove) | 215.208 | 4.126.998 |
| Control plots | | |
| Type 1 (PT): toxic sludge not removed | | |
| PT2 | 210.320 | 4.127.203 |
| Type 2 (PC): located at the southernmost end of the Corridor (Doñana National Park) | | |
| PCD1 (Oak Meadow) | 193.503 | 4.113.287 |
| PCD2 (Eucalyptus Grove) | 191.050 | 4.112.132 |
| PCD3 (Pine forest) | 188.476 | 4.104.626 |
| PCD4 (River forest) | 179.893 | 4.119.246 |

populations (Grümm 1980; de los Santos et al. 1982). A series of eight pit-falls traps were placed at each sampling-site. The traps consisted of cylindrical recipients with a capacity of 1000 cc and were buried up to the top end and partially covered to prevent flooding. The traps were baited with different substances to attract specimens: commercial acetic acid (two traps), meat (two traps), carnivorous (two traps) or herbivorous (two traps) excrements. Ethylic alcohol (70%) was used to preserve the specimens.

### *Statistical methods*

In order to characterize and compare the communities, the Richness ($R_1$, Margalef), Diversity ($H'$, Shannon) and Evenness($E_1$, Pielou) indices were calculated. Although recent literature (Belaoussoff et al. 2003) analyses the effectiveness of these indices in assessing disturbance on assemblages of ground beetles and concludes that they are not always useful, no information is available regarding the validity or rejection of these methods for assessing changes in dung beetle and darkling beetle communities. Hence, we decided to apply the diversity indices that are most frequently used by ecologists when studying insect populations (Southwood 1991).

We also calculated the rarefraction curve (Hurlbert) in order to compare our results with those of the adjacent areas. This procedure is commonly used when sample sizes vary as a result of applying different methodologies. Details about description and application of this statistical method can be found in Ludwig and Reynolds (1988).

The affinity of the communities was determined by calculating the Euclidean distance coefficients and applying a dissimilarity function (Ludwig & Reynolds, *op. cit.*) to the quantitative matrix in order to obtain the corresponding dendrograms. This procedure considers both presence–absence and abundance of species.

### *Taxonomical nomenclature and biogeographical criterion*

Regarding the taxonomical treatment of the species we followed to Serrano (2003) for Carabidae, Baraud (1992) for Scarabaeoidea and those of Porta (1923), Gebien (1938–42, 1942–44) and Español (1954) for Tenebrionidae.

Considering that the current ecological situation of a faunistic territory can be investigated on a mesoscale by the biogeographical analysis of its components (Vargas 1993; Adrian & Scholtz, 2001) and that comparisons from a biogeographical point of view allow to evidence significant changes in the faunistic composition of disturbed areas, a biogeographical analysis of the different research areas was also performed.

The term 'biogeographical element' refers to the taxonomic units which are characteristic of a territory. The Holdhaus (1929) criteria was used to assign the carabids and tenebrionids to different biogeographical categories. The distributions proposed by La Greca (1964) were used to classify the scarabaeids.

## Results

### *Faunistic characterization*

The spatial distribution of the edaphic coleoptera species captured in the research area is displayed in Appendix 1.

An initial assessment of the results included in these tables indicates that the edaphic coleoptera community of the Guadiamar basin is not very complex. We recorded 61 Carabidae species, 55 Scarabaeoidea species and 28 Tenebrionidae species and they can be largely considered scarce, accidental or only occasionally present.

Diversity H', Richness (number of species and $R_1$) and Evenness $E_1$ indexes were calculated to detect differences between the fauna collected from the Guadiamar basin stations (PS) – and therefore more affected by the spill – and the fauna collected in the Biocenters.

The results in Table 2 indicate that the Biocenters lodge more diverse Scarabaeoidea and Tenebrionidae communities with greater specific richness. In contrast, the Carabids were more abundant in the basin stations due to the hygrophilous nature of the species of this group and the humid conditions of the sample plots (PS).

Evenness values were coherent with those of Diversity, with the exception of the Scarabaeids, whose communities also show the dominant effect of some

*Table 2.* Values of N° (species number), Richness ($R_1 > 1$), Diversity ($H'$) and Evenness ($E_1$) Indices for the sampling sites located in the Guadiamar River basin: PS, PT and Biocenters.

| | PS1 | PS2-3 | PT2 | PS4 | PS6-7 | BDV | BOA | BPA | BEA |
|---|---|---|---|---|---|---|---|---|---|
| Carabidae | | | | | | | | | |
| N° | 9 | 22 | 22 | 11 | 33 | 10 | 10 | 9 | 7 |
| $R_1$ | 2.77 | 4.34 | 3.48 | 2.03 | 5.54 | 1.56 | 2.26 | 1.61 | 1.16 |
| $H'$ | 1.71 | 2.45 | 1.23 | 1.23 | 2.83 | 1.28 | 1.58 | 0.74 | 1.14 |
| $E_1$ | 0.78 | 0.79 | 0.40 | 0.51 | 0.81 | 0.55 | 0.69 | 0.34 | 0.58 |
| Scarabaeoidea | | | | | | | | | |
| N° | 7 | 10 | 4 | 10 | 13 | 29 | 19 | 13 | 22 |
| $R_1$ | 1.60 | 2.10 | 1.44 | 2.96 | 3.73 | 4.23 | 3.22 | 2.21 | 3.48 |
| $H'$ | 1.38 | 1.63 | 1.32 | 1.72 | 2.30 | 2.31 | 1.72 | 1.06 | 1.95 |
| $E_1$ | 0.71 | 0.71 | 0.95 | 0.75 | 0.90 | 0.68 | 0.58 | 0.41 | 0.63 |
| Tenebrionidae | | | | | | | | | |
| N° | 4 | 4 | 5 | 7 | 7 | 8 | 12 | 6 | 14 |
| $R_1$ | 1.00 | 0.83 | 0.75 | 1.61 | 1.21 | 1.47 | 3.61 | 2.08 | 2.15 |
| $H'$ | 1.16 | 1.13 | 1.08 | 1.23 | 0.62 | 1.43 | 2.33 | 1.67 | 1.29 |
| $E_1$ | 0.84 | 0.82 | 0.67 | 0.63 | 0.32 | 0.69 | 0.94 | 0.93 | 0.49 |

very abundant species such as *Typhaeus momus*, *Onthophagus punctatus* or *Aphodius castaneus*.

These general conclusions can be examined in greater depth by applying the same indices to the quantitative data obtained from the various sample sites.

The results obtained for the **Carabidae** fauna (Table 2) indicate that station **PS1**, which is adjacent to the spill, and thus the most affected, is also the poorest in species. In contrast, **PS7** yields the highest diversity and richness values, as it lodges more than 50% of the species recorded in all the studied area. The fauna found in this station is very heterogeneous, with typically riparian species (*Paranchus albipes*, *Polystichus connexus*, *Chlaenius spoliatus* or *Agonum marginatum*) cohabiting alongside species that are associated with croplands (*Carterus fulvipes* or *Harpalus distinguendus*) or irrigation canals (genus *Brachinus*).

The Diversity, Richness and Evenness indices (Table 2) for the **Scarabaeoidea** indicate, as expected, that a richer and better structured coprophagous community dwells in the BDV station (meadow).

In the sample plots (PS), the taxocenosis of the dung beetles is still very simple, revealing a dominance of generalist species with high ecological plasticity such as *Onthophagus taurus*, *O. furcatus*, *Aphodius ghardimaouensis* or *A. castaneus*. Besides coprophagous Scarabaeids, the occasional presence of floricolous species has been detected (*Netocia morio*, *N. oblonga*, *N. cuprea*, *Tropinota squalida*, *Valgus hemipterus*) as well as rhizophagous (*Anomala ausonia*, *A. quadripunctata*, *Pentodon algerinum* or *P. idiota*), saprophagous (*Pleurophorus caesus*, *P. mediterranicus*) or sapro-necrophagous (*Omorgus suberosus*).

In sample plot **PT**, where sludge has not been removed, the colonization process is still in the beginning stages.

The **Tenebrionidae** fauna is poor in terms of Richness and Diversity (Table 2), especially in the **PS** plots, where only *Gonocephalum rusticum*, *Scleron armatum* and *Tentyria platyceps* are well represented. Paradoxically, they are even more abundant in the plot containing toxic sludge (**PT**).

The specific composition of the Biocenters differs greatly. The **BEA** station, where dense populations of conspicuous species such as *Erodius goryi*, *Tentyria platyceps* and *Pimelia costata* were detected, appears to be more favorable to Tenebrionids and could be considered the source of dispersion towards the Corridor. The highest levels of Richness, Diversity and Evenness were recorded at the Olive Grove Biocenter, although populations of different species were not as numerous.

Interestingly, the presence of another peculiar darkling beetle, *Akis granulifera*, was detected in the Meadow Biocenter and has been found to be strictly associated with this type of environment in the study area.

### *Temporal course of the edaphic community*

The temporal course of the edaphic community varies for the three Taxa studied (Appendix 2). The Tenebrionidae are clearly uniform in their temporal dimension due to the wide phenology of the majority of its species and the low specific turnover rate that keeps the populations at stable richness and diversity levels. Tenebrionids are less abundant only during the cold season when recorded species are accidental.

Nevertheless, the seasonal nature that characterizes the biological rhythm of the majority of carabids and scarabaeids leads to fluctuations in the temporal dynamics, spring being the most favorable season.

Dung beetles can be considered as typically spring fauna, comprising species of the *Aphodius* genus (*A. hydrochaeris*, *A. baraudi* and *A. merdarius*) and other groups with a broad phenology but which intensify their activity during specific phases of the annual cycle (i.e. species of the *Onthophagus* genus in summer or species of the Geotrupidae family in autumn and/or winter). It should also be noted that some ephemeral species populations increase during the summer–autumn interval (i.e. *Aphodius castaneus* and *A. annamariae*).

During spring and summer, the ground beetle community is dominated by typically riparian species such as Bembidiinae and Platyninae, while in autumn and winter silvicolous (*Calathus granatensis* or *Steropus globosus*) or generalist species (*Pseudophonus rufipes*) acquire greater significance.

When comparing the specific annual abundance (Appendix 2) of each taxa (i.e. spring 2000 with spring 2001), no noticeable differences were observed and can seldom be explained. No clear increase in terms of abundance was observed during the sampling period, suggesting that the edaphic communities are not undergoing a significant recovery process.

However, the reverse happens for certain species such as *P. rufipes* or *S. globosus* whose abundance values almost doubled in the second sampling year.

### *Comparison with the adjacent areas*

Due to differences in the time and the sampling methods used to collect data from Sierra Morena, an in-depth comparative analysis is not viable. Instead, we have provided a general overview by calculating the corresponding rarefraction curves.

Figure 1 shows the rarefraction curves obtained for Carabidae, Tenebrionidae and Scarabaeoidea in the study sites. The results indicate that when examining the Guadiamar basin as a whole, the Carabidae display intermediate values, the Scarabaeoidea have proximate values and Tenebrionidae have higher values than in adjacent areas (Doñana and Sierra Morena), thus suggesting that the toxic spill has not had a significant impact on the area.

In the dendrograms shown in Figure 2, data from the PS stations (damaged area) and the Biocenters (B) are considered independently. The Sierra Morena is segregated from the other sites for all three taxonomical groups. There was a strong affinity observed between Doñana, the PS and the Biocenters for Tenebrionids. The Carabidae from the Basin (PS) and close areas (B) result associated in the dendrogram, the absence of species determining the degree of association.

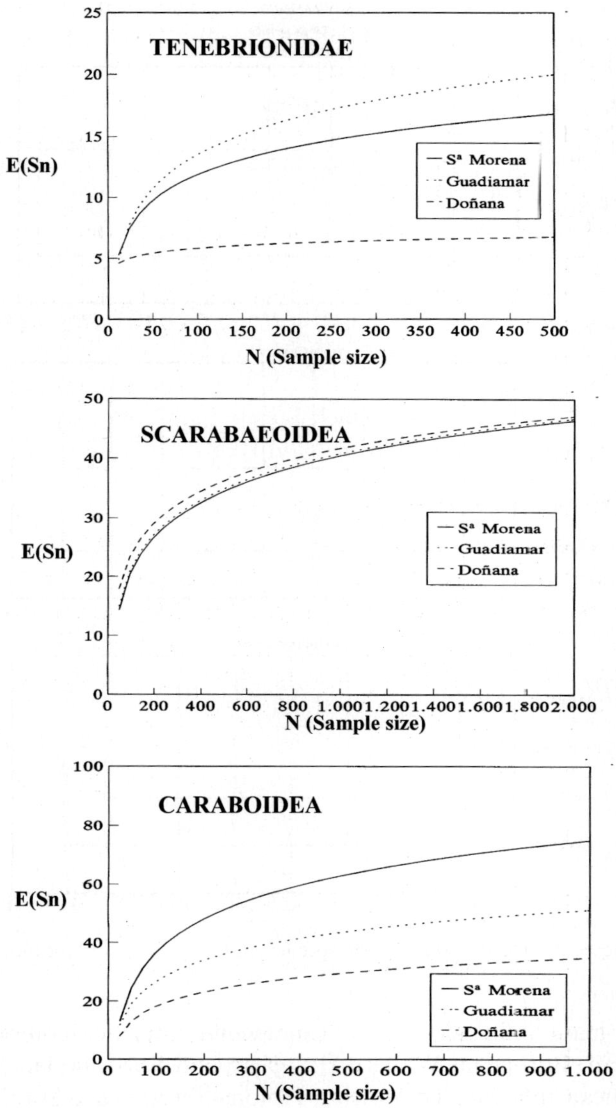


*Figure 1.* Rarefraction curves obtained for the different studied Taxa.

As for the Scarabaeoidea, the Basin (PS) is separate from Doñana and the Biocenters, which form a group with high specific affinity.

*Biogeographical analysis*

The percentage of different biogeographical categories for the Carabid collected from the research areas is shown in Table 3. An analysis of the data

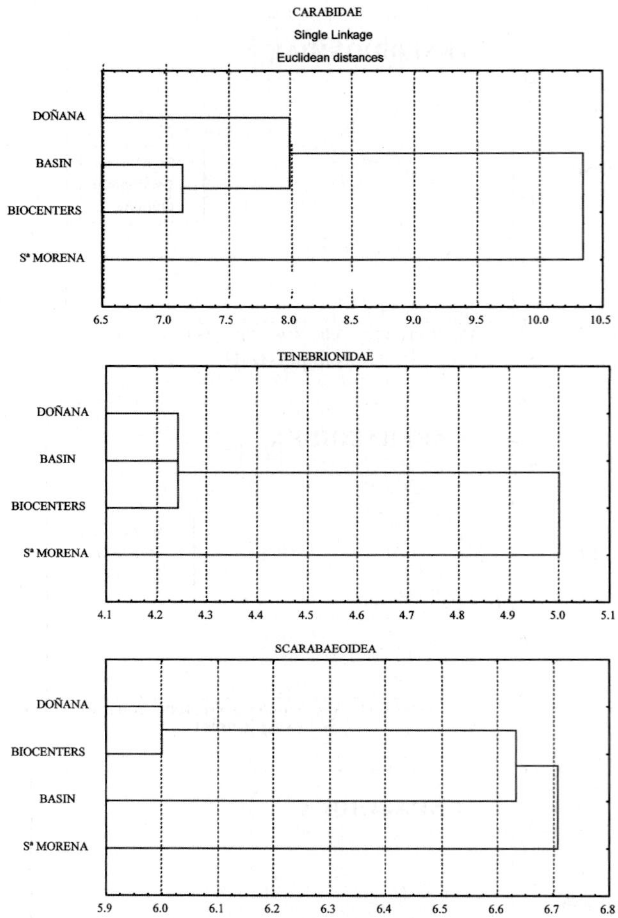


*Figure 2.* Dendrograms of the specific clustering of studied localities.

reveals two main influences: (a) a **Continental** influence comprising Euro-Mediterranean, European, Western-European and Eurosiberian elements; (b) a **Mediterranean** influence comprising Holomediterranean, South-Mediterranean and West-Mediterranean elements. The remaining elements have either a fairly restricted or '**Endemic**' distribution (i.e. Iberian-Maghrebi, Betic-Riff, Iberian, Lusitanian) or **wide distribution** patterns, as in the case of Cosmopolitan, Palaeartic or Western-Palaeartic elements.

The Carabid fauna of the Guadiamar displays a predominantly Mediterranean influence (57.4% of the species), a moderate abundance of endemic (19.7%) and widely distributed elements (18%) and a low continental (European) component (4.9%).

The data from Doñana National Park also show a marked Mediterranean influence (59.7% of the species) as compared to Continental elements (3.2%).

*Table 3.* Percentage of each biogeographical category for Carabidae, Tenebrionidae and Scarabaeoidea in the Guadiamar River basin and the adjacent areas: Doñana National Park and Sierra Morena mountains.

| Biogeographical category | Guadiamar | Doñana | Sierra Morena |
|---|---|---|---|
| | % | % | % |
| Carabidae | | | |
| Cosmopolitan | 1.6 | 1.6 | 5.2 |
| Palaeartic | 18.0 | 17.7 | 13.4 |
| European | 4.9 | 3.2 | 6.1 |
| Mediterranean | 57.4 | 59.7 | 51.1 |
| Iberian-Maghrebi | 8.2 | 9.7 | 10.0 |
| Iberian endemism | 11.5 | 8.1 | 14.2 |
| Tenebrionidae | | | |
| Cosmopolitan | 7.1 | | |
| Palaeartic | 3.6 | | 10.5 |
| European | 3.6 | 12.5 | |
| Mediterranean | 21.4 | 12.5 | 10.5 |
| Iberian-Maghrebi | 10.7 | | 15.8 |
| Iberian endemism | 53.5 | 75.0 | 63.2 |
| Scarabaeoidea | | | |
| Cosmopolitan | 3.6 | 5.1 | 3.6 |
| Palaeartic | 7.1 | 11.9 | 10.7 |
| European | 16.0 | 8.6 | 17.8 |
| Mediterranean | 44.5 | 46.4 | 46.3 |
| Iberian-Maghrebi | 7.1 | 8.5 | 8.4 |
| Iberian endemism | 21.5 | 18.6 | 12.4 |

The remaining fauna is comprised of endemisms (17.8%) and widely distributed elements (19.3%).

Our results were similar to those found by Cárdenas and Bach (1988b). In both cases, the Mediterranean component is the largest with values greater than 50%, followed by elements with restricted distribution (near 20% are endemisms). However, the continental influence is lower in the Guadiamar than in the Sierra Morena.

Table 3 also shows the biogeographical results for Tenebrionidae species. A predominance of elements with a restricted distribution was observed, making up 64.2% of the species in the Guadiamar basin, 75% in Doñana and as high as 79% in the Sierra Morena. An in-depth analysis of the endemic component, which is the most significant for this Taxa, shows that of all the Iberian elements found in the Guadiamar basin, there is a slight predominance of strictly Iberian elements as compared to those with a more restricted distribution, such as the Betic elements (25%). This result is also found, even intensified, in both Doñana and Sierra Morena.

The biogeographical analysis for the Scarabaeid fauna in the Guadiamar basin reveals a clear proliferation of Mediterranean elements comprising 45% of the species, with a noticeable presence (28.6%) of species having restricted distribution (Iberian endemism and Iberian-Maghrebi elements). These results are in line with those obtained for the surrounding reference areas: the patterns observed in Doñana and the Sierra Morena also show a clear dominance of the Mediterranean component (Table 3).

## Discussion

The results showed above provide enough information on the current state of the edaphic system in the Guadiamar basin to evaluate the impact of the toxic spill by comparing the damaged zone with other non-affected (Biocenters) or well-preserved nature areas (the Doñana National Park or Sierra Morena mountains).

The initial basis for our research was the general assumption that faunistic data, both quantitative and qualitative, for concrete habitats enables us to examine the structure and functioning of animal communities and the impact of drastic changes on them, either due to natural causes or human intervention (Southwood 1991). Furthermore, in practical studies, to evaluate biodiversity and conservation strategies, supplementary information can be used such as the richness of an indicative taxa or the endemic condition (with a fairly restricted geographical distribution) of its components (Jaarsveld et al. 1998). Hence, we have based our conclusions on the comparative analysis of the different indices (mainly diversity and richness) and on the biogeographical composition of the communities in the different areas.

Carabids, Scarabaeids and Tenebrionids are considered valid indicators of the state of an edaphic system. The populations studied from the Guadiamar basin and adjacent areas do not differ significantly from those found in the Doñana or Sierra Morena ecosystems, meaning that the Guadiamar basin can act as a connecting area or ecological corridor. Thus, the restoration and conservation of this type of corridor would enable the re-colonization of fragmented habitats and the stabilization of the metapopulations (Vermeulen 1994).

The rarefraction curves obtained for Carabidae, Tenebrionidae and Scarabaeoidea indicate that, overall, the situation in the Guadiamar basin has not become poorer in species. Instead, intermediate (Carabidae), proximate (Scarabaeoidea) and even higher values (Tenebrionidae) were found in the basin in comparison to the other study sites, suggesting that the toxic spill has had a low impact on this area.

However, a detailed analysis of the stations closest to the river, and thus more affected by the spill, reveals an environmental heterogeneity manifested in the faunistic poverty and structural simplicity, a process which becomes increasingly more evident closer to the site of the toxic spill.

On the basis of species absence, we can infer that the Carabidae of the Bembidiini and Tachyni tribes, which are usually dominant in riparian habitats, were most affected by the ecological disaster and subsequent interventions. There are some exceptions, however, such as *Asaphidion curtum* and *Bembidion ambiguum*, which appear to tolerate high levels of toxicity and continue to be abundant in the control plots containing sludge. The PS1 station, which is closest to the site of the spill, was the poorest in species due to the greater environmental impact on this area and the fluctuating level of the Guadiamar River, an additional factor of instability that prevents the establishment of more complex communities. In contrast, the parameters that define the ground beetle community in PS6-7 show high values, which respond to the distance from the disaster site and proximity to the borders of Doñana and other unaffected areas, as well as croplands that are clearly within their area of influence. This suggests that this plot is a potential source area for the dispersion of carabids to other areas of the Green Corridor that have been more disturbed.

Both the dominant Carabidae in the Basin (*Pseudophonus rufipes*, *Harpalus subsinuatus, Poecilus kugelanni* and *Paranchus albipes)* and in the Biocenters (*Steropus globosus, Orthomus velocissimus* and *Calathus granatensis*) are species with low environmental requirements, making them potential agents for the re-colonization of the Ecological Corridor. In fact, this process appears to have already begun in some cases (i.e. *P. rufipes* or *S. globosus*), although a more extensive follow-up study is needed to confirm this assertion.

The Tenebrionidae were the least affected group according to all of the parameters considered in the study. Certain generalist species (*Gonocephalum rusticum, Scleron armatum* or *Tentyria platyceps*) are not only present in the sample plots but are also abundant where the toxic sludge remains (PT). These species have a high environmental plasticity and are capable of colonizing environments that other species are unable to tolerate. Furthermore, their limited dispersal capacity (apterism) may explain why they survived (by seeking refuge) during the clean-up operations in the only area where toxic sludge has not been removed. These ubiquitous elements should be the first agents for the re-colonization of rehabilitated areas in the Corridor. In the dendrogram, the darkling beetles communities from Guadiamar (Basin and Biocenters) and Doñana were the most closely linked owing to the high proportion of common species belonging to the genera *Tentyria*, *Gonocephalum*, *Erodius* and *Pimelia*, which share a thermo-xerophilous condition and preference for sandy substrates.

As expected, only the Biocenters and in particular the Oak Meadows, lodge rich, well-structured dung beetle communities while re-colonization in the sample plots (PS) is still in the early stages. This can be explained by the fact that the spill and subsequent clean-up operations have made it difficult for populations of vertebrates to get established and pasture, thus reducing the available trophic resources. However, in the surrounding area (Biocenters), there are abundant species that are generally of high ecological valence, such as *Onthophagus punctatus, O. opacicollis, O. furcatus, Aphodius hydrochaeris, A. baraudi, A. castaneus* and *A. annamariae*, which should be taken into account when determining possible colonizing agents.

Biogeographical analysis reveals that the current composition of edaphic coleoptera communities in the Guadiamar has not undergone significant changes in comparison to external areas (Sierra Morena and Doñana National Park), the Mediterranean component being predominant in all the groups. Furthermore, there is a marked proliferation of Iberian and even Betic endemisms and a notable scarcity of widely distributed elements (Holopaleartic and Cosmopolitan).This could be interpreted as a good indication of the conservation of autochthonous fauna, which is always more sensitive to alterations in the habitat.

Biogeographical affinity between the Carabidae fauna in the Guadiamar and the adjacent ecosystems is to be expected given the environmental similarities between the Guadiamar and Bembézar basins (Central Sierra Morena) and the proximity between the study area and the plots located in Doñana National Park. From a biogeographical point of view, the current composition of the Guadiamar fauna is quite similar to other Carabidae communities of the southern peninsula.

As for Tenebrionidae, a high affinity has been observed between all three study areas. This affinity is due to the fact that the majority of the representatives of this Family are apterous or brachipterous, suggesting the marked predominance of geographically restricted elements (i.e. Iberian or Betic endemisms). Moreover, the peculiar features of the habitat (arid environments and sandy substrates) favor the existence of a particular fauna to the detriment of elements with a greater biogeographical distribution (paleartic or cosmopolitan species).

The predominance of the Mediterranean component gives rise to fairly homogeneous Scarabaeid fauna in the Guadiamar basin, Sierra Morena and Doñana National Park, which do not differ significantly. Therefore, it could be asserted that, just as with the Carabidae and Tenebrionidae, the biogeographical composition of Scarabaeoidea in the Guadiamar basin has not been substantially altered as a result of the ecological disaster, in spite of the discrepancies observed at a specific level.

## Conclusions

In short, following our assessment of the current state of the edaphic coleoptera in the Guadiamar River basin, we conclude that, overall, the soil fauna has not been significantly affected by the Aznalcóllar mine toxic spill. However a gradient, on the north–south axis, is observed with increasing diversity the farther away from the site of the accident. A similar gradient has also been observed by Solà et al. (2003) for the fresh-water macro-invertebrates and by Tejedo and Reques (2003) for the amphibian community from the Guadiamar River.

Temporal evolution suggests that the most affected areas could undergo a process of rehabilitation, given that several potential re-colonizing species have been identified; species that could be examined in greater depth in subsequent follow-up studies. The conservation of this natural heritage will chiefly depend upon the environmental measures taken, including the consolidation of the Green Corridor as a bridge between the Sierra Morena and Doñana National Park.

## Appendix 1

Spatial distribution for the species recorded in the research area. Symbol key as in the text.

| Species | PS1 | PS2-3 | PT2 | PS4 | PS6-7 | BDV | BOA | BPA | BEA |
|---|---|---|---|---|---|---|---|---|---|
| **Carabidae** | | | | | | | | | |
| *Calosoma maderae* (Fabricius) | | | 2 | 4 | 35 | | 2 | | |
| *Carabus rugosus* Fabricius | 1 | | | | | | 4 | | |
| *Carabus lusitanicus* Fabricius | | | | | | 16 | | 5 | 13 |
| *Distichus planus* (Bonelli) | | | 3 | 1 | 4 | | | | |
| *Scarites cyclops* Crotch | | | | | | | | 2 | 5 |
| *Apotomus rufus* (Rossi) | | | | | 4 | | | | |
| *Trechus obtusus* Erichson | | 1 | | | | | | | |
| *Asaphidion curtum* (Heyden) | | 2 | 6 | | 3 | | | | |
| *Bembidion iricolor* Bedel | | | | | 1 | | | | |
| *Bembidion vicinus* Lucas | | | 2 | | 1 | | | | |
| *Bembidion ambiguum* Dejean | | 2 | | | 3 | | | | |
| *Bembidion tethys* Netolitzky | | | | | | 1 | | 1 | |
| *Percus politus* (Dejean) | | | | | | | 1 | | |
| *Ancholeus gisellae* (Csiki) | | | 1 | | 2 | | | | |
| *Orthomus velocissimus* (Waltl) | 1 | 3 | | | | 14 | 21 | 2 | 16 |
| *Poecilus quadricollis* (Dejean) | | 4 | | | 3 | | | | |
| *Poecilus kugelanni* (Panzer) | 1 | 32 | 56 | 16 | 12 | | | | |
| *Poecilus purpurascens* (Dejean) | 1 | 1 | 9 | | | | | | |
| *Steropus globosus* (Fabricius) | 9 | 4 | | | | 139 | 19 | 117 | 115 |
| *Amara similata* (Gyllenhall) | | 1 | 1 | 1 | 1 | | | | |
| *Amara aenea* DeGeer | | 3 | | 1 | | 3 | 1 | | |
| *Amara metallescens* Zimmermann | | | 4 | | 50 | | | | |
| *Olisthopus fuscatus* Dejean | | | | | 1 | | | | |
| *Agonum marginatum* Linnaeus | | | | | 1 | | | | |
| *Anchomenus dorsalis* (Pontoppidan) | | 2 | | 2 | | | | | |
| *Paranchus albipes* Fabricius | | 9 | 1 | | 20 | | | | |
| *Platyderus emblema* Marseul | | | | | | | | | 1 |
| *Calathus granatensis* Vuillefroy | | 1 | | | | 134 | | 12 | 20 |
| *Calathus mollis* (Marsham) | | | 2 | | | | | | |
| *Laemostenus complanatus* (Dejean) | | 23 | 3 | 11 | 8 | | | | |
| *Amblystomus mauritanicus* (Dejean) | 1 | | | | | | | | |
| *Carterus fulvipes* Latreille | | | | | 16 | | | | |
| *Harpalus distinguendus* Duftschmid | | 5 | 11 | | 23 | | | | |
| *Harpalus oblitus* Dejean | | | 2 | | 1 | | | | |
| *Harpalus attenuatus* Stephens | | | 2 | | | | | | |
| *Cryptophonus tenebrosus* (Dejean) | | | 10 | | 3 | | | | |

*Appendix 1.* (*Continued*)

| Species | PS1 | PS2-3 | PT2 | PS4 | PS6-7 | BDV | BOA | BPA | BEA |
|---|---|---|---|---|---|---|---|---|---|
| *Ophonus sahlbergianus* Lutshnik | | | 1 | | | | | | |
| *Ophonus subsinuatus* Rey | | 1 | | | 44 | | | | |
| *Ophonus ardosiacus* Lutshik | | | | 4 | 6 | | | | |
| *Ophonus opacus* (Dejean) | | | | | 11 | | | | |
| *Pseudophonus rufipes* DeGeer | | 1 | 292 | 92 | 30 | | | | |
| *Pseudophonus griseus* (Panzer) | | | | 1 | | | | | |
| *Acupalpus ibericus* Jaeger | | | | | | | 1 | | |
| *Licinus punctatulus* (Fabricius) | | | | | 1 | 1 | 3 | | |
| *Chlaenius chrysocephalus* Rossi | | | 1 | | | | | | |
| *Chlaenius velutinus* (Duftschmid) | | 5 | | | | | | | |
| *Chlaenius spoliatus* (Rossi) | | 2 | | | 2 | | | | |
| *Dinodes decipiens* (Dufour) | | | | | 1 | | | | |
| *Platytarus faminii* (Dejean) | | | | | 2 | | | | |
| *Paradromius linearis* (Olivier) | | 1 | 1 | | | | 1 | | |
| *Mesolestes scapularis* (Dejean) | 2 | 6 | | | | | | 1 | |
| *Microlestes corticalis* (Dufour) | | | | | 10 | | | | |
| *Microlestes luctuosus* Holdhaus | | | | | | 4 | | | |
| *Microlestes abeillei* (Brisout) | 1 | | | | | | | | |
| *Syntomus fuscomaculatus* Motschulsky | | | | | | 1 | | 1 | |
| *Syntomus foveatus* (Geffroy) | | | | | | 11 | | | |
| *Syntomus foveolatus* (Dejean) | | | | | | | 1 | 1 | |
| *Polystichus connexus* (Fourcroy) | | | | | 1 | | | | |
| *Brachinus immaculicornis* Dejean | | | 3 | | 4 | | | | 3 |
| *Brachinus sclopeta* (Fabricius) | 1 | 17 | | 4 | 5 | | | | |
| *Brachinus humeralis* Ahrens | | | 1 | | 12 | | | | |
| **Scarabaeoidea** | | | | | | | | | |
| *Omorgus suberosus* (Fabricius) | | | | | 1 | | | | |
| *Trox perlatus* (Geoffroy) | | | | | | 1 | | 24 | 70 |
| *Bolbelasmus gallicus* (Mulsant) | | | | | | 1 | | | |
| *Typhaeus momus* (Olivier) | | | | | | 9 | | 21 | 105 |
| *Ceratophyus hoffmannseggi* Fairm. | | | | | | 1 | | | |
| *Sericotrupes niger* (Marsham) | | | | | | 3 | | | |
| *Jekelius nitidus* (Jekel) | | | | | | | | | 1 |
| *Ochodaeus inermis* Reitter | | | | | | | 1 | | |
| *Aphodius immundus* Creutzer | | | | | | 1 | | | |
| *A. hydrochaeris* (Fabricius) | | | | | | 110 | 3 | | 5 |
| *A. annamariae* Baraud | | | | | | 89 | | | |
| *A. castaneus* Illiger | 13 | 25 | | 1 | 2 | 64 | 150 | | 1 |
| *A. foetidus* (Herbst) | | 1 | | | | 16 | | | |
| *A. ghardimaouensis* Balthasar | 18 | 1 | | 1 | 6 | 41 | 21 | 2 | 8 |
| *A. mayeri* Pilleri | | | | | | | 3 | | |
| *A. lineolatus* Illiger | | 1 | | | | 2 | 4 | 1 | 3 |
| *A. erraticus* (Linnaeus) | | | | | | | 1 | | |
| *A. merdarius* (Fabricius) | | | | | | 5 | 6 | | 6 |
| *A. baraudi* Villarreal | | | | | | | | 7 | 40 |
| *A. tersus* Erichson | | | | | | | | 1 | |
| *A. striatulus* Waltl | | | | | | | | | 1 |
| *A. villarreali* Baraud | | | | | | | 14 | | |

*Appendix 1.* (*Continued*)

| Species | PS1 | PS2-3 | PT2 | PS4 | PS6-7 | BDV | BOA | BPA | BEA |
|---|---|---|---|---|---|---|---|---|---|
| *Heptaulacus brancoi* Baraud | | | | | | | | | 1 |
| *Pleurophorus caesus* (Creutzer) | 7 | | 2 | | 1 | | | | |
| *P. mediterranicus* Pittino & Mariani | | | 2 | | | | | | |
| *Scarabaeus sacer* Linnaeus | 1 | | | | | | | | |
| *Copris hispanus* (Linnaeus) | | | | | | 6 | 1 | | 1 |
| *Euoniticellus pallipes* (Fabricius) | | | | | | 1 | | | |
| *Bubas bison* (Linnaeus) | | | | 1 | | 7 | | | 1 |
| *B. bubalus* (Olivier) | | | | | | | 1 | | |
| *Caccobius schreberi* (Linnaeus) | | | | | | 3 | 1 | 1 | 4 |
| *Onthophagus furcatus* (Fabricius) | 1 | 21 | 3 | 2 | 5 | 202 | 26 | 1 | 2 |
| *O. taurus* (Schreber) | | 14 | | 11 | 1 | 5 | 10 | 1 | 2 |
| *O. ruficapillus* Brullé | | 1 | | | 1 | 1 | | | |
| *O. opacicollis* Reitter | | 4 | | 1 | 1 | 72 | 7 | | 5 |
| *O. similis* (Scriba) | | | | | | 18 | | | 1 |
| *O. vacca* (Linnaeus) | | | | | | | 3 | | |
| *O. punctatus* (Illiger) | | 3 | | | 1 | 75 | 7 | 165 | 134 |
| *O. maki* (Illiger) | | | | | 1 | | | | |
| *Rhizotrogus sp.* | | | | | | 1 | | | |
| *Chasmatopterus villosulus* (Illiger) | | | | | | 1 | | | |
| *Euserica mulsanti* (Brenske) | | | | | | 2 | 1 | | |
| *E. mutata* (Gyllenhal) | 1 | | | | | 5 | | 1 | |
| *Anomala ausonia* Erichson | | | | 1 | | | | | |
| *A. quadripunctata* (Olivier) | | | | 1 | | | | | |
| *Oryctes nasicornis* (Linnaeus) | | | | | | | | | 3 |
| *Pentodon algerinum* (Herbst) | | | | | 2 | | | | |
| *P. idiota* (Herbst) | | | | 1 | 1 | | | | |
| *Valgus hemipterus* (Linnaeus) | | | | | 2 | | | | |
| *Cetonia carthami* Gory & Percheron | | | | | | 1 | | | |
| *Netocia morio* (Fabricius) | | | | 1 | | 4 | 5 | 3 | 12 |
| *Netocia oblonga* (Gory & Percheron) | 1 | | | | | | | | |
| *Netocia cuprea* (Fabricius) | | 1 | | | | | | | |
| *Tropinota squalida* (Scopoli) | | | 1 | | | | | | |
| *Oxythyrea funesta* (Poda) | | | | | | | | | 12 |
| **Tenebrionidae** | | | | | | | | | |
| Erodius goryi Allard | | | | | | | | | 96 |
| *Pachychila hispanica* Solier | | 1 | 5 | 1 | 5 | | 2 | | |
| *Tentyria platyceps* Stevens | 10 | 7 | 116 | 16 | 1 | 38 | 3 | 1 | 32 |
| *Tentyria sp.* | | | | 1 | | | | | |
| *Asida goudoti* Solier | | | | | | | | | 1 |
| *Stenosis hispanica* Solier | | | | | | 7 | 2 | | 5 |
| *Dichillus subcostatus* Solier | | | | | | 1 | 1 | | |
| *Sepidium bidentatum* Solier | | | | | | 1 | | 1 | 29 |
| *Akis granulifera* Sahlberg | | | | | | 46 | 1 | | |
| *Scaurus punctatus* Fabricius | | | | 1 | | 4 | 1 | | 1 |
| *Pimelia costata* Waltl | | | | | | | 1 | 3 | 241 |
| *Blaps hispanica* Solier | 5 | | | | | 19 | 4 | 2 | 3 |
| *B. waltli* Seidlitz | | | | | | | 1 | 3 | 3 |
| *Dendarus elongatus* Mulsant | | | | | | | | | 1 |
| *Gonocephalum rusticum* Olivier | 4 | 17 | 53 | 2 | 118 | | | 1 | |
| *G. pusillum* Fabricius | | | | | | | | | 2 |

*Appendix 1.* (*Continued*)

| Species | PS1 | PS2-3 | PT2 | PS4 | PS6-7 | BDV | BOA | BPA | BEA |
|---|---|---|---|---|---|---|---|---|---|
| *Scleron armatum* Waltl | | 12 | 26 | 19 | 13 | | | | |
| *Litoborus planicollis* Waltl | | | 3 | | 1 | | | | |
| *Micrositus longulus* Mulsant & Rey | | | | | | | | | 2 |
| *Cnemeplatia atropos* Kosz | | | | | | | 1 | | |
| *Probaticus granulatus* Allard | | | | | | | | | 1 |
| *Tribolium castaneum* Herbst | | | | | 1 | | 1 | | |
| *Alphitobius diaperinus* Panzer | 1 | | | | | | | | |
| *Belopus elongatus* Herbst | | | | | 1 | | | | |
| *Misolampus gibbulus* Herbst | | | | | | | | | 2 |
| *Oochrotus unicolor* Lucas | | | | | | 1 | | | |
| *Nalassus skopini* Español | | | | 1 | | | | | |
| *Boromorphus tegenioides* Lucas | | | | | | | 3 | | |

## Appendix 2

Seasonal distribution and total number of specimens for the species recorded in the research area.

| Species | Spring 2000 | Summ. 2000 | Autumn 2000 | Winter 2000 | Spring 2001 | Summ. 2001 | Autumn 2001 | Total |
|---|---|---|---|---|---|---|---|---|
| **Carabidae** | | | | | | | | |
| *Calosoma maderae* (Fabricius) | 2 | 13 | | | 23 | 5 | | **43** |
| *Carabus rugosus* Fabricius | | | | 1 | 2 | 1 | 1 | **5** |
| *Carabus lusitanicus* Fabricius | 1 | | 2 | 14 | 10 | | 7 | **34** |
| *Distichus planus* (Bonelli) | 1 | 1 | | 1 | 1 | 1 | 3 | **8** |
| *Scarites cyclops* Crotch | | 2 | | | 2 | 3 | | **7** |
| *Apotomus rufus* (Rossi) | | | | | 2 | | 2 | **4** |
| *Trechus obtusus* Erichson | | | | | 1 | | | **1** |
| *Asaphidion curtum* (Heyden) | 6 | | | 2 | 1 | 1 | 1 | **11** |
| *Bembidion iricolor* Bedel | | | | | 1 | | | **1** |
| *Bembidion vicinus* Lucas | 1 | | | 2 | | | | **3** |
| *Bembidion ambiguum* Dejean | | | | | 2 | 2 | 1 | **5** |
| *Bembidion tethys* Netolitzky | | | | | 1 | 1 | | **2** |
| *Percus politus* (Dejean) | | | | | | 1 | | **1** |
| *Ancholeus gisellae* (Csiki) | | | | 2 | 1 | | | **3** |
| *Orthomus velocissimus* (Waltl) | 19 | 2 | 3 | 8 | 6 | 7 | 12 | **57** |
| *Poecilus quadricollis* (Dejean) | 1 | | | 3 | 1 | 1 | 1 | **7** |
| *Poecilus kugelanni* (Panzer) | 2 | | 4 | 6 | 68 | 29 | 8 | **117** |
| *Poecilus purpurascens* (Dejean) | 1 | | | 2 | 4 | 3 | 1 | **11** |
| *Steropus globosus* (Fabricius) | 66 | 20 | 6 | 4 | 120 | 140 | 47 | **403** |
| *Amara similata* (Gyllenhall) | 1 | | | 1 | 1 | 1 | | **4** |
| *Amara aenea* DeGeer | 4 | 1 | | | | 3 | | **8** |
| *Amara metallescens* Zimmermann | | | | | 5 | | 49 | **54** |

*Appendix 2.* (*Continued*)

| Species | Spring 2000 | Summ. 2000 | Autumn 2000 | Winter 2000 | Spring 2001 | Summ. 2001 | Autumn 2001 | Total |
|---|---|---|---|---|---|---|---|---|
| *Olisthopus fuscatus* Dejean | | | | | | | 1 | **1** |
| *Agonum marginatum* Linnaeus | | 1 | | | | | | **1** |
| *Anchomenus dorsalis* (Pontoppidan) | | | | 3 | 1 | | | **4** |
| *Paranchus albipes* Fabricius | 16 | | | 9 | 1 | | 4 | **30** |
| *Platyderus emblema* Marseul | | | | 1 | | | | **1** |
| *Calathus granatensis* Vuillefroy | 45 | 10 | 26 | 6 | 31 | 20 | 29 | **167** |
| *Calathus mollis* (Marsham) | | | 1 | 1 | | | | **2** |
| *Laemostenus complanatus* (Dejean) | 3 | 2 | 11 | 7 | 10 | 2 | 10 | **45** |
| *Amblystomus mauritanicus* (Dejean) | | | | | | | 1 | **1** |
| *Carterus fulvipes* Latreille | | | 16 | | | | | **16** |
| *Harpalus distinguendus* Duftschmid | 2 | | 2 | 13 | 17 | 5 | | **39** |
| *Harpalus oblitus* Dejean | | | | | 3 | | | **3** |
| *Harpalus attenuatus* Stephens | | | | | | | 2 | **2** |
| *Cryptophonus tenebrosus* (Dejean) | | | | | | | 13 | **13** |
| *Ophonus sahlbergianus* Lutshnik | | | | | | | 1 | **1** |
| *Ophonus subsinuatus* Rey | | | 1 | | | 4 | 40 | **45** |
| *Ophonus ardosiacus* Lutshik | | | | | | | 10 | **10** |
| *Ophonus opacus* (Dejean) | | | | | | | 11 | **11** |
| *Pseudophonus rufipes* DeGeer | 1 | | 76 | 8 | 45 | 140 | 145 | **415** |
| *Pseudophonus griseus* (Panzer) | | | 1 | | | | | **1** |
| *Acupalpus ibericus* Jaeger | | | | | | 1 | | **1** |
| *Licinus punctatulus* (Fabricius) | 1 | | 1 | 1 | | | 2 | **5** |
| *Chlaenius chrysocephalus* Rossi | | | | | | | 1 | **1** |
| *Chlaenius velutinus* (Duftschmid) | | | | 5 | | | | **5** |
| *Chlaenius spoliatus* (Rossi) | | | 1 | 3 | | | | **4** |
| *Dinodes decipiens* (Dufour) | | | | | | 1 | | **1** |
| *Platytarus faminii* (Dejean) | | | | | | | 2 | **2** |
| *Paradromius linearis* (Olivier) | 1 | | | | 1 | | 1 | **3** |
| *Mesolestes scapularis* (Dejean) | 2 | 2 | | | | 4 | 1 | **9** |
| *Microlestes corticalis* (Dufour) | | | 1 | | | 9 | | **10** |
| *Microlestes luctuosus* Holdhaus | | 2 | 1 | 1 | | | | **4** |
| *Microlestes abeillei* (Brisout) | | | | | | | 1 | **1** |
| *Syntomus fuscomaculatus* Motschulsky | | 1 | 1 | | | | | **2** |
| *Syntomus foveatus* (Geoffroy) | | 3 | 5 | | 1 | 2 | | **11** |
| *Syntomus foveolatus* (Dejean) | | 1 | | | | 1 | | **2** |
| *Polystichus connexus* (Fourcroy) | 1 | | | | | | | **1** |
| *Brachinus immaculicornis* Dejean | 3 | | 2 | | 1 | 1 | 3 | **10** |
| *Brachinus sclopeta* (Fabricius) | 2 | | | 21 | 4 | | | **27** |
| *Brachinus humeralis* Ahrens | 12 | | 1 | | | | | **13** |
| **Scarabaeoidea** | | | | | | | | |
| *Omorgus suberosus* (Fabricius) | | | | | | 1 | | **1** |
| *Trox perlatus* (Geoffroy) | 7 | | 27 | 31 | 11 | | 19 | **95** |

*Appendix 2.* (*Continued*)

| Species | Spring 2000 | Summ. 2000 | Autumn 2000 | Winter 2000 | Spring 2001 | Summ. 2001 | Autumn 2001 | Total |
|---|---|---|---|---|---|---|---|---|
| *Bolbelasmus gallicus* (Mulsant) | 1 | | | | | | | **1** |
| *Typhaeus momus* (Olivier) | 11 | | 13 | 79 | 25 | | 7 | **135** |
| *Ceratophyus hoffmannseggi* Fairm. | | | | 1 | | | | **1** |
| *Sericotrupes niger* (Marsham) | | | 2 | | | | 1 | **3** |
| *Jekelius nitidus* (Jekel) | | | | | | | 1 | **1** |
| *Ochodaeus inermis* Reitter | | | | | | | 1 | **1** |
| *Aphodius immundus* Creutzer | 1 | | | | | | | **1** |
| *A. hydrochaeris* (Fabricius) | 109 | | 1 | 8 | | | | **118** |
| *A. annamariae* Baraud | | | 83 | | | | 6 | **89** |
| *A. castaneus* Illiger | | | 175 | | | 2 | 79 | **256** |
| *A. foetidus* (Herbst) | 15 | | 1 | | 1 | | | **17** |
| *A. ghardimaouensis* Balthasar | 15 | 3 | 50 | | 19 | 1 | 10 | **98** |
| *A. mayeri* Pilleri | | | | 3 | | | | **3** |
| *A. lineolatus* Illiger | 3 | | 7 | 1 | | | | **11** |
| *A. erraticus* (Linnaeus) | | | | | | 1 | | **1** |
| *A. merdarius* (Fabricius) | 14 | 2 | | | 1 | | | **17** |
| *A. baraudi* Villarreal | 37 | | | 9 | 1 | | | **47** |
| *A. tersus* Erichson | | | | 1 | | | | **1** |
| *A. striatulus* Waltl | 1 | | | | | | | **1** |
| *A. villarreali* Baraud | | | 5 | 9 | | | | **14** |
| *Heptaulacus brancoi* Baraud | 1 | | | | | | | **1** |
| *Pleurophorus caesus* (Creutzer) | 7 | 1 | | 1 | | | 1 | **10** |
| *P. mediterranicus* Pittino & Mariani | 2 | | | | | | | **2** |
| *Scarabaeus sacer* Linnaeus | | | | | | 1 | | **1** |
| *Copris hispanus* (Linnaeus) | 6 | | | 2 | | | | **8** |
| *Euoniticellus pallipes* (Fabricius) | 1 | | | | | | | **1** |
| *Bubas bison* (Linnaeus) | | | | 8 | | | 1 | **9** |
| *B. bubalus* (Olivier) | 1 | | | | | | | **1** |
| *Caccobius schreberi* (Linnaeus) | 2 | 7 | | | | | | **9** |
| *Onthophagus furcatus* (Fabricius) | 37 | 97 | 27 | | 4 | 89 | 9 | **263** |
| *O. taurus* (Schreber) | 8 | 4 | | | 2 | 26 | 4 | **44** |
| *O. ruficapillus* Brullé | 1 | | | | 1 | 1 | | **3** |
| *O. opacicollis* Reitter | 69 | 1 | 7 | 3 | 1 | 7 | 2 | **90** |
| *O. similis* (Scriba) | 11 | | 6 | | 1 | 1 | | **19** |
| *O. vacca* (Linnaeus) | 3 | | | | | | | **3** |
| *O. punctatus* (Illiger) | 208 | 102 | 20 | 1 | 19 | 17 | 18 | **385** |
| *O. maki* (Illiger) | | | | | 1 | | | **1** |
| *Rhizotrogus sp.* | 1 | | | | | | | **1** |
| *Chasmatopterus villosulus* (Illiger) | | | | | 1 | | | **1** |
| *Euserica mulsanti* (Brenske) | 1 | 2 | | | | | | **3** |
| *E. mutata* (Gyllenhal) | 1 | 2 | | | 3 | 1 | | **7** |
| *Anomala ausonia* Erichson | | | | | | 1 | | **1** |
| *A. quadripunctata* (Olivier) | | | | | | 1 | | **1** |
| *Oryctes nasicornis* (Linnaeus) | | 2 | | | | 1 | | **3** |
| *Pentodon algerinum* (Herbst) | 2 | | | | | | | **2** |
| *P. idiota* (Herbst) | | | | | | | 2 | **2** |
| *Valgus hemipterus* (Linnaeus) | | | | | 2 | | | **2** |

*Appendix 2.* (*Continued*)

| Species | Spring 2000 | Summ. 2000 | Autumn 2000 | Winter 2000 | Spring 2001 | Summ. 2001 | Autumn 2001 | Total |
|---|---|---|---|---|---|---|---|---|
| *Cetonia carthami* Gory & Percheron | | | | | | 1 | | **1** |
| *Netocia morio* (Fabricius) | 3 | 19 | | | 1 | 2 | | **25** |
| *Netocia oblonga* (Gory & Percheron) | 1 | | | | | | | **1** |
| *Netocia cuprea* (Fabricius) | | | | | | 1 | | **1** |
| *Tropinota squalida* (Scopoli) | | | | | | 1 | | **1** |
| *Oxythyrea funesta* (Poda) | | 6 | | | | 5 | 1 | **12** |
| **Tenebrionidae** | | | | | | | | |
| *Erodius goryi* Allard | 48 | 3 | | | | 44 | 1 | **96** |
| *Pachychila hispanica* Solier | | 2 | | | 1 | 5 | 6 | **14** |
| *Tentyria platyceps* Stevens | 5 | 17 | 47 | 1 | 8 | 125 | 25 | **224** |
| *Tentyria sp* | | | | | | | 1 | **1** |
| *Asida goudoti* Solier | 1 | | | | | | | **1** |
| *Stenosis hispanica* Solier | 5 | 5 | 1 | | 1 | 2 | | **14** |
| *Dichillus subcostatus* Solier | 2 | | | | | | | **2** |
| *Sepidium bidentatum* Solier | 26 | 2 | | | 1 | 2 | | **31** |
| *Akis granulifera* Sahlberg | 20 | 9 | 3 | | 6 | 8 | 1 | **47** |
| *Scaurus punctatus* Fabricius | 3 | | | | 1 | | 3 | **7** |
| *Pimelia costata* Waltl | 57 | 41 | 13 | | 41 | 49 | 44 | **245** |
| *Blaps hispanica* Solier | 7 | 1 | 13 | | 1 | 6 | 5 | **33** |
| *B. waltli* Seidlitz | 1 | | 5 | | | | 1 | **7** |
| *Dendarus elongatus* Mulsant | | 1 | | | | | | **1** |
| *Gonocephalum rusticum* Oliv. | 17 | 7 | 5 | 1 | 28 | 21 | 116 | **195** |
| *G. pusillum* Fabricius | | | | | | 2 | | **2** |
| *Scleron armatum* Waltl | 6 | | | | 35 | 7 | 22 | **70** |
| *Litoborus planicollis* Waltl | | | | | | | 4 | **4** |
| *Micrositus longulus* Mulsant & Rey | | | 2 | | | | | **2** |
| *Cnemeplatia atropos* Kosz | 1 | | | | | | | **1** |
| *Probaticus granulatus* Allard | | | 1 | | | | | **1** |
| *Tribolium castaneum* Herbst | | 1 | | | | 1 | | **2** |
| *Alphitobius diaperinus* Panzer | 1 | | | | | | | **1** |
| *Belopus elongatus* Herbst | 1 | | | | | | | **1** |
| *Misolampus gibbulus* Herbst | | 1 | 1 | | | | | **2** |
| *Oochrotus unicolor* Lucas | 1 | | | | | | | **1** |
| *Nalassus skopini* Español | | | | | 1 | | | **1** |
| *Boromorphus tegenioides* Lucas | 1 | | 1 | 1 | | | | **3** |

## References

Aguilar J., Bellver R., Dorronsoro C., Fernández J., Fernández E., García I., Ariarte A., Martín F., Ortiz I. and Simón M. 2003. Estudio de la contaminación del suelo y técnicas de tratamiento de inmovilización de metales en la cuenca del Guadiamar. In: Junta de Andalucía. (ed.), PICOVER: Ciencia y Restauración del Río Guadiamar. Consejería de Medio Ambiente, pp. 52–59.

Anguas J.M., Carmona J.J., Isendahl N., Robles P. and Schmidt G. 2002. Directorio de organizaciones ambientales de la comarca de Doñana. (ed.), WWF/ADENA. Madrid, Spain.

Baraud J. 1992. Coléoptères Scarabaeoidea carabidae, Faune de France, 78. Ed. Société Linnéenne de Lyon.

Belaoussoff S., Kevan P.G., Murphy S. and Swanton C. 2003. Assessing tillage disturbance on assemblages of ground beetles (Coleoptera: Carabidae) by using a range of ecological indices. Biodiversity and Conservation 12: 851–882.

Boscaini A., Franceschini A. and Maiolini B. 2000. River ecotones: carabid beetles as a tool for quality assessment. Hydrobiologia 422/423: 173–181.

Cárdenas A.M. and Bach C. 1988a. Contribución al conocimiento de los Carábidos (Col. Carabidae) de Sierra Morena Central. 1ª Parte. Boletín Asociación española Entomología 12: 9–25.

Cárdenas A.M. and Bach C. 1988b. Estudio de la comunidad de Carábidos de la Sierra de Hornachuelos. Stvdia Oecologica 5: 335–351.

Cárdenas A.M. and Bach C. 1989. Contribución al conocimiento de los Carábidos (Col. Carabidae) de Sierra Morena Central. 2ª Parte. Boletín Asociación española Entomología 13: 9–19.

Cárdenas A.M. and Bujalance J.L. 1985. Sobre los Tenebrionidae de los alrededores del embalse del Bembézar. Boletim Sociedade Portuguesa Entomologia 3: 497–506.

Cárdenas A.M. and Hidalgo J.M. 1998. Oviposition patterns in *Steropus globosus* (Fabricius, 1792) (Coleoptera, Carabidae). Miscel. lània Zoològica 21(1): 9–19.

Castro H. 2003. Environmental connectivity and conservation in the Mediterranean Basin. In: Protected areas in the Mediterranean basin (ed.), Dirección General de la RENP. Consejería de Medio Ambiente.Sevilla (Spain): 3–9.

Davis A.L.V. and Scholtz C.H. 2001. Historical vs. ecological factors influencing global patterns of scarabaeinae dung beetle diversity. Diversity and Distributions. 7: 161–174.

de los Santos A., Montes C. and Ramírez-Díaz L. 1982. Modelos espaciales de algunas poblaciones de coleópteros terrestres en dos ecosistemas del bajo Guadalquivir (S.W. España). Mediterranea 6: 65–92.

Español F. 1954. Los Tenebriónidos (col.) de Baleares. Trabajos del Museo de Ciencias Naturales de Barcelona. Nueva Serie Zoológica 1(5): 84.

García-Canseco V. 2002. Parque Nacional de Doñana. (ed.), Canseco. Talavera de la Reina (Spain).

García M.R. and Montes C. 2003. Linkages in the Mediterranean landscape: the role of protected areas in the territorial context. Consejería de Medio Ambiente, Junta de Andalucía.

Gebien H. 1938–42. Katalog der Tenebrioniden. Mitteilungen der Müncher Entomologidche Gesellschaft.

Gebien H. 1942–44. Katalog der Tenebrioniden. Mitteilungen der Müncher Entomologidche Gesellschaft.

Grümm L. 1980. Methods of approximate estimation of energy flow through carabid populations. Ekologia Polska 28(1): 129–149.

Hidalgo J.M., Bach C. and Cárdenas A.M. 1998a. Los Scarabaeoidea coprófagos de las comarcas naturales de la provincia de Córdoba: I. Aphodiidae. Boletín Asociación española Entomología 22(3–4): 9–35.

Hidalgo J.M., Bach C. and Cárdenas A.M. 1998b. Los Scarabaeoidea coprófagos de las comarcas naturales de la provincia de Córdoba: II. Trogidae, Geotrupidae y Scarabaeidae. Boletín Asociación española Entomología 22(3–4): 203–230.

Holdhaus K. 1929. Die geographische Verbreitung der Insekten. Schröder Handbuch Entomologie 2: 592–1058.

Jaarsveld van A.S., Freitag S., Chown S.L., Muller C., Koch S., Hull H., Bellamy C., Krüger M., Endrödy-Younga S., Mansell M.W. and Scholtz C.H. 1998. Biodiversity assessment and conservation strategies. Science 279: 2106–2108.

La Greca M. 1964. Le categorie corologiche degli elementi faunistici italiani. Memorie della Societá Entomologica Italiana 43: 147–165.

Ludwig J.A. and Reynolds J.F. 1988. Statistical Ecology (ed.). John Wiley & Sons, New York.

Montes C. 1999. Corredores ecológicos. El Corredor Verde del Guadiamar: Conexión entre espacios naturales. Medioambiente 31: 48–53.

Montes C., Arenas J.M. and Borja F. 2003. Iniciativa científico técnica del PICOVER. *In* PICOVER. Ciencia y Restauración del Río Guadiamar. Junta de Andalucía. (ed.), Consejería de Medio Ambiente, pp. 14–41.

Naiman R.S. and Roger K.H. 1997. Large animal and the maintenance of system level characteristics in river corridors. Bioscience 47: 521–528.

Naiman R.S., Decamps H. and Pollock M. 1993. The role of riparian corridors in maintaining regional biodiversity. Ecological Application 3: 209–212.

Porta A. 1923. Fauna Coleopterorum Italica 2: 90–165.

PICOVER 1999. Programa de Investigación del Corredor Verde. Junta de Andalucía. (ed.), Consejería de Medio Ambiente, Sevilla.

PICOVER. 2003. Ciencia y Restauración del Río Guadiamar. Junta de Andalucía. (ed.), Consejería de Medio Ambiente, Sevilla.

Serrano J. 2003. Catálogo de los Carabidae (Coleoptera) de la Península Ibérica. Monografías, Sociedad Entomológica Aragón 9: 130.

Scholtz C.H. and Holm E. 1995. Insects of Southern Africa. (ed.), Butterworth, Durban, South Africa.

Solà C., Plans M. and Prat N. 2003. Los macroinvertebrados de los ríos Agrio y Guadiamar: evolución de la comunidad, concentración de metales y toxicidad. *In* PICOVER. Ciencia y Restauración del Río Guadiamar. Junta de Andalucía. (ed.), Consejería de Medio Ambiente: 110–125.

Southwood T.R.E. 1991. Ecological Methods, with particular reference to the study of insect populations. (ed.), Chapman & Hall, London.

Tejedo M. and Reques R. 2003. Evaluación de los efectos del vertido tóxico de las minas de Aznalcóllar sobre la comunidad de Anfibios del río Guadiamar. *In* PICOVER. Ciencia y Restauración del Río Guadiamar. Junta de Andalucía. (ed.), Consejería de Medio Ambiente: 156–169.

Toja J., Alcalá E., Burgos M.D., Martín G., Plazuelo A., Schutter de T., Prat N., Plans M. and Solá C. 2003. Efecto del vertido tóxico en las comunidades de plancton y perifiton del río Guadiamar. *In* PICOVER. Ciencia y Restauración del Río Guadiamar. Junta de Andalucía. Ed. Consejería de Medio Ambiente: 94–109.

Vargas J.M. 1993. Siete pecados capitales en Biogeografía. Zoológica Baetica 4: 39–56.

Vermeulen H.J.W. 1994. Corridor function of a road verge for dispersal stenotopic heath land ground beetle Carabidae. Biological Conservation 69: 339–349.

Biodiversity and Conservation (2006) 15:371–379
DOI 10.1007/s10531-005-3432-2


# Diversity of the scuttle fly (Diptera: Phoridae) communities in the plantations of moist pine forests of the Białowieża Primeval Forest and the Tuchola Forest (Poland)

EWA DURSKA
*Museum and Institute of Zoology, PAS, Wilcza 64, 00-679 Warszawa, Poland (e-mail: edurska@miiz.waw.pl; fax: +48-22-6296-302)*

Received 22 July 2004; accepted in revised form 28 February 2005

**Key words:** Diversity, Phenology, Phoridae, Pine plantations

**Abstract.** Scuttle fly communities of pine plantations were investigated in 1986 and 1987 in two sites, Białowieża Primeval Forest and Tuchola Forest. Flies were collected using yellow plastic bowls placed on the ground. Up to now 108 species were identifined, and 46 occurred on both sites. Although the number of species was very similar in Tuchola Forest (77) and Białowieża Primeval Forest (75) species diversity was considerably lower on the former site ($p < 0.05$). Ten dominant species were common: *Megaselia brevicostalis, M. giraudii, M. manicata, M. nigriceps, M. pleuralis, M. pulicaria*-complex, *M. pumila, M. verralli, Metopina oligoneura* and *Triphleba opaca*. Five of these were characteristic of both communities (*Megaselia verrali, M. brevicostalis, M. pumila, Metopina oligoneura* and *Triphleba opaca*). Similarity of qualitative composition for dominants was rather high (Sø = 0.67), but the quantitative similarity was low (Mo = 0.26). During two study seasons in the community of the Białowieża Primeval Forest the dominance structure did not change markedly and *Megaselia verralli* was the dominant (over 20%). *M. verralli* was also the dominant (ca. 30%) in the communities in the Tuchola Forest in 1986, but next year *M. pulicaria*-complex (ca.50%) dominated. Most of the dominant species are multivoltines and generalists. In Tuchola Forest the disturbances caused by the chemical treatment against *Neodiprion sertifer* might be the main factor changing the dominance structure in the phorid community in 1987 (extremely high dominance of *Megaselia pulicaria*-complex).

## Introduction

In intensively managed forests, the clear cut of mature stand and establishment of a new plantation initiate the secondary succession of the ecosystem. In the Polish climate, the secondary succession of moist pine forest lasts from 120 to 150 years (Szujecki 1980). Newly planted culture is colonized by early successional, open area species, mostly generalists with good dispersal powers and pressed for time (Brown and Southwood 1983; Prinzing 2003). There is little information on how disturbances affect Phoridae diversity in the earliest phase of succession. Only my investigations (Durska 2001a) in the pine plantations in the Białowieża Primeval Forest and Prescher et al. (2002) in the chestnut belt of the Alps after wildfires, provide comparable data.

Tscharntke and Brandl (2004) commented on plant–insect interactions thus: 'Communities of insects within a patch are a complex mixture of species that function as a metapopulation...'. Indeed, habitat heterogeneity resulting from various disturbances, like cutting or wildfires, may create very similar conditions for plant-insect interactions. Pine plantations, small, surrounded by older pine stands and patchily distributed, could be analysed by island ecology (Simberloff 1988; Saunders et al. 1991).

The aim of the study was to compare species composition and phenology of the phorid communities of pine plantations established after clear cuts in natural (Białowieża Primeval Forest) and managed (Tuchola Forest) moist pine forests. For this purpose the abundance of the species, dominance structure, similarity indices, index of species diversity and fidelity formula (comparing to the old-growth phase stands) were compared.

## Materials and methods

### *Study areas and sampling method*

Both, the Białowieża Primeval Forest (52°30′–52°50′ N, 23°40′–24°00′ E) and the Tuchola Forest (53°30′–53°50′ N, 18°15′–18°40′ E) are large forest complexes situated in the Polish Lowland. The former mostly consists of natural forest vegetation while in the latter managed tree stands predominate. Plant communities of the moist pine forests are in the Białowieża Primeval Forest represented by the association Peucedano-Pinetum Material in its subboreal variety, and in the Tuchola Forest by the Coastal-Silesian variety of the Leucobryo-Pinetum Material (Matuszkiewicz et al. 1993).

In each site, three several-hectare plots of even-aged pine plantations were selected. In individual plots the age of pines ranged between 3 and 8 years. Distances between plots in a given site did not exceed 5 km.

Phorids were collected in 1986 and 1987 using yellow plastic bowls (diameter 18 cm) with 75% ethylene glycol and detergent. Five traps per each plot were installed on the ground.

The trapping lasted from April through October in the Białowieża Primeval Forest and to mid-November in the Tuchola Forest. Traps were emptied fortnightly.

### *Data analysis*

The analysis of phorid communities was based only on male individuals because most females of the genera *Megaselia* and *Phora* are unidentifiable to species (in our present state of knowledge) and therefore they have been identified to genus only. Moreover, number of species is probably higher, because it was not possible to identify a lot of *Megaselia* males.

To describe phorid communities in the two sites, the abundance of males of individual species and the dominance structures were determined.

Species diversity of the communities was estimated using Shannon–Weaver index. Significance of differencers in the Shannon–Weaver index was checked by chi-squared test (Magurran 1988).

To perform between-site comparisons of the communities, two similarity indices were calculated: (1) Sørensen quotient (Sø) (Sørensen 1948) – comparing presences and absences of the species and (2) Morisita's index (Mo) modified by Horn (1966) to calculate the similarity of dominance structures of the communities.

The fidelity formula ($F$) expressed quantitatively was used to distinguish species characteristic of particular habitats ($F = a/b \times 100$: where $a$ = number of specimens of a given species in a given habitat, and $b$ = total number of specimens of a given species in all compared habitats). Phorid communities of pine plantations were compared with the communities of the old-growth phase stands from the respective sites (Durska 1996; Durska 2001a).

## Results

### *Species composition*

A total of 15,178 adult scuttle flies individuals (Table 1) ware caught in both study areas, but almost 60% of them in the second study season in the Tuchola Forest. The determined specimens belonged to 108 species, which constitute over 30% of the Polish fauna (Disney 1991; Durska 2001b; Durska and Disney unpublished).

For comparison only determined males (6607 specimens) were used. Qualitative similarity (Sørensen's quotient) calculated for dominants was rather high (Sø = 0.67), but the quantitative similarity was low (Mo = 0.26) (Table 2).

Although the species number was slightly higher in the Tuchola Forest, the diversity value ($H'$) there was significantly lower than in Białowieża Primeval Forest ($\chi^2 = 6.249$; $k = 1$; $p < 0.05$) (Tables 1). Of 108 species, 46 occured on

*Table 1.* Parameters of abundance and diversity of Phoridae in pine plantations of Białowieża Primeval Forest and Tuchola Forest.

| | Białowieża Primeval Forest | | | Tuchola Forest | | |
|---|---|---|---|---|---|---|
| | 1986 | 1987 | 1986–87 | 1986 | 1987 | 1986–87 |
| No. of individuals | 2032 | 3087 | 5119 | 1374 | 8685 | 10,059 |
| No. of males | 972 | 1493 | 2465 | 840 | 4175 | 5015 |
| No. of determined males | 875 | 1197 | 2072 | 791 | 3744 | 4535 |
| No. of determined species | 42 | 59 | 75 | 43 | 59 | 77 |
| Shannon–Weaver index | 3.95 | 3.86 | 3.87 | 3.05 | 2.78 | 2.91 |

*Table 2.* Similarity of phorid communities in pine plantations from Białowieża Primeval Forest and Tuchola Forest.

| | Sørensen quotient | Morisita quotient |
|---|---|---|
| All species | 0.61 | 0.36 |
| Common dominants | 0.67 | 0.26 |

*Table 3.* Trophic group placement, percentage contribution and fidelity of phorid species dominating in pine plantations; s = saprophagous, p = polyphagous, ? = unknown.

| Species | Trophic groups | Pine plantations | | | |
|---|---|---|---|---|---|
| | | Białowieża Primeval Forest | | Tuchola Forest | |
| | | % | *F* | % | *F* |
| *Megaselia verralli* (Wood, 1910) | ? | 22.3 | 99.4 | 5.3 | 97.4 |
| *Megaselia brevicostalis* (Wood, 1910) | s | 14.8 | 95.3 | 8.2 | 92.9 |
| *Megaselia nigriceps* (Loew, 1866) | s | 11.0 | 96.4 | 3.4 | 40.8 |
| *Megaselia manicata* (Wood, 1910) | ? | 7.4 | 81.8 | 2.2 | 28.1 |
| *Triphleba opaca* (Meigen, 1830) | s | 5.2 | 86.3 | 9.7 | 95.0 |
| *Megaselia pleuralis* (Wood, 1909) | s | 4.7 | 87.9 | 6.4 | 19.9 |
| *Megaselia pumila* (Meigen, 1830) | ? | 3.1 | 94.4 | 4.1 | 99.5 |
| *Metopina oligoneura* (Mik, 1867) | s | 2.2 | 90.4 | 1.5 | 97.4 |
| *Megaselia giraudii*-complex (Egger, 1862) | p | 1.3 | 3.3 | 2.2 | 7.0 |
| *Megaselia pulicaria*-complex (Fallen, 1823) | p | 1.0 | 19.1 | 46.2 | 81.5 |
| Total | | 73.0 | | 89.2 | |

both sites, resulting in Sørensen quotient of similarity of 0.61 (Table 2). Ten species (*Megaselia brevicostalis*, *M. giraudii*, *M. manicata*, *M. nigriceps*, *M. pleuralis*, *M. pulicaria*-complex, *M. pumila*, *M. verralli*, *Metopina oligoneura* and *Triphleba opaca)* occurred on both pine plantations with dominances ≥ 1% (Table 3). The dominants made up ca. 90% of each community and fidelity (comparing to the old-growth stands) of most of the species exceeded 80%. Five species with the highest fidelity (*Megaselia verrali*, *M. brevicostalis*, *M. pumila*, *Metopina oligoneura* and *Triphleba opaca*) were considered as characteristic of the phorid communities in pine plantations (Table 3).

In the both phorid communities, the number of species was almost the same in the first and the second study seasons (Table 1). The dominant species with the known diet are saprophagous or polyphagous species (Table 3).

In the community of the Białowieża Primeval Forest the dominance structure was very similar during two years and the pioneer species with a high colonization potential – *Megaselia verralli* dominanted (over 20%) (Figure 1a). *M. verralli* was also the dominant (ca. 30%) in the Tuchola Forest in 1986 but the next year (1987) the dominance structure changed rapidly and the polyphagous *M. pulicaria*-complex dominated (ca.50%) (Figure 1b).

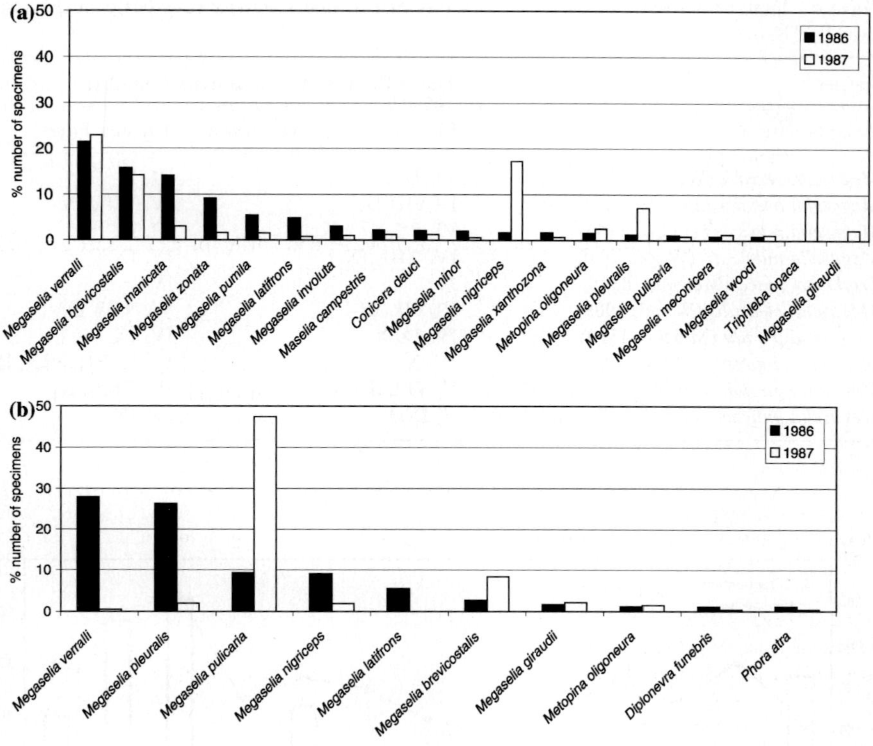


*Figure 1.* Dominance structure of phorid communities of pine plantation in the Białowieża Primeval Forest (a) and the Tuchola Forest (b) (pooled values from 1986–1987; male abundance ≥ 1%).

### *Phenology*

In both sites the majority of the both pine plantation species reached their abundance peaks in spring (April–May; the lower peak) and autumn (September; the higher peak) (Figure 2) *Megaselia brevicostalis, M. giraudii, M. manicata, M. nigriceps, M. pleuralis, M. pulicaria*-complex, *M. pumila, M. verralli*, and *Metopina oligoneura* are mulivoltine, with spring (IV–VI) and autumn (IX) maximum of male activity. *Triphleba opaca* is univoltine with one maximum in spring (IV) (Table 4).

## Discussion

Using yellow plastic bowls it is possible to collect material from the forest canopy as well as the forest floor (Durska 1996, 2001a). Su and Woods (2001) studied the vertical distribution of insect faunas in managed forests and

*Table 4.* Dominant species of both pine plantations and males activity (number of males individuals ≥ 1%).

| Species | Time of peaks of male activity (months) | |
|---|---|---|
| Pine plantations: | Białowieża Primeval Forest | Tuchola Forest |
| *Megaselia verralli* (Wood, 1910) | VI, IX | VI, X |
| *Megaselia brevicostalis* (Wood, 1910) | IV,VIII,IX | IV, VIII, IX |
| *Megaselia nigriceps* (Loëw, 1866) | VI, IX | VI, IX |
| *Megaselia manicata* (Wood, 1910) | IV, VIII, IX | IV, IX |
| *Triphleba opaca* (Meigen, 1830) | IV | IV |
| *Megaselia pleuralis* (Wood, 1909) | VI/VII, IX | VI, VIII, IX |
| *Megaselia pumila* (Meigen, 1830) | IV, IX/X | IV, IX |
| *Metopina oligoneura* (Mik, 1867) | IV, X | V, VI, VII, VIII, IX |
| *Megaselia giraudii*-complex (Egger, 1862) | V, VI/VII, IX | IV, VIII, IX |
| *Megaselia pulicaria*-complex (Fallen, 1823) | V, IX/X | IV, IX |

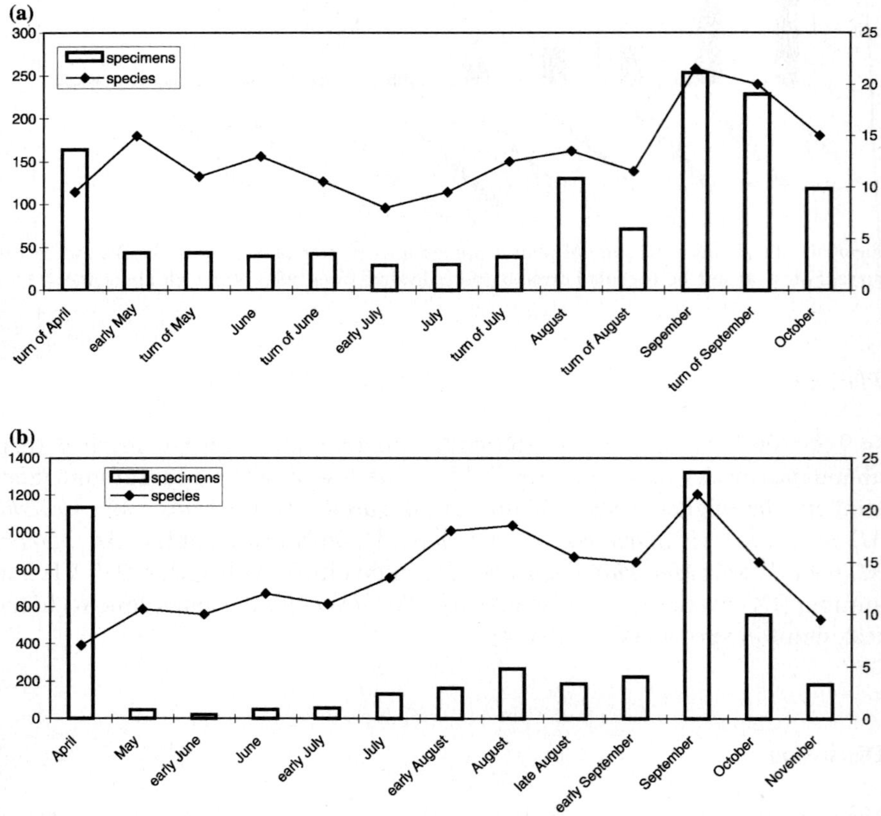


*Figure 2.* Phenology of phorid communities in pine plantation of the Białowieża Primeval Forest (a) and the Tuchola Forest (b) (mean values of specimens and species from 1986–1987).

reported that the lower traps adequately sampled the canopy fauna. Disney (2004) reported that yellow traps were selective for particular species. Thunes et al. (2004) using a motorized conopy fogger received only 101 phorid specimens that belonged to 16 determined species. Nevertheless *Megaselia* was the most numerous genus (over 80% of collected specimens) and Phoridae 'showed an extraordinary high number of new faunistical records'. The yellow bowls used in my study confirmed the results of the previous studies that the contribution of the genus *Megaselia* to all Phoridae caught is about 70%. (Disney 1994; Durska 1996, 2001a). Studies by Goos (1975) on flying insects in the sugar–beet plantations and also my investigations (Durska 1996, 2001a) indicated that Phoridae are one of the most abundant families of Diptera that are caught in yellow plastic bowls.

The structure of insect communities most of all is connected with the phase of succession and degree of anthropopressure including the use of chemicals against forest pests. In the Tuchola Forest the disturbances caused by the chemical treatment against *Neodiprion sertifer* might be the most important factor changing the dominance structure in the phorid community in 1987 (Sawoniewicz 1999). The lower values of diversity (*H'*) recorded in that year were associated with the extremely high dominance of polyphagous *Megaselia pulicaria*-complex (Table 1, Figure 1b). After chemical treatment against aphids the considerable increase of the abundance of Phoridae was observed by Goos (1975) in the sugar–beet plantations suggesting that the quantitative environmental differences may affect community structure.

The high number of species in the phorid communities of both investigated Forests (over 75 species in each site) (Durska 1996, 2001a), as well as of chestnut forests in the Swiss Alps after wildfires (Prescher et al. 2002), is probably an effect of habitat heterogeneity. Small habitats colonized by plants and animals during the secondary succesion are like islands re-colonized by characteristic species. *Megaselia verralli* and probably some other species dominating in both pine plantations and the chestnut belt in the Swiss Alps, seem to prefer habitats under high stress or disturbance. The new enviromental conditions, after patchy disturbances, provide opportunities for those open-area species to reach dominant position in the community.

Insect species typical of rather unstable habitat of pine plantation would be expected to be generalists with multivoltine life cycle, and tolerant of abiotic stress (Southwood 1988; Durska 2002; Prinzing 2003). Data on the life cycle strategies of the dominant *Megaselia* species are consistent with this assumption. The different phenology has only one of the dominants univoltine *Triphleba opaca* which is known as cold-adapted species (Durska 1996, 2001a, 2003; Soszyńska and Durska 2002; Soszyńska 2004).

The composition of a phorid community depends, first of all, on microlimatic factors and the physical structure of the habitat and seems to be little affected by plant species composition. It appeared that different kinds of disturbance, i.e. cutting and wildfire, similarly affect phorid communities. Also spider and carabid faunas have been found to response to grazing and burning in a parallel way (Gibson et al. 1992; Zulka et al. 1997; Moretti et al. 2002; Fernandez and Costas 2004)

The high diversity of the Phoridae communities as a result of stress or disturbance endorses the intermediate disturbance hypothesis (IDH) (Connell 1978; Huston 1979). IDH is 'a promoter of species coexistance' and 'broader in scope and richer in detail than has previously been recognized' (Roxburgh et al. 2004).

## Acknowledgements

I especially thank Dr Piotr Ceryngier and the anonymous referee for their advice and helpful comments on drafts of the manuscript. I wish to thank Dr R. Henry L. Disney for his useful comments and improvement of the English presentation of this paper. I am grateful to MSc Krzysztof Gagla for his invaluable assistance in the segregation of the material.

## References

Brown V.K. and Southwood T.R.E. 1983. Tropic diversity, niche breadth and generation times of exopterygote insects in a secondary succession. Oecologia 56: 220–235.

Connell J.H. 1978. Diversity in tropical rain forests and coral reefs. Science 199: 1302–1310.

Disney R.H.L. 1991. Family Phoridae. In: Soós A. and Papp L. (eds), Catalogue of Palaearctic *Diptera*. Volume 7 (Dolichopodidae – Platypezidae). Akademiai Kiado, Budapest pp. 143–204.

Disney R.H.L. 1994. Scuttle flies: The Phoridae. Chapman and Hall, London, 467pp.

Disney R.H.L. 2004. Species preferences for white versus yellow water traps for scuttle flies (Diptera: Phoridae). Entomologist's Monthly Magazine 140: 31–35.

Durska E. 1996. The species composition and structure of scuttle fly communities (Diptera: Phoridae) in mature tree stands in pine forests at different stages of habitat degradation. Fragmenta faunnistica 39: 267–285.

Durska E. 2001a. Secondary succession of scuttle fly (Diptera: Phoridae) communities in moist pine forest in Białowieża Forest. Fragmenta faunistica 44: 81–130.

Durska E. 2001b. Zadrowate (Diptera: Phoridae). In: Gutowski J.M. and Jaroszewicz B. (eds), Catalogue of the fauna of Białowieża Primeval Forest. IBL, Warszawa, pp. 286–289.

Durska E. 2002. The phenology of dominant scuttle-fly (Diptera: Phoridae) species in the Białowieża Forest. Entomologica Fennica 13: 123–127.

Durska E. 2003. The phenology of *Triphleba* Rondani species (Diptera: Phoridae) in moist pine forests in the Białowieża Forest. Entomologica Fennica 14: 177–182.

Fernández Fernández M.M. and Salgado Costas J.M. 2004. Recolonization of burnt pine forest (*Pinus pinaster*) by Carabidae (Coleoptera). European Journal of Soil Biology 40: 47–53.

Gibson C.W.D., Hambler C. and Brown V.K. 1992. Changes in spider (Araneae) assemblages in relation to succession and grazing management. Journal of Applied Ecology 29: 132–142.

Goos M. 1975. Influence of aphicides used in sugar–beet plantations on arthropods. IV. Studies on the flying insects captured in yellow receptacles. Polskie Pismo Entomologiczne Bulletin entomologique de Pologne, Wrocław 45: 423–444. (in Polish, summary in English).

Horn H.S. 1966. Measurement of "overlap" in comparative ecological studies. American Naturalist 100: 419–424.

Huston M. 1979. A general hypothesis of species diversity. American Naturalist 113(1): 81–101.

Magurran A.E. 1988. Ecological Diversity and its Measurement. Princeton University Press, Princeton, New Jersey, 179pp.

Matuszkiewicz J., Degórski M. and Kozłowska A. 1993. Description of the plant association structure and soils of pine forest stands situated in five regions of Poland. In: Species composition and structure of pine forests fauna in Poland. Part I. Fragmenta faunnistica vol. 36. pp. 13–36.

Moretti M., Conedera M., Duelli P. and Edwards P.J. 2002. The effects of wildfire on ground-active spiders in deciduous forests on the Swiss southern slope of the Alps. Journal of Applied Ecology 39: 321–336.

Prescher S., Moretti M. and Duelli P. 2002. Scuttle flies (Diptera, Phoridae) in *Castanea sativa* forests in the southern Alps (Ticino, Switzerland), with thirteen species new to Switzerland. Mitteilungen der Schweizerischen Entomologischen Gesellschaft. Bulletin de la Société Entomologique Suisse 75: 289–298.

Prinzing A. 2003. Are generalists pressed for time? An interspecific test of the the time-limited dispersal model. Ecology 84(7): 1744–1755.

Roxburgh S.H., Shea K. and Wilson J.B. 2004. The intermediate disturbance hypothesis: patch dynamics and mechanisms of species coexistence. Ecology 85(2): 359–374.

Saunders D., Hobbs R.J. and Margules C.R. 1991. Biological consequences of ecosystem fragmentation a review. Conservation Biology 5: 18–32.

Sawoniewicz J. 1999. Development dynamics of Ichneumonidae (Hymenoptera) communities of *Pinus sylvestris* L. canopies in fresh pine forest in the three health zones of Poland's forests. Thesis, Faculty of Forestry, Warsaw Agricultural University, Poland.

Simberloff D. 1988. The contribution of population and community biology to conservation science. Annual Review of Ecology and Systematics 19: 473–51.

Soszyńska A. and Durska E. 2002. Coldadapted scuttle flies species of *Triphleba* Rondani (Diptera: Phoridae). Annales Zoologici 52(2): 279–283.

Soszyńska A. 2004. The influence of environmental factors on the supranivean activity of flies (Diptera) in Central Poland. European Journal of Entomology 101: 481–489.

Southwood T.R.E. 1988. Tactics, strategies and templets. Oikos 52: 3–18.

Sørensen T. 1948. A method of establishing groups of equal amplitude in plant sociology based on similarity of species content and its application to analyses of the vegetation on Danish commons. Biologiske Skrifter 5: 1–34.

Su J.C. and Woods S.A. 2001. Importance of sampling along a vertical gradient to compare insect fauna in managed forests. Environmental Entomology 30(2): 400–408.

Szujecki A. 1980. Ekologia owadów leśnych. PWN, Warszawa, 603pp.

Thunes K.H., Skartveit J., Gjerde I., Starý J., Solhøy T., Fjellberg A., Kobro S., Nakahara S., zur Strassen R., Vierbergen G., Szadziewski R., Hagan D.V., Grogan W.L.Jr., Jonassen T., Aakra K., Anonby J., Greve L., Aukema B., Heller K., Michelsen V., Haenni J.-P., Emeljanov A. F., Douwes P., Berggren K., Franzen J., Disney R.H.L., Prescher S., Johanson K. A., Mamaev B., Podenas S., Andersen S., Gaimari S.D., Nartshuk E., Søli G. E. E., Papp L., Midtgaard F., Andersen A., von Tschirnhaus M., Bächli G., Olsen K. M., Olsvik H., Földvári M., Raastad J. E., Hansen L. O. and Djursvoll P. 2004. The arthropod community of Scots pine (*Pinus silvestris* L.) canopies in Norway. Entomologica Fennica 15: 65–90.

Tscharntke T. and Brandl R. 2004. Plant–insect interactions in fragmented landscapes. Annual Review of Entomology 49: 405–430.

Zulka K.P., Milasowszky N. and Lethmayer Ch. 1997. Spider biodiversity potential of an ungrazed and grazed inland salt meadow in the National Park "Neusiedler See-Seewinkel" (Austria): implications for management (Arachnida: Araneae). Biodiversity and Conservation 6: 75–88.

Biodiversity and Conservation (2006) 15:381–395
DOI 10.1007/s10531-005-3435-z 

# Spatial distribution of ground beetles (Coleoptera: Carabidae) and moths (Lepidoptera) in the Mrtvý luh bog, Šumava Mts (Central Europe): a test of habitat island community

ALEŠ BEZDĚK[1,2,*], JOSEF JAROŠ[1] and KAREL SPITZER[1]
[1]*Institute of Entomology, Academy of Sciences of the Czech Republic, Branišovská 31, CZ-370 05 České Budějovice, Czech Republic;* [2]*Faculty of Biological Sciences, University of South Bohemia in České Budějovice, Branišovská 31, CZ-370 05 České Budějovice, Czech Republic;* **Author for correspondence (e-mail: bezdek@entu.cas.cz; phone: +420-387775270; fax: +420-385310354)*

Received 3 November 2004; accepted in revised form 28 February 2005

**Key words:** Carabidae, Central Europe, Community ecology, Conservation, Ecological gradient, Habitat island, Lepidoptera, Paleorefugium, Peat bog, Relicts

**Abstract.** Spatial distribution of ground beetles and moths in the isolated Central European Mrtvý luh bog was analyzed. The most stenotopic tyrphobiontic species (relicts restricted to the peat bog) of ground beetles (Coleoptera: Carabidae) and moths (Lepidoptera) are distributed according to a distinct ecological gradient between the bog margin (lagg) and the bog centre. The degree of habitat preference between the bog margin and centre is taxonomically specific and significant. A list of stenotopic species of high conservation value is given. Several tyrphobiontic species occur in the treeless bog centre only. The tyrphophilous and tyrphoneutral species are distributed in the peat bog mostly randomly, some of such species prefer bog margins. The migratory highly opportunistic moths from habitats outside the bog usually cross the treeless centre. Most of the ubiquitous tyrphoneutral and migratory moths represent the faunal component, which is a very temporary phenomenon only, not associated with the peat bog permanently. The conservation of insect biodiversity associated with isolated peat bogs depends on complete preservation of the edaphic ecological conditions of the fragile paleorefugial habitat island and its spatial structure.

## Introduction

Most of the Bohemian landscape (Czech Republic) is man-made and was repeatedly modified over centuries. Recently, only few types of habitats represent isolated fragments of natural semi-virgin biotopes more or less unaffected by human impact (e.g., glacial stony scree communities, remnants of primary forests and some wetlands). The best examples of such biotops are 'habitat islands' of peat bogs in the Šumava Mts (South Bohemia, Central Europe see e.g., Spitzer 1981, 1994, 2004). These Central European isolated peat bogs could be classified as ancient paleorefugia (sensu Nekola 1999) and as relictual habitats of some formerly widespread biota of forest-tundra biomes of the early Holocene (Jankovská 1980, 1995; Mikkola and Spitzer 1983) with the endemic plant community Pino rotundatae–Sphagnetum

(Neuhäusl 1972). Holocene history, local microclimate and constant edaphic peatland environment have been the basic factors favouring survival of cold-adapted and stress-tolerant relict insects and their unique association with the bog (Spitzer 1981, 2004). Our aim in this publication is to answer the following questions (1) Are all the ground beetles (Coleoptera: Carabidae) and all phototactic moths (Lepidoptera) distributed within the bog habitat island randomly or are their communities specifically associated with the central and marginal habitats of the bog? (2) What are the basic ecological groups of the bog insect species (sensu Peus 1932; Roubal 1934; Mikkola and Spitzer 1983) and how are they associated with plant communities along a mesoclimatic gradient between the bog centre and margins (lagg)? (3) Is the investigated bog Mrtvý luh (one of the largest peat bogs in Central Europe) a true paleorefugial habitat island differing from other Central European peatlands in insect biodiversity?

## Material and methods

### *Study area*

The investigated bog locality was the Mrtvý luh Nature Reserve in the Šumava Mts, SW Bohemia (48°52′ N, 13°52′ E, 740 m a.s.l.), which is a large montane oligotrophic valley peat bog covering an area of 350 ha bordered by the confluence of the rivers Studená Vltava and Teplá Vltava, a part of the core zone of the Šumava National Park (Figures 1 and 2). The Mrtvý luh bog was already an important locality of ecological and entomological faunistic research (see e.g., Holubičková 1960; Novák and Spitzer 1972; Šula and Spitzer 2000; Spitzer et al. 2003). The carabid beetles and phototactic Lepidoptera were monitored in two different habitats within the bog:

(A) *Marginal open elfin pine forest associated with the lagg*. The open pine forest forms the marginal belt of the peat bog. The tree layer is dominated by polycormic bog pine *Pinus rotundata* Link (= *P.* '*mugo* Turra s. lat.'), the ground layer by various species of *Vaccinium*. The herb coverage is about 40% and the moss layer is strongly dominated by *Sphagnum angustifolium* (Jensen ex Russow), with scattered *S. fallax* (Klingg.) and *S. flexuosum* Dozy & Moll. (see Table 1, cf. also Novák and Spitzer 1972; Spitzer et al. 2003). The arrangement of botanical data in Table 1 are based on the method of visual estimation after Mueller Dombois and Ellenberg (1974).

(B) *Treeless centre*. The treeless centre forms most of the central part of the peat bog with some scattered shrubby formations of *Pinus rotundata, Vaccinium uliginosum* L., *Eriophorum vaginatum* L. and *Calluna vulgaris* (L.) being the most abundant plant species. The low shrubby plant cover forms about 50% of the total area. The moss layer is dominated by *Sphagnum fuscum* (Schimp.), followed by *S. angustifolium* and *Polytrichum stictum* Brid. (Table 1).

*Figure 1.* Aerial photograph of the Mrtvý luh bog and adjacent peatbogs.

The habitats A and B strongly differ in plant composition and structure (Table 1), and also in unequal microclimatic (thermal) characteristics – for minimum and maximum temperatures measured near the ground see Table 2.

*Sampling*

Ten unbaited pitfall traps were used in both sites for collecting carabid beetles. Each trap consisted of a 1000 ml glass bottle (diameter 60 mm) with a plastic cover to protect the traps from rainfall and litter. Traps were partially filled with 4% formaldehyde as a preservation liquid. The traps were arranged in two straight lines of five traps at 5 m intervals. Pitfall traps have been operated in each site for three years (2000–2002), periodically during the whole vegetation season from the end of April to the end of September. One pitfall trap sample is consisted of four days' catch. Additionally, five pitfall traps were placed in a mixed forest closely adjacent to the peat bog in 2002. The species composition and abundance of Lepidoptera were monitored by BL–Pennsylvania light traps (8 W black light tube). One light trap was operated at each site, one sample consists of a four nights' catch. The pitfall and light traps were operated simultaneously ones or twice per month, and they were emptied on the same dates. Eleven samples were taken in 2000, nine samples in 2001 and six samples in 2002.

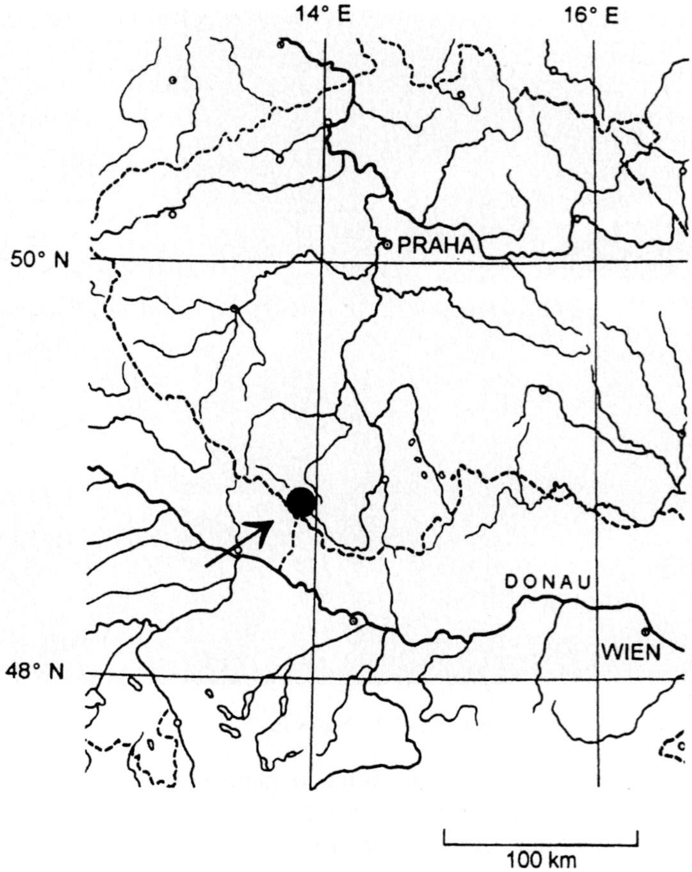


*Figure 2.* Geographical position of the Mrtvý luh bog in Central Europe.

All the ground beetles and moths were identified and divided in to the following ecological categories according to their habitat preferences: (1) tyrphobionts are species obligatorily associated with peat bogs, and restricted to the bogs within the temperate and southern boreal zones and represent relict taxa, (2) tyrphophilous species are peat bog characteristic, but not restricted to bogs, and are usually much more abundant on peat bogs in comparison with other habitats (conifer forests, heathlands, other types of wetlands), (3) tyrphoneutral species have no specific preference for peat bogs, and are usually distributed in various habitats around the bog locality, (4) migratory species are only a very temporary foreign component of the peat bog fauna not permanently living in peatlands except for a short period of adult migratory behaviour (e.g., sometime including those searching for nectar sources). The ecological classification of these four categories and explanation of the terminology follow several already published sources (e.g., Peus 1932; Roubal 1934;

*Table 1.* Plant community (phytocenological releves) of the Mrtvý luh peat bog of the study sites.

| Plant species | % |
|---|---|
| Mrtvý luh–margin | |
| $E_2$ (20%) | |
| *Pinus rotundata* Link | 20 |
| $E_1$ (40%) | |
| *Andromeda polifolia* L. | 0.1 |
| *Calluna vulgaris* (L.) Hull | 1 |
| *Oxycoccus palustris* Pers. | 0.5 |
| *Vaccinium myrtillus* L. | 10 |
| *Vaccinium uliginosum* L. | 10 |
| *Vaccinium vitis-idaea* L. | 20 |
| $E_0$ (90%) | |
| *Aulacomnium palustre* (Hedw.) Schwägr. | 0.1 |
| *Dicranum bergeri* Blandow ex Hoppe | 0.1 |
| *Polytrichum juniperinum* Hedw. | 2 |
| *Sphagnum angustifolium* (C.E.O. Jensen ex Russow) C.E.O. Jensen | 60 |
| *Sphagnum fallax* (H. Klinggr.) H. Klinggr. | 20 |
| *Sphagnum flexuosum* Dozy & Molk. | 10 |
| *Sphagnum magellanicum* Brid. | 1 |
| Mrtvý luh–centre | |
| $E_1$ (50%) | |
| *Andromeda polifolia* L. | 5 |
| *Calluna vulgaris* (L.) Hull | 15 |
| *Eriophorum vaginatum* L. | 15 |
| *Oxycoccus palustris* Pers. | 0.5 |
| *Vaccinium uliginosum* L. | 15 |
| $E_0$ (80%) | |
| *Cephalozia connivens* (Dicks.) Lindb. | 0.1 |
| *Cladonia* sp. | 1 |
| *Dicranum bergeri* Blandow ex Hoppe | 0.5 |
| *Mylia anomala* (Hook.) Gray | 2 |
| *Pleurozium schreberi* (Brid.) Mitt. | 1 |
| *Polytrichum strictum* Brid. | 15 |
| *Sphagnum angustifolium* (C.E.O. Jensen ex Russow) C.E.O. Jensen | 5 |
| *Sphagnum flexuosum* Dozy & Molk. | 1 |
| *Sphagnum fuscum* (Schimp.) H. Klinggr. | 50 |
| *Sphagnum magellanicum* Brid. | 0.1 |
| *Sphagnum rubellum* Wilson | 3 |

*Table 2.* Maximum and minimum temperatures (°C) measured near ground in the centre and in the margin of Mrtvý luh bog in 2000.

| | May | | June | | July | | August | |
|---|---|---|---|---|---|---|---|---|
| | Centre | Margin | Centre | Margin | Centre | Margin | Centre | Margin |
| Max. temperature | 32.2 | 31.4 | 36.5 | 33.7 | 31.1 | 29.2 | 35.7 | 32.2 |
| Average max. temperature | 24.4 | 23.5 | 26.9 | 26.3 | 22.2 | 20.8 | 27.5 | 24.9 |
| Min. temperature | −5.8 | −4.9 | −5.8 | −3.5 | −2.2 | −0.6 | −4.0 | −2.7 |
| Average min. temperature | 0.3 | 0.9 | 2.0 | 2.9 | 5.0 | 5.4 | 5.4 | 5.9 |

Spitzer 1981; Roháček and Máca 1982; Mikkola and Spitzer 1983; Spitzer and Jaroš 1993; Dapkus 2001; Spitzer 2004).

### *Statistical analysis*

CCA–(partial) canonical correspondence analysis method using CANOCO ver. 3.12 program package (Ter Braak 1987) was used for evaluation of the

*Table 3.* List of Carabidae collected outside the Mrtvý luh bog (mixed forest near the bog border) in 2002. Tn–tyrphoneutral.

| Species | Category | Number of specimens |
|---|---|---|
| *Abax parallelepipedus* (Piller & Mitterpacher) | tn | 21 |
| *Amara ovata* (Fabricius) | tn | 1 |
| *Carabus auronitens* Fabricius | tn | 2 |
| *Carabus convexus* Fabricius | tn | 1 |
| *Carabus hortensis* Linnaeus | tn | 4 |
| *Carabus vioiaceus* Linnaeus | tn | 3 |
| *Nebria brevicollis* (Fabricius) | tn | 1 |
| *Pterostichus aethiops* (Panzer) | tn | 1 |
| *Pterostichus oblongopunctatus* (Fabricius) | tn | 3 |
| *Pterostichus rhaeticus* Heer | tn | 1 |
| *Trichotichnus laevicollis* (Duftschmid) | tn | 1 |

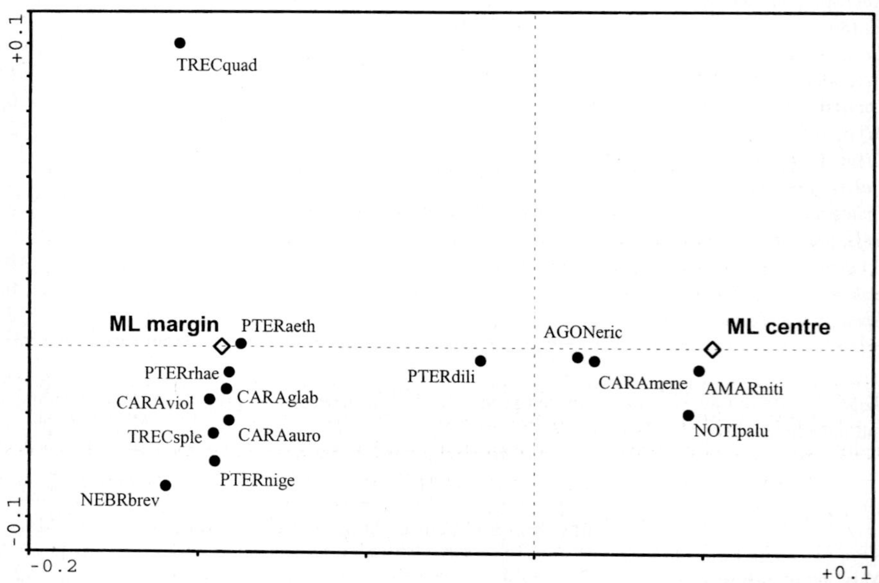


*Figure 3.* CCA ordination biplot representing the habitat preferences of peat bog Carabidae. Abbreviations of species names are composed from the first four letters of genera and species names (see Appendix A).

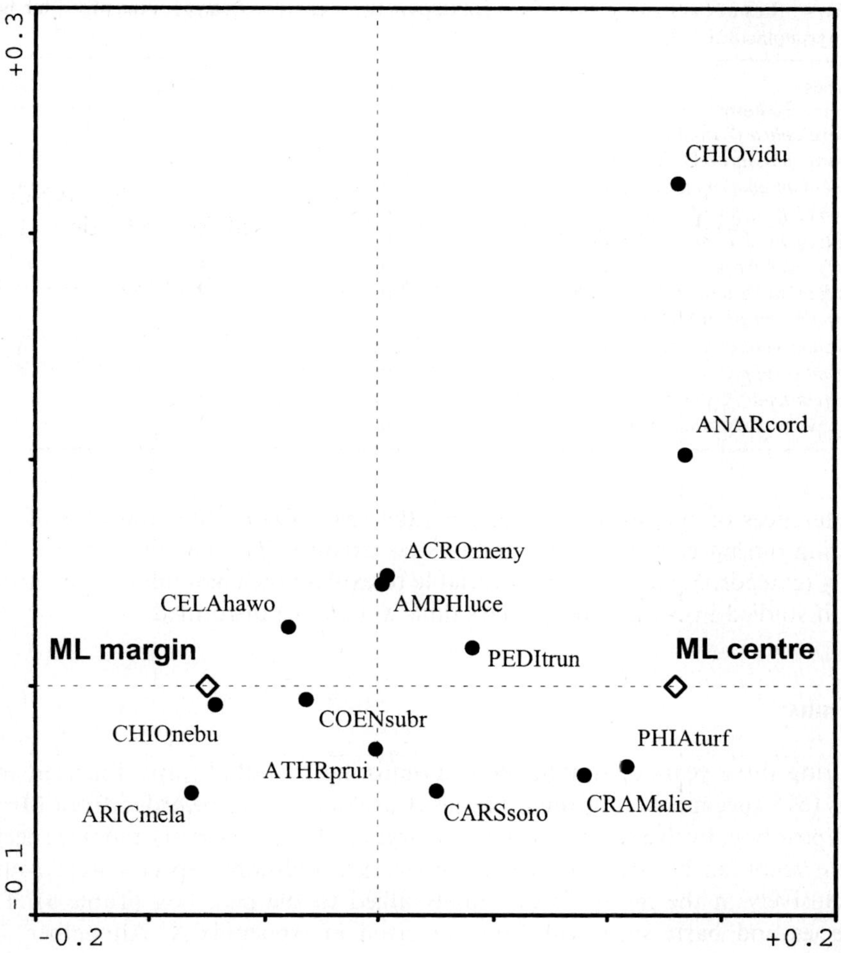


*Figure 4.* CCA ordination biplot representing the habitat preferences of tyrphobiontic Lepidoptera. Abbreviations of species names are composed from the first four letters of genera and species names (see Appendix A).

habitat preferences of ground beetles and phototactic Lepidoptera to the particular habitat type. Constrained ordinations roughly correspond to regressions, where both the explanatory and response variables are multivariate (although weighted averaging is used instead of least squares in CCA). Traditionally, the explanatory variables are called environmental variables in constrained ordinations. The CCA method is based on unimodal response of species abundance to environmental gradient. For more details, see Jongman et al. (1987).

The pitfall trap data and light trap data were analyzed separately. To disentangle the differences in species composition between years and seasonal

*Table 4.* List of Carabidae collected by light traps in two successional stages of the Mrtvý luh bog. Tn–tyrphoneutral.

| Species | Category | Margin | Centre |
|---|---|---|---|
| *Amara aulica* (Panzer) | tn | | 1 |
| *Amara consularis* (Duftschmid) | tn | | 1 |
| *Bembidion quadrimaculatum* (Linnaeus) | tn | | 1 |
| *Bembidion varium* (Olivier) | tn | | 2 |
| *Bradycellus caucasicus* (Chaudoir) | tn | | 1 |
| *Bradycellus harpalinus* (Audinet–Serville) | tn | 4 | 11 |
| *Bradycellus verbasci* (Duftschmid) | tn | 2 | |
| *Harpalus froelichii* Sturm | ? | | 1 |
| *Harpalus calceatus* Duftschmid | ? | | 1 |
| *Harpalus rufipes* (De Geer) | tn | | 1 |
| *Tachyta nana* (Gyllenhal) | tn | 1 | |
| *Trechus quadristriatus* (Schrank) | tn | 1 | 2 |

preferences of species during the year, the year of collecting and date of collecting during each year were used as covariables. The habitat type was the only (categorial) environmental variable to explain the community structure of both studied insect groups. Species data were log-transformed.

## Results

During three years of simultaneous monitoring by pitfall traps, fourteen species (515 specimens) of ground beetles (Carabidae) were recorded from Mrtvý luh peat bog, twelve species at site A (marginal elfin pine forest) and five species were recorded at site B (treeless centre). Six additional species were found exclusively in the mixed forest closely allied to the peat bog (Table 3). The species and basic statistical data are given in Appendix A. Altogether 346 species (6914 specimens) of phototactic Lepidoptera were recorded in both sites, 292 species were discovered at site A and 199 species at site B.

The results provide evidence that tyrphobiontic and tyrphophilous carabid beetles are distributed in both monitored sites, but their preference for treeless centre is evident. Of the 130 specimens of ground beetles found by pitfall traps on the treeless centre, almost 95% belonged to only three species: *Carabus menetriesi* Hummel, *Agonum ericeti* (Panzer) (both are tyrphobionts) and one tyrphophilous species *Pterostichus diligens* (Sturm). All three species were also found in the marginal elfin pine forest site, but in lower abundancies. Solely forest and/or eurytopic carabids are distributed in the mixed forest allied to the peat bog near the lagg beyond the margins.

Similar results were obtained for the phototactic Lepidoptera data set, but a preference of tyrphobiontic and tyrphophilous species for the treeless centre is less demonstrable (Figures 4 and 5). Tyrphobionts like *Crambus alienellus* (Germar & Kaulfuss) and *Phiaris turfosana* (Herrich-Schäffer) were recorded

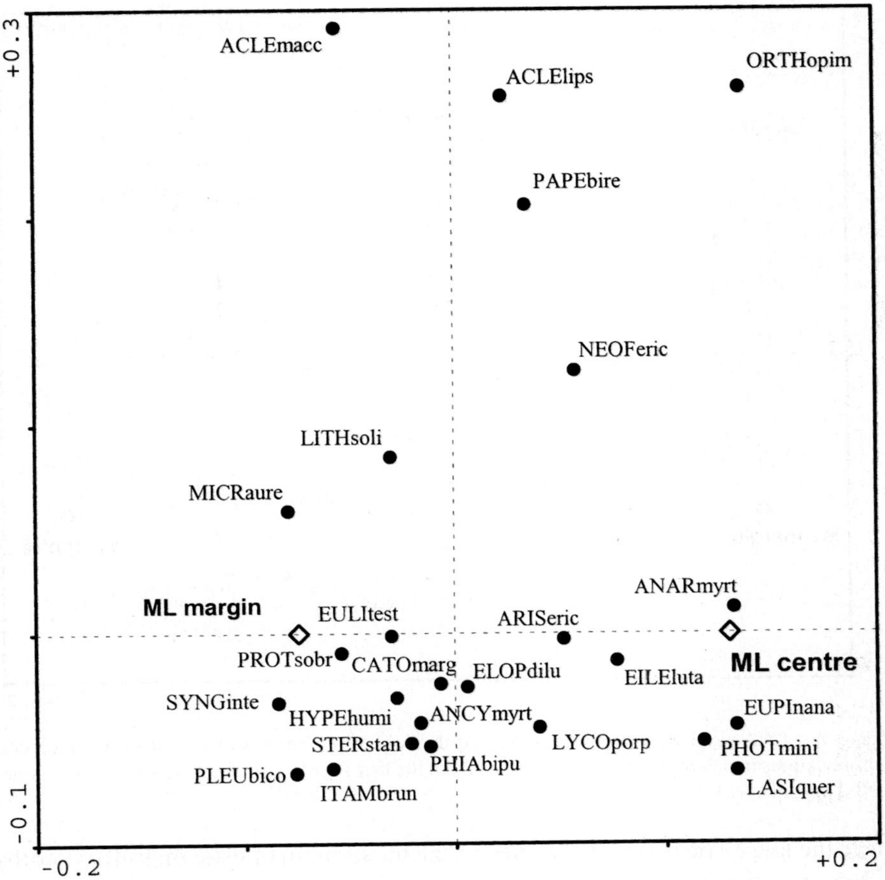


*Figure 5.* CCA ordination biplot representing the habitat preferences of tyrphophilous Lepidoptera. Abbreviations of species names are composed from the first four letters of genera and species names (see Appendix A).

predominantly from the centre, but several local stenotopic tyrphobiontic lepidopteran species are also characteristic for the marginal elfin pine forest, e.g. *Chionodes nebulosella* (Heinemann) and *Arichanna melanaria* (L.). The tyrphophilous moths of the centre seem to be predominantly associated with the heath (*Calluna vulgaris*). On the other hand, typical migratory species not associated with the bog at all were mostly recorded migrating through the treeless centre (Figure 6, Appendix A).

The distribution of ground beetles and moths within the bog depends on the ecological category of species–the tyrphobiontic species mostly prefer the bog centre, but some of them are distributed randomly in the habitat island (Figure 3, Appendix A) but not dispersing outside the bog (with the exception of some "erratic flight"– cf. Novák and Spitzer 1972; Mikkola and Spitzer 1983). The tyrphophilous taxa are distributed widely, mostly within the lagg or

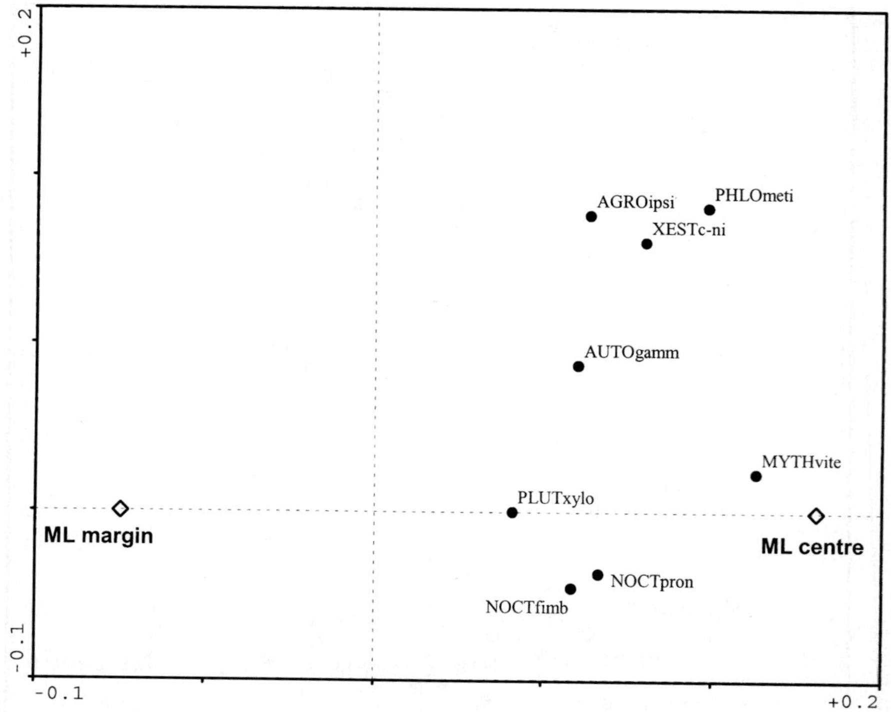


*Figure 6.* CCA ordination biplot representing the habitat preferences of migratory Lepidoptera. Abbreviations of species names are composed from the first four letters of genera and species names (see Appendix A).

near the lagg. The migratory routes of the most opportunistic migratory moths often cross the treeless centre during a very temporary long distance migration Figure 6, Appendix A). Such behaviour is characteristic for many eurytopic tyrphoneutral and migratory moths from outside of the peat bog. Few single opportunistic ground beetles behave also by a "migratory" way, attracted by light traps from outside of the bog (e.g. *Harpalus calceatus* Duftschmid and *H. froelichii* Sturm – see Table 4).

## Discussion

The isolated peat bogs of the temperate zone represent an important unique habitat island with associated relict insect and other biota (e.g., Peus 1932; Spitzer 1981, 1994; Roháček and Máca 1982; Mikkola and Spitzer 1983). Such paleorefugia (Nekola 1999) seem to be very important for testing basic hypotheses about habitat islands (= ancient isolated biotops) and their specific insect biodiversity for applications in conservation biology (Spitzer et al. 1999, 2003). The Mrtvý luh is one of the most valuable large montane peatlands in Central Europe for studies of isolated biotops and their specific insect

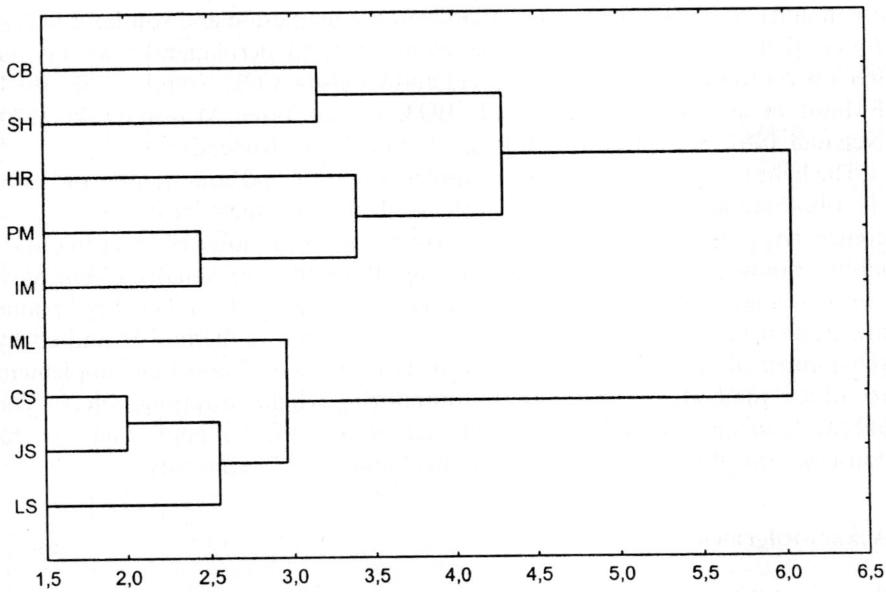


*Figure 7.* Similarity of tyrphobiontic moth communities (presence–absence data, single linkage, Euclidean distance) among Central European peat bogs: CB–Červené blato (South Bohemia), CS–Chalupská slat' (Šumava Mts), HR–Hradčanské rybníky (North Bohemia), IM–Ibmer Moor (Upper Austria), JS–Jezerní slat' (Šumava Mts), LS–Luzenská slat' (Šumava Mts), ML–Mrtvý luh (Šumava Mts), PM–Pürgschachen Moor (Styria), SH–Suchá hora (North Slovakia).

communities (Novák and Spitzer 1972; Šula and Spitzer 2000). The entomofaunistic position of the Mrtvý luh is unique among the other peat bog paleorefugia in Central Europe – see Figure 7 based on data of distribution of relict tyrphobiontic Lepidoptera (see Spitzer and Jaroš 1993, 2001).

Our results provide evidence that the majority of tyrphobiontic species obligatorily associated with the peat bog are distributed within the habitat islands with high preference for the bog centre (e.g., the ground beetles *Carabus menetriesi* and *Agonum ericeti* and the moths *Crambus alienellus* and *Phiaris turfosana*). The limited dispersal power of the carabid *Agonum ericeti* is also described by De Vries and Den Boer (1990) for an application in habitat conservation. Small dispersal power is also recorded in tyrphobiontic populations of the noctuid moth *Coenophila subrosea* (Stephens) (Šula and Spitzer 2000). The higher abundance of some cold adapted tyrphobiontic moths (e.g., *Coenophila subrosea*) near the bog margins (lagg) is not clear. Further studies of tyrphobiontic insect dispersal (mainly larval associations with food plants and microclimate) within the "bog island" are badly needed. A majority of the tyrphophilous species is associated with the margins or are distributed in the bog randomly like the opportunistic tyrphoneutral taxa. For monitoring studies of ground beetles (Carabidae), the method of pitfall traps seems to be the most useful and effective way for investigations along an ecological gradient and for ecological community studies. Our results conform to other community data from isolated peatlands in the temperate and southern boreal zones, that only a few carabid species are able to permanently inhabit the treeless centres of the bogs (Butterfield and Coulson 1983; Främbs 1988, 1994; Främbs et al. 2002, Holmes et al. 1993; Hůrka 1960; Mossakowski 1970; Nenadál 1987; Spitzer et al. 1999)–see Figure 3 and Appendix A.

The light trap data are basic for complete faunistic and community studies of the phototactic moths associated with peatlands and their local habitat preference (Appendix A). But for "fine grain" investigations of microhabitats along ecological gradients, better ecological results are usually obtained if selected guilds of moths (Lepidoptera) associated with relict bog plants are studied (see Spitzer et al. 2003). Such an elaborate method is limited by the number of investigated species, however it can be an important supplement to other methods of community monitoring (light trapping, etc.). The entomofaunistic monitoring is fundamental for holistic appr oach to the conservation of fragile bog habitat islands and their biodiversity.

**Acknowledgements**

We would like to thank to Hana Zikmundová for technical assistance and to Blaine A. Mathison for linguistic help. Jan Kučera kindly provided us with some botanical data. We are gratefull to Jan Ševčík for an air photo of the Mrtvý luh bog, and to the two anonymous referees for helpful comments. This research was supported by the Grant Agency of the Academy of Sciences of the Czech Republic–grant No. IAA600070501.

## Appendix

Appendix A. List of all Carabidae and Lepidoptera species in two successional stages of the Mrtvý luh bog with their CCA scores on first axis. Tb–tyrphobiont, tf–tyrphophil, tn–tyrphoneutral (10 most abundant species only), m–migratory species.

| Species | Category | Margin | Center | CCA–score |
|---|---|---|---|---|
| COLEOPTERA: CARABIDAE | | | | |
| *Agonum ericeti* (Panzer) | tb | 65 | 253 | 0.0186 |
| *Carabus menetriesi* Hummel | tb | 17 | 84 | 0.0256 |
| *Pterostichus diligens* (Sturm) | tf | 26 | 32 | −0.0228 |
| *Amara nitida* Sturm | tn | | 1 | 0.0694 |
| *Carabus auronitens* Fabricius | tn | 2 | | −0.1293 |
| *Carabus glabratus* Paykull | tn | 4 | | −0.1300 |
| *Carabus violaceus* Linnaeus | tn | 3 | | −0.1377 |
| *Nebria brevicollis* (Fabricius) | tn | 1 | | −0.1564 |
| *Notiophilus palustris* (Duftschmid) | tn | | 6 | 0.0654 |
| *Pterostichus aethiops* (Panzer) | tn | 1 | | −0.1245 |
| *Pterostichus niger* (Schaller) | tn | 1 | | −0.1351 |
| *Pterostichus rhaeticus* Heer | tn | 5 | | −0.1293 |

*Appendix A.* Continued.

| Species | Category | Margin | Center | CCA–score |
|---|---|---|---|---|
| *Trechus quadristriatus* (Schrank) | tn | 1 | | −0.1506 |
| *Trechus splendens* Gemminger & Harold | tn | 3 | | −0.1358 |
| LEPIDOPTERA | | | | |
| *Acronicta menyanthidis* (Esper) | tb | 65 | 18 | −0.0106 |
| *Amphipoea lucens* (Freyer) | tb | 15 | 14 | 0.0037 |
| *Anarta cordigera* (Thunberg) | tb | | 2 | 0.1364 |
| *Arichanna melanaria* (Linnaeus) | tb | 23 | | −0.0818 |
| *Athrips pruinosella* (Liennig & Zeller) | tb | 260 | 108 | −0.0013 |
| *Carsia sororiata* (Hübner) | tb | 35 | 20 | 0.0253 |
| *Celaena haworthii* (Curtis) | tb | 39 | 6 | −0.0392 |
| *Coenophila subrosea* (Stephens) | tb | 268 | 20 | −0.0319 |
| *Crambus alienellus* (Germar & Kaulfuss) | tb | 3 | 16 | 0.0911 |
| *Chionodes viduella* (Fabricius) | tb | | 1 | 0.1329 |
| *Chionodes nebulosella* (Heinemann) | tb | 5 | | −0.0706 |
| *Pediasia truncatellus* (Zetterstedt) | tb | 9 | 13 | 0.0416 |
| *Phiaris turfosana* (Herrich-Schäffer) | tb | 3 | 35 | 0.1101 |
| *Acleris maccana* (Treitschke) | tf | 13 | 1 | −0.0565 |
| *Acleris lipsiana* (Denis & Schiffermüller) | tf | 1 | 1 | 0.0232 |
| *Anarta myrtilli* (Linnaeus) | tf | | 9 | 0.1342 |
| *Ancylis myrtillana* (Treitschke) | tf | 8 | 3 | −0.0173 |
| *Apotomis sauciana* (Frölich) | tf | 1 | | −0.0710 |
| *Aristotelia ericinella* (Zeller) | tf | 6 | 18 | 0.0516 |
| *Bryotropha boreella* (Douglas) | tf | 1 | | −0.0710 |
| *Catoptria margaritella* (Denis & Schiffermüller) | tf | 81 | 22 | −0.0070 |
| *Coleophora glitzella* Hofmann | tf | 2 | | −0.0659 |
| *Crambus ericella* (Hübner) | tf | 2 | | −0.0673 |
| *Eilema lutarella* (Linnaeus) | tf | 1 | 7 | 0.0775 |
| *Elachista alpinella* Stainton | tf | 1 | | −0.0929 |
| *Elophos dilucidaria* (Denis & Schiffermüller) | tf | 146 | 59 | 0.0050 |
| *Ematurga atomaria* (Linnaeus) | tf | 1 | | −0.0929 |
| *Eulithis testata* (Linnaeus) | tf | 197 | 21 | −0.0302 |
| *Eupithecia plumbeolata* (Haworth) | tf | 13 | 1 | −0.0422 |
| *Eupithecia nanata* (Hübner) | tf | | 7 | 0.1344 |
| *Eurois occulta* (Linnaeus) | tf | 13 | 4 | −0.0191 |
| *Hypena crassalis* (Fabricius) | tf | 2 | | −0.0626 |
| *Hypenodes humidalis* Doubleday | tf | 7 | 2 | −0.0286 |
| *Hyppa rectilinea* (Esper) | tf | 15 | 2 | −0.0414 |
| *Itame brunneata* (Thunberg) | tf | 43 | 1 | −0.0522 |
| *Lasiocampa quercus* (Linnaeus) | tf | | 5 | 0.1355 |
| *Lithomoia solidaginis* (Hübner) | tf | 72 | 10 | −0.0309 |
| *Lycophotia porphyrea* (Denis & Schiffermüller) | tf | 635 | 1039 | 0.0403 |
| *Micropterix aureatella* (Scopoli) | tf | 1 | | −0.0789 |
| *Neofaculta ericetella* (Geyer) | tf | 12 | 36 | 0.0578 |
| *Neofaculta infernella* (Herrich-Schäffer) | tf | 17 | | −0.0771 |
| *Orthosia opima* (Hübner) | tf | | 2 | 0.1378 |
| *Papestra biren* (Goeze) | tf | 12 | 10 | 0.0345 |
| *Phiaris micana* (Denis & Schiffermüller) | tf | 3 | | −0.0721 |
| *Phiaris bipunctana* (Fabricius) | tf | 67 | 9 | −0.0125 |
| *Photedes minima* (Haworth) | tf | | 1 | 0.1190 |

*Appendix A.* Continued.

| Species | Category | Margin | Center | CCA–score |
|---|---|---|---|---|
| *Pleurota bicostella* (Clerck) | tf | 13 | | −0.0767 |
| *Protolampra sobrina* (Duponchel) | tf | 270 | 7 | −0.0548 |
| *Sterrhopterix standfussi* (Wocke) | tf | 6 | 1 | −0.0222 |
| *Stictea mygindiana* (Denis & Schiffermüller) | tf | 1 | | −0.0691 |
| *Syngrapha interrogationis* (Linnaeus) | tf | 3 | | −0.0807 |
| *Alcis repandata* (Linnaeus) | tn | 107 | 1 | −0.0699 |
| *Apamea monoglypha* (Hufnagel) | tn | 41 | 79 | 0.0530 |
| *Arctia caja* (Linnaeus) | tn | 32 | 46 | 0.0312 |
| *Cybosia mesomella* (Linnaeus) | tn | 67 | 13 | −0.0303 |
| *Eilema depressa* (Esper) | tn | 57 | 8 | −0.0301 |
| *Eulithis populata* (Linnaeus) | tn | 98 | 1 | −0.0719 |
| *Exoteleia dodecella* (Linnaeus) | tn | 55 | 12 | −0.0177 |
| *Pempelia palumbella* (Denis & Schiffermüller) | tn | 42 | 57 | 0.0535 |
| *Rhopobota naevana* (Hübner) | tn | 137 | 5 | −0.0627 |
| *Xestia baja* (Denis & Schiffermüller) | tn | 112 | 66 | 0.0076 |
| *Agrotis ipsilon* (Hufnagel) | m | 6 | 21 | 0.0640 |
| *Autographa gamma* (Linnaeus) | m | 2 | 5 | 0.0608 |
| *Mythimna vitellina* (Hübner) | m | | 1 | 0.1134 |
| *Noctua pronuba* (Linnaeus) | m | 13 | 61 | 0.0671 |
| *Noctua fimbriata* (Schreber) | m | 2 | 4 | 0.0588 |
| *Phlogophora meticulosa* (Linnaeus) | m | 2 | 22 | 0.0987 |
| *Plutella xylostella* (Linnaeus) | m | 2 | 2 | 0.0413 |
| *Xestia c-nigrum* (Linnaeus) | m | 4 | 29 | 0.0805 |

## References


Butterfield J. and Coulson J.C. 1983. The carabid communities on peat bog and upland grasslands in northern England. Holarctic Ecology 6: 163–174.

Dapkus D. 2001. Macrolepidoptera of Lithuanian peatbogs. Norwegian Journal of Entomology 48: 161–166.

De Vries H.H. and Den Boer P.J. 1990. Survival of populations of *Agonum ericeti* Panz. (Col., Carabidae) in relation to fragmentation of habitats. Netherlands Journal of Zoology 40: 484–498.

Främbs H. 1988. Habitatpräferenz von Carabiden (Coleoptera) auf Palsmooren in Torne–Lapmark, Schweden. Mitteilungen der deutschen Gesselschaft für allgemeine and angewandte Entomologie 6: 379–390.

Främbs H. 1994. The importance of habitat structure and food supply for carabid beetles (Coleoptera, Carabidae) in peat bogs. Memoirs of the Entomological Society of Canada 169: 145–159.

Främbs H., Dormann W. and Mossakowski D. 2002. Spatial distribution of carabid beetles on Zehlau bog. Baltic Journal of Coleopterology 2: 7–13.

Holmes P.R., Fowles A.P., Boyce D.C. and Reed D.K. 1993. The ground beetle (Coleoptera, Carabidae) fauna of Welsh peatland biotopes–species assemblages in relation to peatland habitats and management. Biological Conservation 65: 61–67.

Holubičková B. 1960. Studie o vegetaci blat. I. (Mrtvý luh). [Studies on peatland vegetation. I. (Mrtvý luh)]. Sborník VŠZ Praha 1960: 129–149 (in Czech).

Hůrka K. 1960. Die Carabidenfauna des Sooser Moores in Westböhmen (Col. Carabidae). Acta Universitatis Carolinae–Biologica, suppl. 1960: 59–82.

Jankovská V. 1980. Paläogeobotanische Rekonstruktion der Vegetations–entwicklung im Becken Třeboňská pánev während des Spätglazials und Holozäns. Academia, Praha, 151

Jankovská V. 1995. Relationship between the Late Glacial and Holocen vegetation and the animal component of their ecosystems. Geolines (Praha) 2: 11–16.

Jongman R.H., Ter Braak C.J. and van Tongeren O.F.R. 1987. Data Analysis in Community and Landscape Ecology. Pudoc, Wageningen, 298.

Mikkola K. and Spitzer K. 1983. Lepidoptera associated with peatlands in central and northern Europe: a synthesis. Nota Lepidopterologica 6: 216–229.

Mossakowski D. 1970. Ökologische Untersuchungen an epigäischen Coleopteren atlantischer Moor- und Heidestandorte. Zeitschrift für wissenschaftliche Zoologie 181: 233–316.

Mueller-Dombois D. and Ellenberg H. 1974. Aims and Methods of Vegetation Ecology. J. Wiley & Sons, New York, 547.

Nekola J.C. 1999. Paleorefugia and neorefugia: the influence of colonization history on community pattern and process. Ecology 80: 2459–2473.

Nenadál S. 1987. Die Biotopenbindung der Laufkäfer (Carabidae) an irgend unterschieden Sukzessionsstufen der Torfgrunde und Torfmoore im Landschaftschutzgebiet Žd'árské vrchy. Vlastivědný Sborník Vysočiny 8: 193–202 (in Czech, German abstr.).

Neuhäusl R. 1972. Subkontinentale Hochmoore und ihre Vegetation. Studie ČSAV (Praha) 13: 11–21.

Novák I. and Spitzer K. 1972. Výsledky faunisticko-ekologického studia Lepidopter (Noctuidae a Geometridae) rašeliniště Mrtvý luh u Volar a okolí. Ergebnisse des faunistisch-ökologischen Studiums der Lepidopterenfauna (Noctuidae und Geometridae) des Hochmoores Mrtvý luh bei Volary und dessen Umgebung. Sborník Jihočeského muzea v Českých Budějovicích, Přírodní vědy 12 (suppl. 1): 3–63 (in Czech, German abstr.).

Peus F. 1932. Die Tierwelt der Moore unter besonderer Berücksichtigung der europäischen Hochmoore. Handbuch der Moorkunde (Berlin) 3: 1–277.

Roháček J. and Máca J. 1982. Acalypterate Diptera of the peat bogs in North Moravia (Czechoslovakia) Part 2. Ecological classification Opomyzidae, Anthomyzidae, Asteiidae, Diastatidae, Drosophilidae. Časopis Slezského muzea, série A, Vědy přírodní (Opava) 31: 193–213.

Roubal K. 1934. Die Coleopterenwelt (Tyrphobionte, Tyrphophile, Tyrphoxene, etc.) der Treboner (Wittingauer) Moore. Folia Zoologica et Hydrobiologica (Riga) 7: 56–97.

Spitzer K. 1981. Ökologie und Biogeographie der bedrohten Schmetterlinge der südböhmischen Hochmoore. Beihefte zu den Veröffentlichungen für Naturschutz und Landschaftspflege in Baden-Württemberg 21: 125–131.

Spitzer K. 1994. Biogeographical and ecological determinants of the central European peat bog Lepidoptera: The habitat island approach to conservation. Nota Lepidopterologica, suppl. 5: 45–49.

Spitzer K. 2004. Insect biodiversity of Central European peat bogs: the habitat island concept of conservation. International Peat Journal 12: 113–117.

Spitzer K., Bezděk A. and Jaroš J. 1999. Ecological succession of a relict Central European peat bog and variability of its insect biodiversity. Journal of Insect Conservation 3: 97–106.

Spitzer K. and Jaroš J. 1993. Lepidoptera associated with the Červené Blato bog (Central Europe): conservation implications. European Journal of Entomology 90: 323–336.

Spitzer K. and Jaroš J. 2001. Moths and butterflies of the Chalupská slat' bog, Šumava Mountains. Sborník Jihočeského muzea v Českých Budějovicích, Přírodní vědy 44: 43–55.

Spitzer K., Jaroš J. and Bezděk A. 2003. Leaf-spinning moths (Lepidoptera) feeding on boreal *Vaccinium uliginosum* L. along an ecological gradient of central European peat bogs. Entomologica Fennica 14: 46–52.

Šula J. and Spitzer K. 2000. Allozyme polymorphism in isolated populations of the moth *Coenophila subrosea* (Lepidoptera: Noctuidae) from three Central European peat bogs. European Journal of Entomology 97: 7–12.

Ter Braak C.J.F. 1987. CANOCO–a FORTRAN Program for Canonical Community Ordination. TNO Institute of Applied Computer Science, Wageningen, 95.

Biodiversity and Conservation (2006) 15:397–414
DOI 10.1007/s10531-005-5399-4 

# Surrogate habitats demonstrate the invasion potential of the African pugnacious ant

PIA ADDISON[1,2,*] and MICHAEL J. SAMWAYS[2]
*[1]ARC Infruitec-Nietvoorbij, Pest Management Division, Private Bag X5026, Stellenbosch, 7599, South Africa; [2]Current address: Department of Entomology and Centre for Agricultural Biodiversity, Stellenbosch University, Private Bag Xl, Matieland 7602, South Africa; *Author for correspondence (e-mail: pia@sun.ac.za; phone: +27-21-808-4671; fax: +27-21-808-4807)*

Received 21 July 2004; accepted in revised form 5 April 2005

**Key words:** Agriculture, *Anoplolepis custodiens*, Ant, Cover crops, Invasive, Mealybug mutualism, Vineyards

**Abstract.** Many ant species are highly invasive and are a significant component of disturbed ecosystems. They can have a major suppressive effect upon indigenous invertebrates, including other ants. Despite overwhelming circumstantial evidence for the ecological resourcefulness of many ants, there appears to be no experimental evidence illustrating the habitat breadth of a potentially invasive ant species. We demonstrate here that a particularly opportunistic and locally dominant ant *Anoplolepis custodiens*, which is a major indigenous African pest, overrides habitat structure to maintain its population level. We compared *A. custodiens* activity, morphology, foraging behaviour and ant species diversity in artificially established surrogate habitats (cover crops) in a vineyard containing an ample food resource in the form of the honeydew-producing mealybug *Planococcus ficus*. These cover crops were chosen so as to create highly altered habitats. The ant's ability to overcome these potentially suppressive habitat conditions hinged on its tight mutualism with the mealybug, and on its chasing away mealybug parasitoids. This ant species is predicted to be a latent invasive beyond Africa. It is unlikely to be impeded once it has established a foothold in a variety of novel habitats. It could locally invade to obtain food resources in a wide range of habitat types. Furthermore, in agricultural systems, cover crops are unlikely to control such an ant. Potential invasives such as this ant should be flagged as important quarantine suspects.

## Introduction

Invasive alien ants have invaded and altered many ecosystems throughout the world (Holway et al. 2002). *Pheidole megacephala* (Fabricius), for example, has had a major impact on indigenous insects (Wilson 1996) and spiders (Gillespie 1999) in Hawaii. *Solenopsis invicta* Buren locally excluded four endemic ant species in Texas (Cook 2003). *Linepithema humile* (Mayr) in South Africa has excluded indigenous ant species (Donnelly and Giliomee 1985) and pre-empts seed burial by indigenous ants (Bond and Slingsby 1983). Although *L. humile* tends to follow watercourses and lines of disturbance, particularly in Mediterranean-type ecosystems (Holway 1995; Human et al. 1998), it is now penetrating indigenous evergreen forest (Ratsirarson et al. 2002). During the pioneering invasive stage colonies can go through a genetic bottleneck that reduces genetic diversity, mitigates intraspecific aggression among spatially separate nests, and leads to the formation of supercolonies which dominate other, sympatric ant species (Tsutsui et al. 2003).

In the artificial ecosystem Biosphere 2, the ant *Paratrechina longicornis* (Latreille) eventually became dominant, feeding almost entirely on homopteran honeydew. Among *Anoplolepis* species, *A. gracilipes* (F. Smith) (formerly known as *A. longipes*) in the Seychelles has had a major impact on indigenous invertebrates both on the ground and in the tree canopy (Hill et al. 2003). *Anoplolepis gracilipes* has also been highly invasive in many other tropical countries, invading various ecosystems (Fowler et al. 1990). A second *Anoplolepis* species, *A. custodiens* (F. Smith), aptly named the pugnacious ant, is another very aggressive ant. In its African home, it can outcompete many other indigenous ant species when honeydew food resources are available to sustain it (Samways 1999). It is also a major predator of indigenous insects (Steyn 1954; Löhr 1992). At the spatial scale of the landscape, it is highly invasive. *Anoplolepis custodiens* has exhibited extreme dominance over other ant species in citrus and guava orchards (Samways 1990) in north-eastern South Africa, in citrus orchards in south-eastern South Africa (Myers 1957), and in vineyards in south-western South Africa (Addison and Samways 2000) and once established has proven difficult to control (Addison 2002; Samways 1990). This ant also exhibits aggressive behaviour towards some other animal groups, as was found in the central parts of South Africa where chicken runs need to be protected against *A. custodiens* (Prins et al. 1990). Its current distribution in the rest of sub-Saharan Africa needs confirmation, although it has been recorded from Zanzibar and Tanzania (Way 1953), Democratic Republic of Congo (Wheeler 1922), Angola and the 'Ethiopian region' (www.antbase.org).

Observations have indicated that *A. custodiens* does not appear to exhibit any intra-specific aggression between colonies (Way 1953; Steyn 1954), which lends itself to invasive expansions (Tsutsui et al. 2003). Arguably, it is a serious potential invader of countries outside Africa, and should receive quarantine attentiveness by non-African nations (Samways 1999). But how invasive is it really? We cannot undertake classical experiments in order to study this aspect. We can however, create artificial habitats to determine its invasiveness in surrogate habitats with different compositional and structural characteristics. This can be undertaken by planting cover crops within an agroecosystem, where adequate food resources are available to encourage it to forage, or not, in the surrogate habitats. In a vineyard ecosystem, for example, *A. custodiens* feeds on honeydew from mealybugs and so reduces the effectiveness of mealybug parasitoids (Kriegler and Whitehead 1962).

In its natural habitat in the southern Karoo, this ant nests in open, well insolated soil and feeds on dead and live animal matter as well as honeydew and nectar (Dean 1992). Should the ant not be able to survive in the surrogate habitats, it is postulated that there would be a change in food web structure. The change would come about because the ants are no longer chasing away the parasitoids, which in turn, reduce the ant's mealybug food resource (honeydew).

In other words, there would be a collapse in the ant-mealybug mutualism. In contrast, should the ants be invasive, they are likely to override habitat type and quality in favour of a generalized food source (honeydew). We test here the ability of *A. custodiens* to maintain its population level in a range of habitat structures designed principally to suppress it and break the ant-mealybug mutualism.

## Site and methods

### *Site and surrogate habitats*

The experimental site was at Bonnievale, South Africa (33.26 S 20.01 E) in a 9 year old, trellised Chenin blanc vineyard infested with vine mealybug *Planococcus ficus* (Signoret) and the common pugnacious ant *A. custodiens*. Irrigation was by micro-jets with supplementary irrigation during winter dry spells. A middle section of the vineyard was selected for the trial, with the experimental area being 1.5 ha. A 4×4 latin square was used with each plot 950 $m^2$ (11 rows × 30 vines). Each data area was 144 $m^2$ (5 rows × 10 vines) in the centre of each 950 $m^2$ plot.

Surrogate habitats (cover crops) were chosen to thickly cover the ground, as well as to create major changes in habitat conditions and therefore be potentially inhibitory to ant activity. These cover crops were equivalent to a wide range of foreign habitats. Two cover crops, creeping vetch *Vicia dasycarpa* (Fabaceae) and triticale *Triticale v.* Usgen 18 (Graminae) were used as habitats that were seasonally part green and part trash (litter). A third cover crop was permanently green, consisting of a seed mixture during the first season (permanent dwarf fescue, creeping red Harold, SR-4–200 and Santiago medic) and a pure stand of dwarf fescue *Festuca* sp. (Poaceae) during the second season. These three surrogate habitats were compared to an almost open soil control (an ideal habitat, as indicated by very high ant densities) in which weed cover was suppressed using herbicides and a tractor-drawn rotary mower. Seed planting densities were: Vetch (50 kg $ha^{-1}$) and triticale (100 kg $ha^{-1}$), while the permanent seed mixture consisted of a permanent dwarf fescue (16 kg $ha^{-1}$) and creeping red Harold, SR-4–200 and Santiago medic (all at 8 kg $ha^{-1}$). During the second year, a pure stand of fescue was sown at 30 kg $ha^{-1}$.

Cover crops were sown during April 2001 and again in April 2002. Soil preparation was one month prior to sowing, with disking to a depth of about 15 cm. Immediately after sowing, the seeds were raked into the soil using a tractor-drawn furrow plough. The vetch, triticale and control plots were sprayed with herbicide (glyphosate SL 360 g/l water at 6 l/ha) during September 2001 and 2002. This resulted in a mat of litter cover during the hot, dry summer in the case of vetch and triticale, while keeping the ground largely free of vegetation in the control. One month after spraying, the control plots were mechanically mowed with a rotary mower, while the fescue plots were mowed to a height of 30 cm above the soil surface to slash high-growing weeds.

During the first week in April 2002, the control plots were again slashed with a rotary mower.

*Soil temperature and moisture measurements*

One 2-channel soil temperature and moisture logger (MCS 486-TSM, Mike Cotton Systems, Cape Town, South Africa), which uses gypsum blocks to measure soil moisture (Toome 2002), was placed into each of the cover crop and control plots. Sensors were moved alternately to 10 cm and 30 cm below the soil surface on a monthly basis. Monitoring was from June 2001 to March 2002, with hourly readings. Accumulated degree-day units (Baskerville and Emin 1969) were calculated in each cover crop and the control using temperatures above 9.7 °C, which is the lower threshold of development for *A. custodiens* medium worker pupae in laboratory colonies (Steyn 1954). For comparing temperatures between cover crops and the control, the following formula was used: Accumulated °D $=\sum[(\text{hourly temperature reading} - 9.7)/24]$, where: °D = Degree Days.

*Ant activity and species diversity sampling*

Foraging activity of epigaeic ants was monitored using pitfall traps of test tubes ($18 \times 150$ mm) containing 10 ml of seven parts 70% ethyl alcohol and three parts glycerol (Majer 1978). An outer case, consisting of irrigation pipe, 160 mm in length, was permanently sunk into the ground and used as a trap sleeve to facilitate changing of traps. Test tubes were sunk into the casing, and the ground levelled so that the lip was even with the soil surface. A total of 64 traps were used, four in each of 16 cover crop and control plots. Traps were changed every 2 weeks, except when unfavourable weather caused trap changing to be postponed. Sampling was from 19 June 2001 to 18 March 2003. All ants were identified to species.

*Ant nest entrance counts and foraging distance*

All active nest entrances were counted in a 1 m $\times$ 1 m area directly opposite pitfall traps. This was done during March 2001 (pre-treatment counts) and subsequently during November 2001 and 2002 and March 2002 and 2003. To establish whether cross-infestation of ants between cover crop treatments was likely, the average distance that 120 ants (30 per treatment) travelled from 120 randomly selected nest entrances to their foraging end point was recorded during April 2003. Each observation was for 2 min, as ants usually turned back in the direction of the nest or ended at a honeydew food source by this time.

*Ant and mealybug population levels*

*A. custodiens* and *P. ficus* population levels were monitored in the vine canopy during April 2001 (pre-treatment counts), February 2002, November 2002 and March 2003. A total of 40 vines in the data area in each of the 16 cover crop and control plots were designated for monitoring, which was done by inspecting the leaves (10 per vine), stems and fruit of vines. Vines were classified as infested or not infested.

*Mealybug parasitoid densities*

Parasitoid monitoring was with visually-attractant yellow, sticky Bug Traps™ (Agribiol, Vlaeberg, South Africa), 200 mm × 100 mm. One trap was placed in the data area of each cover crop and control, making a total of 16 traps. Traps were fixed onto cross wires in the vine canopy, which were 1.5 m above the ground. Traps were therefore situated 36 m apart in each row, and 26 m apart between each replicate. Parasitoid identifications were to species level where possible. Monitoring started during June 2001 with traps being changed once a month thereafter until March 2003.

Data from pitfall trap catches, nest entrance counts, ant and mealybug infestations in the vine canopy, and yellow Bug Traps for parasitoids were analysed using Analysis of Variance (ANOVA) and Least Significant Differences (LSD) calculated to compare habitat types. Data for parasitoid counts were transformed ($\log \{x + 1\}$) to normalize means.

*Head capsule measurements*

Sampling was in the vineyard, and in a section of natural vegetation 900 m away from the vineyard which was dominated by succulents (Acocks 1988). Traps were changed every 2 weeks during March and April 2003. In the natural vegetation, 16 pitfall traps were sunk into the soil close to nests. In the vineyard, the same traps were used to sample ant diversity and activity. A total of 200, randomly selected *A. custodiens* workers per cover crop, control and natural vegetation were used. A further 50 of the smallest workers from cover crops and the control were also collected. Length and width of head capsules were measured using a digital camera with measuring function and software (PhotoLib 3.03) mounted onto a stereomicroscope. Ants were decapitated and positioned on sticky slides. The formula of Southwood (1978) was used to determine whether the sample size was sufficient: $n = (\text{Standard Deviation}/0.05 \text{ times Mean})^2$, where 0.05 is a predetermined standard error of the mean.

Head capsule size was calculated by multiplying head width by head length. Head capsule size between workers from different surrogate habitats in the vineyard and natural vegetation was compared using a *t*-test. The 200 randomly selected workers were analysed separately from the 50 smallest workers.

## Results

### *Cover crops and environmental conditions*

Triticale grew to a height of 1 m. After spring herbicide treatment, the plants broke at the base and formed a thick, dry layer on the soil surface during summer. Vetch grew to 20 cm and spread along the soil surface to form a dense, leafy layer with little of the soil surface exposed. After the herbicide treatment, vetch formed a dense, dry layer over the entire soil surface during summer. Both triticale and vetch out competed weeds. The permanent mixture and fescue were less competitive with weeds, and together, formed a dense green layer throughout the year. Control plots were largely free of weeds during summer, although 40% of the soil surface was covered with weeds by the end of winter. Control plots had weeds such as yellow sorrel (*Oxalis pescaprae* L.), wild radish (*Raphanus raphanistrum* L.), wild mustard (*Rapistrum rugosum* L.), sowthistle (*Sonchus olearceus* (L.)), small mallow (*Malva parviflora* L.), red pigweed (*Amaranthus thunbergii* Moq.) and white goosefoot (*Chenopodium album* L.).

Maximum and minimum soil temperatures were significantly higher in the control plot than in the cover crops at both soil depths (Table 1). A maximum reduction in soil temperature of 3 °C was brought about by the green cover (fescue) at 30 cm depth (Table 1). Accumulated heat units, calculated at 10 cm and 30 cm, were highest in the control (1078 °D and 1370 °D, respectively), intermediate in triticale (880 °D and 933 °D) and vetch (856 °D and 859 °D), and lowest in fescue (801 °D and 787 °D). Mean percentage soil moisture was significantly lower in control plots than in fescue or vetch plots, but not in triticale plots (Table 1). A maximum difference of 5.21% soil moisture at 10 cm depth was found between control and fescue.

### *Ant behaviour*

*Anoplolepis custodiens* activity on the soil surface was significantly higher in the triticale plots than in the other treatments, which did not differ significantly from each other (Figure 1). Ants could have utilized seeds from triticale as an additional food source. During March of both years, ant activity in triticale plots declined relative to the other treatments, possibly as a result of seeds being buried during soil preparation (disking). In each habitat type, *A. custodiens* was significantly more prolific than any of the other ant species (ANOVA, LSD, $p \leq 0.05$), thereby exhibiting extreme dominance (Figure 2),

*Table 1.* Mean difference of soil temperature (°C) and soil moisture (%) between ground cover treatments.

| Cover crop treatments | Triticale | | Vetch | | Permanent mix | |
|---|---|---|---|---|---|---|
| Soil depth | 10 cm | 30 cm | 10 cm | 30 cm | 10 cm | 30 cm |
| Maximum soil temperature (°C) | | | | | | |
| Control | 2.20 ($p$ = 0.01) | 1.87 ($p$ = 0.01) | 2.02 ($p$ = 0.01) | 2.49 ($p$ = 0.01) | 2.25 ($p$ = 0.01) | 3.00 ($p$ = 0.01) |
| Triticale | | | −0.19 ($p$ = 0.58) | 0.61 ($p$ = 0.01) | 0.05 ($p$ = 0.88) | 1.12 ($p$ = 0.01) |
| Vetch | | | | | 0.24 ($p$ = 0.02) | 0.51 ($p$ = 0.01) |
| Minimum soil temperature (°C) | | | | | | |
| Control | 1.49 ($p$ = 0.01) | 1.13 ($p$ = 0.01) | 1.80 ($p$ = 0.01) | 1.98 ($p$ = 0.01) | 2.25 ($p$ = 0.01) | 2.60 ($p$ = 0.01) |
| Triticale | | | 0.32 ($p$ = 0.05) | 0.85 ($p$ = 0.01) | 1.11 ($p$ = 0.01) | 1.46 ($p$ = 0.01) |
| Vetch | | | | | 0.79 ($p$ = 0.01) | 0.62 ($p$ = 0.01) |
| Soil moisture (%) | | | | | | |
| Control | −0.57 ($p$ = 0.07) | 3.81 ($p$ = 0.01) | −3.26 ($p$ = 0.01) | −5.15 ($p$ = 0.01) | −5.21 ($p$ = 0.01) | −4.63 ($p$ = 0.01) |
| Triticale | | | −2.68 ($p$ = 0.05) | −8.97 ($p$ = 0.01) | −4.63 ($p$ = 0.01) | −8.44 ($p$ = 0.01) |
| Vetch | | | | | −1.95 ($p$ = 0.02) | −0.52 ($p$ = 0.54) |

Positive numbers next to means indicate that the readings of treatments in the left column are larger than those in the top row ($p$ = 0.05).

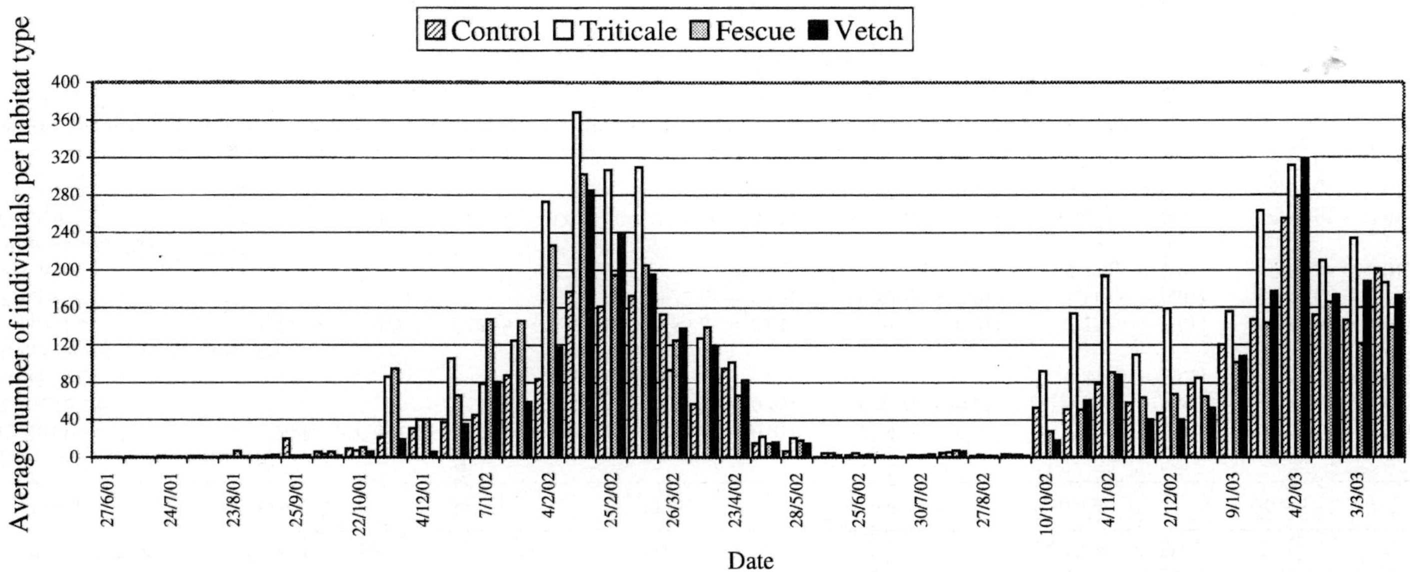


*Figure 1.* Average number of the ant *Anoplolepis custodiens*, as measured using pitfall traps, for four habitat types. Mean number of ants were as follows ($\pm$standard error): Triticale ($404 \pm 39.4$a), fescue ($299 \pm 2.9$b), vetch ($273 \pm 39.4$b) and control ($246 \pm 26.3$b). Numbers with the same letter do not differ significantly ($p \leq 0.05$).

which is characteristic of this ant. Mean number of species (±standard error) per habitat type were: fescue (3.43 ± 0.18a), vetch (3.06 ± 0.23a), control (3.00 ± 0.32a) and triticale (2.31 ± 0.24b), where numbers followed by the same letter do not differ significantly (ANOVA, LSD, $p \leq 0.05$). The lower species richness in triticale plots was associated with increased *A. custodiens* activity in this cover crop.

Seasonal foraging activity of *A. custodiens* is shown for the four habitat types (Figure 3). *Anoplolepis custodiens* was least active from June to September, although maximum temperatures remained above 11.3 °C, the temperature at which Dean (1992) found this ant to cease being active. Activity increased greatly in November 2001 and October 2002, when average maximum temperatures reached 26 °C and visibility and activity of mealybugs on the

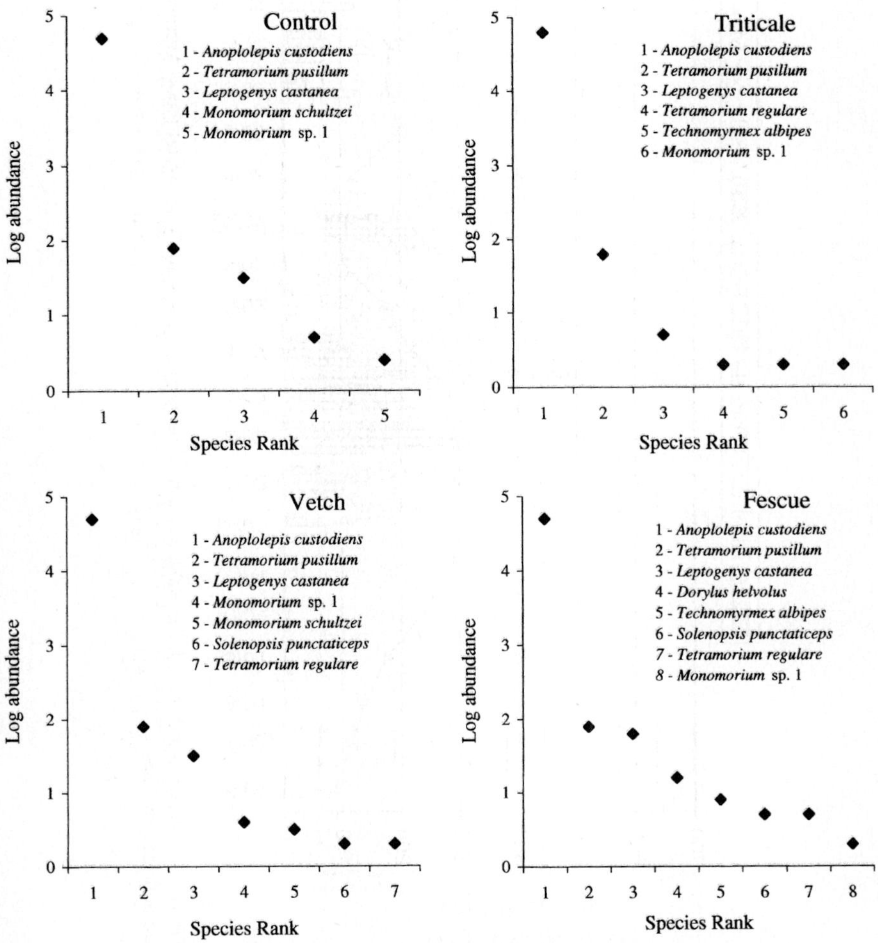


*Figure 2.* Abundance/rank plots of the ant assemblages found in four habitat types.

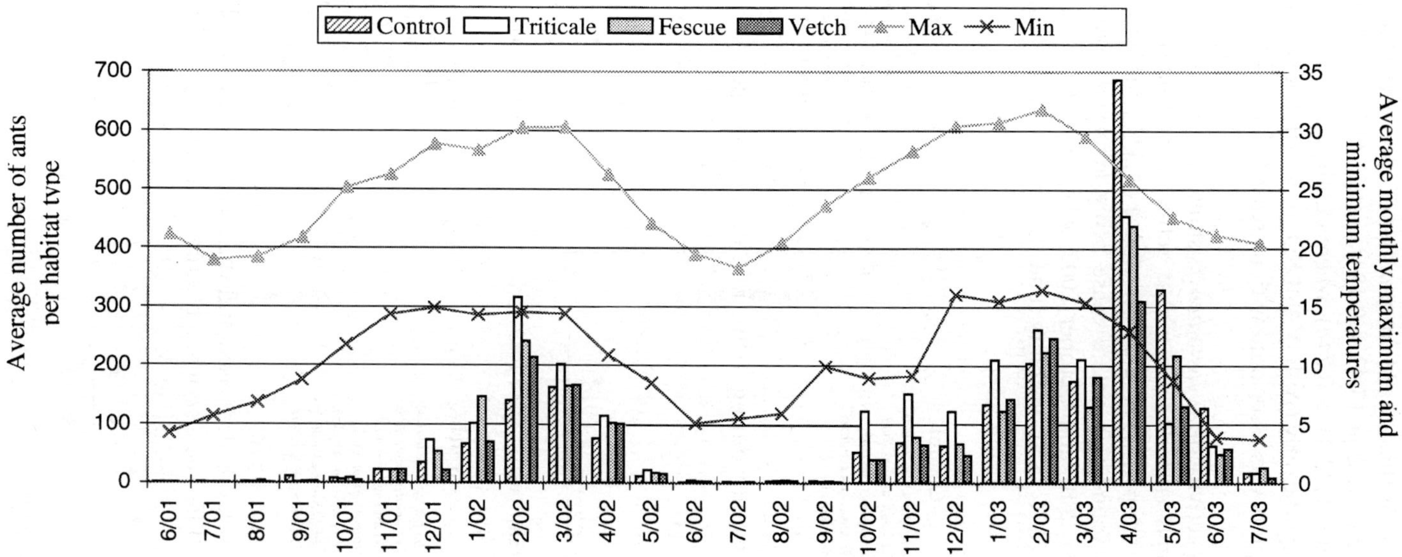


*Figure 3.* Average number of *A. custodiens*, measured using pitfall traps, for four habitat types. Average monthly minimum and maximum temperatures are indicated by lines.

*Table 2.* Mean number of *Anoplolepis custodiens* nest entrances (±standard error) on five sampling dates (the first being pre-treatment counts) where four ground cover treatments were compared.

| Cover crop treatment | March 2001 | November 2001 | March 2002 | November 2002 | March 2003 |
|---|---|---|---|---|---|
| Control | 4.00 ± 0.60a | 0.50 ± 0.23a | 9.81 ± 1.10a | 0.87 ± 0.22a | 2.78 ± 0.54a |
| Triticale | 3.25 ± 0.54a | 0.65 ± 0.19a | 11.0 ± 1.46a | 0.84 ± 0.25a | 2.21 ± 0.65a |
| Fescue | 4.18 ± 0.70a | 0.81 ± 0.22a | 8.75 ± 1.01a | 0.75 ± 0.19a | 3.68 ± 0.87a |
| Vetch | 2.68 ± 0.40a | 0.18 ± 0.08a | 8.37 ± 0.85a | 0.53 ± 0.17a | 2.87 ± 0.67a |

Numbers in a column followed by the same letters do not differ significantly (ANOVA, LSD, $p \leq 0.05$).

stems, bunches and leaves started to increase. Mealybugs are inactive and hide under the bark of vines during winter.

No significant differences in the number of nest entrances were found between habitat types on any of the sampling dates (Table 2). As Myers (1957) found that *A. custodiens* readily moved their nests as citrus trees grew larger and shaded the soil surface, it was predicted that if the ants found the cover crops unfavourable, they would move their nests out of the vineyard during the 2 years of the study. In the triticale plots, ants travelled an average distance of 34 cm, in fescue plots 32 cm, in vetch plots 25 cm and in control plots 26 cm. Shortest foraging distance during a two-min observation period was 10 cm, and the longest 79 cm, indicating that cross-infestation of ants between plots was unlikely, as the data areas were located 24 m apart in each row, and 19 m apart in each column of the Latin square.

### *Ant and mealybug population levels*

Ant population levels in the vine canopy in the pre-treatment count were variable between treatments, with the fescue plots having a significantly higher infestation than the control plots (Figure 4a). During November 2002, the ant infestation of the vines in the vetch plots was significantly lower than in the other treatments (Figure 4a). As vetch plants formed a dense, dry layer by November, movements of the ant between its nest and vines was restricted. By March the following year, there were no significant differences. Percentage mealybug population levels in fescue and triticale plots were significantly higher than in vetch and control plots during pre-treatment counts (Figure 4b). Mealybug population levels decreased and then increased in March 2003, but with no significant difference between treatments (Figure 4b). In summary, cover crops had no effect on either ant or mealybug population levels.

### *Mealybug parasitoids*

Three species of parasitoids were recorded: *Coccidoxenoides peregrinus* (Timberlake) (Hymenoptera: Encyrtidae), *Anagyrus* sp. (Hymenoptera: Encyrtidae)

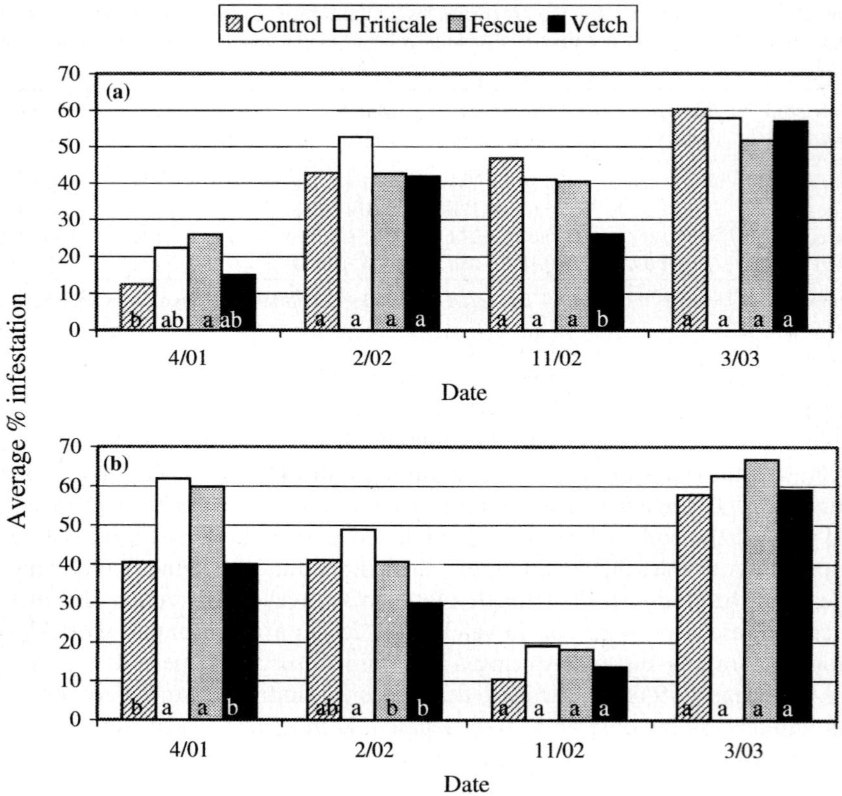


*Figure 4.* Average percentage of the ant *Anoplolepis custodiens* (a) and the mealybug *Planococcusficus* (b) occurrence in vines on four sampling dates with four habitat types. April 2001 represents the pre-treatment sampling date. Letters that differ on each column indicate a significant difference, analysed for each date separately ($p \leq 0.05$).

and *Leptomastix dactylopii* Howard (Hymenoptera: Encyrtidae). Highest mean number of parasitoids was in control plots, although this was only significantly different for *C. peregrinus* in vetch plots and *L. dactylopii* in triticale and vetch plots (Table 3). The higher numbers of parasitoids in control plots were pos-

*Table 3.* Transformed means ($\log\{x + 1\} \pm$ standard error) of *Planococcus ficus* parasitoids caught in sticky yellow Bug Traps™ where four ground cover treatments were compared.

| Cover crop treatment | *Anagyrus* sp. | *Coccidoxenoides peregrinus* | *Leptomastix dactylopii* |
|---|---|---|---|
| Control | 0.60 ± 0.10a | 1.00 ± 0.09a | 0.26 ± 0.07a |
| Triticale | 0.53 ± 0.09a | 0.91 ± 0.09ab | 0.14 ± 0.041 |
| Fescue | 0,48 ± 0.05a | 0.96 ± 0.10a | 0,18 ± 0.05ab |
| Vetch | 0.50 ± 0.09a | 0.75 ± 0.09b | 0.11 ± 0.051 |

Numbers in a column followed by the same letters do not differ significantly (ANOVA, LSD, $p \leq 0.05$).

sibly due to the larger variety of weeds found there during winter, which could have provided the parasitoids with a greater variety of habitats to utilize as refuges. Parasitoid population levels remained relatively constant throughout the experiments, with no significantly large increase that would be expected should they have been released from ant pressure.

### *Head capsule measurements*

Workers (all-sizes and smallest) had significantly larger head capsules in both natural vegetation and in control plots, than in triticale and vetch plots (Tables 4 and 5). No significant difference was found in head capsule size between workers from natural vegetation, control or fescue plots. Size-frequency distribution (Figure 5) indicates that *A. custodiens* had continuous polymorphism (Oster and Wilson 1978), with minors, medias and majors occurring together. Species expand their physical polymorphism when they occupy habitats with few competitors, as was the case for *A. custodiens* here. Size-frequency distribution of workers in natural vegetation was due to more of the larger individuals occurring here than in cover crops or control plots, while control plots contained more of the larger individuals than in cover crops (indicated by the median). Following Oster and Wilson (1978) and Brian and Brian (1951), *A. custodiens* size-frequency distribution approximates size-frequency distribution of its prey, with individuals in natural vegetation making use of a broader range of food sources, such as honeydew from scale insects and termites, than

*Table 4*. Comparison between all-sized worker head capsule size (width × length) from various habitat types (df = 398).

| Habitat type (means in mm) | Natural vegetation (1.52) | | | Control (1.56) | | |
|---|---|---|---|---|---|---|
| | *t*-value | *p*-value | SE of difference | *t*-value | *p*-value | SE of difference |
| Control (1.56) | 0.48 | 0.62 | 0.076 | – | – | – |
| Triticale (1.28) | 3.85 | 0.0001* | 0.062 | 3.46 | 0.0006* | 0.08 |
| Fescue (1.53) | 0.13 | 0.89 | 0.076 | 0.31 | 0.76 | 0.09 |
| Vetch (1.37) | 2.00 | 0.05* | 0.073 | 2.08 | 0.04* | 0.09 |

*Indicates means that are significantly different.

*Table 5*. Comparison between smallest worker head capsule size (width × length) from various habitat types associated with vineyards (df = 98).

| Habitat type (means in mm) | Control (0.632) | | |
|---|---|---|---|
| | *t*-value | *p*-value | SE of difference |
| Triticale (0.716) | 3.72 | 0.0003* | 0.022 |
| Fescue (0.631) | 0.07 | 0.941 | 0.015 |
| Vetch (0.683) | 3.06 | 0.0029* | 0.016 |

*Indicates means that are significantly different.

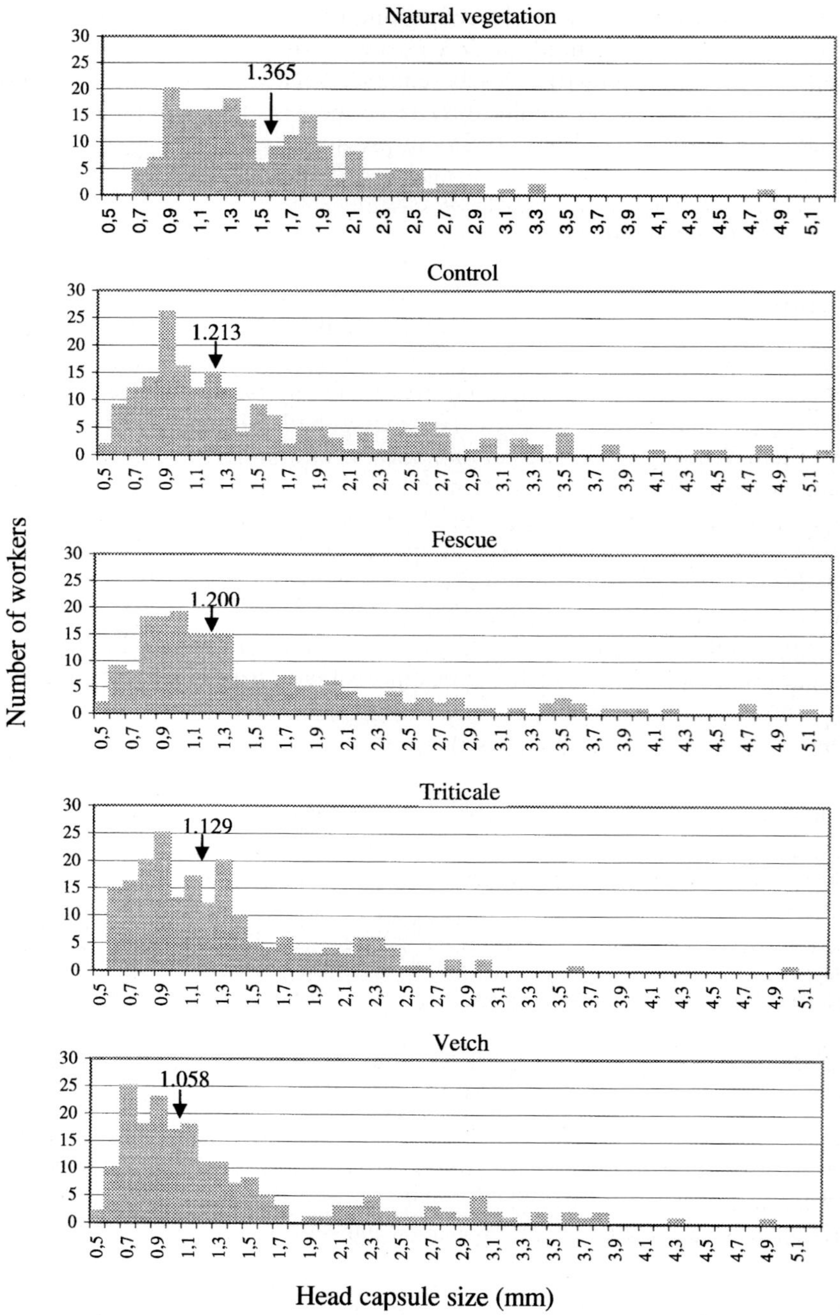


*Figure 5.* The size-frequency distribution of 200 all-sized *A. custodiens* workers from different structural habitats associated with vineyards. Arrows indicate the median.

in a vineyard with mealybugs. Control plots in vineyards resembled natural vegetation in that there were a great variety of microhabitats due to various weeds. This contrasted with the pure stand of cover crop, such as vetch and triticale. Workers in cover crops were smaller than those in control plots, which suggests that prey type is not the only determinant of worker size, but that temperature could also play a role since soil temperatures in cover crop plots were significantly lower than in control plots.

## Discussion

*Anoplolepis custodiens* is a local, highly aggressive invasive species, which reduces population levels of many other ants. It is also highly predaceous on a wide range of insects. It can reach high population levels and locally can be a keystone species. Nevertheless, *A. custodiens* is sensitive to tree canopy cover and will move its nest away from increasingly dense canopy shade, returning when the canopy thins (Myers 1957). Despite this, the results here strongly suggest that the ant could gain a foothold in bushy, herbaceous or grassy habitats and that it is unlikely to be prevented from spreading. In all likelihood, this would be triggered by a mutualism with honeydew-producing insects such as mealybugs. This is likely to happen whether or not mealybug parasitoids are present. Once the mutualistic spiral between ant and honeydew-producer has started, the parasitoids, not surprisingly, would probably not be able to stop it.

The pugnacious ant's potential invasive success hinges on the crucial fact that food resources drive it to maintain high population levels in a wide range of habitat types, with the exception of dense tree canopies. This is further substantiated by its morphological adaptation to that food resource as indicated by the variation in head capsule measurements for the different habitats. Both a limitation in the food source and a reduction in soil temperature have been ascribed to causing smaller workers in a study conducted in Scotland on *Myrmica rubra* Linnaeus (Brian and Brian 1951). Therefore, it would appear that *A. custodiens* would sacrifice size for a lucrative food source as it is highly tolerant of soil temperature changes, litter presence and grassy or herbaceous ground covers, so long as there is ample food and sunlight. In summary, these results indicate that this ant has the ability to spread and dominate a variety of open-canopy subtropical habitats. As such habitats are widespread globally, there is no reason why this species should not penetrate these areas, given time and human agency. But how likely is this to become a reality?

The ant is unlikely to be transported with export fruit or grapes as these are picked from the tree or vine and individually packed. Far more likely is the possibility of being transported in soil or some sort of ground cover or litter. Furthermore, as *A. custodiens* is polygynous, there is no reason why it should not go through the same genetic bottleneck process as *Linepithema humile*, with the formation of supercolonies through reduced intraspecific competition (Tsutsui et al. 2003).

*A. custodiens* is not the only potentially invasive ant in southern Africa. *Pheidole megacephala*, which is also competitively superior given abundant food resources, is already a well-known tramp species. *Anoplolepis steingroeveri* is another possible invasive, but it is far less common and less geographically widespread than *A. custodiens* and perchance is less likely to be an invasive candidate.

From this study, we recommend cataloguing, in different countries, potentially invasive ants and other insects, and indeed other organisms. While this is done on a regular basis for export agricultural produce, it has not yet been a sufficiently widespread practice in the invasive alien insect sector, although methodologies are being developed for screening potential pest plants (Daehler et al. 2004).

As a corollary, *A. custodiens* is not just a potential invader of open, natural or disturbed indigenous ecosystems, but also of agroecosystems. The cover crops used here are not likely to be a pest management option, either in African or overseas systems. The main driver regulating ant abundance in this study was the food source in the form of honeydew from mealybugs, which dominated over any possible inhibitions imposed by the cover crops. In addition, seeds from triticale could have been utilized, as indicated by the abrupt decline in ant activity, relative to the other treatments, when seeds were buried during soil preparation. Samways (1983) likewise found that habitat modification would be unsuitable as a primary method of ant management for dominant *Pheidole* spp. in South African citrus, and that trunk barriers are an ecologically more appropriate method of management. *A. custodiens* maintains its notoriety as an insect to monitor and control immediately, should it arrive in a new country.

## Acknowledgements

Thanks to staff of ARC Infruitec-Nietvoorbij, in particular E.C. du Toit for technical assistance, J. Fourie for advice on cover crops, V.M. Walton and L. Williams for parasitoid identifications and F. Calitz for assistance with the statistical analyses. Mr. F. van der Merwe of Morgentsond is thanked for the use of his vineyard for the trial. Funding was from the Agricultural Research Council and Winetech.

## References

Acocks J.P.H. 1988. Veld types of South Africa. Memoires of the Botanical Survey of South Africa 57: 1–146.

Addison P. 2002. Chemical stem barriers for the control of ants (Hymenoptera: Formicidae) in vineyards. South African Journal of Enology and Viticulture 23: 1–8.

Addison P. and Samways M.J. 2000. A survey of ants (Hymenoptera: Formicidae) in vineyards in the western Cape Province, South Africa. African Entomology 8: 251–260.

Baskerville G.L. and Emin P. 1969. Rapid estimation of heat accumulation from maximum and minimum temperatures. Ecology 50: 514–517.

Bond W.J. and Slingsby P. 1983. Seed dispersal by ants in shrublands of the Cape Province and its evolutionary implications. South African Journal of Science 79: 231–233.
Brian M.V. and Brian A.D. 1951. Insolation and ant populations in the West of Scotland. Transactions of the Royal Entomological Society of London 102: 303–330.
Cook J.L. 2003. Conservation of biodiversity in an area impacted by the red imported fire ant *Solenopsis invicta* (Hymenoptera: Formicidae). Biodiversity and Conservation 12: 187–195.
Daehler C.C., Denslow J.S., Ansari S. and Kuo H.-C. 2004. A risk-assessment system for screening invasive pest plants from Hawaii and the Pacific Islands. Conservation Biology 18: 360–368.
Dean W.R.J 1992. Temperatures determining activity patterns of some ant species in the southern Karoo, South Africa. Journal of the Entomological Society of southern Africa 55: 149–156.
Donnelly D. and Giliomee J.H. 1985. Community structure of epigaeic ants (Hymenoptera: Formicidae) in fynbos vegetation in the Jonkershoek Valley. Journal of the Entomological Society of southern Africa 48: 247–257.
Fowler H.G., Bernardi J.V.E., Delabie J.C., Forti L.C. and Pereira-Da-Silva V. 1990. Major ant problems of South America. In: Vander Meer R., Jaffe K. and Cedeno A. (ed.), Applied Myrmecology: A World Perspective. Westview Press, Boulder USA, pp. 3–14.
Gillespie R.G. 1999. Naiveté and novel perturbations: conservation of native spiders on an oceanic island system. Journal of Insect Conservation 3: 263–272.
Hill M., Holm K., Vel T., Shah N.J. and Matyot P. 2003. Impact of the introduced crazy ant *Anoplolepis gracilipes* on Bird Island, Seychelles. Biodiversity and Conservation 12: 1969–1984.
Holway D.A. 1995. Distribution of the Argentine ant (*Linepithema humile*) in northern California. Conservation Biology 9: 1634–1637.
Holway D.A., Lack L., Suarez A.V., Tsutsui N.D. and Case T.J. 2002. The causes and consequences of ant invasions. Annual Review of Ecology and Systematics 33: 181–233.
Human K.G., Weiss A., Sandler B. and Gordon D.M. 1998. Effect of abiotic factors on the distribution and activity of the invasive Argentine ant (Hymenoptera: Formicidae). Environmental Entomology 27: 822–833.
Kriegler P.J. and Whitehead V.B. 1962. Notes on the biology and control of *Crematogaster peringueyi* var *angustior* Arnold on grape vines (Hymenoptera: Formicidae). Journal of the Entomological Society of southern Africa 25: 287–290.
Löhr B. 1992. The pugnacious ant, *Anoplolepis custodians* (Hymenoptera: Formicidae), and its beneficial effect on coconut production in Tanzania. Bulletin of Entomological Research 82: 213–218.
Majer J.D. 1978. An improved pitfall trap for sampling ants and other epigaeic invertebrates. Journal of the Australian Entomological Society 17: 261–262.
Myers N.J. 1957. Studies on the Biology of Ants Associated with Citrus Trees. M.Sc. dissertation, Department of Entomology, Rhodes University, Grahamstown, South Africa.
Oster G.F. and Wilson E.O. 1978. Caste and Ecology in the Social Insects (Monographs in Population Biology, no. 12). Princeton University Press, Princeton, NJ.
Prins A.J., Robertson H.G. and Prins A. 1990. Pest ants in urban and agricultural areas of southern Africa. In: Vander Meer R., Jaffe K. and Cedeno A. (ed.), Applied Myrmecology: A World Perspective. Westview Press, Boulder, USA.
Ratsirarson H., Robertson H.G., Picker M.D. and van Noort S. 2002. Indigenous forests versus exotic eucalypt and pine plantations: a comparison of leaf litter invertebrate communities. African Entomology 10: 93–99.
Samways M.J. 1983. Community structure of ants (Hymenoptera: Formicidae) in a series of habitats associated with citrus. Journal of Applied Ecology 20: 833–847.
Samways M.J. 1990. Ant assemblage structure and ecological management in citrus and subtropical fruit orchards in southern Africa. In: Vander Meer R., Jaffe K. and Cedeno A. (ed.), Applied Myrmecology: A World Perspective. Westview Press, Boulder, USA.
Samways M.J. 1999. Managing insect invasions by watching other countries. In: Sandlund O.T., Schei P.J. and Viken A. (ed.), Invasive Species and Biodiversity Management. Kluwer, Dordrecht, The Netherlands.

Southwood T.R.E. 1978. Ecological Methods 2nd edn. Chapman & Hall, London.

Steyn J.J. 1954. The pugnacious ant *Anoplolepis custodiens* (F. Smith) and its relation to the control of the citrus scales at Letaba. Memoires of the Entomological Society of southern Africa 3: 1–96.

Toome P. 2002. Soil moisture monitoring with gypsum blocks. Part 1: Gypsum blocks as an irrigation scheduling tool. The Australian and New Zealand Grapegrower and Winemaker November: 82–84.

Tsutsui N.D., Suarez A.V. and Grosberg R.K. 2003. Genetic diversity, asymmetrical aggression, and recognition in a widespread invasive species. Proceedings of the National Academy of Sciences 100: 1078–1083.

Way M.J. 1953. The relationship between certain ant species with particular reference to biological control of the coreid, *Theraptus* spp. Bulletin of Entomological Research 44: 669–691.

Wheeler W.M. 1922. The ants of the Belgian Congo. Bulletin of the American Museum of Natural History 45: 1–1139.

Wilson E.O. 1996. Hawaii: a world without social insects. Bishop Museum Occasional Papers 45: 3–7.

Biodiversity and Conservation (2006) 15:415–427
DOI 10.1007/s10531-005-6319-3


# Impacts of catastrophic earthquakes on the insect communities in estuarine mangroves, northern Taiwan

CAN-JEN W. MAA[1], HUEI-JIN WANG[1] and HSING-JUH LIN[2,3,*]
*[1]Research Center for Biodiversity, Academia Sinica, Taipei, Taiwan 115, ROC; [2]Department of Life Sciences, National Chung Hsing University, Taichung, Taiwan 402, ROC; [3]Institute of Marine Environmental Chemistry and Ecology, National Taiwan Ocean University, Keelung 202, Taiwan, ROC*
**Author for correspondence (e-mail: hjlin@dragon.nchu.edu.tw; phone: +886-4-22840416; fax: +886-4-22874740)*

Received 23 September 2004; accepted in revised form 11 April 2005

**Key words:** Amplified shaking, Danshui River, Non-biting midge, Rove beetle, The 3–31 earthquake, The Chi–Chi earthquake

**Abstract** As a catastrophic earthquake is unpredictable and occurs only occasionally, impacts on biotic communities are seldom known. Monthly changes in insect communities in the mangroves along the Danshui Estuary were monitored for more than 3 years from before and after two catastrophic earthquakes in Taiwan the Chi–Chi earthquake ($M_L$ = 7.3) of September 21, 1999 and the 3–31 earthquake ($M_L$ = 6.8) of March 31, 2002. Here we show that the Chi–Chi earthquake not only caused large declines in total individual number but also total species number of insects. It also resulted in greater variability among samples, and shifts in insect communities. Non-biting midges and rove beetles, whose immatures inhabited the riparian underground or aquatic sediments, were most severely affected. By 7 months after the Chi–Chi earthquake, the insect communities had recovered to a level comparable to that before the earthquake. However, the influence of the 3–31 earthquake on the insect communities was less severe. It is concluded that the more-severe impacts of the Chi–Chi earthquake than the 3–31 earthquake can be attributable to differences in ground shaking, occurrence time, biodiversity, and growing conditions of insects at those times.

## Introduction

The extent to which biotic communities are affected by natural catastrophes has long been of ecological and biogeographic interest (Sousa 1984; Brown and Lomolino 1998; Spiller et al. 1998; Vandermer 2000). As a catastrophic earthquake is unpredictable and of low frequency, monitoring studies of communities before an earthquake and the immediate responses are often absent. Hence, knowledge of impacts of earthquakes on communities is very limited and little documented (Wells et al. 2001; Castilla 1988; Fang et al. 2002).

Coastal wetlands are regarded as being among the most productive ecosystems because of the tremendous volumes and varieties of inhabiting plants (Teal 1962). Plants have been shown to correlate with large numbers of insects in wetlands (Lin et al. 2003). Long-term changes in the insect communities were monitored monthly in the mangroves along the estuary of Danshui River in the Taipei basin of northern Taiwan for more than 3 years from April 1999 to

August 2002. During the study period, two catastrophic earthquakes hit Taiwan. The Chi–Chi earthquake, rated 7.3 on the Richter scale ($M_L = 7.3$), occurred in central Taiwan (23° 85′ N, 120° 78′ E) on 21 September 1999 with its epicenter identified at a distance of 155 km from the study sites. Peak horizontal accelerations of 70–116 *gal* were recorded in the Danshui Estuary. The 3–31 earthquake, rated 6.8 on the Richter scale ($M_L = 6.8$), occurred in the sea off northeastern Taiwan (24° 24′ N, 122° 17′ E) on 31 March 2002 with its epicenter located 132 km from the study sites. Peak horizontal accelerations of 43–63 *gal* were recorded in the estuary. There were 2413 and 5 human fatalities in the Chi–Chi and 3–31 earthquakes, respectively. These long-term observations before and after catastrophic earthquakes provide an opportunity to examine repeatedly their impacts on insect communities in the estuarine mangroves.

Community-level conservation approaches have been suggested for non-charismatic, little-known taxa like insects (Hughes et al. 2000). Herein this study focused on: (1) the immediate impacts of two catastrophic earthquakes on insect communities; (2) the species which were most affected; (3) the recovery time of the affected communities; and (4) differences in the impacts caused by the two earthquakes.

## Materials and methods

### *Study site and sampling*

Mangroves, composed of homogeneous forests of *Kandelia candel*, line the Danshui Estuary of subtropical northern Taiwan. Insects in three mangrove forests, Bali (25° 10′ N, 121° 25′ E), Chuwei (25° 09′ N, 121° 27′ E), and Guandu (25° 07′ N, 121° 28′ E) were collected monthly (Figure 1). The areas of the mangrove forests are 12 ha in Bali, 24 ha in Guandu, and 50 ha in Chuwei. Mean air temperature ranged from 15.2 °C in February to 28.9 °C in July with an annual rainfall of 2600 mm. These forests are subjected to a semidiurnal tidal regime with a tidal amplitude of about 1–2 m. Salinities of the overlying waters ranged from 7 to 12 psu at low tide and might reach 25 psu at high tide.

Abiotic variables of sediment, including pH, grain size, silt/clay content, sorting coefficient of sediment, total organic content (TOC), and total nitrogen (TN) immediately before the Chi–Chi earthquake were available from the Guandu mangroves collected on July 15, 1999 (Lin et al. 2003). For comparisons, sediment samples were collected from the same site on September 29, 1999 immediately after the earthquake. For the collection of sediment, PVC corers with an inner diameter of 2.6 cm were pushed 2 cm deep into the sediment. One of the 2 sediment samples collected was used for analysis of granulometry, while the other was refrigerated and used for analysis of organic content. Granulometry of the sediment was determined following Buchanan and Kain (1971) and Hsieh (1995). The TOC content of the sediment, expressed as percent dry

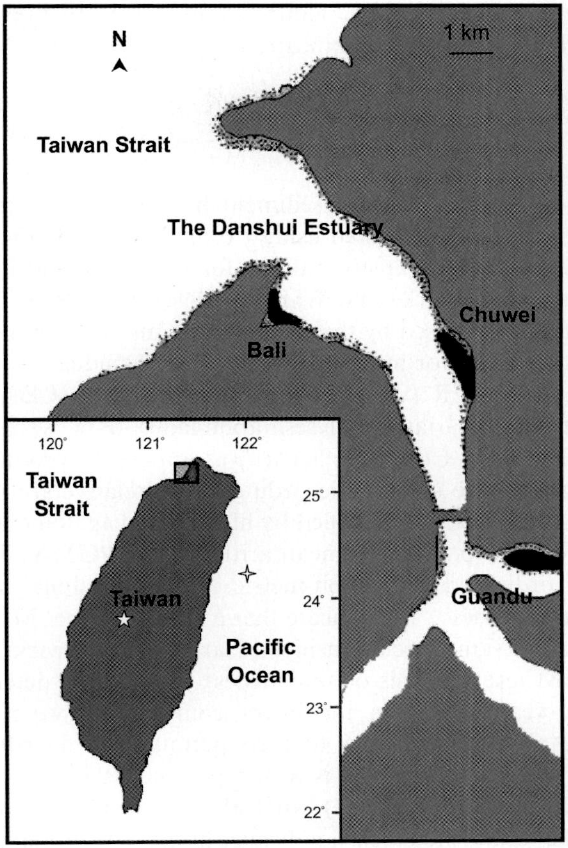


*Figure 1.* Locations of three study mangroves of Bali, Chuwei, and Guandu along the Danshui Estuary in northern Taiwan. ☆ Epicenter of the Chi–Chi earthquake, ☆ Epicenter of the 3–31 earthquake.

weight, was measured by adding 1 N HCl to remove inorganic carbon. The TOC and TN contents were then measured using an elemental analyzer (Perkin Elmer EA 2400 Series II).

Light traps and sticky paper were used for insect collection. A light trap was composed of an 8-W ultraviolet light-tube, a 12-V storage battery, and a water container measuring 18 cm × 28 cm × 10 cm (width × length × depth). Within the container, 500 ml of a 1% sodium dodecyl sulfate solution was added for trapping insects. Sticky paper surrounded the light trap would catch those insects which landed around the light trap instead of dropping into the solution. Each light trap was supported on a wooden stand. These stands were 4 m high in the mangroves. Hence, the light traps would stand taller than the mangroves and above sea level at high tide. These UV light traps at 320–340 nm were set for a period of 12 h from dusk of the first day to dawn of the following day to collect insects. Collected insects were preserved in 75% alcohol for enumeration and identification in the laboratory.

### *Data analyses*

Differences in abiotic variables of sediment between two times (July 15 and September 29, 1999) were detected using *t* tests. Before the analyses, the data were tested by the power transformations for normality and homogeneity of variance assumptions (Clarke and Warwick 1994). In order to reveal changes in insect communities caused by the earthquakes, monthly abundance data for each species at each site for more than 3 years were studied using multivariate analyses in the PRIMER (vers. 5.2) computer package (Clarke and Gorley 2001). Prior to multivariate analyses, abundance data were $\log_{10}(n + 1)$-transformed. The Bray–Curtis coefficient was applied to produce a similarity matrix between any two samples according to the relative abundance of each species. The data matrix was classified by hierarchical agglomerative clustering using the unweighted pair group mean arithmetic (UPGMA) linking method and was then ordinated using non-metric multidimensional scaling (MDS) techniques. Stress values $<0.2$ indicate that a 2-dimensional MDS plot gives a usable summary of sample relationships (Clarke and Warwick 1994).

The ANOSIM test (analysis of similarities) was used to determine whether site or year-to-year effects on the insect communities were significant by comparing the observed statistics to their permutation distributions for the absence of differences. ANOSIM is a non-parametric analog of MANOVA (multivariate analysis of variance) without the assumption of multivariate normality. If the results indicated significance at the 0.05 probability level, both pairwise *R* values and the Bonferroni correction for the significance level were used to determine which levels differ. Insect communities were well separated when $R > 0.75$, overlapped but clearly different when $R > 0.50$, and barely separable at all when $R < 0.25$. The SIMPER (similarity of percentages) analysis was employed to reveal the most-representative species for each insect community within a site or a year. Warwick and Clarke (1993) found that the variability among samples collected from stressed areas is generally much greater than that from control sites. Hence, the MVDISP (multivariate dispersion) analysis was then used to measure the variability of insect communities among samples from the three study sites after the two earthquakes.

## Results

### *Abiotic variables*

There were no distinct differences in abiotic variables of sediment immediately before and after the Chi–Chi earthquake in the Guandu mangroves (Table 1).

*Table 1.* Abiotic variables in sediments (mean ± SE, $n = 4$) of the Guandu mangroves recorded before and after the Chi–Chi earthquake of 21 September 1999.

| Variable | Before (15 July 1999, Lin et al. 2003) | After (29 September 1999) | *t*-test (*p* value) |
|---|---|---|---|
| Grain size (mm) | 0.020 ± 0.001 | 0.030 ± 0.008 | 0.27 |
| Silt/clay content (%) | 83.8 ± 1.2 | 65.9 ± 7.8 | 0.06 |
| Sorting coefficient | 1.78 ± 0.15 | 2.48 ± 0.36 | 0.13 |
| TOC (%) | 2.07 ± 0.24 | 2.11 ± 0.10 | 0.89 |
| TN (%) | 0.18 ± 0.03 | 0.16 ± 0.01 | 0.49 |

The Guandu mangroves had silt sediments. Silt/clay content of sediment tends to be lower after the Chi–Chi earthquake ($p = 0.06$). However, differences in grain size and sorting coefficient could not be shown to be statistically significant.

### *Abundance changes*

Approximately 300 insect species belonging to 104 families and 18 orders were collected at the three study sites during the 41-month study. Generally, at all three study sites, total individual number (Figure 2) and total species number (Figure 3) of insects per effort showed a clear bimodal seasonal pattern with one peak in April (spring) and a second peak in August (summer). Total individual numbers at the three study sites in September 1999 immediately after the Chi–Chi earthquake declined by 90% as compared to those in September of 2000, whereas species number declined by 75%. However, impacts of three typhoons (the Bilis typhoon, the Xangsane typhoon, and the Nari typhoon) on the species number and the total individual number were relatively minor as compared those of the Chi–Chi earthquake. There was only a small decrease in total individual number of insects immediately after the 3–31 earthquake (Figure 2). The total species number appears not to be affected by the 3–31 earthquake (Figure 3).

### *Community changes*

The insect communities collected before the Chi–Chi earthquake at all three study sites were barely separable at all (ANOSIM test, $R < 0.25$, $p > 0.05$, Figure 4a). However, insect communities from 1999 of the Chi–Chi earthquake could be separated from those collected in 2000, 2001 and 2002 (ANOSIM test, $R > 0.50$, $p < 0.01$, Figure 4b). The variability of insect communities among samples from the three study sites was found to increase immediately after the Chi–Chi earthquake (MVDISP analyses, Figure 5). Community similarities (30–40%), including insect community samples in September 1999 just after the Chi–Chi earthquake, decreased when compared to other month-pairs

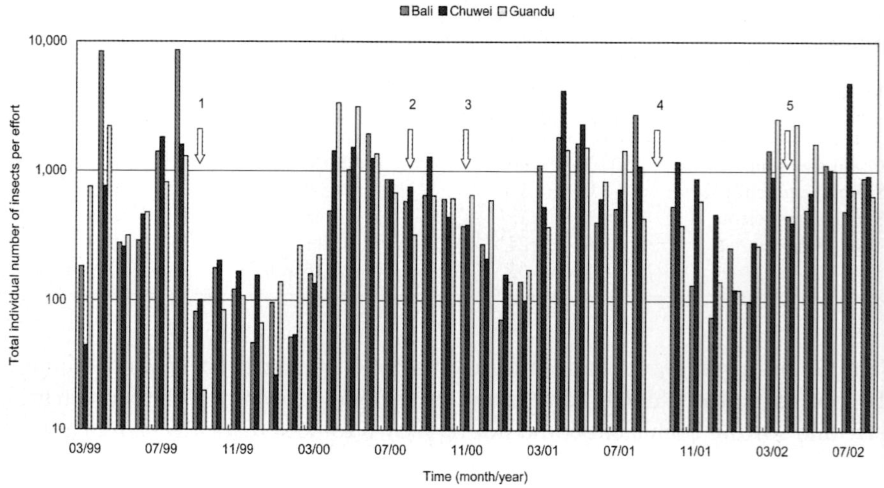


*Figure 2.* Seasonal pattern of total individual numbers of insects per effort in the mangroves of Bali, Chuwei, and Guandu along the Danshui Estuary. (1. the Chi–Chi earthquake; 2. the Bilis typhoon; 3. the Xangsane typhoon; 4. the Nari typhoon, no data available; 5. the 3–31 earthquake).

(55–65%). Shifts in species composition could be attributed to large decreases in individual numbers of non-biting midges and rove beetles (SIMPER analyses). The primary species of non-biting midges were *Tanytarsas formosanus*, *Cricotopus* (*Isocladius*) *sylvestris*, *Microchironomus tener*, *Harnischia curtilamellata*, and *Polypedilum nubifer*. The primary species of rove beetles were possibly *Trogophloeus* spp.

Generally, the insect communities in the mangroves of the Danshui Estuary can be classified into a winter group (from November to February, Figure 6) and a summer group (from March to October, Figure 6). However, insect communities had been shifted in September 1999 after the Chi–Chi earthquake from the summer group to the winter group and remained so for 7 months until April 2000 (MDS ordination, Figure 7a) as compared to the gradation of insect communities from April 2000 to April 2001 (MDS ordination, Figure 7b). This suggests a rapid recovery time of 7 months after the impacts of the Chi–Chi earthquake. Despite the increase in the variability among samples (Figure 5), no clear shifts in insect communities occurred in April 2002 just after the 3–31 earthquake (Figure 7c) as occurred after the Chi–Chi earthquake (Figure 7a).

## Discussion

Biotic communities are characterized by continuous changes in species composition in response to various biotic and abiotic factors over time (Greene and

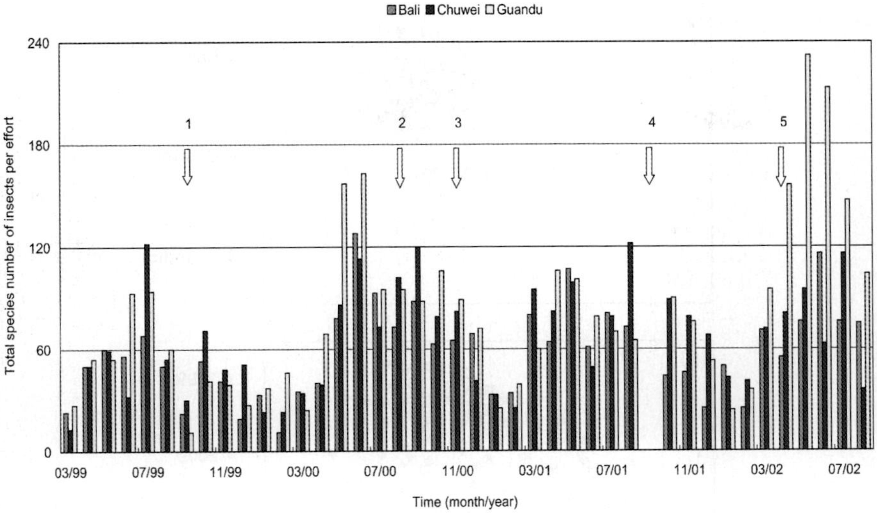


*Figure 3.* Seasonal pattern of total species numbers of insects per effort in the mangroves of Bali, Chuwei, and Guandu along the Danshui Estuary (1. the Chi–Chi earthquake; 2. the Bilis typhoon; 3. the Xangsane typhoon; 4. the Nari typhoon, no data available; 5. the 3 – 31 earthquake).

Schoener 1982). The damage resulting from an earthquake is influenced in a number of ways by the characteristics of the soil in the affected area. In addition to ground shaking, the Chi–Chi earthquake has been reported to induce soil liquefaction in central Taiwan (Wen and Ray 2001). Hengesh and Bachhuber (1996) indicated that the spatial distributions of amplified ground-shaking intensity and the occurrence of liquefaction are not random, but generally are restricted to alluvial basins that contain thick deposits of low-strength clay (amplified shaking), or shallow layers of low-density, saturated, granular sediments (liquefaction). Mangroves in the Danshui Estuary of the Taipei basin are, thus, susceptible to both of these hazards (Teng and Liao 2000; Chien and Loh 2002). It is conceivable that the shifts in the insect communities in 1999 could be linked to ground shaking and/or soil liquefaction from the Chi–Chi earthquake.

However, soil characteristics of the Guandu mangroves show little evidence of induction of soil liquefaction by the Chi–Chi earthquake. Wen and Ray (2001) found that the most important factor for soil liquefaction susceptibility during earthquakes is uniformed size distribution of deposits, which can be indexed by the sorting coefficient. They indicated that the mean grain size for soil liquefaction susceptibility is about 0.15 mm, fine sand sediments. However, the Guandu mangroves had silt sediments with a mean grain size of 0.02 mm (Table 1). Based on the equation suggested by Wen and Ray (2001), the threshold of sorting coefficient for soil liquefaction susceptibility during an earthquake at the magnitude of the Chi–Chi earthquake (70 – 116 *gal*) is about 1.34. The higher mean sorting coefficient of 1.78 in the Guandu mangroves

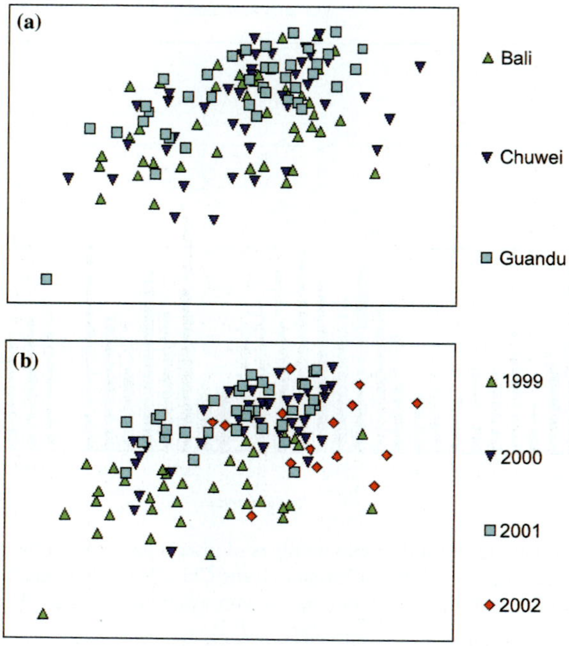


*Figure 4.* Two-dimensional MDS ordination of insect communities collected from April 1999 to August 2002 using Bray–Curtis similarities on log-transformed abundance for: (a) three study sites: Bali, Chuwei, and Guandu; (b) four study periods: 1999, April 1999 to March 2000; 2000, April 2000 to March 2001; 2001, April 2001 to March 2002; and 2002, April 2002 to August 2002 (stress = 0.14).

suggests that the sediments were poorly sorted and the induction of soil liquefaction was unlikely.

Ground shaking would be much larger at sites located on the soft layers of alluvial basins. Huang (2004) demonstrated that the Taipei basin amplified the seismic waves of the Chi–Chi earthquake and the 3–31 earthquake by factors of 4.0–9.5 at frequencies of 0.6–0.8 Hz (1.0–1.7 s periods), with the largest amplification being at lower frequencies. The strongest shaking lasted much longer on the Taipei basin than at the surrounding bedrock areas. It is clear that the Taipei basin increases low frequency shaking during the two earthquakes. Deposits of loose granular soils might be compacted by ground vibrations induced by earthquakes, resulting in large and differential settlement of the ground surface (Seed and Idriss 1982). The process of the shaking effects and the motions of soil cause the destruction of soil structure, which may disturb underground or soil surface organisms. However, there were no distinct changes in soil structure of the Guandu mangroves immediately after the Chi–Chi earthquake, although silt/clay content of sediment tends to be lower (Table 1). This suggests that soil in the mangroves enriched with water suffered less structure damage, but might experience more pressure wave during earthquakes. It is more likely

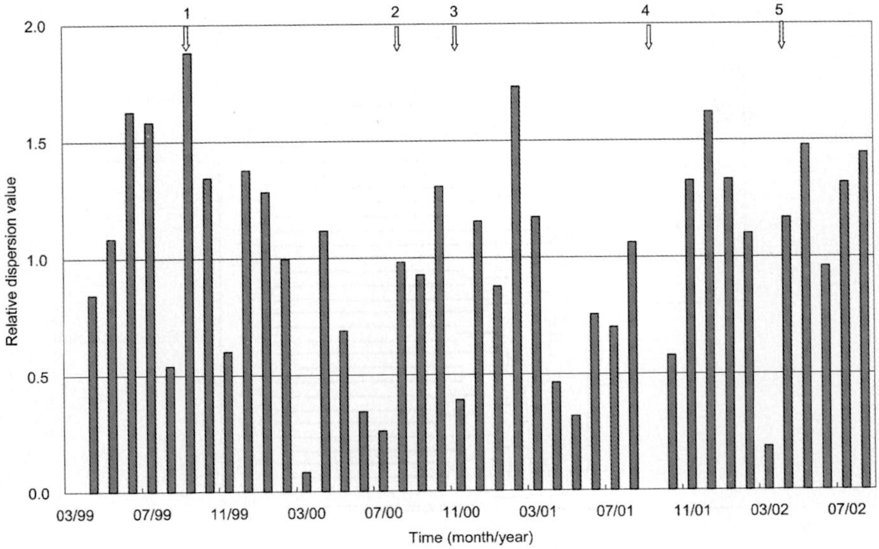


*Figure 5.* MVDISP analyses of the relative dispersion of insect communities among samples collected at Bali, Chuwei, and Guandu along the Danshui Estuary (1. the Chi–Chi earthquake; 2. the Bilis typhoon; 3. the Xangsane typhoon; 4. the Nari typhoon, no data available; 5. the 3–31 earthquake).

that the individual number and species number of rove beetles and non-biting midges, whose immatures inhabited the riparian underground or aquatic sediments, were drastically reduced due to pressure wave right after a catastrophic earthquake. The mechanism needs to be further examined.

However, our results showed that a catastrophic earthquake does not necessarily exert a severe impact on the abundance and structure of insect communities. Changes in communities caused by any force vary from negligible to extreme, depending on the intensity of the force and the vulnerability of the target organisms (Sousa 1984). Therefore, the cause of the more-severe impacts on insect communities by the Chi–Chi earthquake might be the stronger intensity, longer duration, and higher frequency of ground shaking (Huang 2004). In the Chi–Chi earthquake, the maximum ground acceleration reached a value of 116 *gal*; the duration of strong shaking lasted more than 1 min; and the motion continued with 76 aftershocks with magnitudes exceeding 5.0 $M_L$, including 8 with magnitudes exceeding 6.0 $M_L$. In contrast, for the 3–31 earthquake, the recorded maximum ground acceleration reached 63 *gal* with 7 aftershocks having magnitude exceeding 5.0 $M_L$, among which none exceeded a magnitude of 6.0 $M_L$.

The second possibility for the cause of the more-severe impacts on insect communities by the Chi–Chi earthquake is the different occurrence time between the two earthquakes. Earthquake occurred at different time might have different impacts on insect communities. Our 41-month observations showed

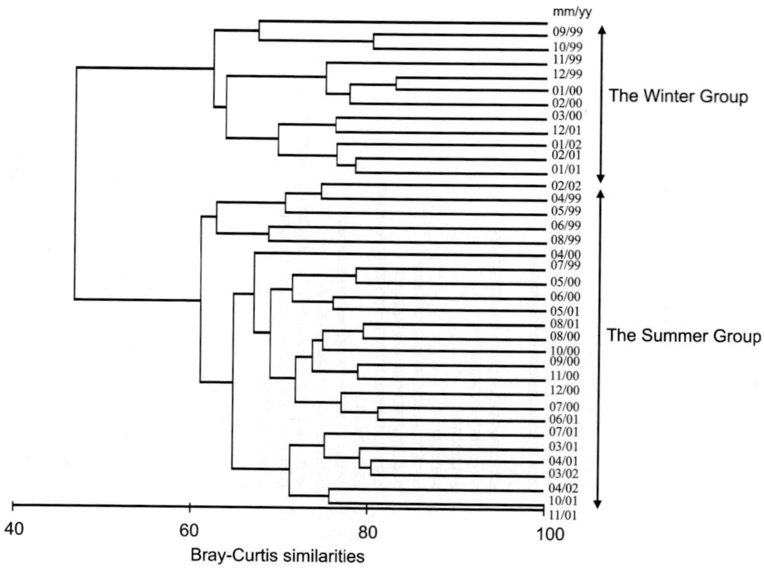


*Figure 6.* Dendrogram for hierarchical clustering of monthly insect communities, using group-average linking of Bray–Curtis similarities calculated on log-transformed abundance data.

that the insect growing seasons of the Danshui Estuary occurred in spring and summer. For example, rove beetles and non-biting midges are the dominant insects in the mangroves. The eggs are laid in a mass on the surface of the water which is usually attached to the edge of the river or twigs in contact with the water (authors' personal observations). The eggs usually hatch in 2–3 days. On the second or third day after hatching, the larvae leave the mass, burrow into the mud to build small tubes in which they live. The larval stage can last 7–9 days. The larvae transform into pupae while still in the tubes. The pupal stage normally lasts 2–5 days. The pupae leave the tube and actively swim to the surface a few hours before the adult emerges. The adults which emerge mate during swarming at night. The adults do not feed during their adult existence and only live for 3–5 days. The entire life cycles can be completed in 15–20 days, depending on temperature. The adult emergence begins in March and shows a clear bimodal seasonal pattern with one peak in April and a second peak in August (Figure 2). With the Chi–Chi earthquake occurring at the end of summer with slowed insect growth and the 3–31 earthquake occurring in spring when insects were actively growing and the biodiversity was high, it is likely that the greater biological potential and biodiversity of spring insects would fuel a speedier and stronger bounce back and made the impacts of the 3–31 earthquake hardly to detect.

The third possibility is the higher air temperature when the 3–31 earthquake occurred. The occurrence and development of insects are related to ambient temperatures. Mean monthly air temperatures in April of 2002 immediately after

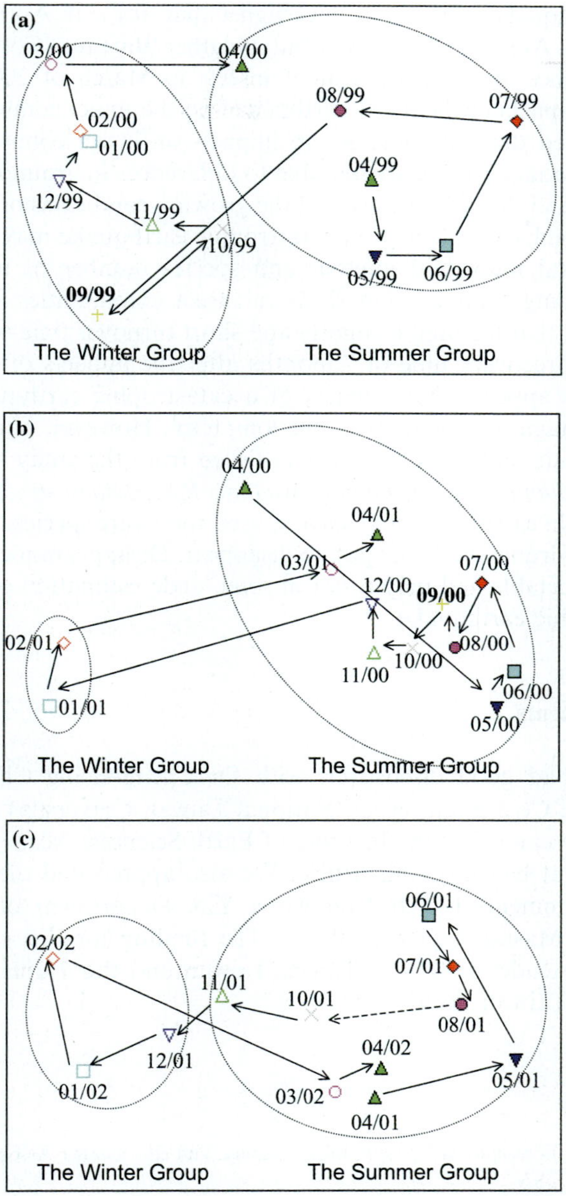


*Figure 7.* Two-dimensional MDS ordination of shifts of insect communities (a) from April 1999 to April 2000 (stress = 0.04), (b) from April 2000 to April 2001 (stress = 0.03), and (c) from April 2001 to April 2002 (data on September 2001 were not available due to the Nari typhoon, stress = 0.04) in the mangroves along the Danshui Estuary. The two classified groups, the summer group and the winter group, are indicated for each period.

the 3–31 earthquake were 1.5–1.9 °C higher than those in April 2001 (Climatological Data Annual Report, Central Weather Bureau). Consequently, the earlier occurrence and development of insects in March of 2002 might have lessened the impacts of the 3–31 earthquake on the insect communities.

We concluded that the more-severe impacts on insect communities by the Chi–Chi earthquake can be attributable to differences in ground shaking, time of the quake, and the biodiversity and the growing conditions of insects at that time. The immediate impacts of a catastrophic earthquake may result in large declines in total individual number and species number of insects, greater variability among samples, and shifts in insect communities. However, our results showed that the high fecundity and short turnover time of insects led to a rather rapid recovery time of 7 months after the impacts of a catastrophic earthquake. It appears that impacts of a catastrophic earthquake on insect communities might be negligible in the long term. However, after the Chi–Chi earthquake, four midge species have vanished from the study sites, including *Limnophyes minimus*, *Polypedilum convictum*, *Polypedilum* sp., and *Cricotopus* sp. The Chi–Chi earthquake may have caused some rare species, which are very sensitive to environmental changes, to disappear. Disappearance of rare species is often undetectable and may result in some underestimation of the influence of a catastrophic earthquake.

## Acknowledgements

The authors had good discussions with Dr Cheng-Hsing Chen (Professor, Department of Civil Engineering, National Taiwan University) and Dr Hung-Chie Chiu (Research Fellow, Institute of Earth Sciences, Academia Sinica) on this subject that benefited this article. We also appreciated the very effective and useful comments by Dr Chih-Ming Yin, Department of Entomology, University of Massachusetts, Amherst. The funding for this study was provided by the Academia Sinica, Taipei, Taiwan and the Taipei City Government, Taipei, Taiwan, ROC.

## References

Brown J.H. and Lomolino M.V. 1998. Biogeography. 2nd ed.. Sinauer Associates, Sunderland, Massachusetts, USA.

Buchanan J.B. and Kain J.M. 1971. Measurement of the physical and chemical environment. In: Holme N.A. and McIntyre (eds), Methods for the Study of Marine Benthos. IBP Handbook no. 16. Blackwell Scientific Publications, Oxford, UK, pp. 30–58.

Castilla J.C. 1988. Earthquake-caused coastal uplift and its effects on rocky intertidal kelp communities. Science 242: 440–443.

Chien W.-Y. and Loh C.-H. 2002. Characteristics of ground shaking by the 3–31 earthquake in the Taipei Basin. Modern Construction and Civil Engineering 273: 12–15. (in Chinese).

Clarke K.R. and Gorley R.N. 2001. PRIMER v5: user manual/tutorial. PRIMER-E. Plymouth Press, UK.

Clarke K.R. and Warwick R.M. 1994. Changes in Marine Communities: An Approach to Statistical Analysis and Interpretation. Plymouth Press, UK, Natural Environment Research Council.

Fang L.S., Chen I.-S., Tew K.S., Han C.-C., Lee T.-F. and Chen I.-M. 2002. Impact of the 1999 earthquake on mountain stream fishes in Taiwan. Fisheries Science 68: 446–448.

Greene C.H. and Schoener A. 1982. Succession on marine hard substrata: a fixed lottery. Oecologia 55: 289–297.

Hengesh J.V. and Bachhuber J.L. 1996. Seismic hazard zonation of greater San Juan area northern Puerto Rico coastal plain. USGS Award No.:1434-HQ-96-GR-02765.

Hsieh H.-L. 1995. Spatial and temporal patterns of polychaete communities in a subtropical mangrove swamp: influences of sediment and microhabitat. Marine Ecology Progress Series 127: 157–167.

Huang H.C. 2004. Site effect of soft soil layers using the 921 Chi–Chi earthquake sequence. NSC 91–2119-M-194–004, National Science Council, Taipei, Taiwan ROC, pp. 7. (in Chinese).

Lin H.-J., Shao K.-T., Chiou W.-L., Maa C.-J.W., Hsieh H.-L., Wu W.-L., Severinghaus L.L. and Wang Y.-T. 2003. Biotic Communities of Freshwater Marshes and Mangroves in Relation to saltwater incursions: Implications for Wetland Regulation. Biodiversity and Conservation 12: 647–665.

Seed H.B. and Idriss I.M. 1982. Ground Motions and Soil Liquefaction during Earthquakes. Earthquake Engineering Research Institute, California, USA.

Sousa W.P. 1984. The role of disturbance in natural communities. Annual Review of Ecology and Systematics 15: 353–391.

Spiller D.A., Losos J.B. and Schoener T.W. 1998. Impact of a catastrophic hurricane on island populations. Science 281: 695–697.

Teal J.M. 1962. Energy flow in the salt marsh ecosystem of Georgia. Ecology 43: 614–624.

Teng C.-J. and Liao W.-Y. 2000. Study on focusing of seismic wave energy in basins. News Report of National Center for Research on Earthquake Engineering 33: 1–5. (in Chinese).

Vandermer J., dela Cerda I.G., Boucher D., Perfecto I. and Ruiz J. 2000. Hurricane disturbance and tropical tree species diversity. Science 290: 788–791.

Warwick R.M. and Clarke K.R. 1993. Increased variability as a symptom of stress in marine communities. Journal of Experimental Marine Biology and Ecology 172: 215–226.

Wells A., Duncan R.P. and Stewart G.H. 2001. Forest dynamics in Westland, New Zealand: the importance of large, infrequent earthquake-induced disturbance. Journal of Ecology 89: 1006–1018.

Wen S.-B. and Ray D.-T. 2001. The Geological Criteria of Soil Liquefaction Area in Earthquake Induction. NSC 90–2611-E-006–030.National Science Council, Taipei, Taiwan ROC, pp. 4. (in Chinese).

Biodiversity and Conservation (2006) 15:429–438
DOI 10.1007/s10531-005-6320-x 

# Countryside Stewardship Scheme and butterflies: a study of plant and butterfly species richness

R.G. FIELD[1,*], T. GARDINER[1], C.F. MASON[2] and J. HILL[3]
*[1]Centre for Environment and Rural Affairs, Writtle College, Writtle, Chelmsford, Essex; [2]Department of Biological Sciences, University of Essex, Wivenhoe Park, Colchester, Essex, CO4 3SQ; [3]Department of Science, Writtle College, Writtle, Chelmsford, Essex, CM1 3RR; *Author for correspondence: Address: 166 Sherwood Avenue, Northampton, NN2 8TE, 01604–844927 (e-mail: rfield8633@aol.com)*

Received 24 March 2005; accepted in revised form 11 April 2005

**Key words:** Agri-environment schemes, Butterflies, Butterfly species richness, Countryside Stewardship Scheme, Field margins, Habitats, Nectar sources, Plant species richness

**Abstract.** Butterfly and plant species richness were recorded from 1997 to 2000 on 2 and 6 m grass margins created at three farms in Essex which had entered the Countryside Stewardship Scheme (CSS) in October 1996. On both the 2 and 6 m margins there was a significant relationship between the length of hedgerow and the number of plant species observed on the margins, but the seed mixtures used may not have been ideal and natural regeneration should not have been used on the clay soils of Essex. Butterfly species richness was significantly greater on the 2 m margins than on the control sections, and was greater when a higher number of grass species were included in the original seed mixture. Plant species richness was greater on the 6 m margins when established by natural regeneration. CSS grass margins could be improved as butterfly habitats if they are linked to existing habitats such as hedgerows, are sown with a better range of native grasses and herbs and are managed in a way more conducive to wildlife. These changes to the policy of establishment of CSS margins could help combat habitat loss and fragmentation within the countryside.

## Introduction

Britain, although heavily populated has 76% (MAFF 2000) of it's land surface devoted to agriculture. This landscape has altered dramatically, the main reasons being the increased reliance on winter sown crops, increased farm specialisation and the loss of semi-natural habitats (Marshall 1998). The loss of the semi-natural habitats has resulted in an increase in habitat fragmentation, an important cause of species decline (English Nature 1995). For example, the agricultural improvement of an area of semi-natural grassland to one of grass and clover ley can reduce butterfly species richness from 17 to 23 species to one to three species (Thomas 1984).

The Countryside Stewardship Scheme (CSS) grass margins (2 & 6 m) are a method of replacing lost habitats and providing connectivity (MAFF 1999). However de Snoo and Udo de Haes (1994) suggest that grass only margins have no beneficial effect on insect populations. The range of grasses recommended for the CSS grass margins were reviewed by Marshall (1998), who identified those suitable for a range of different soil types. Natural regeneration was also suggested as one option for developing 6 m margins, though West and Marshall (1996) found that natural regeneration was most successful on lighter soils, where the local seed bank was still diverse (Marshall 1998; West et al. 1999) and weed pressure was low (Marshall and Moonen 1998).

Studies of grass margin establishment observed the lowest plant species diversity on clay soils and in margins sown with grass only mixtures (West and Marshall 1996). The greatest cost benefit was provided by sowing a simple grass and wildflower mix. Other studies relating to the inclusion of wildflower seeds in grass margins have demonstrated positive results for invertebrate biodiversity (Feber et al. 1994; Smith et al. 1994; Hopkins and Feber 1997; Barker and Reynolds 1999; Kirkham et al. 1999; Smith et al. 1999).

Dover (1999) suggested that butterflies are important indicators of farmland biodiversity. Common species were positively associated with sheltered areas, such as field corners, and with flower rich sites, whilst few butterflies were observed on margins where the boundary vegetation consisted of a simple grass bank. Butterfly species richness was highest on margins established from wildflower and grass mixtures (Buys 1995; Buys et al. 1996; Feber et al. 1996; Carreck et al. 1999; Kirkham et al. 1999). Furthermore recent studies have suggested that larval host plants may not be the limiting factor on butterfly abundance, and that sub-optimal adult resources may be of greater importance (Dover 1999). The lack of nectar sources was found to be a significant factor in the potential fecundity and longevity of adult butterflies (Watt et al. 1974; Murphy et al. 1983; Dover 1994; Feber et al. 1996), and may be a limiting factor for butterflies in today's arable landscape (Dover 1999).

The size of habitats is a key factor. Thomas (1984) identified the minimum habitat requirement for each butterfly species. Many common butterflies have 'closed populations' (Warren 1992) and of these, 34 species have a minimum breeding habitat of two hectares or less. Thus if the grass margins are not linked to other suitable habitats or are smaller than the minimum area, they are of little use.

The aim of this study was to investigate the development of 2 and 6 m CSS grass margins over a 4 year period and to assess changes to butterfly and plant species richness over that period.

## Materials and methods

The three farms chosen for this study were Writtle College Farm, Writtle, Chelmsford (NGR: TL 670070); Greenstead Green Farm, Greenstead Green, Halstead (NGR: TL 810280); and Gorrells Farm, Highwood, Chelmsford (NGR: TL 630036). All entered the Countryside Stewardship Scheme (CSS) on the 1st October 1996. The agreements with MAFF encompassed various features such as: hedgerow coppicing, hedge planting, 2 and 6 m grass margins and pond clearance, but in this study only the 2 and 6 m grass margins were evaluated.

A number of 2 and 6 m grass margins and control sections were established (Table 1) using a range of different grass mixtures (Tables 2 and 3) and establishment methods (Table 1). With the exception of one margin at Writtle all the 6 m grass margins and control sections were adjacent to running water. All the available 6 m grass margins were included in the study, while the number of control sections was limited at each site to the few suitable field edges which had no grass margins on them. This in the case of Writtle was two (one 2 and one 6 m), Highwood one 2 m and Greenstead Green two (one 2 and one 6 m). The vegetation on the CSS grass margins was monitored during the period 1998 – 2000. All species present in each grass margin and hedgerow were recorded and then allocated an abundance classification using DAFOR (Bullock 1996). The monitoring took place during July and early August of each year and provided data on plant species richness and abundance.

*Table 1.* Attributes of the margins at the three sites.

| Code | Width of margin (m) | Section length (m) | Aspect | Hedgerow length (m) | Sown | Riverside |
|---|---|---|---|---|---|---|
| Writtle | | | | | | |
| W2.1 | 2 | 274 | NE/SW | 150 | Yes | No |
| W2.2 | 2 | 274 | NW/SE | 274 | Yes | No |
| W2.3 | 2 | 270 | NW/SE | 270 | Yes | No |
| WN2.4 | No margin | 133 | NE/SW | 100 | No | No |
| W6.1 | 6 | 631 | E/W | 310 | NR | Yes |
| W6.2 | 6 | 701 | E/W | 350 | NR | Yes |
| W6.3 | 6 | 720 | NNE/SSW | 200 | NR | Yes |
| W6.4 | 6 | 190 | E/W | 0 | Yes | No |
| WN6.5 | No margin | 450 | E/W | 400 | No | Yes |
| Greenstead Green | | | | | | |
| G2.1 | 2 | 450 | E/W | 390 | Yes | No |
| G2.2 | 2 | 141 | E/W | 141 | Yes | No |
| G2.3 | 2 | 250 | E/W | 150 | Yes | No |
| G2.4 | 2 | 320 | NE/SW | 320 | Yes | No |
| G2.5 | 2 | 285 | NE/SW | 0 | Yes | No |
| GN2.6 | No margin | 180 | E/W | 160 | No | No |
| G6.1 | 6 | 417 | NW/SE | 417 | Yes | Yes |
| G6.2 | 6 | 322 | NW/SE | 322 | Yes | Yes |
| G6.3 | 6 | 166 | NW/SE | 166 | Yes | Yes |
| G6.4 | 6 | 345 | NW/SE | 345 | Yes | Yes |
| GN6.5 | No margin | 250 | NW/SE | 250 | No | Yes |
| Highwood | | | | | | |
| H2.1 | 2 | 200 | N/S | 200 | Yes | No |
| H2.2 | 2 | 762 | E/W-N/S | 450 | Yes | No |
| H2.3 | 2 | 467 | N/S-E/W | 467 | Yes | No |
| H2.4 | 2 | 500 | NE/SW | 400 | Yes | No |
| H2.5 | 2 | 285 | ENE/WSW | 0 | Yes | No |
| HN2.6 | No margin | 343 | ENE/WSW | 300 | No | No |

NR – Natural regeneration.

*Table 2.* Grass mixtures used on the 2 m and 6 m grass margins.

| Date sown | 2 m | | 6 m | |
|---|---|---|---|---|
| | Number of margins sown | Number of grass species sown (mixture number) | Number of margins sown | Number of grass species sown (mixture number) |
| Writtle | | | | |
| Oct 1997 | 3 | 5 (3) | 1 | 9 |
| Greenstead Green | | | | |
| Oct 1996 | 5 | 6 (1) | 4 | 6 (1) |
| Oct 1997 | 2 | 4 (2) | | |
| Highwood | | | | |
| Oct 1997 | 3 | 4 (2) | | |
| Oct 1998 | 2 | 4 (2) | | |

Butterfly species richness was recorded between 1997 and 2000 by monitoring butterfly observations at the height of the flight period for grassland butterflies on all of the 2 and 6 m grass margins and the control section using the transect method (Pollard 1977). This involved recording butterfly numbers once a week when the weather conditions were suitable (Pollard and Yates 1993). The lengths of the grass margins and the number of visits were standardised to produce the abundance and species richness of butterflies observed per km per visit for each section.

Non-parametric statistical analysis techniques were applied to the data collected. Where only two samples were compared Mann–Whitney *U*-test (unmatched samples) was used. Where more than two samples were compared the Kruskal – Wallis statistic was applied. If significance was identified Dunn's multiple comparison procedure was used to identify differences between samples (Dunn 1964). The Spearman rank correlation coefficient was used when investigating for correlation between variables.

*Table 3.* Seed mixtures used on the grass margins.

| | Writtle | Greenstead Green | | Highwood |
|---|---|---|---|---|
| Date established | Oct 1997 | Oct 1996 | Oct 1997 | Oct 97–Oct 98 |
| Length in research transect | 818 m | 1020 m | 426 m | 2214 m |
| Seed mix | Mix 3 | Mix 1 | Mix 2 | Mix 2 |
| *Dactylis glomerata* | 50% | | 50% | 50% |
| *Festuca pratensis** | 10% | | 25% | 25% |
| *Festuca arundinacea* | | | 10% | 10% |
| *Poa pratensis** | | 7.5% | 15% | 15% |
| *Festuca ovina** | 20% | 25% | | |
| *Cynosurus cristatus* | 15% | 7.5% | | |
| *Trisetum flavescens** | 5% | | | |
| *Festuca rubra* ssp. *commutata* | | 30% | | |
| *Agrostis tenuis** | | 5% | | |
| *Festuca rubra* | | 25% | | |

* Not suitable for soil type (Marshall 1998).

## Results

Butterfly species richness was significantly greater ($U = 257.5$, $p < 0.01$) on the 2 m grass margins than on the control sections Table 4). The mean plant species richness on the 2 m grass margins declined over the period of research but the decline was not significant (Table 4). There was significantly greater plant ($U = 0$, $p < 0.05$) and butterfly ($U = 0$, $p < 0.01$) species richness on 2 m grass margins with hedgerows than 2 m grass margins without hedgerows. Butterfly species richness was greater numerically on the grass margins sown with Mixture 1 than on either of the other two mixtures, but the difference was not significant (Table 5).

There was a significant relationship between the number of woody hedgerow species and the number of plant species found on the adjacent 2 m grass margins ($r_s = 0.8$, $p < 0.01$). A similar relationship was found between the length of hedgerow and the number of plant species found on the adjacent 2 m grass margins ($r_s = 0.677$, $p < 0.05$). At Writtle and Greenstead Green there was also a significant relationship between the number of grass species sown in the 2 m margins and the butterfly species richness on those margins ($r_s = 0.694$, $p < 0.05$). There was also a significant relationship between plant species richness and butterfly species richness ($r_s = 0.762$, $p < 0.05$).

The range of butterflies observed at all three sites comprised of a range of breeding and vagrant species. No uncommon species were recorded. Breeding species included: *Maniola jurtina*, *Pyronia tithonus*, *Aphantopus hyperantus*,

*Table 4.* Plant and butterfly species richness on 2 m margins and control sections.

| Date | July 1997 | July 1998 | July 1999 | July 2000 |
|---|---|---|---|---|
| Mean plants species richness 2 m margins (Range) | | 38.2 (49–30) | 32.3 (49–18) | 33.6 (39–22) |
| Mean butterfly species richness 2 m margins (Range) | 7.6 (11–5) | 8.53 (12–6) | 8.22 (13–4) | 8.24[1] (12–3) |
| Mean butterfly species richness control (Range) | 7.39 (10–4) | 7.00 (9–3) | 6.67 (9–4) | 4.67[1] (5–4) |

[1] Means significantly different over the four year period ($p < 0.01$, Mann–Whitney $U$ test).

*Table 5.* Butterfly species richness on 2 m margins established using different seed mixtures.

| | Mixture 1 | Mixture 2 | Mixture 3 |
|---|---|---|---|
| Number of grasses in mixture | 6 | 4 | 5 |
| July 1997 | 7.6 | | |
| July 1998 | 8.7 | 6 | 8 |
| July 1999 | 10 | 5 | 7.7 |
| July 2000 | 10.7 | 8 | 7.3 |

*Ochlodes venata*, *Thymelicus* spp. and *Pieris* spp., while vagrants included: *Inachis io*, *Aglais urticae*, and *Vanessa* spp.

Plant species richness was significantly greater on 6 m margins established by natural regeneration ($U = 0$, $p < 0.05$) and seeded grass margins ($U = 1$, $p < 0.05$) than on the margin established from a grass ley (Table 6). However, there was no significant difference in butterfly species richness between the 6 m grass margins and the control sections. Furthermore, there was no significant relationship between the areas of the 6 m grass margins or number of woody hedgerow species and plant species richness, but there was a significant relationship between the length of the hedgerow and the number of plant species found in the adjacent 6 m grass margin ($r_s = 0.8$, $p < 0.05$).

Where *Dactylis glomerata* and *Festuca arundinacea* were sown they became either **D**ominant or **A**bundant (using DAFOR) by 2000. No other species sown became so prevalent (Table 7). Of the five main species recorded (Table 7) *Dactylis glomerata* is a larval foodplant for six species of grassland butterflies, *Festuca rubra* a foodplant for three and *Agrostis tenuis* four (Asher et al. 2001). Whether *Festuca rubra* ssp. *commutata* is a larval foodplant for any grassland butterfly is unknown. The key nectar sources for grassland butterflies such as *Leucanthemum vulgare*, *Centaurea* spp. and *Knautia arvensis* (Feber et al. 1994; Feber and Smith 1995; Smith et al. 1999) were not included within the original seed mixtures and only a few *Leucanthemum vulgare* plants were recorded in one 2 m grass margin at one of the three sites. No other margins contained the main nectar sources of grassland butterflies except for the occasional *Cirsium* spp.

*Table 6.* Mean (Range) plant species richness on 6 m margins.

| Type of establishment | Number of replicates | July 1998 | July 1999 | July 2000 |
|---|---|---|---|---|
| Natural regeneration | 2 | 43 (44–42) | 48 (48) | 45[a] (52–38) |
| From grass ley | 1 | 22 | 25 | 24[b] |
| Seeded | 5 | | 32.8 (49–24) | 36.2[a] (46–28) |
| Mean 6 m | 8 | 36 (44–22) | 35.6 (49–24) | 36.9 (52–24) |

Means followed by differing superscripts are significantly different at $p < 0.05$, Mann–Whitney *U* test.

*Table 7.* Frequency in 2000 using DAFOR of the main grasses used in the margin mixtures.

| Grasses | Used in | Sown % margins | D | A | F | O | R | nr |
|---|---|---|---|---|---|---|---|---|
| *Dactylis glomerata* | 11 | 61 | 6 | 5 | | | | |
| *Festuca arundinacea* | 6 | 33 | | 6 | | | | |
| *Cynosurus cristatus* | 11 | 61 | | 5 | 3 | 1 | 2 | |
| *Festuca rubra.* ssp.*commutata* | 7 | 39 | | 2 | 3 | 1 | | 1 |
| *Agrostis tenuis* | 8 | 44 | | | 2 | 4 | 2 | |

D = Dominant, A = Abundant, F = Frequent, O = Occasional, R = Rare, nr = not recorded.

**Discussion**

Many of the observations noted in this paper are similar to findings of other authors in Northern Europe. In UK, France and the Netherlands, Marshall et al. (1996) identified a relationship between the number of hedgerow species and the number of plant species found on margins. In this study, a similar set of findings were observed for the 2 m but not 6 m grass margins. Marshall and Moonen (1998) suggested this was due to the complex structure of hedgerows and the variety of habitats they consisted of.

In a series of experiments at ADAS Boxworth on similar soil types and also in the East Anglian region, Kirkham et al. (1999) observed that butterfly species richness correlated with the number of grass species sown, with an average of 7.8 species on a basic mix, 7.7 on a tussocky mix (7.64 in this study) and 9.6 on a diverse mix (9.22 in this study), with 11.0 being observed on a diverse grass and wildflower mix. Similar results could have been expected in this study had a diverse grass and wildflower mix been used as the recording was undertaken during a similar period.

On the 2 m grass margins where *Dactylis glomerata* was sown, it soon became dominant, as had been found in other studies (Cowling and Lockyer 1968; Spedding and Diekmahns 1972; Smith et al. 1997). On twelve out of the thirteen 2 m grass margins monitored, plant species richness reduced over the period of observation. West et al. (1999); Marshall and Nowakowski (1995) and Luken (1990) have shown that the aggressive nature of certain grass species sown in a margin reduces species diversity and restricts vegetation change. Of the list of 11 grass species sown, five were on Marshall's (1998) list of grasses unsuitable for the soil type at the three research sites. By 2000, of those five, three were not recorded and the other two only very rarely noted.

As the majority of the 6 m grass margins and their associated control sections were established next to running water, with an associated bank side vegetation, there was no difference in butterfly species richness between 6 m grass margins and control sections. In the early years of the experimental period, butterfly species richness was higher on the 6 m grass margins established by natural regeneration than on the sown 6 m grass margins. This related to high densities of *Cirsium* spp. on the natural regeneration margins. In 1998 and 1999 these margins were treated to reduce *Cirsium* spp. Once the density of *Cirsium* spp. had been reduced, butterfly species richness reduced also. The 6 m grass margins established from a grass ley had the lowest plant species richness by the end of the research period, but had the highest butterfly species richness out of any of the 6 m grass margins. This could have been due to a more open sward which was no more than 30 cm high (Field 2002).

The required management for the 6 m grass margins was a summer cut with the cuttings removed (MAFF 1999). This had been identified as the worst option for butterflies in a range of treatment experiments reported by Feber et al. (1996). The highest numbers of butterflies were found on uncut margins, or those cut in spring and autumn.

It is widely accepted that the greater the width of the grass margin the better for conservation purposes. However, this study suggests that there is actually very little difference in butterfly and plant species richness between 2 and 6 m grass margins. The main problems with both types of margins were the lack of nectar sources, the same problem also identified by other authors (Feber et al. 1994; Feber and Smith 1995; Smith et al. 1999) and area of suitable habitat as identified by Thomas (1984). No 2 m grass margins were large enough for species to form distinct colonies on, while certain of the 6 m were large enough they possibly lacked the suitable nectar sources. The management of 2 m grass margins was more butterfly friendly (Feber et al. 1996) than that of the 6 m grass margins, thus possibly negating the better initial seed mixture used and the larger area of the 6 m margins.

The use of natural regeneration for 6 m grass margins should be discouraged on clay soils, as identified in this study and by Marshall (1998), as all that is produced is a weedy strip which would be unacceptable to farmers. The use of a tussocky type of grass mixture for 2 m grass margins recommended for the CSS (MAFF 1999) should also be discouraged unless it includes butterfly nectar sources because at the moment all that is produced is a strip of improved grassland with a reduced plant and butterfly fauna (Smith et al. 1997; Kirkham et al. 1999; Field 2002).

In conclusion, the 2 and 6 m grass margins as established under the rules of CSS in October 1996 were often no better than having no grass margin at all. The results from the research suggest that a major opportunity has been missed with regard to creating habitat. What has been developed is a 'monoculture' of improved grassland, which Thomas (1984) rates as supporting at best only one to three butterfly species. With a little thought and reference to previous research such as that of Smith et al. (1993), an unimproved type patchy pasture/tall grassland that has the potential to support 23–28 butterfly species (Thomas 1984) could have been created.

## References


Asher J., Warren M., Fox R., Harding P., Jeffcoate G. and Jeffcoate S. 2001. The Millennium Atlas of Butterflies in Britain and Ireland. Oxford University Press, Oxford.

Barker A.M. and Reynolds C.J. 1999. The value of planted grass field margins as a habitat for sawflies and other chick-food insects. In Aspects of Applied Biology 54: Field Margins and Buffer Zones: Ecology, Management and Policy. Association of Applied Biologists, Wellesbourne pp.109–116.

Bullock J. 1996. Plants. In: Sutherland W.J. (ed), Ecological Census Techniques: A Handbook. Cambridge University Press, Cambridge, pp. 111–137.

Buys J.C. 1995. Kansen voor natuur bij braakegging van perceelranden. Akkerranden in Nederland, In: de Snoo G.R., Rotteveel A.J.W. and Heemsbergen H. (eds), Wageningen, pp. 43–51.

Buys J.C., Oosterveld E.B. and Ellenbroek F.M. 1996. Kansen voor natuur bij braak II verslag van een tweejarig praktijkonderzoek. (No CLM 253–1996). Centrum voor Landbouw en Milieu.

Carreck N.L., Williams I.H. and Oakley J.N. 1999. Enhancing farmland for insect pollinators using flower mixtures. In Aspects of Applied Biology 54: Field Margins and Buffer Zones: Ecology, Management and Policy. Association of Applied Biologists, Wellesbourne, pp. 101–108.

Cowling D.W. and Lockyer D.R. 1968. A comparison of the yield of three grass species at various levels of nitrogenous fertiliser grown alone or in a mixture. Journal of Agricultural Science 71: 127–136.

de Snoo G.R. and Udo de Haes H.A. 1994. Onbespoten akkerranden voor natuur, milieu en bedrijf. Landschap 11(4): 17–32.

Dover J.W. 1994. Arable Field Margins: factors affecting butterfly distribution and abundance. In: Boatman N. (ed.), BCPC Monograph no 58, Field Margins: Integrating agriculture and conservation. British Crop Protection Council, Warwick, pp. 109–116.

Dover J.W. 1999. Butterflies and field margins. In Aspects of Applied Biology 54: Field Margins and Buffer Zones: Ecology, Management and Policy. Association of Applied Biologists, Wellesbourne pp.109–116.

Dunn O.J. 1964. Multiple comparisons using rank sums. Technometrics 6: 241–252.

English Nature 1995. Rebuilding the English Countryside. English Nature, Peterborough.

Feber R.H. and Smith H. 1995. Butterfly conservation on arable farmland. In: Pullen A.S. (ed.), Ecology and Conservation of Butterflies. Chapman and Hall, London.

Feber R.H., Smith H. and Macdonald D.W. 1994. The effects of field margin restoration on the Meadow Brown butterfly (*Maniola jurtina*). In: Boatman N. (ed.), BCPC Monograph no 58, Field Margins: Integrating agriculture and conservation. British Crop Protection Council, Warwick, pp. 295–300.

Feber R.H., Smith H. and MacDonald D.W. 1996. The effects on butterfly abundance of the management of uncropped edges of arable fields. Journal of Applied Ecology 33: 1191–1205.

Field R.G. 2002. A study of butterflies on Farmland. Unpublished Ph.D thesis, University of Essex, Colchester.

Hopkins A. and Feber R.E. 1997. Management for plant and butterfly species diversity on organically farmed grassland field margins. In Management for grassland diversity. Proceedings of International. Symposium of the European Grassland Federation, pp.69–73

Kirkham F.W., Sherwood A.J., Oakley J.N. and Fielder A.G. 1999. Botanical composition and invertebrate populations in sown grass and wildflower margins. In Aspects of Applied Biology 54, Field Margins and Buffer Zones: Ecology, Management and PolicyAssociation of Applied Biologists, Wellesbourne, pp.291–298

Luken J.O. 1990. Directing Ecological Succession. Chapman and Hall, London.

MAFF 1999. The Countryside Stewardship Scheme: Information and how to apply. MAFF, London.

MAFF 2000. Agricultural Statistics in Great Britain in 1999. MAFF, London.

Marshall E.J.P 1998. Guidelines for the siting, establishment and management of arable field margins, beetle banks, cereal conservation headlands and wildlife seed mixtures. Issue no 2. IACR, Long Ashton.

Marshall E.J.P., Baudry J., Moonen C., Fevre E. and Thomas C.F.G. 1996. Factors affecting the floral diversity in European field margins networks. In: Simpson I. and Dennis P. (eds), The Spatial Dynamics of Biodiversity Proceedings 5th Annual IALE (UK) Conference. IALE, UK, pp. 97–104.

Marshall E.J.P. and Moonen C. 1998. A Review of Field Margin Conservation Strips in Europe. IACR, Long Ashton.

Marshall E.J.P. and Nowakowski M. 1995. Successional changes in the flora of a sown field margin strip managed by cutting and herbicide application. Brighton Crop Protection Conference – Weeds. BCPC, Farnham, pp. 973–978

Murphy D.D., Launer A.E. and Ehrlich P.R. 1983. The role of adult feeding in egg production and population dynamics of the checkerspot butterfly (*Euphydryas editha*). Oecologica 56: 257–263.

Pollard E. 1977. A method for assessing the abundance of butterflies. Biological Conservation 12: 115–134.

Pollard E. and Yates T. 1993. Monitoring Butterflies for Ecology and Conservation. Chapman and Hall, London.

Smith H., Feber R.E., Johnson P.J., McCallum K., Jensen S.P., Younes M. and MacDonald D.W. 1993. The Conservation Management of Arable Field Margins. English Nature, Peterborough.

Smith H., Feber R.E. and MacDonald D.W. 1994. The role of wildflower seed mixtures in field margin restoration. In: Boatman N. (ed.), BCPC Monograph no 58, Field Margins: Integrating agriculture and conservation. British Crop Protection Council, Warwick, pp. 289–294.

Smith H., Feber R.E. and MacDonald D.W. 1999. Sown field margins: why stop at grass? In: Aspects of Applied Biology 54: Field Margins and Buffer Zones Ecology, Management and Policy. Association of Applied Biologists, Wellesbourne, pp.291–298.

Smith H., McCallum K. and MacDonald D.W. 1997. Experimental comparison of the nature conservation value, productivity and ease of management of a conventional and a more species-rich grass ley. Journal of Applied Ecology 34: 53–64.

Spedding C.R.W. and Diekmahns E.W. 1972. Grasses and Legumes in British Agriculture. Bulletin of the Commonwealth Bureau of Pasture and Field Crops No. 49. Royal Commonwealth Agriculture Bureau, Farnham.

Thomas J.A. 1984. The Conservation of Butterflies in Temperate Countries: Past efforts and lessons for the future. In: The Biology of Butterflies. Sypm. of the Royal Entomological Society of London 11: 333–353.

Warren M.S. 1992. Butterfly populations. In: Dennis R.L.H. (ed.), The Ecology of Butterflies in Britain. Oxford University Press, Oxford. 73–92.

Watt W.B., Hoch P.C. and Mills S.G. 1974. Nectar resource use by *Colias* butterflies. Oecologia 14: 353–374.

West T.M. and Marshall E.J.P. 1996. Managing sown flower margin strips on contrasted soil types in three Environmentally Sensitive Areas. In Aspects of Applied Biology, 44, Vegetation Management in Forestry, Amenity and Conservation Areas. Association of Applied Biologists, Wellesbourne, pp. 269–276

West T.M., Marshall E.J.P, Westbury D.B. and Arnold G.M. 1999. Vegetation development on sown and unsown field boundary strips established in three Environmentally Sensitive Areas. In Aspects of Applied Biology 54: Field Margins and Buffer Zones: Ecology, Management and Policy. Association of Applied Biologists, Wellesbourne, pp. 235–240

Biodiversity and Conservation (2006) 15:439–449
DOI 10.1007/s10531-005-6971-7 

# Habitat specificity and variation of coleopteran assemblages between habitats in a Southern African (Swaziland) agricultural landscape

C.N. MAGAGULA
*Department of Biological Sciences, University of Swaziland, P/Bag 4, Kwaluseni, Swaziland (e-mail: Cebisile@uniswacc.uniswa.sz; phone: +268-518-4011; fax: +268-518-5276)*

Received 17 August 2004; accepted in revised form 25 April 2005

**Key words:** Agroecosystems, Biodiversity, Carabidae, Coleoptera, Scarabaeidae

**Abstract.** Assessment of Coleopteran diversity and abundance was carried out in the lowveld region of Swaziland by pitfall trapping in five distinct habitats. Additionally, the study was to ascertain if any of the families collected illustrated habitat fidelity. 18 coleopteran families, comprising 2903 individuals were collected, with an additional 29 unidentified specimens, totalling 2932 beetles. Numerically, the top five families were the Scarabaeidae (2425), Carabidae (211), Tenebrionidae (149), Nitidulidae (37) and Curculionidae (22). 66.6% of all specimens collected were harboured within the pine windbreak while the lowest number (2.6%) was collected from the mature citrus orchard. The highest proportion (77%) of scarabaeid beetles collected occurred within the pine windbreak. Habitat specificity was illustrated by one family, Meloidae, which was found only in the young orchard. Community analysis was further carried out to ascertain distribution patterns of the dominant coleopteran families. Three scarabaeid species were found only in the pine windbreak while three carabid species occurred only in the indigenous savanna. Significant differences between habitats were observed in the number of beetles collected while no significant differences were observed in terms of the number of families observed in each habitat ($p < 0.05$). Results indicate that conservation of the various habitat patches within the mosaic studied could facilitate conservation of whole communities rather than individual species thus facilitating effective conservation of the agricultural landscape.

## Introduction

Spatial distribution of insects within an ecosystem is the result of both behavioural and environmental factors. With increased pressure from anthropogenic activities, habitat alteration and associated changes in biodiversity within agroecosystems is inevitable. While vulnerable to modification, habitat mosaics in fragmented landscapes are essential for ecosystem function. Fragmentation and loss of habitats are among the most important causes of species decline worldwide (Varchola and Dunn 1999; Heliölä et al. 2001; Zilihona and Nummelin 2001).

Communities are poorly described by species alone and better understanding can be achieved if the ecology of species assemblages is used as a basis of classification and applied to predict effects of environmental disturbances. The challenge is to be able to identify the correct species assemblages that will provide key ecological services such as biological control, nutrient recycling, soil conservation through their biological synergisms (Altieri and Nicholls 1999; Cole et al. 2002).

With this in mind, this study was initiated to assess coleopteran diversity within a southern African agroecosystem. Coleopteran diversity was assessed due to the diversity of this insect group and the various important roles in ecosystem functioning attributed to this group (Hanski and Cambefort 1991). The study intended to determine habitat characteristics and management practises important in influencing habitat selection by these insects. Addressing these is important since abundance and diversity of fauna in agroecosystems is considered to enhance sustainability (Kromp 1999).

## Materials and methods

### *Study area and sampling method*

The study was carried out within Tambuti Estate (26.43° S, 31.43° E), which had 931.6 ha under citrus and the rest being 97.8 ha indigenous savanna and 400 ha sugar cane fields. The estate is bordered by the Great Usuthu river on the western and southern borders while the northern end is indigenous savanna bush dominated by *Acacia* and Tamboti. The estate thus has various natural, semi-natural and man-made habitat mosaics.

Stratified random sampling was used to select five distinct habitats, i.e. indigenous savanna, pine windbreak, riparian border, mature and young orchards. Pitfall trapping was used at three randomly selected sampling points within each habitat, using ethylene glycol as a preservative. This method was chosen since it is one of the most consistently reliable methods for sampling epigaeic macroinvertebrates (Samways 1990; Doube 1991; French and Elliot 1999). Each point habitat was sampled using a set of four pitfall traps representing the corners of a 1 $m^2$ quadrant. Each set of four traps represent a point sample replicated three times. Each of the five sites thus had three sampling point habitats, giving a total of 12-point habitats per sampling site, with 60 pitfall traps in total.

Traps were serviced fortnightly and sampling was carried out over a period of 12 months. Species were sent to the National Flagship Institute (Transvaal museum, South Africa) for identification.

Environmental variables measured were the soil moisture, percent insolation and percent litter cover. Percent insolation and litter cover were measured as a percentage of a 1 $m^2$ quadrant exposed to sunlight for % insolation or under leaf litter for % litter cover. The percentage means were then scored on a point scale, e.g. 0–5% was given a score of 0 while 96–100% was given a score of 10 (Table 1).

*Table 1.* Summary of environmental data for the five sampling sites. Scores for environmental variables as explained in the text.

| Site | Habitat type (Dominant bush/tree) | % Soil moisture | Litter cover | Insolation |
|---|---|---|---|---|
| 1 | Riparian border (Lantana) | 0.313 ± 0.218 | 10 | 0 |
| 2 | Mature orchard (> 3 years) (Citrus) | 2.440 ± 0.1174 | 6 | 2 |
| 3 | Pine windbreak (Pine) | 2.448 ± 0.526 | 10 | 3 |
| 4 | Young orchard (< 3 years) (Citrus) | 2.342 ± 0.267 | 1 | 6 |
| 5 | Indigenous savanna (Acacia, Tamboti) | 3.785 ± 0.713 | 9 | 4 |

The data was analysed at family and species level for the families Carabidae and Scarabaeidae using PRIMER v5, where various biodiversity indices were calculated. Shannon diversity index was used to determine Coleopteran family diversity at each site. Species evenness and richness within each habitat were also calculated to determine the effect of habitat type on the distribution of these families. Community analyses comprising Non-Metric Dimensional Scaling (MDS) and Principal Component Analysis (PCA) were then carried out to ascertain coleopteran distribution patterns within the agricultural ecosystem. NMDS was to ascertain similarities in coleopteran distribution between the habitats while PCA was carried out in order to determine the extent to which the environmental variables measured are related to the coleopteran assemblages. In order to exclude vagrant species, habitat specificity was assessed for families and species which had more than one specimen collected within each habitat (Clarke and Warwick 1994).

## Results

2932 beetles were collected, comprising 18 identified coleopteran families. Significant differences in numerical assemblage of families between sites (ANOSIM, global $R = 0.886$, $p = 0.001$) was observed. Numerically, the most common families were Scarabaeidae, Carabidae, Tenebrionidae, constituting 82.7, 7.2 and 5.1% respectively (Table 2). All other families combined contributed only about 5% of the samples collected.

The pine windbreak had the highest number of beetles, where 66.6% of the beetles sampled were collected. This site was dominated by Scarabaeidae, with this family constituting 77% of beetles collected from this site. In terms of the total numbers collected, the young orchard and indigenous savanna were comparable with 16.3% and 10.8% respectively while the lowest numbers were in the mature orchard (2.6%) and riparian border (3.7%) (Table 2). With respect to coleopteran assemblages, the three most abundant families from each site are shown in Table 2. As can be seen, Scarabaeidae, Carabidae and Tenebrionidae were the most abundant overall and dominated in three of the

*Table 2.* Relative abundance of coleopteran families at Tambuti Estate.

| Family | Riparian | Mature | Windbreak | Young | Savanna | Total | # of sites occurring |
|---|---|---|---|---|---|---|---|
| Carabidae | 35[1] | 16[2] | 26[2] | 75[2] | 59[3] | 211[2] | 5 |
| Curculionidae | 0 | 0 | 4 | 11 | 7 | 22 | 3 |
| Tenebrionidae | 33[2] | 12 | 23[3] | 17 | 64[2] | 149[3] | 5 |
| Scarabaeidae | 30[3] | 24[1] | 1869[1] | 326[1] | 176[1] | 2425[1] | 5 |
| Coccinellidae | 0 | 0 | 2 | 2 | 0 | 4 | 2 |
| Chrysomelidae | 0 | 0 | 1 | 0 | 1 | 2 | 2 |
| Meloidae | 0 | 0 | 0 | 20[3] | 0 | 20 | 1 |
| Paussidae | 0 | 0 | 0 | 1 | 0 | 1 | 1 |
| Nitidulidae | 1 | 15[3] | 6 | 15 | 0 | 37 | 4 |
| Staphylinidae | 0 | 1 | 3 | 4 | 0 | 8 | 3 |
| Elateridae | 1 | 0 | 0 | 1 | 0 | 2 | 1 |
| Trogidae | 0 | 2 | 4 | 0 | 0 | 6 | 2 |
| Hydrophylidae | 1 | 0 | 0 | 0 | 4 | 5 | 2 |
| Alleculidae | 0 | 1 | 0 | 0 | 1 | 2 | 2 |
| Histeridae | 2 | 0 | 0 | 0 | 0 | 2 | 1 |
| Anthribidae | 0 | 2 | 0 | 2 | 0 | 4 | 2 |
| Buprestidae | 0 | 1 | 0 | 0 | 0 | 1 | 1 |
| Scolytidae | 0 | 0 | 1 | 2 | 0 | 3 | 2 |
| Unidentified | 5 | 3 | 14 | 1 | 4 | 27 | 5 |
| N | 108 | 77 | 1953 | 477 | 316 | 2931 | |
| % of N | 3.67 | 2.63 | 66.64 | 16.27 | 10.78 | | |

The three most abundant families from each site are shown in superscript numbers at each site.

five habitats, while Nitidulidae, Meloidae and Curculionidae were the next abundant families overall. The latter families, however, occurred mainly in the managed habitats, i.e. the orchards and were absent or very few in the other habitats. With the exception of the latter three families, there was no distinct

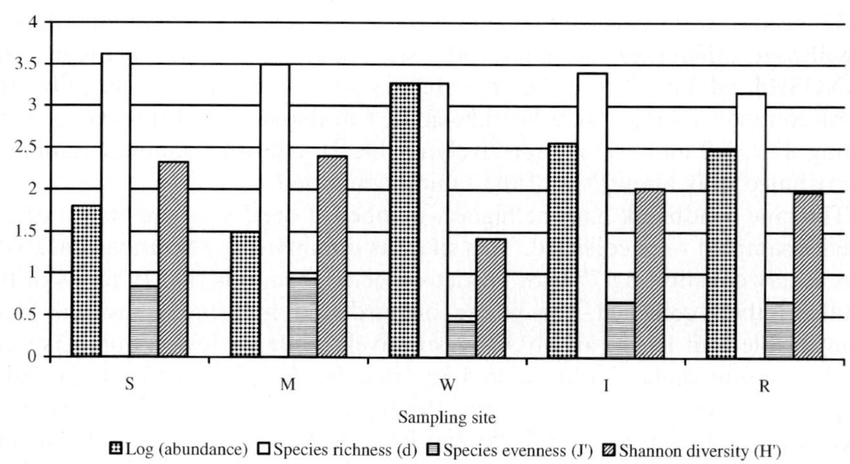


*Figure 1.* Species abundance ($\log_{10}$), richness and evenness as calculated for all habitats.

coleopteran assemblage characteristic of any habitat. Analyses of the various biodiversity indices indicated that there were no significant differences between the five habitats (Figure 1).

PCA analysis was used to ascertain the effects of the measured environmental variables on the biotic patterns observed. PC1 was dominated by decreasing insolation while the second axis was dominated by increasing litter cover. Soil moisture had moderate effect on both axes. The PCA explained 96.6% of the variability in these variables (Figure 2).

*Analyses of Scarabaeidae and Carabidae*

Further analyses were carried out on the two families which were consistently present in all sites and occurred in sufficiently high numbers for analyses, i.e. Scarabaeidae and Carabidae. Tenebrionids are a poorly studied group in the region and their habits are poorly known. Consideration of the two coleopteran families was deemed appropriate because they represent a wide array of taxonomic, functional and habitat diversity (McGeoch 1998) and thus considered to be beneficial in agroecosystems. A total of 36 species were collected from the two families (Table 3).

Total abundance of both families from the pine windbreak was significantly higher than numbers collected from the other habitats (ANOSIM, global $R = 0.702$, $p = 0.001$). Differences in population abundances, however, did not translate to any significant differences in species richness (Table 3, Figure 3).

Although total abundance was lower in managed habitats like the orchards, these had comparable species diversity as in the unmanaged habitats. Despite having the highest abundance and species diversity, the pine windbreak had the

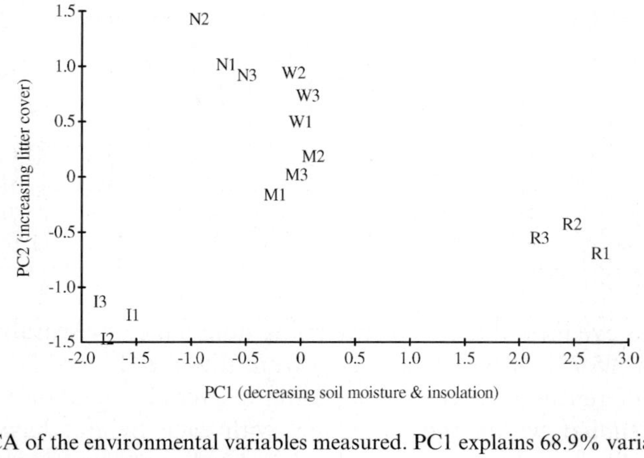


*Figure 2.* PCA of the environmental variables measured. PC1 explains 68.9% variability and PC2 explained 27.7% variability observed.

*Table 3.* Species distribution of Scarabaeidae and Carabidae in all five habitats.

| Species | Riparian | Mature | Windbreak | Young | Savanna | Total | Relative species abundance |
|---|---|---|---|---|---|---|---|
| *Onthophagus* sp. 1 | 0 | 0 | 14 | 5 | 1 | 20 | 0.767 |
| *O. bicafrons* | 8 | 2 | 980 | 35 | 27 | 1052 | 40.34 |
| *Anachalcos convexus* | 0 | 0 | 250 | 32 | 90 | 372 | 14.26 |
| *Aphodius* sp. 1 | 1 | 1 | 3 | 2 | 0 | 7 | 0.268 |
| *Onthophagus* sp. 2 | 5 | 6 | 19 | 15 | 4 | 49 | 1.879 |
| *Sisyphus* sp. 1 | 0 | 0 | 1 | 0 | 1 | 2 | 0.077 |
| *Onthophagus* sp. 3 | 0 | 4 | 121 | 14 | 93 | 232 | 8.896 |
| *O.* near *pullus* | 0 | 0 | 2 | 0 | 0 | 2 | 0.077 |
| Scarabaeid 1 | 5 | 0 | 2 | 3 | 0 | 10 | 0.383 |
| Scarabaeid 2 | 0 | 0 | 14 | 1 | 0 | 15 | 0.575 |
| *Schizonycha* sp. | 0 | 3 | 6 | 2 | 2 | 13 | 0.498 |
| Scarabaeid 3 | 0 | 0 | 4 | 0 | 0 | 4 | 0.153 |
| *O. vinctus* | 6 | 0 | 2 | 27 | 8 | 43 | 1.649 |
| *Onthophagus* sp. 4 | 0 | 0 | 32 | 4 | 10 | 46 | 1.764 |
| Scarabaeid 4 | 0 | 0 | 23 | 0 | 0 | 23 | 0.882 |
| *Phoxomela umbrosa* | 1 | 0 | 0 | 0 | 0 | 1 | 0.038 |
| *Plaesiorrhinella trivittata* | 0 | 3 | 6 | 3 | 1 | 13 | 0.498 |
| *O. interstitialis* | 0 | 2 | 352 | 162 | 0 | 516 | 19.79 |
| Carabid 1 | 0 | 1 | 0 | 0 | 0 | 1 | 0.038 |
| *Tefflus delagorguei* | 2 | 3 | 13 | 25 | 17 | 60 | 2.301 |
| *Haplotrachelus* sp. | 2 | 1 | 1 | 1 | 20 | 25 | 0.959 |
| Carabid 2 | 0 | 1 | 0 | 6 | 2 | 9 | 0.345 |
| Carabid 3 | 2 | 3 | 2 | 15 | 7 | 29 | 1.112 |
| Carabid 4 | 1 | 0 | 0 | 0 | 0 | 1 | 0.038 |
| *Cypholoba* sp. | 0 | 0 | 2 | 0 | 8 | 10 | 0.383 |
| Carabid 5 | 1 | 0 | 2 | 0 | 0 | 3 | 0.115 |
| Carabid 6 | 0 | 0 | 0 | 1 | 0 | 1 | 0.038 |
| *Dromica ambitiosa* | 0 | 0 | 3 | 1 | 1 | 5 | 0.192 |
| Carabid 7 | 3 | 0 | 0 | 0 | 0 | 3 | 0.115 |
| *Graphipterus* sp. | 1 | 1 | 1 | 0 | 2 | 5 | 0.192 |
| Carabid 8 | 0 | 0 | 1 | 0 | 0 | 1 | 0.038 |
| Carabid 9 | 1 | 0 | 1 | 1 | 2 | 5 | 0.192 |
| Carabid 10 | 0 | 0 | 1 | 4 | 0 | 5 | 0.192 |
| Carabid 11 | 19 | 0 | 0 | 0 | 0 | 19 | 0.729 |
| Carabid 12 | 0 | 0 | 0 | 0 | 1 | 1 | 0.038 |
| Carabid 13 | 5 | 0 | 0 | 0 | 0 | 5 | 0.192 |
| Carabidae | 10 | 6 | 10 | 8 | 9 | 18 | 7.209 |
| Scarabaeidae | 6 | 7 | 17 | 13 | 10 | 18 | 92.79 |
| Species total | 16 | 13 | 27 | 21 | 19 | 36 | 100 |
| Individuals | 63 | 31 | 1858 | 359 | 297 | 2608 | |

lowest species evenness due to overwhelming dominance by scarabaeid beetles that made up 95.7% of beetles collected from this site (Figure 3).

Despite differences in species composition, overall abundance of carabid beetles was similar within the landscape, with each habitat having its own species assemblage and the young orchard and indigenous savanna having the

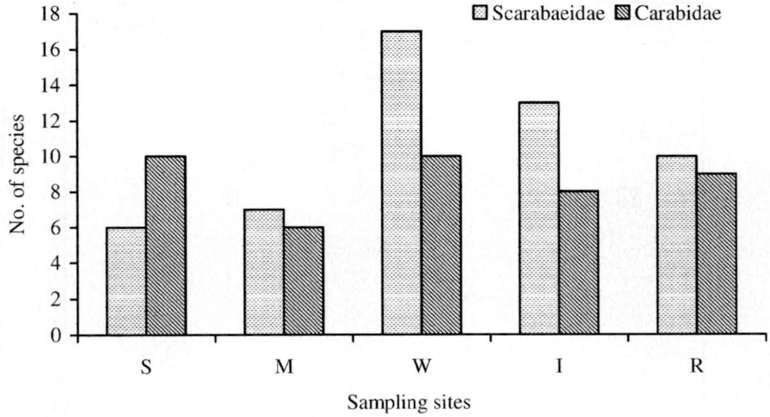


*Figure 3.* Species distribution between sites at Tambuti estate. S = indigenous savanna, M = mature orchard, W = pine windbreak, I = Young orchard, R = riparian border.

highest abundance of these beetles as shown in Figure 4, where these two habitats are shown to have the largest bubbles. On the other hand, the abundance of scarabaeid beetles varied between habitats. Figure 5 indicates that the windbreak had the highest numbers, with 77% of beetles in this family collected from the windbreak. Three species, i.e. *Onthophagus* near *pullus*, scarabaeid 3 and 4 occurred only in the pine windbreak while three unidentified carabid species (7, 11 and 13) were restricted to the riparian border.

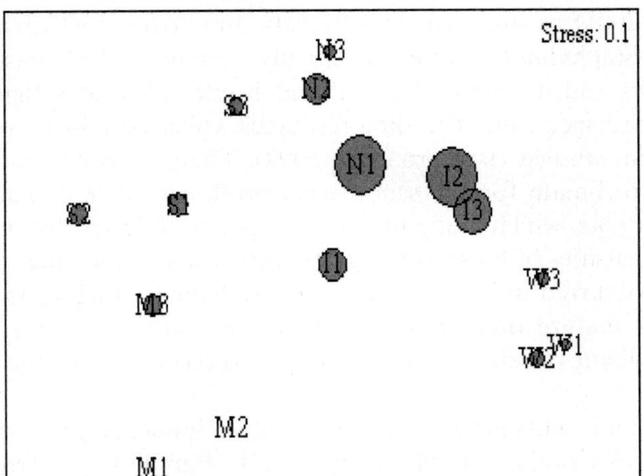


*Figure 4.* MDS with carabid distribution across sites sampled. Larger bubbles indicate greater abundance.

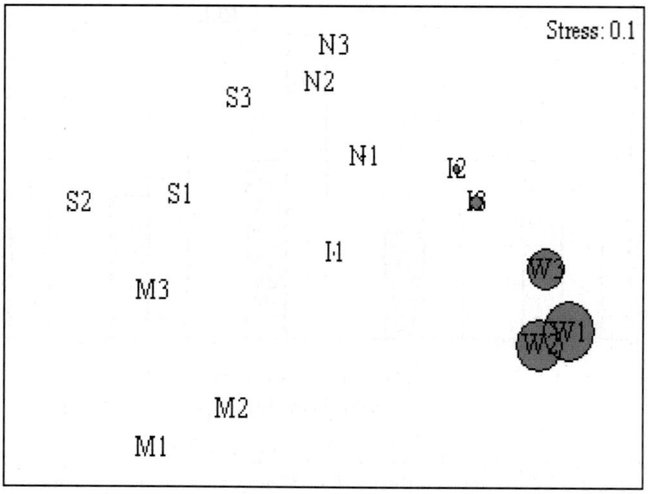


*Figure 5.* MDS with scarabaeid distribution across sites sampled. Larger bubbles indicate greater abundance.

## Discussion

Coleopteran diversity varied with respect to habitat analysed. Dominance by the family Scarabaeidae, both numerically and in terms of diversity was observed in all habitats except the riparian border. At species level, this was due to high abundance of *Onthophagus* spp., the most speciose genus in the family Scarabaeidae in Africa (Cambefort and Walter 1991). Although in a different habitat structure, scarabaeid dominance was also observed by Zilihona and Nummelin (2001) as opposed to Anderson and Ashe (2000) where Curculionidae and Staphylinidae were numerically dominant. The presence of such high numbers and diversity of scarabaeid beetles illustrates that this family utilises a broad spectrum of feeding resources which were likely present in the agroecosystem studied (Barbero et al. 1999). The pine windbreak provided a suitable microclimate for the scarabaeids beetles, with dominance by *Onthophagus bicafrons*, which made up 40% of species collected from this habitat. Population densities of these were significantly lower in the other habitats. The scarabaeids occurred in low densities in dense habitats such as the indigenous savanna and mature orchard where the dense canopy had the potential to reduce their flight capabilities and searching success (Steenkamp and Chown 1996).

Land use and management was observed to influence the composition of coleopteran assemblages (Cole et al. 2002). Families occurring in lower abundances illustrated habitat specificity. Using single site occurrence as a criterion, only the Meloidae illustrated habitat specificity, since they occurred only in the young orchard in moderate numbers. Members of this family are known to be economically significant in agriculture as pests, thus explaining their restricted occurrence in this habitat. Nitidulidae were found mainly in the managed orchards and culturally managed pine windbreak. The mature orchard had the lowest abundance probably due to accumulative effects of management practises in this habitat (Paoletti 1999). Additionally this habitat is at least 10 years old and can be considered a unique habitat fragment within the indigenous savanna. Diversity in such fragments is known to be lower than in continuous natural habitats.

Management practises may have either detrimental or advantageous effects on coleopteran fauna (Clarke et al. 1997). At species level, habitat specificity was illustrated by six species, with three carabid species found only in riparian border while another three scarabaeid species occurred only in the pine windbreak. Compared to the orchards, both the pine wind break and riparian border had minimal or no disturbance due to agricultural activities and this likely provide suitable microclimate for stenotopic species. Such habitat specificity emphasised that all habitat mosaics are ecologically important and conservation worthy within the landscape studied.

In addition to being less dense, windbreaks act as margins and as important refugia for various species, whereby the insects invade from surrounding fragments and surrounding vegetation. This also contributes towards improving diversity within the agroecosystem studied as well as enhancing the functioning of biodiversity components (Altieri and Nicholls 1999; Paoletti 1999). All biodiversity components of the mosaic studied provided suitable habitat for the maintenance of biotic diversity. The species occurring within the landscape studied are likely to exploit the variety of resources available within the habitat patches assessed (Doube 1991). The significant differences in coleopteran assemblages between sites could indicate that none of the habitats can be regarded as a substitute for the other (Zilihona and Nummelin 2001). Maintenance of the different habitats should thus be considered for effective management of coleopteran diversity.

Habitat isolation and fragmentation cause disruption in biological processes that maintain biodiversity and ecosystem functioning, which to a large degree are mediated by insects (Didham et al. 1996). While species in both families illustrated habitat specificity, there was no coincidence in habitats they were specific to, i.e. stenotopic carabids species were restricted to the riparian border while scarabaeids were restricted to the pine windbreak. The lack of correlation in habitat specificity could be explained by that the species may have different ecological requirements, e.g. moisture, soil preference, etc. (Rainio and Niemela 2003). Stenotopic species are more appropriate for conservation since they are usually more effective indicators than eurytopic species (Buchs 2003) probably due to their specific habitat requirements.

Since species diversity is affected by more than one factor, no one group can be used as predictor of biodiversity. In order to maximise biodiversity, even on structural and functional qualities, it is best to use several groups with different ecological requirements, as observed in this study. Enhancement and conservation of the coleopteran assemblages could help determine conditions under which conservation of these beetles could be incorporated into agroecosystems, within which they are deemed to be beneficial (Clarke et al. 1997; Buchs et al. 2003). There is a need to understand the effects of fragmentation on ecosystem function in these fragmented habitats as well as how well these effects represent the response of other species not investigated.

## References

Altieri M.A. and Nicholls C.I. 1999. Biodiversity, ecosystem function and insect pest management in agricultural ecosystems. In: Collins W.W. and Qualset C.O. (eds), Biodiversity in Agroecosystems. CRC Press, Florida.

Anderson R.S. and Ashe J.S. 2000. Leaf litter inhabiting beetles as surrogates for establishing priorities for conservation of selected tropical montane cloud forests in Honduras, Central America (Coleoptera; Staphylinidae, Curculionidae). Biodiversity and Conservation 9(5): 617–653.

Barbero E., Palestrini C. and Rolando A. 1999. Dung beetle conservation: effects of habitat and resource selection (Coleoptera: Scarabaeoidea). Journal of Insect Conservation 3(2): 75–84.

Buchs W. 2003. Biotic indicators for biodiversity and sustainable agriculture – introduction and background. Agriculture, Ecosystems and Environment (2003) 98: 1–16.

Buchs W., Harenberg A., Zimmermann J. and Weiß B. 2003. Biodiversity, the ultimate agri-environmental indicator? Potential and limits for the application of faunistic elements as gradual indicators in agroecosystems. Agriculture, Ecosystems and Environment 98: 99–123.

Cambefort Y. and Walter P. 1991. Dung beetles in tropical forests in Africa. In: Hanski I. and Cambefort Y. (eds), Dung Beetle Ecology. Princeton University press, New Jersey.

Clarke M.S., Stuart H.G. and Spence J.R. 1997. Habitats and management associated with common ground beetles (Coleoptera: Carabidae) in a Michigan agricultural landscape. Environmental Entomology 26: 519–527.

Clarke K.R. andWarwick R.M. 1994. Change in marine communities: an approach to statistical analysis and interpretation. Natural Environmental Research Council, United Kingdom.

Cole L.J., McCracken D.J., Dennis P., Downie I.S., Griffin A., Foster G.N., Murphy K.J. and Waterhouse T. 2002. Relationships between agricultural and ecological groups of ground beetles (Coleoptera: Carabidae) in Scottish farmland. Agriculture, Ecosystems and Environment 93: 323–336.

Didham R.K., Ghazoul J., Stork N.E. and Davis A.J. 1996. Insects in fragmented forests: a functional approach. Trends in Evolutionary Ecology 11: 255–260.

Doube B.M. 1991. Dung beetles of Southern Africa. In: Hanski I. and Cambefort Y. (eds), Dung Beetles Ecology. Princeton University press, New Jersey.

French B.W. and Elliot N.C. 1999. Spatial and temporal distribution of groundbeetles (Cloleoptera: Carabidae) assemblages in riparian strips and adjacent wheat fields. Environmental Entomology 28: 597–607.

Hanski I. and Cambefort Y. 1991. Dung beetle ecology. Princeton University Press, New Jersey.

Heliölä J., Koivula M. and Niemela J. 2001. Distribution of carabid beetles (Coleoptera: Carabidae) across a Boreal Forest-Clearcut Ecotone. Conservation Biology 15: 370–377.

Kromp B. 1999. Carabid beetles in sustainable agriculture: a review on pest control efficiency, cultivation impacts and enhancement. Agriculture, Ecosystems and Environment 74: 187–228.

McGeoch M. 1998. The selection, testing and application of terrestrial insects as indicators. Biological Reviews 73: 181–201.

Paoletti M.G. 1999. Using bioindicators based on biodiversity to assess landscape sustainability. Agriculture, Ecosystems and Environment 74: 1–8.

Rainio J. andNiemela J. 2003. Groundbeetles (Coleoptera: Carabidae) as bioindicators. Biodiversity and Conservation 12: 487–506.

Samways M.J. 1990. Species temporal variability: epigaiec ant assemblages and management for abundance and scarcity. Oecologia 84: 482–490.

Steenkamp H.E. andChown S.L. 1996. Influence of dense stands of an exotic tree *Prosopis glandulosa* Benson, on a savanna dung beetle (Coleoptera: Scarabaeinae) assemblage in southern Africa. Biological Conservation 78: 305–311.

Varchola J.M andDunn J.P. 1999. Changes in ground beetle (Coleoptera: Carabidae) assemblages in farming systems bordered by complex or simple roadside vegetation. Agriculture, Ecosystems and Environment 73: 41–49.

Zilihona I.J. E. andNummelin M. 2001. Coleopteran diversity and abundance in different habitats near Kihansi waterfall in the Udzungwa mountains, Tanzania. Biodiversity and Conservation 10: 769–777.

Biodiversity and Conservation (2006) 15:451–472
DOI 10.1007/s10531-005-0306-6 

# Effects of habitat disturbance can be subtle yet significant: biodiversity of hawkmoth-assemblages (Lepidoptera: Sphingidae) in Southeast-Asia

JAN BECK[1,3,*], IAN J. KITCHING[2] and K. EDUARD LINSENMAIR[1]
*[1]Department of Animal Ecology and Tropical Biology, Biocenter, University of Würzburg, Am Hubland, 97074 Würzburg, Germany; [2]Department of Entomology, The Natural History Museum, Cromwell Road, London SW7 5BD, U.K.; [3]Kuala Belalong Field Studies Centre, Dept. of Biology, Universiti Brunei Darussalam, Tungku Link, Gadong BE1410, Brunei; *Author for correspondence (e-mail: jbeck @biozentrum.uni-wuerzburg.de)*

Received 21 January 2005; accepted in revised form 23 June 2005

**Key words:** Borneo, Community ecology, Habitat disturbance, Life history, Moths, Multidimensional scaling, Species diversity

**Abstract.** Sphingid biodiversity was compared in a large number of light-trapping samples on Borneo and elsewhere in the Indo-Australian tropics, using our own quantitative light-trapping samples supplemented by other collectors' published and unpublished data. No effects of anthropogenic habitat disturbance on the within-habitat diversity (measured as Fisher's $\alpha$) were observed, but the faunal composition of assemblages differs significantly under varying degrees of disturbance. Altitude, year of sampling and sampling regime (full night vs. part of the night) were identified as additional parameters that influence the composition of local samples. The frequency of subfamilies in samples varies under different disturbance regimes: Smerinthinae decline along a gradient from primary habitats to heavily disturbed sites, while Macroglossinae show the reverse trend. Connections between the reactions of subfamilies to disturbance and altitude and potential life-history differences between the subfamilies are discussed. Hypothetically, *capital breeding* Smerinthinae might be commoner and more speciose in stable primary habitats, whereas many *income breeding* Macroglossinae are probably adapted to thrive in ephemeral, disturbed habitats. Furthermore, we show that estimates from local samples fall short of the total known species richness of Borneo by about 10%.

## Introduction

Biodiversity research is to a large extent concerned with the documentation and understanding of the influence of habitat disturbance on the species diversity and composition of biological communities (e.g. Lawton et al. 1998). To understand how biological communities react to human habitat destruction or fragmentation is not only academically interesting, but also of vital interest to ecosystem management, which will undoubtedly be of increasing concern to human societies in the future (e.g. Hector et al. 2001), especially considering the immense damage that is being done to today's tropical ecosystems (Bowles et al. 1998; Sodhi et al. 2004). However, time and manpower-constrained research usually involves investigation of one or a number of 'handpicked' taxonomic groups, which are inferred to reflect the reactions of other groups of organisms (Hammond 1994). This assumption has only rarely been tested within the same sampling sites (e.g. Lawton et al. 1998; Schulze et al. 2004), and although local diversity for a majority of taxa diminishes with increasing disturbance, reactions of different groups are often quite dissimilar in detail (e.g. termites: Gathorne-Hardy et al. 2002, scarabid beetles: Holloway et al. 1992; chrysomelid beetles: Wagner 1999, leaf litter ants: Brühl 2001, canopy invertebrates: Simon and Linsenmair 2001; Floren et al. 2001, vertebrates: Johns 1992; Lambert 1992, mantids: Helmkampf et al., in press, butterflies: Hamer et al. 1997; Fermon et al. 2005). Each taxon has certain specific habitat requirements (e.g. Dennis 2003) that often lead to species loss with the disturbance of tropical rainforest habitat, but may sometimes and for some taxa also lead to inverted patterns or reactions that cannot be explained by disturbance alone (e.g. Beck et al. 2002).

Lepidopteran diversity and its change under different disturbance regimes or habitat gradients have been intensively investigated in northern Borneo (e.g. Holloway et al. 1992; Chey et al. 1997; Beck et al. 2002; Willott 1999; Hamer and Hill 2000; Fiedler and Schulze 2004) and elsewhere in the Indo-Australian tropics (e.g. Holloway 1998; Fermon et al. 2005). While for some groups (e.g. Geometridae, Pyralidae, Arctiinae, fruit-feeding butterflies) clear negative effects of habitat disturbance on species diversity and community composition were found (particularly when changing from forests to heavily disturbed or open agricultural land, e.g. Willott 1999; Schulze 2000; Beck et al. 2002), night-active hawkmoths were found to be apparently unresponsive to habitat disturbance, with regard to both within-habitat disturbance and community composition (Schulze and Fiedler 2003).

Hawkmoths are an attractive 'model group' for ecological investigations (e.g. Pearson 1994) due to the availability of a comparatively large amount of background information (taxonomy, host plants, distribution; e.g. Beck and Kitching 2004 and references therein) that is matched for tropical invertebrates only by butterflies. Reactions of their within-habitat diversity and community composition to anthropogenic disturbance and other habitat gradients (e.g. altitude) are a crucial point for the understanding of ecological processes on a community level. We re-evaluated these reactions for hawkmoths in Borneo and elsewhere in Southeast-Asia, using a larger dataset and somewhat different methods than Schulze and Fiedler (2003). Particular questions posed to the data set were:

(1) Are the within-habitat diversity and faunal composition of sphingid assemblages in Borneo and elsewhere in Southeast-Asia really not influenced by habitat disturbance?
(2) How do other environmental gradients, such as altitude or geographic position, influence local hawkmoth assemblages?
(3) Are there differences between taxonomic subgroups of Sphingidae in their reactions to environmental gradients? Such effects were found in an analysis of an altitudinal gradient on Mt. Kinabalu (north-eastern Borneo; Schulze 2000, see also Schulze et al. 2000) and might also play a role in the

biogeographical patterns of hawkmoths throughout the Malesian archipelago (Beck 2005; see also Beck and Kitching 2004).

## Methods

### *Field methods and data sources*

During an extensive light-trapping program Sphingidae were quantitatively recorded at various sites across Southeast-Asia. Moths were attracted to a generator-driven 125 W mercury-vapour (MV) lamp that was placed inside a white gauze cylinder about 1.7 m in height. Three combined 15 W blacklight tubes (*Sylvania blacklight-blue*, powered by 12 V 'dry-fit' batteries) were used at the few sites where logistic conditions precluded the use of a generator. Moths were individually marked with a waterproof pen and stored inside the gauze cylinder until dawn, when they were released. Individual marking ensured that pseudoreplicates, which could be caused by re-catches in following nights, were avoided. Only if species identification was unsure (<10 percent of specimens) were the moths killed and stored or digital photos taken for later determination. Each site was sampled for 3–9 consecutive complete nights (depending on logistic feasibility), which yielded an average of more than 3/4 of the estimated total species richness at each site (based on various estimators of 'true' local species richness, such as *Chao1*; see Beck 2005, Beck and Linsenmair, in press, for details). Most sites were chosen to allow sampling from open airspace, in open landscapes or in the forest canopy (accessed either by platforms or steep slopes or cliffs), as Sphingidae are known to avoid flying in dense undergrowth (Schulze and Fiedler 1997). Sampling sites were situated as deep as logistically possible (at least 1/2 km) inside a habitat type in order to minimize the overlap of faunas from neighbouring habitats.

Published as well as unpublished data of other collectors were used to supplement our own data. We used all data available to us, provided that we were reasonably confident of the quantitative completeness of collections. Generally, only sites with a minimum of 20 individuals were considered for analysis. Sampling by others was mostly carried out in similar short-term, high intensity light trapping sessions as described above, but light sources, sampling schedule and duration differed between sources. All data were corrected for a unified taxonomy, following an updated version of Kitching and Cadiou (2000) and more recently published taxonomic literature, and species identifications were checked wherever possible. Data for mainly day-active genera (i.e., *Macroglossum, Cephonodes* and *Sataspes*) were generally excluded if they were occasionally recorded at light.

From own observations or site descriptions of other authors, habitats were grouped in three disturbance classes: (1) *Primary habitats* without any significant human disturbance, usually primary rainforests; (2) *Secondary habitats*, ranging from selectively logged forests through secondary forests to sites that were at least partly forested; and (3) *Heavily disturbed* sites, consisting of anthropogenically opened landscapes, often near villages, agricultural sites or plantations. Complete habitat descriptions could not be obtained for all sampling sites. Smaller sample sizes compared to the total number of sites in some tests are due to missing values for altitude or disturbance class for some samples.

*Biodiversity statistics*

Species richness or diversity in a habitat cannot be measured directly as the number of observed species if samples are incomplete, which is a normal condition in entomology, particularly where tropical taxa are concerned (Gotelli and Colwell 2001). Furthermore, absolute abundance of specimens at light is influenced by non-habitat related variables (e.g. weather, moonlight; Yela and Holyoak 1997) and cannot therefore be used directly in analyses. An appropriate measure must be employed that is largely independent of the sampling effort or success and gives a reliable, comparable estimate of diversity. For this purpose, Fisher's $\alpha$ was calculated for every site. This well-established index of diversity has proven robust and suitable for comparisons of biodiversity in a number of comparative studies and is considered to be the best index of within-habitat diversity (Hayek and Buzas 1997; Southwood and Henderson 2000). The underlying assumption of this index, a fit of the species-abundance relation of the data to the log-series distribution, was met in 89 of 93 sites (KS-tests, $p > 0.05$), but Fisher's $\alpha$ has also proven relatively robust if this assumption is violated (Hayek and Buzas 1997). To assess the reliability of $\alpha$-values, 95% confidence intervals were computed (Kenney and Krebs 1998).

NESS($m_{max} = 10$)-indices of faunal similarity (Grassle and Smith 1976) were used to investigate differences between local species assemblages. This measure considers quantitative data and is not biased by incomplete samples (Grassle and Smith 1976). By changing its parameter $m$, rare species can be weighted lower (low $m$) or higher (high $m$). NESS-indices were used to produce non-metric Multidimensional Scaling (MDS)-plots, which allow display and testing of distance data in a reduced number of dimensions (see Legendre and Legendre 1998 for details on MDS). In a recent comparison, Brehm and Fiedler (2004) suggested that non-metric MDS plots based upon NESS with the highest possible $m$ (in our case $m_{max} = 10$) are superior to other ordination methods for displaying quantitative ecological data. Dimension values can be tested for the influence of habitat parameters by standard statistical methods (Legendre and Legendre 1998). NESS-values were calculated using a computer program provided by S. Messner (pers. com.); MDS and all standard statistics were computed with *Statistica 6.1* (StatSoft 2003).

Multiple statistical tests from the same data set can lead to spurious results and were controlled for by sequential adjustment of $p$-values according to the method of Hochberg (1988). All results fulfil these conditions, but re-tests of the same topic (e.g. non-parametric checks of parametric test results, tests on data subsets with more homogeneous data) were not controlled.

## Results

Our data compilation led to quantitative light-trapping data for 93 sites from the Indo-Australian region (Table 1: 17,676 specimens, 159 night-active species) and includes data from the 16 lowland sites on Borneo used in Schulze and Fiedler (2003). For Borneo alone, 57 sites (12,333 specimens, 77 species) were analysed.

### *Within-habitat diversity*

A comparison of disturbance classes did not indicate an influence of anthropogenic disturbance on within-habitat diversity of Sphingidae in 57 samples from Borneo (Figure 1). A comparison of median values of Fisher's $\alpha$ among the three classes confirms this conclusion (Kruskal–Wallis ANOVA: $H_{df=2} = 0.395$, $p = 0.825$). Similarly, no clear and significant effects of altitude (with data ranging from sea level to 2600 m a.s.l.) could be found (Figure 2), but a fitted curve (negative exponential least squares method) suggests a mid-elevational peak of Fisher's $\alpha$ above 1000 m altitude. A restriction of analysis to data from the 30 sampling sites with more than 80 individuals, or to the 17 sampling sites with more than 150 individuals, did not reveal any clearer patterns of diversity with regard to habitat disturbance or elevation, so small samples can be ruled out as a reason for artefacts.

Diversity measures within other regions (data not shown, see Table 1 for regions) do not give any indication that the unresponsiveness of sphingid diversity to habitat disturbance is specific to Borneo. However, data from other islands cannot be directly compared to the Borneo samples because there are significant differences in median Fisher's $\alpha$ between regions (KW-ANOVA: $N = 93$, $H_{df=10} = 27.091$, $p = 0.003$). There are no effects of latitude in the data, but local diversity decreases with increasing longitude (Spearman rank correlation: $N = 93$, $R = -0.282$, $p = 0.006$), which is probably an effect of regional biogeography in the Malesian archipelago (Beck 2005, see maps in Beck and Kitching 2004).

### *Effects of habitat parameters on the faunal composition of communities*

NESS($m_{max} = 10$)-indices of faunal similarity from sampling sites in Borneo were used to display similarity of sites as proximity in an MDS-plot. Figure 3 shows a 2-dimensional MDS for easier graphic display, while a 3-dimensional MDS with lower Stress-values was used for further analyses (Stress $= 0.145$;

*Table 1.* Sources of quantitative light-trapping data for Sphingidae in Southeast-Asia.

| Region | No. sites | No. specimens | No. own sample sites | Sources (published) | Sources (pers. com.) |
|---|---|---|---|---|---|
| Borneo | 57 | 12.333 | 23 | Chey (1994), Chey (2002), Holloway (1976), Schulze (2000), Tennent (1991), and Zaidi and Chong (1995) | G. Martin (NHM London), J.D. Holloway[a] (NHM London) |
| Peninsula Malaysia | 3 | 284 | 1 | | Azmi M. (FRIM Kepong) |
| Northern Vietnam | 1 | 3.223 | | | T. Larsen |
| Flores | 3 | 324 | 3 | | |
| Lombok | 1 | 29 | | | U. Buchsbaum (ZSM München) |
| Luzon | 2 | 45 | | | W. Mey (NKM Berlin) |
| Negros | 1 | 36 | | | W. Mey (NKM Berlin) |
| New Guinea | 6 | 480 | | | H.v. Mastrigt, U. Buchsbaum (ZSM München) |
| Seram | 11 | 650 | | | J.D. Holloway[b] (NHM London) |
| Sulawesi | 5 | 147 | | | J.D. Holloway[c] (NHM London) |
| Taiwan | 3 | 125 | | | W. Mey (NKM Berlin), U. Buchsbaum (ZSM München) |

Classification of 'regions' mostly refers to separate islands as well as the climatically distinct continental areas of Malaya and Northern Vietnam. [a]See Holloway (1984), [b]Holloway (1993), and [c]Holloway et al. (1990).

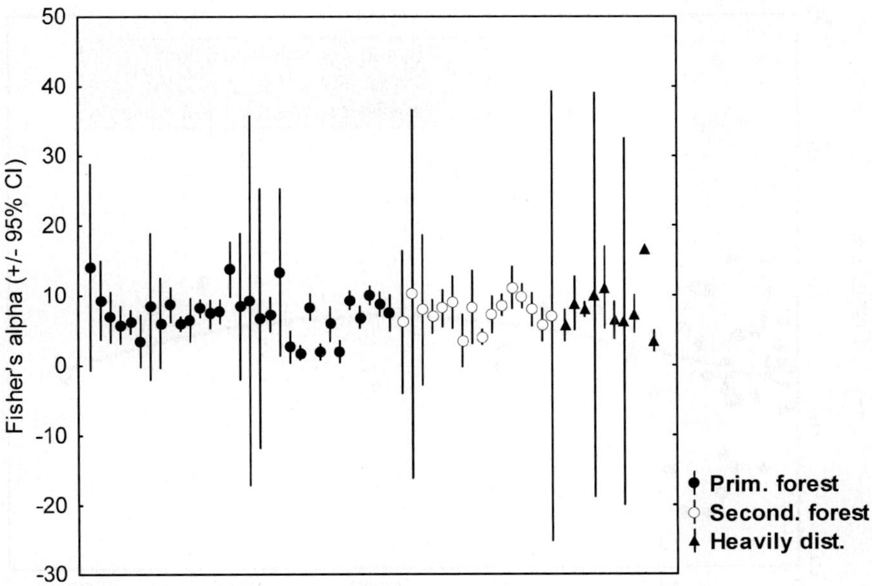


*Figure 1.* Fisher's α (±95% confidence intervals) for 57 sites on Borneo. No significant differences between sites can be observed (see text). See Table 1 for data sources.

Stress is a measure of information loss in MDS; a Shepard diagram did not reveal deviations from general assumptions of the model; see Legendre and Legendre 1998, StatSoft 2003 for details). Preliminary analyses identified three potentially influential variables: moderate effects of habitat disturbance and elevation and – unexpectedly – a strong effect of the source of the data. As 'data source' is not a satisfying natural variable, a Generalized Linear Model (GLM: Guisan et al. 2002, StatSoft 2003) was used to identify how values of the three MDS-dimensions are influenced by the potentially important parameters of disturbance, elevation, sampling procedure [full night (39 sites) vs. not full night (18 sites)], light source [MV (47 sites), blacklight (5 sites), kerosene lamp (5 sites)] and 'half-decade of sampling year' [(5–21 sites per half-decade), assuming data were collected 1–2 years prior to publication if not otherwise stated]. All suspected parameters except 'lamp type' have a significant influence on MDS-values in the multivariate design (Figure 4, Table 2), and several direct influences on dimension values are suggested from univariate tests: habitat disturbance is influencing dimension 2, whereas altitude of the sampling site has an effect on dimension 3 of the MDS. Of the 'data source'-related parameters, the 'year of sampling' is influencing dimension 1, but also has an effect on dimension 2 (the 'disturbance-axis'). 'Full night'-sampling affects dimensions 1 and 3 (the 'altitude-axis'); thus, samples from higher altitudes are associated with incomplete sampling nights. Effects of 'full night' sampling were expected due to preferred flight times during the night in various species (Beck and Linsenmair, in press). To confirm further that the effect of

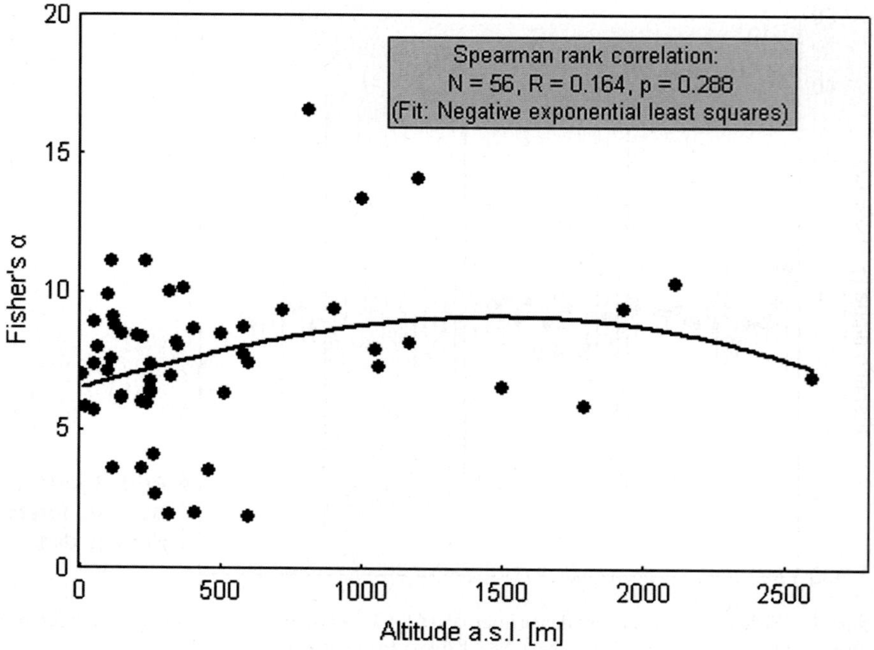


*Figure 2.* Plots Fisher's α of 56 sampling sites on Borneo as a function of altitude. No significant effects can be observed, although the fitted curve suggests an increase of diversity in medium elevations (see Discussion).

disturbance is not an artefact of the data source (e.g. *via* year of sampling), a GLM was used to analyse MDS-data based only on our own sampling on Borneo (18 sites, always full night sampling, sampled between 2001 and 2003). The model is significant only for dimension 2 of three dimensions ($R^2 = 0.433$, $F = 3.563$, $p = 0.042$), which is based solely on the effect of habitat disturbance (univariate test: $F = 4.994$, $p = 0.023$).

## *Effects of habitat parameters on relative abundance of subfamilies*

Sphingid subfamilies differ in their life histories and have been reported to vary in relative abundance under different conditions of disturbance and elevation (Holloway 1987). Across 93 sites from Southeast-Asia, relative abundances of three subfamilies (as *specimens/total catch*) were compared for effects of disturbance class, elevation and geographic position (latitude/longitude) with a GLM. Results (Table 3) indicate that only disturbance has a significant effect on subfamily frequency, while trends ($p < 0.10$) for an influence of elevation and latitude were found. Univariate tests indicate an effect of latitude on Sphinginae frequency and effects of disturbance on Smerinthinae and Macroglossinae frequencies. GLMs are robust to deviations of data from

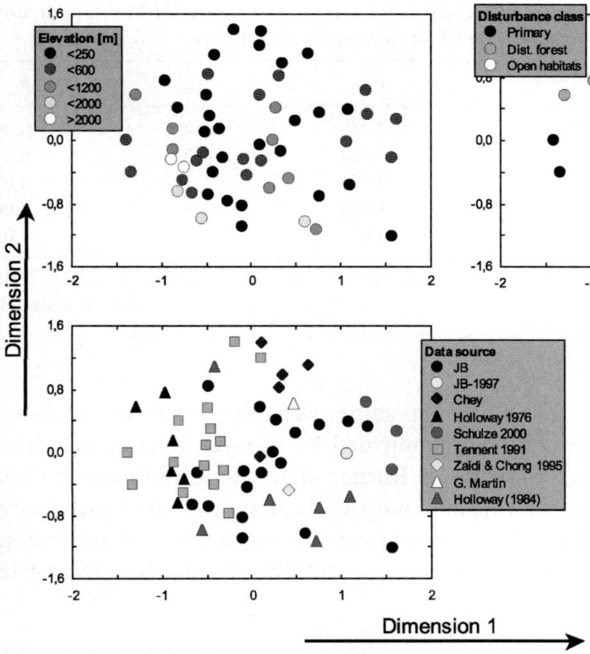


*Figure 3.* 2-dimensional non-metric MDS-plots for 57 Bornean sampling sites (Stress = 0.237), displaying differences in elevation, habitat disturbance and data source.

*Table 2.* Results for a standard Generalized Linear Model (GLM; StatSoft 2003), analysing potentially influential factors on dimension-values of a MDS.

| | Multivariate significance test | | | Univariate results | | | | | |
|---|---|---|---|---|---|---|---|---|---|
| | | | | Dim1 | | Dim2 | | Dim3 | |
| | 1-Wilks $\lambda$ | $F_{df=3}$ | $p$ | $F_{df=1}$ | $p$ | $F_{df=1}$ | $p$ | $F_{df=1}$ | $p$ |
| Elevation | **0.304** | **7.147** | **0.0004** | 0.663 | 0.4191 | 3.783 | 0.0573 | **13.707** | **0.0005** |
| Disturbance | **0.205** | **4.210** | **0.0100** | 2.808 | 0.0999 | **11.319** | **0.0015** | 0.722 | 0.3993 |
| Sampling year | **0.407** | **11.218** | **<0.0001** | **23.757** | **<0.0001** | **6.599** | **0.0132** | 0.192 | 0.6629 |
| Lamp type | 0.024 | 0.407 | 0.7487 | 0.428 | 0.5158 | 0.298 | 0.5878 | 0.245 | 0.6225 |
| Full night | **0.383** | **10.148** | **<0.0001** | **22.387** | **<0.0001** | 1.151 | 0.2883 | **4.161** | **0.0466** |
| Constant | 0.398 | 10.782 | <0.0001 | 22.999 | <0.0001 | 6.173 | 0.0163 | 0.201 | 0.6562 |

The model is significant for all three dimensions (Dim1: $R^2_{multiple} = 0.418$, $F_{df=5} = 7.321$, $p < 0.0001$; Dim2: $R^2_{multiple} = 0.333$, $F_{df=5} = 5.103$, $p < 0.001$; Dim3: $R^2_{multiple} = 0.396$, $F_{df=5} = 6.695$, $p < 0.0001$). Multivariate significance tests identify all suspected factors except 'lamp type' as influential (1-Wilks $\lambda$ can be interpreted as a measure of explained variance, analogous to $R^2$ in univariate tests; StatSoft 2003). Significant effects (bold print) in univariate tests suggest influences of a factor on respective dimensions (see also Figure 5).

*Table 3.* Multiple significance tests of influential parameters in a standard GLM (StatSoft 2003) on the proportion of sphingid subfamilies in 93 samples in Southeast-Asia.

| | 1-Wilks $\lambda$ | $F$ | df | $p$ |
|---|---|---|---|---|
| Latitude | 0.062 | 2.467 | 2 | 0.092 |
| Longitude | 0.017 | 0.654 | 2 | 0.523 |
| Elevation | 0.065 | 2.571 | 2 | 0.083 |
| Disturbance | **0.179** | **3.828** | **4** | **0.005** |
| Constant | 0.077 | 3.079 | 2 | 0.052 |

The model gives significant predictions for Sphinginae ($R^2_{\text{multiple}} = 0.143$, $F_{\text{df} = 5} = 2.509$, $p = 0.037$) and Smerinthinae ($R^2_{\text{multiple}} = 0.156$, $F_{\text{df} = 5} = 2.767$, $p = 0.024$), while it is (barely) non-significant for Macroglossinae ($R^2_{\text{multiple}} = 0.120$, $F_{\text{df} = 5} = 2.046$, $p = 0.082$).

normality (Guisan et al. 2002), which some variables exhibited (KS-test: $p < 0.01$). Furthermore, results were confirmed by non-parametric univariate tests (see Figure 4). Similar analyses of Borneo data alone (not shown) produced no significant multivariate results, but trends along the same pattern (for disturbance). Similarly, analyses of the relative species richness of subfamilies (as *species/total species richness*) show no significant results, but follow the

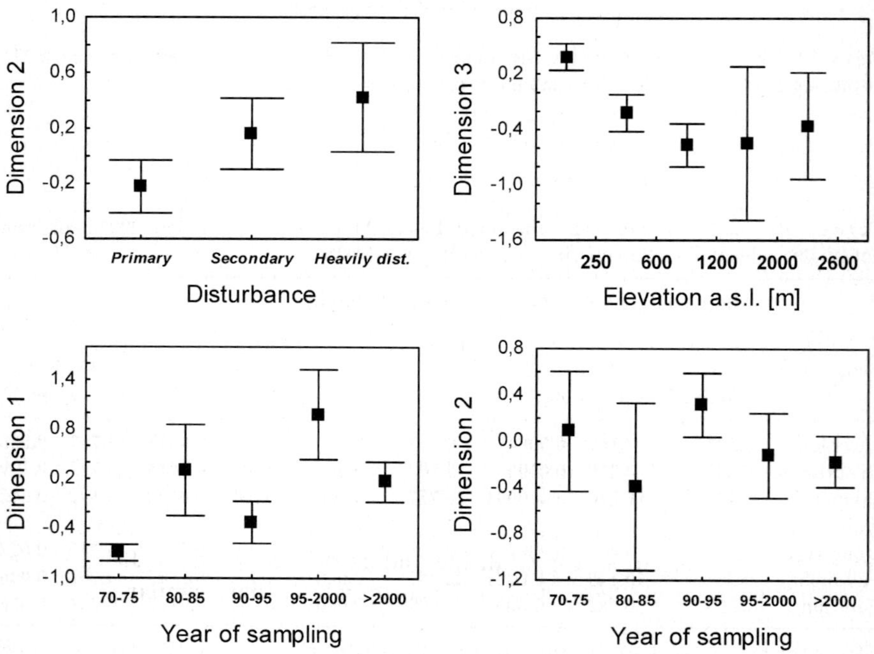


*Figure 4.* Mean dimension values (±95% CI) from a 3-dimensional MDS-plot for 57 sampling sites on Borneo. Sites were grouped by disturbance, elevation class and half-decade of sampling year (see text). Significance of the effect of parameters on dimension values was established with a GLM (Table 2) and was also confirmed with non-parametric, univariate tests (not shown).

same pattern as specimen frequencies. Subfamily frequency changes with increasing longitude (less Smerinthinae, more Macroglossinae) are not significant, but their direction matches results of biogeographical analyses of island faunas across the region (Beck 2005, Beck and Kitching 2004).

## Discussion

### *Habitat disturbance, altitude and life history traits*

No influence of habitat disturbance on the within-habitat diversity of hawkmoths in Borneo was found, hereby confirming the conclusion Schulze and Fiedler (2003) reached with their considerably smaller data set. Furthermore, among data from other regions (albeit on a smaller number of sites) no trends of any influence of disturbance were found. This result is in striking contrast to the reaction of a number of other Lepidoptera groups (e.g. Nymphalidae, Geometridae, Pyralidae: Hamer and Hill 2000; Beck et al. 2002; Fiedler and Schulze 2004), which decrease considerably in diversity at sites of high anthropogenic habitat disturbance. However, when taking a closer look at the data it became evident that taxonomic subgroups within the Sphingidae do react to habitat disturbance, but seem to compensate for each others effect with regard to total diversity: The frequency of Smerinthinae specimens (and species) decreases with disturbance, whereas that of Macroglossinae increases. Generally, it must be expected that the ecological similarity of species, and therefore their habitat choice, be correlated with their phylogeny (Webb 2000), so effects of habitat parameters on higher taxon frequency are not surprising. Many Smerinthinae species (i.e., the tribe Smerinthini) have a reduced, non-functional proboscis that does not allow for adult feeding (Miller 1997; Kitching and Cadiou 2000). This implies a *capital breeding* life history, where only larval resources are used for egg production and adult energy expenditure (see e.g. Boggs 1997 and references therein for terminology), and which can have significant impacts on ecological characteristics of Lepidoptera species (Tammaru and Haukioja 1996). Presumably associated with this life history are a shorter adult life span and greater sexual dimorphism (e.g. Janzen 1984). Macroglossinae, on the other hand, have a well-developed proboscis. Their *income breeding* life history implies that adult resources are used for reproduction and body maintenance, potentially resulting in longer adult life and associated features (Janzen 1984, see also Kaitala et al. 2002). For Sphinginae, which share similar life history traits as Macroglossinae, no changes in their generally low frequency were observed. There is an indication that Smerinthinae are less efficient dispersers (Beck 2005), possibly due to their shorter adult life span or lesser flight abilities. Data indicate the trend that the partly *capital breeding* Smerinthinae are better adapted to stable primary habitats, whereas *income breeding* Macroglossinae thrive in disturbed sites that were probably mostly ephemeral

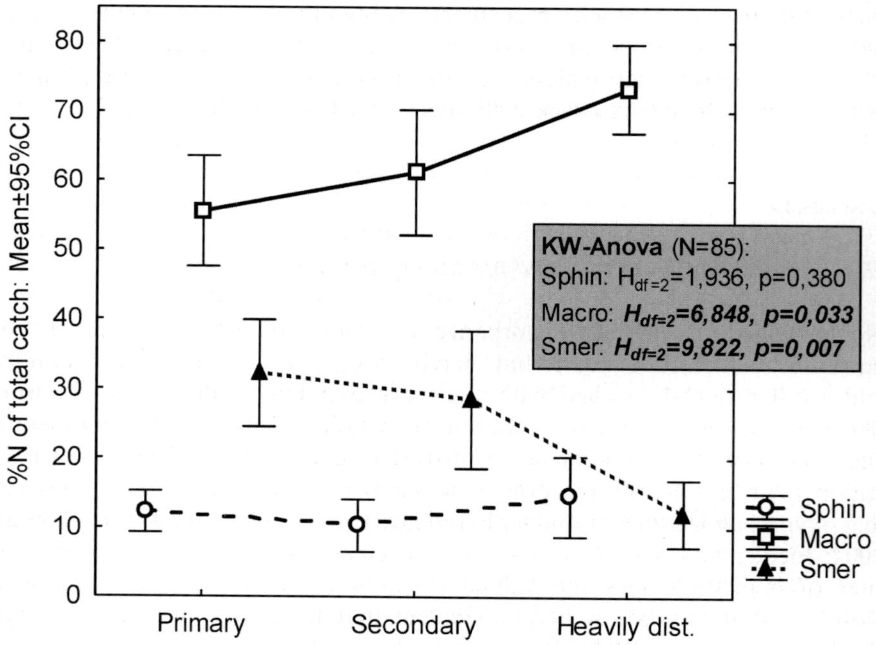


*Figure 5.* The influence of habitat disturbance on the relative abundance of subfamilies in 93 local samples across Southeast-Asia. Non-parametric univariate test values are given in the graph, which were largely confirmed in a multivariate analysis (see Table 3).

prior to the recent, massive anthropogenic habitat conversion in the Indo-Australian tropics (Sodhi et al. 2004; Bowles et al. 1998). Further support of the idea that life history influences the response of taxa to habitat disturbance comes from a comparison to species and specimen numbers of the mostly *capital breeding* family Saturniidae, which follow similar patterns as Smerinthinae, albeit with greatly reduced sample sizes (in Borneo; Beck 2005). However, this hypothesis is weakened by the Smerinthinae tribe Ambulycini, which are adult flower feeders (as indicated by pollen found on their proboscises, Beck 2005) yet exhibit the same reactions to habitat disturbance as the confirmed non-feeding Smerinthini (not shown).

Faunal composition of sphingid assemblages on Borneo is significantly influenced by habitat disturbance (Figure 5). This finding is in contrast to results from Schulze and Fiedler (2003) who found no influence of disturbance on sphingid $\beta$-diversity despite similar analysis techniques (i.e., NESS-index, non-metric MDS). Two differences between Schulze and Fielder (2003) and our study might be responsible for this difference. A large sample size might have made it possible to find community changes that were not visible in Schulze and Fiedler's smaller data set. Furthermore, the classification of habitat disturbance differed between the studies. Whereas Schulze and Fiedler (2003) dichotomously compared strictly primary habitat with habitats of any degree of disturbance, we, due to a larger number of samples, were able to classify habitats more finely by also differentiating between secondary, degenerated forests and heavily disturbed, open landscapes. Several studies on Lepidoptera diversity indicated that this stage of habitat conversion might create the greater change in communities than a primary forest to secondary forest conversion (e.g. Willott 1999; Schulze 2000; Beck et al. 2002).

Most studies on habitat disturbance investigate a gradient (e.g. Beck et al. 2002) or differently disturbed sites (e.g. Willott 1999) in close proximity to each other, thereby avoiding the influence of biogeographical or regional differences in species composition. Data in this study, on the other hand, were compiled from different sources, times and regions, and are therefore less controlled for additional influences besides the 'target' variables (i.e., habitat disturbance), which probably 'blur' effects to a certain degree. Multivariate analysis becomes necessary to filter out relevant effects and sometimes leaves some doubt if the right parameters of a large number of possibilities were chosen (McNally 2000). The resulting large sample sizes, on the other hand, allow results and conclusions on a larger regional scale, as (at least on Borneo) all major habitat types were covered by data.

In preliminary analyses 'data source' proved to be the most influential predictor of community composition. This is an important call for careful analysis of multi-source data, but it is a very unsatisfying result if one looks for biologically relevant habitat parameters. Data sources not only varied in methodological aspects (light source, schedule of nightly sampling), but also in sampling regions within Borneo (which differed due to geographic autocorrelation of samples, Beck 2005), the year of sampling, altitudinal zone and habitat type (e.g., Chey 1994 sampled almost exclusively on lowland softwood plantations in the south-east of Sabah). The parameters that were finally chosen for analysis (see Table 2) could all be reasonably expected to cover a portion of the 'data source' variability and succeeded in a significant multivariate model that filtered out effects that are readily interpretable. Elevation and habitat disturbance are major environmental parameters that influence almost all investigated biological communities (e.g. McCoy 1990; Lawton et al. 1998). The sampling schedule (full night vs. part of the night) was expected *a priori* to influence samples (see above), whereas the influence of the year of sampling is an unexpected, yet possibly important and interpretable finding (see below). However, it should not be forgotten that at least some of these parameters could co-vary with as yet unknown variables that influence sphingid assemblages, but are still hidden in the variation of 'data sources' (McNally 2000).

Elevation was found to be a significant predictor of faunal assemblages (Figure 5). However, no statistically significant trends of within-habitat diversity were found, although plots suggest a mid-elevational peak over 1000 m a.s.l., both for Borneo (Figure 2) as well as for pooled data from Southeast-Asia (not shown). Whereas the biodiversity of many taxa generally decreases with increasing altitude in the Indo-Australian tropics (e.g. Wolda

1987; McCoy 1990; Brühl et al. 1998; Häuser et al. 1997), a mid-elevational peak was found for many Lepidoptera groups (Holloway 1993; Holloway et al. 1990). Reasons for this might be an overlap of lowland and montane fauna, Pleistocene extinctions in the lowlands (Holloway et al. 1990; Holloway and Nielsen 1999), or high speciation rates in montane regions (as e.g. in the butterfly genus *Delias*, Parsons 1999). Furthermore, such altitude patterns were associated with the mid-domain effect (Colwell and Lees 2000) in other taxa (e.g. McCain 2004; but see Zapata et al. 2003). Schulze (2000) found a mid-elevational peak for recorded sphingid species richness on an altitudinal gradient on Mt. Kinabalu in northeastern Borneo, which is the highest and best-surveyed mountain on the island (e.g. Schulze et al. 2000). Furthermore, he reports a similar taxonomic dichotomy to that we report here for habitat disturbance (Schulze, 2000): Macroglossinae (and to a lesser degree Sphinginae) are species-rich from the lowland up to the lower montane region, while Smerinthinae are relatively species-poor in the lowland and exhibit a strong rise in recorded species with increasing altitude and reach a peak in the lower montane forest. Both groups show a sharp decline in species richness above 1600 m altitude. Data in this study (not shown) confirm the reported patterns (Schulze 2000) for the species diversity of subfamilies (as Fisher's $\alpha$) for Borneo.

A *capital breeding* life history might be connected to larval food plant choice (Miller 1997) and should lead to the use of stable resources due to limited dispersal abilities (see above with reference to habitat disturbance). Trees are a more stable resource than herbaceous plants and are more commonly taken by Smerinthinae caterpillars than by other subfamilies (Holloway 1987). However, the tree family Dipterocarpaceae, which is dominant in the canopy of Southeast-Asian lowland forests (Whitmore 1990), is only rarely eaten by sphingid caterpillars (Robinson et al. 2001). Schulze (2000) argued that Smerinthinae might be more diverse in montane regions because suitably stable larval resources might be more abundant there, due to a change in plant family composition, and there is also a lower tree diversity in montane forest (e.g. Kitayama 1992), which diminishes the time needed to find a specific resource.

### *Temporal variation in sphingid assemblages: climate, habitat conversion or just 'noise'?*

The year of sampling (measured in half-decades) emerged as an influential predictor of sphingid assemblage composition. If this is not an artefact of an unknown, co-varying parameter of 'data source' (see above), it is an unexpected yet relevant indication of long-term change of a tropical insect community. During three years of our own sampling no effects of seasonality were observed in re-sampled sites on Borneo (Beck and Linsenmair, in press), but species assemblages changed slightly between re-samples. Thus, seasonal effects that were observed on insect taxa in other tropical regions (Novotny and

Basset 1998; Intachat et al. 2001) can probably be excluded as a reason for the observed changes in hawkmoths in Borneo (see also Barlow and Woiwod 1993; Novotny et al. 2002). Long-term changes of Southeast-Asian hawkmoth assemblages are quantitatively documented here for the first time, but were also observed by collectors who regularly sampled at the same sites for decades (e.g. H. Barlow pers. com; H. v. Mastrigt pers. com.).

Long-term temporal changes of biological communities are not an unusual phenomenon, but are mostly well-documented only for plants and vertebrate taxa from temperate regions or for taxa with special relevance, such as game or pest species (see e.g. Maurer 1999; Lawton 2000 for manifold examples and references). Population fluctuations of species can be regular or synchronized by an outside factor (see Selas et al. 2004; Bjørnstad et al. 1998 for examples on moths), or they can be irregular (e.g. Lawton 2000; Maurer 1999). Two potentially influential factors on temporal changes hawkmoth assemblages in Borneo come to mind: deviations from the otherwise very stable and uniform climate during the irregular '-*El Niño* Southern Oscillations' (ENSO, Kitayama et al. 1999), a weather phenomenon that leads to several months of draught every few years and that has potentially far-reaching biological impacts (e.g. on trees: Slik 2004; on butterflies: Cleary and Mooers 2004; Itioka and Yamauti 2004). Furthermore, rapid, large-scale habitat conversion has changed Borneo (as well as many other tropical rainforests, Bowles et al. 1998; Sodhi et al. 2004), which until the beginning of industrial logging in the 1950s (Marsh et al. 1996) was mostly covered with relatively undisturbed forest.

ENSO years since 1970 were identified (source: *http://www.elnino.noaa.gov*) in all of the analysed sampling periods (in 5-year steps) and therefore cannot be associated with the observed sphingid community changes at this temporal resolution. However, the strongest ENSO events were identified in 1983 and 1998, matching the highest values on dimension 1 of the MDS (Figure 5) in the corresponding half-decades. Therefore, the idea that sphingid assemblages are influenced by this global climate phenomenon cannot be ruled out. Temporal changes of the community are mostly projected on dimension 1 of the MDS-plot, but to a lesser degree also on dimension 2, the 'disturbance axis' (see Figure 5). However, values seem to decrease rather than increase with time, thus developing towards 'primary habitat' – a counterintuitive result that might be explained by better accessibility of jungle regions in modern times (allowing easier sampling in primary habitats), or by an increased interest in the ecology of primary habitats.

### *Estimating regional species richness from local samples*

Hawkmoths were exhaustively sampled in northern, non-Indonesian Borneo, which is one of the best-covered regions for this taxon in the Indo-Australian tropics. Despite a significant geographic autocorrelation of quantitative samples (Beck 2005), $\beta$-diversity between regions on the island is apparently not

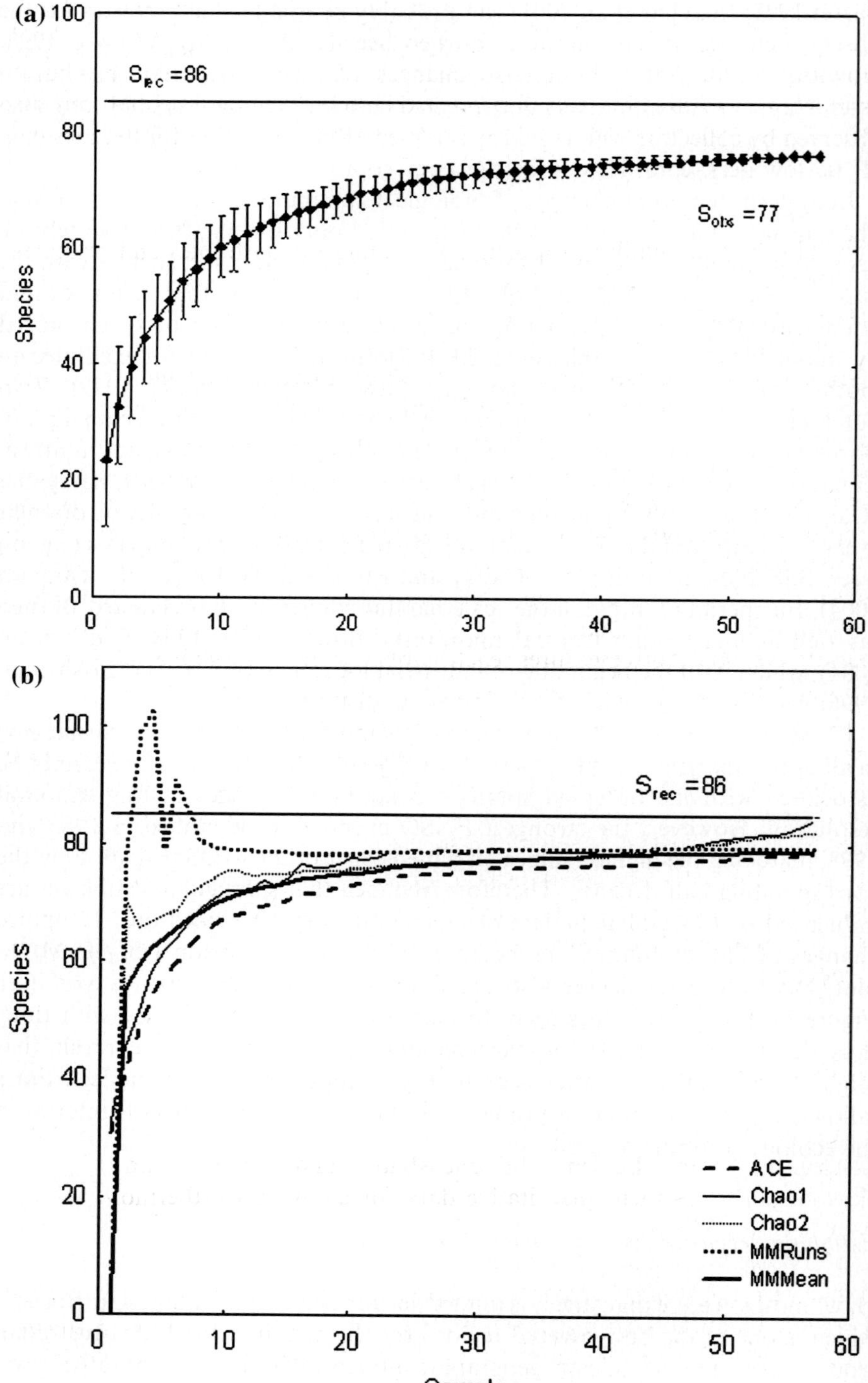
(a)
Species
Srec =86
Sobs =77
0
20
40
60
80
100
0
10
20
30
40
50
60
(b)
Species
Srec =86
ACE
Chao1
Chao2
MMRuns
MMMean
0
20
40
60
80
100
0
10
20
30
40
50
60
Samples

◀

*Figure 6*. Shows a 100-fold randomised species accumulation curve [± SD] (a) and estimators of 'true' species richness (b) from 57 quantitative local samples on Borneo. Randomisations and calculations of richness estimators were executed with *EstimateS 5.01* (Colwell 2000). Total known species richness for Borneo was taken from the Checklist in Beck and Kitching (2004), excluding species of the genera *Macroglossum*, *Cephonodes* and *Sataspes*, which are day-active and can be caught only rarely at light.

very high: Schulze et al. (2000) record two thirds of the known species from Kinabalu Park (as similar proportion was found for butterflies and primates, Häuser et al. 1997; Schulze and Beck 1999). Even though this figure might be misleading because (1) Kinabalu Park is an exceptionally diverse region due to its unmatched altitudinal range and (2) the park is by far the best-sampled region of the island (see above for reference), there is little indication that Borneo's Sphingidae fauna is not (almost) completely known. This gives the unique possibility to compare a large data set of short, intense samples with a relatively complete inventory of 125 years of sampling for this insect taxon (see Checklist in Beck and Kitching 2004). Hammond (1994) suggested using local samples for extrapolation to regional species richness for less well-known groups (see also Krishnamani et al. 2004 and references therein). Figure 6 shows species accumulation curves (Leon-Cortez et al. 1998; Moreno and Halffter 2000) and estimators of 'true' species richness (Colwell 2000) of quantitative data. Both measures reach stable, apparently reliable values at 20–30 local samples, but miss the total known species richness of the island by about 10 percent. Considering (1) that it will rarely be possible to use such a comprehensive set of sampling sites for analysis, (2) that Sphingidae are a relatively species-poor taxon of insects in tropical rainforest regions, and (3) that their $\beta$-diversity and geographic autocorrelation of faunal composition is probably comparatively low due to their high dispersal abilities (Beck 2005), estimates of regional species richness from local samples most probably should generally be corrected upwards to an unknown extent (see also Ugland et al. 2003 for an alternative model, Petersen and Meier 2003 for similar results on European Diptera).

## Acknowledgements

We thank Dr Jeremy D. Holloway, Henk van Mastrigt, Geoff Martin, Azmi Mahyudin, Torben Larsen, Ulf Buchsbaum and Dr Wolfram Mey for allowing us to use their quantitative data for analysis. Furthermore, we are grateful to Prof Dr Konrad Fiedler, Dr Christian H. Schulze and Dr Jeremy D. Holloway for manifold advice on Lepidoptera, statistics, or both. Two anonymous reviewers helped to improve the manuscript. We thank the Economic Planning Unit of Malaysia for granting a research permit for work in Malaysia. We are also grateful to large number of local authorities in Malaysia and Indonesia for allowing JB to sample moths at sites under their authority, as well as providing logistic support – among them Dr Chey Vun Khen (FRC), Dr Jamili Nais (Sabah Parks), Moses Jalawatu, Wilem Kayoi (and Dinas Kehutanan Jayapura) and the Danum Valley Management Committee. The study was financed by scholarships of the 'Graduiertenkolleg' of the German Research Council (DFG) and the Deutscher Akademischer Austauschdienst (DAAD).

## References

Barlow H.S. and Woiwod I.P. 1993. Seasonality and diversity of Macrolepidoptera in two lowland sites in Dumoga-Bone National Park, Sulawesi Utara. In:Knight W.J. (ed.), Insects and the Rain Forests of South-East Asia (Wallacea). Royal Entomological Society, London, pp. 167–172.

Beck J. and Kitching I.J. 2004. The Sphingidae of Southeast-Asia, incl. New Guinea, Bismark and Solomon Islands (including pictures and range-maps for all species and checklists for 114 islands). http://www.sphingidae.biozentrum.uni-wuerzburg.de.

Beck J. and Linsenmair K.E.. Feasibility of light-trapping in community research of moths: attraction radius of light, completeness of samples, nightly flight times and seasonality of Southeast-Asian hawkmoths (Lepidoptera: Sphingidae). J. Res. Lepidoptera 39, in press.

Beck J., Schulze C.H., Linsenmair K.E. and Fiedler K. 2002. From forest to farmland: diversity of geometrid moths along two habitat gradients on Borneo. J. Trop. Ecol. 18: 33–51.

Beck J. 2005. The Macroecology of Southeast-Asian Hawkmoths (Lepidoptera: Sphingidae). PhD thesis. University of Würzburg (Germany). Available online at http://opus.bibliothek.uni-wuerzburg.de/opus/volltexte/2005/1300/.

Bjørnstad O.N., Begon M., Stenseth N.C., Falck W., Sait S.M. and Thompson D.J. 1998. Population dynamics of the Indian meal moth: demographic stochasticity and delayed regulatory mechanisms. J. Anim. Ecol. 67: 110–126.

Boggs C.L. 1997. Reproductive allocation from reserves and income in butterfly species with differing adult diets. Ecology 78: 181–191.

Bowles I.A., Rice R.A., Mittermeier R.A. and da Fonseca G.A.B 1998. Logging and tropical forest conservation. Science 280: 1899–1900.

Brehm G. and Fiedler K. 2004. Ordinating tropical moth ensembles from a elevational gradient: a comparison of common methods. J. Trop. Ecol. 20: 165–172.

Brühl C.A. 2001. Leaf-litter Ant Communities in Tropical Lowland Rain Forest in Sabah, Malaysia: Effects of Forest Disturbance and Fragmentation. Ph.D. thesis, Universität Würzburg, Germany.

Brühl C.A., Gunasalam G. and Linsenmair K.E. 1998. Stratification of ants (Hymenoptera, Formicidae) in a primary rain forest in Sabah, Borneo. J. Trop. Ecol. 14: 285–297.

Chey V.K. 1994. Comparison of Biodiversity between Plantations and Natural Forests in Sabah Using Moths as Indicators. Ph.D. thesis, University of Oxford, UK.

Chey V.K. 2002. Comparison of moth diversity between lightly and heavily logged sites in a tropical rainforest. Malay. Nat. J. 56(1): 23–41.

Chey V.K., Holloway J.D. and Speight M.R. 1997. Diversity of moths in forest plantations and natural forests in Sabah. Bull. Entomol. Res. 87: 371–385.

Cleary D.F.R. and Mooers A.Ø. 2004. Butterfly species richness and community composition in forests affected by ENSO-induced burning and habitat isolation in Borneo. J. Trop. Ecol. 20: 359–367.

Colwell R.K. 2000. EstimateS: statistical estimation of species richness and shared species from samples. Version 5.01. User's Guide and application published at: http://viceroy.eeb.uconn.edu/estimates.

Colwell R.K. and Lees D.C. 2000. The mid-domain effect: geometric constraints on the geography of species richness. Trends Evol. Ecol. 15: 70–76.

Dennis R.L.H. 2003. Towards a functional resource-based concept of habitat: a butterfly biology viewpoint. Oikos 102: 417–426.

Fermon H., Waltert M., Vane-Wright R.I. and Mühlenberg M. 2005. Forest use and vertical stratification in fruit-feeding butterflies of Sulawesi, Indonesia: impacts of conservation. Biodiv. Conserv. 14: 333–350.

Fielder K. and Schulze C.H. 2004. Forest modification affects diversity, but not dynamics of speciose tropical pyraloid moth communities. Biotropica 36: 615–627.

Floren A., Freking A., Biehl M. and Linsenmair K.E. 2001. Anthropogenic disturbance changes the structure of arboreal tropical ant communities. Ecography 24: 547–554.

Gathorne-Hardy F.J., Jones D.T. and Syaukani 2002. A regional perspective on the effects of human disturbance on the termites of Sundaland. Biodiv. Conserv. 11: 1991–2006.

Gotelli N. and Colwell R.K. 2001. Quantifying biodiversity: procedures and pitfalls in the measurement and comparison of species richness. Ecol. Lett. 4: 379–391.

Grassle J.F. and Smith W. 1976. A similarity measure sensitive to the contribution of rare species and its use in investigation of variation in marine benthic communities. Oecologia 25: 13–22.

Guisan A., Edwards T.C.Jr. and Hastie T. 2002. Generalized linear and generalized additive models in studies of species distributions: setting the scene. Ecol. Model. 157: 89–100.

Hamer K.C. and Hill J.K. 2000. Scale-dependent effects of habitat disturbance on species richness in tropical forests. Conserv. Biol. 14: 1435–1440.

Hamer K.C., Hill J.K., Lace L.A. and Langan A.M. 1997. Ecological and biogeographical effects of forest disturbance on tropical butterflies of Sumba, Indonesia. J. Biogeogr. 24: 67–75.

Hammond P. 1994. Practical approaches to the estimation of the extent of biodiversity in specious groups. Phil. Trans. R. Soc. (B) 345: 119–136.

Häuser C.L., Schulze C.H. and Fiedler K. 1997. The butterfly species (Insecta: Lepidoptera: Rhophalocera) of Kinabalu Park, Sabah. Raffles Bull. Zool. 45: 281–304.

Hayek L.-A.C. and Buzas M.A. 1997. Surveying Natural Populations. Columbia University Press, New York.

Hector A., Joshi J., Lawler S.P., Spehn E.M. and Wilby A. 2001. Conservation implications of the link between biodiversity and ecosystem functioning. Oecologia 129: 624–628.

Helmkampf M.E., Schwarz C.J., Maklarim B. Lakim and Beck J.. A first look at the biodiversity of praying mantids (Insecta: Mantodea) in Sabah, Borneo. Sabah Parks Nat. J., in press.

Hochberg Y. 1988. A sharper Bonferroni procedure for multiple tests of significance. Biometrika 75: 800–802.

Holloway J.D. and Nielsen E.S. 1999. Biogeography of the Lepidoptera. In: Kristensen N.P. (issue ed.), Lepidoptera, Moths and Butterflies, Vol. 1: Evolution, Systematics, and Biogeography. Handbuch der Zoologie, Band IV: Arthropoda: Insecta. (ed. M. Fischer), part 35, W. de Gruyter, Berlin, New York, 423–462.

Holloway J.D. 1976. Moths of Borneo with Special Reference to Mt. Kinabalu. Malayan Nature Society, Kuala Lumpur.

Holloway J.D. 1984. The larger moths of the Gunung Mulu National Park – a preliminary assessment of their distribution, ecology, and potential as environmental indicators. Sarawak Museum J. Kuching 51 (New Series), Special Issue 2: 148–191.

Holloway J.D. 1987. The Moths of Borneo, Part 3: Lasiocampidae, Eupterotidae, Bombycidae, Brahmaeidae, Saturniidae, Sphingidae. The Malay Nature Society and Southdene Sdn. Bhd, Kuala Lumpur.

Holloway J.D. 1993. Aspects of the biogeography and ecology of the Seram moth fauna. In: Edwards I.D., MacDonald A.A. and Proctor J. (eds), Natural History of Seram, Maluku, Indonesia, pp. 91–115. Raleigh International .

Holloway J.D. 1998. The impact of traditional and modern cultivation practices, including forestry, on Lepidoptera diversity in Malaysia and Indonesia. In: Newberry D.M., Prins H.H.T. and

Brown N.D. (eds.). Dynamics of Tropical Communities. 37th Symposium of the British Ecological Society, London, 507–597.

Holloway J.D., Kirk-Spriggs A.H. and Chey V.K. 1992. The response of some rain forest insect groups to logging and conversion to plantation. Philos. Trans. R. Soc. (B) 335: 425–436.

Holloway J.D., Robinson G.S. and Tuck K.R. 1990. Zonation in the Lepidoptera of northern Sulawesi. In: Knight W.J.and Holloway J.D. (eds), Insects and the Rain Forests of South East Asia (Wallacea). The Royal Entomological Society of London, pp. 153–166.

Intachat J., Holloway J.D. and Staines H. 2001. Effects of weather and phenology on the abundance and diversity of geometroid moths in a natural Malaysian rain forest. J. Trop. Ecol. 17: 411–429.

Itioka T. and Yamauti M. 2004. Severe drought, leafing phenology, leaf damage and lepidopteran abundance in the canopy of a Bornean aseasonal tropical rain forest. J. Trop. Ecol. 20: 479–482.

Janzen D.H. 1984. Two ways to be a tropical big moth: Santa Rosa saturniids and sphingids. In: Dawkins R. and Ridley M. (eds), Oxford Surveys in Evolutionary Biology. 1: 85–140.

Johns A.D. 1992. Vertebrate responses to selective logging: implications for the design of logging systems. Philos. Trans. R. Soc. (B) 335: 437–442.

Kaitala V., Tesar D. and Ranta E. 2002. Semelparity versus iteroparity and the number of age groups. Evolution. Ecol. Res. 4: 169–179.

Kenney A.J. and Krebs C.J. 1998. Programs for Ecological Methodology, Version 5.2. Published at http://www.zoology.ubc.ca/~krebs.

Kitayama K. 1992. An altitudinal transect of the vegetation on Mount Kinabalu, Borneo. Vegetatio 102: 149–171.

Kitayama K., Lakim M. and Wahab M.Z. 1999. Climate profile of Mount Kinabalu during Late 1995–early 1998 with Special Reference to the 1998 Drought. Sabah Parks Nat. J. 2: 85–100.

Kitching I.J. and Cadiou J.-M. 2000. Hawkmoths of the world. The Natural History Museum, Cornell University Press, london.

Krishnamani R., Kumar A. and Harte J. 2004. Estimating species richness at large spatial scales using data from small discrete plots. Ecography 27: 637–642.

Lawton J.H., Bignell D.E., Bolton B., Bloemers G.F., Eggleton P., Hammond P.M., Hodda M., Holt R.D., Larsoen T.B., MAwdsley N.A., Stork N.E., Srivastava D.S and Watt A.D. 1998. Biodiversity inventories, indicator taxa and effects of habitat modification in tropical forest. Nature 391: 72–76.

Lawton J.H. 2000. Community Ecology in a Changing World. Excellence in Ecology 11. Ecology Institute, Oldendorf/Luhe, Germany.

Legendre P. and Legendre L. 1998. Numerical Ecology. Developments in Environmental Modelling 20, 2nd ed. Elsevier, New York.

Leon-Cortez J.L., Soberon-Mainero J. and Llorente-Bousquets J. 1998. Assessing completeness of Mexican sphinx moths inventories through species accumulation functions. Diversity and Distribution 4: 37–44.

Marsh C.W., Tay J., Pinard M.A., Putz F.E. and Sullivan T.E. 1996. Reduced impact logging: a pilot project in Sabah, Malaysia. In: Schulte A. and Schöne D. (eds), Dipterocarp Forest Ecosystems – Towards Sustainable Management. World Scientific Publishing, Singaporepp. 293–307.

Maurer B.A. 1999. Untangling Ecological Complexity – The Macroscopic Perspective. University of Chicago Press.

McCain C.M. 2004. The mid-domain effect applied to elevational gradients: species richness of small mammals in Costa Rica. J. Biogeogr. 31: 19–31.

McCoy Earl D. 1990. The distribution of insects along elevational gradients. Oikos 58: 313–322.

McNally R. 2000. Regression and model-building in conservation biology, biogeography and ecology: The distinction between – and reconciliation of –'predictive' and 'explanatory' models. Biodiv. Conserv. 9: 644–671.

Miller W.E. 1997. Diversity and evolution of tongue length in hawkmoths (Sphingidae). J. Lepidopterist Soc. 51: 9–31.

Moreno C.E. and Halffter G. 2000. Assessing the completeness of bat biodiversity inventories using species accumulation curves. J. Appl. Ecol. 37: 149–158.

Novotny V. and Basset Y. 1998. Seasonality of sap-sucking insects (Achenorryncha, Hemiptera) feeding on *Ficus* (Moraceae) in a lowland rainforest in New Guinea. Oecologia 115: 514–522.

Novotny V., Miller S.E., Basset Y., Cizek L., Drozd P., Darrow K. and Leps J. 2002. Predictably simple: communities of caterpillars feeding on rainforest trees in Papua New Guinea. Proc. R. Soc. (B) 269: 2337–2344.

Parsons M. 1999. The Butterflies of Papua New Guinea. Their Systematics and Biology. Academic Press, San Diego.

Pearson D.I. 1994. Selecting indicator taxa for the quantitative assessment of biodiversity. Philos. Trans. R. Soc. (B) 345: 75–79.

Petersen F.T. and Meier R. 2003. Testing species richness estimation methods on single-sample collection data using the Danish Diptera. Biodiv. Conserv. 12: 667–686.

Robinson G.S., Ackery P.R., Kitching I.J., Beccaloni G.W. and Hernández L.M. 2001. HOSTS – a database of the hostplants of the world's Lepidoptera. Internet website at http://www.nhm.ac.uk/entomology/hostplants/index.html.

Schulze C.H. and Beck J. 1999. A record of proboscis monkey (*Nasalis larvatus*) (Mammalia: Primates: Ceropithecidae) from Kinabalu Park, Sabah, Malaysia. Sabah Parks Nat. J. 2: 23–26.

Schulze C.H. and Fiedler K. 1997. Patterns of diversity and vertical stratification in hawkmoths of a Bornean rain forest. Mitteilungen der Deutschen Gesellschaft für allgemeine und angewandte Entomologie Band 11/Heft, 1–6.

Schulze C.H. and Fiedler K. 2003. Hawkmoth diversity in northern Borneo does not reflect the influence of anthropogenic habitat disturbance. Ecotropica 9: 99–102.

Schulze C.H. 2000. Auswirkungen anthropogener Störungen auf die Diversität von Herbivoren – Analyse von Nachtfalterzönosen entlang von Habitatgradienten in Ost-Malaysia. Universität Bayreuth, Germany. Ph.D. thesis,

Schulze C.H., Häuser C.L. and Mohamend Maryati 2000. A Checklist of the Hawkmoths (Lepidoptera: Sphingidae) of Kinabalu Park, Sabah, Borneo. Malay. Nat. J. 54: 1–20.

Schulze C.H., Waltert M., Kessler P.J.A., Pitopang R., Shahbuddin, Veddeler D., Mühlenberg M., Gradstein S.R., Leuschner C., Steffan-Dewentter I. and Tscharntke T. 2004. Biodiversity indicator groups of tropical land-use systems: Comparing plants, birds and insects. Ecol. Applicat. 14: 1321–1333.

Selas V., Hogstadt O., Kobro S. and Rafoss T. 2004. Can sunspot activity and ultraviolet-B radiation explain cyclic outbreaks of forest moth pest species? Proc. R. Soc. (London) B 271: 1897–1901.

Simon U. and Linsenmair K.E. 2001. Arthorpods in tropical oaks: differences in their spatial distributions within tree crowns. Plant Ecol. 153: 179–191.

Slik J.W.F. 2004. El Niño droughts and their effects on tree species composition and diversity in tropical raind forests. Oecologia 141: 114–120.

Sodhi N.S., Lian Pin Koh, Brook B.W. and Ng P.K.L. 2004. Southeast-Asian biodiversity: an impending disaster. Trends Ecol. Evol. 19: 654–660.

Southwood T.R.E. and Henderson P.A. 2000. Ecological Methods, 3rd ed. Blackwell Science Ltd, Oxford.

StatSoft Inc. 2003. STATISTICA, Version 6. Computer program and manual at www.statsoft.com.

Tammaru T. and Haukioja E. 1996. *Capital breeders* and *income breeders* among Lepidoptera – consequences to population dynamics. Oikos 77: 561–564.

Tennent W.J. 1991. Notes on some Borneo hawk moths (Lepidoptera: Sphingidae) including *Eupanacra hollowayi* sp. n. and *Macroglossum amoenum* Rothschild and Jordan, new to Borneo. Entomol. Rec. 103: 223–235.

Ugland K.I., Gray J.S. and Ellingsen K.E. 2003. The species-accumulation curve and estimation of species richness. J. Anim. Ecol. 72: 888–897.

Wagner T. 1999. Arboreal Chrysomelid community structure and faunal overlap between different types of forest in central Africa. Adv. Chrysomelidae Biol. I: 247–270.

Webb C.O. 2000. Exploring the phylogenetic structure of ecological communities: an example for rain forest trees. Am. Nat. 156: 145–155.

Whitmore T.C. 1990. An Introduction to Tropical Rainforests. Clarendon Press, Oxford.

Willott S.J. 1999. The effect of selective logging on the distribution of moths in a Bornean rainforest. Philos. Trans. R. Soc. (B) 354: 1783–1790.

Wolda H. 1987. Altitude, habitat and tropical insect diversity. Biol. J. Linn. Soc. 30: 313–323.

Yela J.L. and Holyoak M. 1997. Effects of moonlight and meteorological factors on light and bait trap catches of noctuid moths (Lepidoptera: Noctuidae). Environ. Entomol. 26: 1283–1290.

Zaidi M.I. and Chong M.S. 1995. Moth fauna of Sayap – Kinabalu Park, Sabah: a preliminary survey and assessment. In: Ismail G. and bin Din L. (eds), A Scientific Journey through Borneo. Sayap-Kinabalu Park, Sabah, pp 201–209.

Zapata F.A., Gaston K. J. and Chown S.L. 2003. Mid-domain models of species richness gradients: assumptions, methods and evidence. J. Anim. Ecol. 72: 677–690.

Biodiversity and Conservation (2006) 15:473–487
DOI 10.1007/s10531-005-0602-1 

# The importance of ants and high-shade management to coffee pollination and fruit weight in Chiapas, Mexico

STACY M. PHILPOTT[1,4,*], SHINSUKE UNO[2] and JORGE MALDONADO[3]
[1]*Department of Ecology and Evolutionary Biology, University of Michigan, 830 N. University Ave., Ann Arbor, MI 48109-1048, USA;* [2]*School of Natural Resources and the Environment, University of Michigan, 430 E. University Ave., Ann Arbor, MI 48109-1115, USA;* [3]*c/o El Colegio de la Frontera Sur (ECOSUR), Carretera Antiguo Aeropuerto, Km 2.5, Tapachula, Chiapas 30700, Mexico;* [4]*Present address: Smithsonian Migratory Bird Center, National Zoological Park, 3001 Connecticut Ave NW, Washington DC 20008, USA; *Author for correspondence (e-mail: philpotts@si.edu; phone: +202-633-4206, fax: +202-673-4916)*

Received 12 August 2004; accepted in revised form 24 June 2005

**Key words:** *Coffea arabica*, Ecosystem services, Pollinator and ant diversity, Shade management system

**Abstract.** Recent reports show importance of pollinators to coffee and importance of ants as pollinators or floral protectors in many systems. Arthropod and pollinator diversity, however, declines with management intensification of coffee (*Coffea arabica*) agroecosystems. We investigated influences of both flying pollinators and ants on coffee fruit set and fruit weight in one high-shade (high-biodiversity) and one low-shade (low-biodiversity) coffee farm in Chiapas, Mexico through exclusion experiments. Contradictory to previous reports, flying pollinators alone did not affect coffee fruit set or fruit weight. Individual fruit weights, however, were higher on branches with both ants and flying pollinators (1.78 g ± 0.312 (SE)) compared to branches without ants (1.03 ± 0.029) or branches without ants or flying pollinators (1.05 ± 0.049), but only in the high-shade site. Although the mechanisms producing higher fruit weights are unknown, we discuss how ants or ant-flying pollinator interactions under high-shade coffee management may contribute to increased fruit weight and the implications of high-shade management for both sustainable coffee production and biodiversity conservation.

## Introduction

Pollinators provide ecosystem services (benefits to humans resulting from ecological functions) in agricultural and natural ecosystems especially where pollinator diversity or abundance is high. Studies highlight necessary or positive influences of particular pollinators (Pimentel et al. 1997; Allen-Wardell et al. 1998; Kearns et al. 1998; Norberg 1999; Cunningham et al. 2002; Kremen et al. 2002). Additionally, pollinator (specifically bee) diversity and abundance can increase pollination rates and fruit or seed set (Rathcke and Jules 1993; Aizen and Feinsinger 1994; Steffan-Dewenter and Tscharntke 1999; Kremen et al. 2002; Klein et al. 2003b; Ricketts et al. 2004). Although some argue that pollinator biomass is alone sufficient to maintain pollination services (Myers 1996), there is still much debate regarding the importance of pollinator diversity (Balvanera et al. 2001).

Although ants are not often considered pollinators, and in fact sometimes negatively affect plant reproduction, ants may enhance pollination in some cases. Ants are nectar thieves (Galen 1999; Ghazoul 2001), flower predators (Galen and Cuba 2001), and may reduce pollen viability via antibiotic secretions (Beattie et al. 1984; Ramsey 1995; Wagner 2000). Some plants even have chemical (Willmer and Stone 1997; Wagner and Kay 2002) or mechanical defenses (Guerrant and Fiedler 1981) to deter ants. Many studies, however, show either direct or indirect benefits to plants via their interactions with ants. Ants do visit flowers and act as pollinators (Gomez and Zamora 1992; Garcia et al. 1995; Gomez et al. 1996). Despite that ants are considered less efficient pollinators, ants may pollinate plants as effectively as winged insects especially when considering germination, seedling survival, and growth to reproductive maturity in addition to seed set (Gomez 2000). Ants benefit plant reproduction indirectly by limiting floral predators (Yano 1994; Oliveira 1997; Sporleder and Rapp 1998; Oliveira et al. 1999). Ants also may augment pollination success by attacking pollinators subsequently increasing their movement and thus pollen transfer between flowers (Altshuler 1999).

Arthropod (including ant and flying pollinator) diversity generally declines with increasing management intensification in coffee (*Coffea arabica*) agroecosystems (Perfecto and Snelling 1995; Perfecto and Vandermeer 1996; Klein et al. 2002). Intensified coffee systems are generally characterized by high use of agrochemicals and fertilizers and a reduction or total elimination of shade trees (Moguel and Toledo 1999) and shade composition or cover may affect species richness (e.g. Perfecto and Vandermeer 1996; Calvo and Blake 1998). It is not clear, however, what this decline in richness may mean for coffee pollination.

Arabica coffee is a self-compatible species that may or may not benefit from pollinators. High numbers of visits of one species of pollinator (*Apis mellifera*) correlate to increased coffee fruit set and fruit weight (Raw and Free 1977; Manrique and Thimann 2002; Roubik 2002). Furthermore, some studies have shown the importance of a diverse suite of pollinators (including both social and solitary bees) to coffee pollination (Klein et al. 2003a, b) and pollen deposition (Ricketts 2004). Some researchers, in contrast, have found that coffee does not significantly benefit from insect pollinators (Nogueira-Neto et al. 1959; Sein 1959). No studies, however, specifically separate the effects of bees from other pollinators, possibly including ants, which also visit coffee flowers (Free 1993) and may be coffee pollinators (Klein et al. 2003b).

In order to study the importance of pollinators under two coffee management systems, and the possible importance of ant as pollinators, we set up pollinator exclusion experiments to test the following hypotheses: (1) If flying insects pollinate coffee, and if coffee responds to these visits, fruit set or fruit weight will increase on plants with flying pollinators compared with pollinator exclosures, (2) If ants directly or indirectly beneficially influence coffee pollination, fruit set or fruit weight will be higher on plants with ants than on ant exclosure plants, (3) If a diverse pollinator array benefits coffee, fruit set or fruit weight will be higher where pollinator diversity is also higher. We therefore performed flying pollinator and ant exclosure experiments in both a high-shade farm and a low-shade farm. Previous studies in our study sites confirm that the diversity of ants and some flying pollinators, such as bees, are higher in the high-shade farm compared to the low-shade farm (Ibarra-Núñez et al. 1995; Perfecto and Vandermeer 2002; Armbrecht and Perfecto 2003; Philpott et al. in press).

## Methods

### *Site description and flowering phenology*

We conducted our study in two coffee farms in the Soconusco region of SW Chiapas, Mexico: (1) Finca Irlanda (15°11′ N, 92°20′ W) and (2) Finca Hamburgo (15°10′ N, 92°19′ W) located 40 km NE of Tapachula. Both farms are located from 1000 to 1100 m a.s.l. To investigate differences with respect to pollination between two coffee management systems, we chose two farms differing in shade cover. Finca Irlanda, the high-shade site, has higher tree richness, abundance, percent shade cover, and structural depth than Finca Hamburgo, the low-shade site (Mas and Dietsch 2003; Philpott 2004). Although both farms cultivate multiple varieties of *C. arabica*, all study areas are dominated by var. Typica. According to Moguel and Toledo (1999), Irlanda corresponds to a "commercial polyculture" whereas Hamburgo is a "shaded monoculture". Finca Irlanda is a certified organic farm, and no fungicides or pesticides have been used in Finca Hamburgo for at least 4 years (Perfecto and Vandermeer 2002). Although the two farms do not differ in terms of soil classification or texture, Finca Hamburgo has higher concentrations of some nutrients (Potassium, Phosphorous, and Nitrate) due to chemical fertilizer use and soil acidity is higher in Finca Irlanda (K. Avilés-Vázquez, unpublished data).

In the study region, coffee flowers synchronously between February and April and fruits are harvested from September to December of the same year (personal observation). The main flowering event occurs in the dry season immediately following a rain event, with sporadic flowering (~5–10% of flowers) throughout the year. We conducted our study during the coffee growing season of 2002. During this year, flowering occurred in the study sites from 10 March to 25 April. In general, coffee flowers remain open for approximately 2 days (Free 1993), however, if flowers are not pollinated, they may remain open for at least 5 days (Jiménez-Castano and Castillo-Zapata 1976; Free 1993). Normally, coffee fruits contain two seeds, but occasionally only one ovary develops, a condition known as peaberry (Free 1993).

*Pollinator exclosures and observations*

To test for effects of flying pollinators and/or ants compared to the ability of coffee to self pollinate, we established exclosure experiments in both farms. In each farm, we established 15 replicate blocks, consisting of three coffee plants. Plants were randomly assigned to one of three treatments: (1) open to flying pollinators and ants (open), (2) open to flying pollinators without ants (no-ant), or (3) without flying pollinators or ants (bagged). One branch per plant approximately 1 m above ground was treated. In the no-ant and bagged treatments, we eliminated ants by putting Tanglefoot® around the base of branches and by removing other vegetation making a bridge for ants. Additionally, on bagged branches, we placed bags (0.5×0.5 mm mesh) around entire coffee branches when flower buds were small (~1 month before flowering), and removed bags only after all flowers had fallen. Mesh bags were removed immediately after coffee flowering; however, Tanglefoot® was not completely cleaned off branches until after the coffee harvest. Thus, bagged and no-ant branches contained Tanglefoot®, but mesh bags were placed only on the bagged treatment.

For each experimental branch, we recorded number of flower buds, number of harvestable coffee fruits, weight per coffee fruit, and calculated the final fruit set (# coffee fruits/# flower buds) per branch. We counted flower buds in January 2002 in the high-shade farm and in February 2002 in the low-shade farm, due to slightly later development in the low-shade site. We harvested coffee fruits when the actual harvest began in the high-shade farm (October 2002). We tested for significant differences in numbers of coffee buds, fruits, fruit set, and fruit weight using two-way ANOVA using treatment and site as main factors, and block as a random factor. We used Tukey's post-hoc tests to determine significant differences between treatments in each site. We used raw data for numbers of flower buds and natural logarithm transformed data for number of fruits, fruit set, and fruit weight in all analyses to meet conditions of normality.

We made preliminary studies of flying pollinator and ant communities in study areas. Between 8:00 am and 1:00 pm, we observed coffee plants for 10 min each, recording order or morphospecies of all flying insects visiting coffee flowers during the observation period. We observed flying pollinators in 2002 only in the high-shade site (10 plants) during the height of the flowering event (26 and 27 March) and observed pollinators in both high- (10 plants) and low-shade sites (10 plants) during flowering events the following year (21 and 22 March 2003). Protocol for both years was the same. We conducted ant surveys during July 2002. We placed small (2 g) tuna baits on each coffee plant included in the experiment, and recorded both identity and activity level of ants. We checked tuna baits 30–45 min after they were placed and recorded ant activity per species at each bait with the following index: (1) 1–2 ants, (2) 3–10 ants, (3) > 10 ants. To compare richness of flying pollinators and ants in the two sites, we generated sample-based rarefaction curves (*MaoTao* estimates) using EstimateS Version 7.5 (Colwell and Coddington 1994;

http://www/viceroy/eeb/uconn/.edu/estimates). It is recommended to use sample-based rarefaction curves rescaled as the number of individuals to best compare richness between two sites (Gotelli and Colwell 2001). Yet, because of the social nature of ants, it is advised to use presence/absence (or incidence) data rather than abundance and correspondingly the number of species occurrences rather than number of individuals should be used when graphing species accumulation (Longino et al. 2002). Statistical comparisons of richness are made possible using *MaoTao* estimates as the corresponding 95% confidence intervals both produced using analytical formulas rather than re-sampling techniques. Voucher specimens of all ants and flying pollinators were collected and are stored at the University of Michigan.

## Results

### *Pollination*

There were no significant differences between treatments (open, no-ant, or bagged) in terms of numbers of flower buds, total fruits per branch, or in final fruit set (Table 1). In contrast, fruit weights were significantly higher on open branches than on no-ant or bagged branches, but only in the high-shade site (Table 1). In the high-shade site, fruits on open branches were significantly heavier than fruits on bagged ($p < 0.001$) and no-ant branches ($p < 0.001$) whereas, in the low-shade site, fruit weights on open branches did not differ from bagged ($p = 0.228$) or no-ant ($p = 0.107$) branches. Furthermore, coffee fruits were heavier in the high-shade site than in the low-shade site ($p < 0.001$, $F = 48.6$, df = 1).

### *Pollinator diversity and activity*

Ant surveys revealed a total of 13 ant morphospecies (Table 2). We encountered 10 morphospecies in the high-shade site, and 8 in the low-shade site. Observed sample-based rarefaction curves generated with EstimateS show that we did not sample the majority of the ant community in either site (i.e. our curves did not reach asymptotes) (Figure 1a). Comparing curves and 95% confidence intervals for the high- and low-shade sites showed that although richness is higher in high-shade sites, this is likely not a significant difference. Ant activity did not significantly differ between the two sites ($t$-test, $p = 0.660$, df = 86), although mean activity was somewhat higher in the high-shade site ($2.11 \pm 0.23$ (SE)) than in the low-shade site ($1.97 \pm 0.19$). Additionally, although we did not formally collect data on ants visiting coffee flowers, we saw several ant species (including *Crematogaster* spp., *Myrmelachista* spp., *Pseudomyrmex* spp., and *Brachymyrmex* spp.) on and around coffee flowers (S. Philpott, personal observation).

*Table 1.* Mean (±SE) for numbers of flower buds and fruits, fruit set, and fruit weights in high- (Irlanda) and low-shade (Hamburgo) sites. Statistics are for two-way ANOVA using treatment (open, no-ant, or bagged) and site (high and low-shade) as main factors and block a random factor. Sample sizes were: high-shade open ($N = 15$), no-ants ($N = 15$), bagged ($N = 15$); low-shade open ($N = 14$), no-ants ($N = 13$), and bagged ($N = 13$). Superscripts indicate significant differences between treatments within a site and bold print shows significant differences.

| | | # Flower buds | # Fruits | Fruit set | Fruit weight (g) |
|---|---|---|---|---|---|
| High-shade | Bagged | 91.38 ± 13.25 | 40.92 ± 3.12 | 0.57 ± 0.096 | 1.05 ± 0.049[b] |
| | No-ants | 86.15 ± 13.14 | 46.38 ± 8.22 | 0.54 ± 0.069 | 1.03 ± 0.029[b] |
| | Open | 97.35 ± 11.41 | 45.07 ± 6.70 | 0.51 ± 0.078 | 1.78 ± 0.312[a] |
| Low-shade | Bagged | 90.60 ± 7.66 | 48.46 ± 6.61 | 0.54 ± 0.064 | 0.81 ± 0.031 |
| | No-ants | 131.86 ± 17.65 | 61.87 ± 7.05 | 0.53 ± 0.058 | 0.80 ± 0.038 |
| | Open | 83.87 ± 12.01 | 39.80 ± 4.81 | 0.50 ± 0.045 | 0.90 ± 0.024 |
| Treatment | *p* | 0.3 | 0.204 | 0.686 | **<0.001** |
| | *F* | 1.255 | 1.683 | 0.382 | 13.543 |
| | df | 2 | 2 | 2 | 2 |
| Site | *p* | 0.289 | 0.144 | 0.613 | **<0.001** |
| | *F* | 1.222 | 2.416 | 0.269 | 48.6 |
| | df | 1 | 1 | 1 | 1 |
| Block | *p* | 0.738 | 0.428 | 0.762 | 0.942 |
| | *F* | 0.693 | 1.142 | 0.661 | 0.397 |
| | df | 14 | 14 | 14 | 14 |
| Treatment × site | *p* | 0.054 | 0.313 | 0.742 | **0.035** |
| | *F* | 3.294 | 1.218 | 0.302 | 3.866 |
| | df | 2 | 24 | 2 | 1 |
| Treatment × block | *p* | 0.652 | 0.489 | 0.574 | 0.321 |
| | *F* | 0.86 | 1.015 | 0.933 | 1.208 |
| | df | 28 | 28 | 28 | 28 |
| Site × block | *p* | 0.138 | 0.121 | 0.215 | 0.329 |
| | *F* | 1.654 | 1.72 | 1.433 | 1.212 |
| | df | 13 | 13 | 13 | 13 |

We observed a total of 14 flying pollinator species visiting coffee flowers in surveys; 14 species occurred in the high-shade site and 5 were seen in the low-shade site (Table 3). Hymenoptera were the most frequent visitors accounting for 76% of total visits, with *Trigona* spp. and *Apis mellifera* making up 43.7% and 12.7% of visits, respectively. Dipterans accounted for 18.3% of visits. As for ant surveys, sample-based rarefaction curves generated with EstimateS did not reach asymptotes and also demonstrate that flying pollinator richness in the high- and low-shade sites does not differ (Figure 1b). The number of visits per plant in the high-shade site (3.39 ± 0.61) was more than double that for the low-shade site (1.25 ± 0.25) ($t = -2.272$, df=24, $p = 0.032$).

## Discussion

Surprisingly, fruit set was not higher on open or no-ant branches (both with flying pollinators) than on bagged branches. Nor were coffee fruit weights

*Table 2.* Species list and activity levels for ants on tuna baits in the high-shade (Finca Irlanda) and low-shade (Finca Hamburgo) sites. Activity levels were calculated using an index where 1 = 1 to 2 ants, 2 = 3 to 10 ants, and 3 = more than 10 ants. Total average activity levels were not significantly different between farms.

| Species | High-shade | Low-shade |
|---|---|---|
| *Azteca instabilis* | 0 | 1 |
| *Azteca* sp. 1 | 1 | 0 |
| *Brachymyrmex* sp. 1 | 7 | 4 |
| *Brachymyrmex* sp. 2 | 6 | 3 |
| *Camponotus senex textor* | 1 | 1 |
| *Crematogaster* sp. | 0 | 7 |
| Dolichoderinae sp. 1 | 0 | 1 |
| *Pheidole* sp. 1 | 4 | 0 |
| *Pheidole* sp. 2 | 4 | 0 |
| *Pheidole* sp. 3 | 1 | 0 |
| *Solenopsis geminata* | 2 | 2 |
| *Solenopsis* sp. 1 | 2 | 27 |
| *Wasmannia auropunctata* | 1 | 0 |
| Total species richness | 10 | 8 |
| Average activity level | 2.17 | 1.9 |
| Number of plants sampled | 28 | 60 |

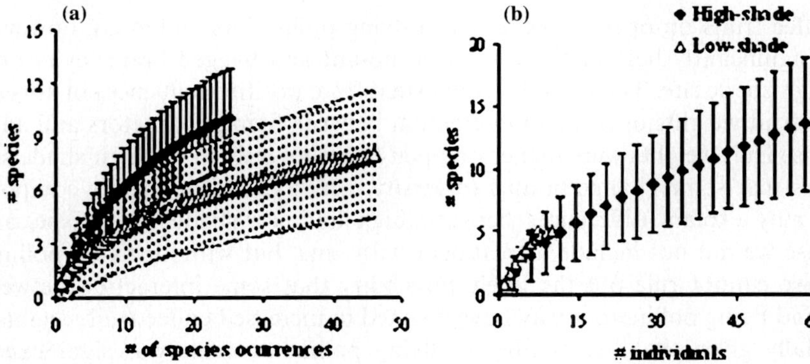


*Figure 1.* Sample-based rarefaction curves for ants (a) and flying pollinators (b) in the high- and low-shade sites. Curves were created using *Mao-Tao* estimates in EstimateS Version 7.5 using incidence data for ants and abundance data for flying pollinators. Ants are plotted by # of individual occurrences and bees against number of individuals to best compare richness between the two management types (see text for full explanation). Error bars are analytically derived 95% confidence intervals. Black diamonds represent the high-shade site and open triangles represent the low-shade site.

higher on branches only with flying pollinators (no-ant) compared to bagged branches in either site. We thus reject our first hypothesis that flying pollinators alone positively influence coffee fruit set or weight, and do not further examine the implications of flying pollinator diversity for coffee pollination thoroughly addressed in other studies (Klein et al. 2003b).

*Table 3.* Species list and activity levels for flying pollinators in the high-shade (Finca Irlanda) and low-shade (Finca Hamburgo) sites. Pollinator observations were based on 10 min observations for each of 20 and 10 coffee plants in the high- and low-shade sites, respectively.

| Order | Species | High-shade | Low-shade |
|---|---|---|---|
| Coleoptera | Coleoptera sp. 1 | 1 | |
| | Coleoptera sp. 2 | 1 | |
| Diptera | Bombyliidae sp. 1 | 1 | |
| | Diptera sp. 1 | 10 | 1 |
| | Dolichopodidae sp. 1 | 1 | |
| Hymenoptera | *Apis mellifera* | 7 | 2 |
| | *Ceratina* sp. 1 | 4 | |
| | *Ceratina* sp. 2 | 3 | 3 |
| | *Trigona* sp. 1 | 28 | 2 |
| | *Trigona* sp. 2 | 1 | |
| | Vespidae sp. 1 | 1 | |
| | Vespidae sp. 2 | 1 | 2 |
| Lepidoptera | Lepidoptera sp. 1 | 1 | |
| Odonata | Odonata sp. 1 | 1 | |
| Total species richness | | 14 | 5 |
| Total number of visits | | 60 | 10 |
| Number of plants observed | | 20 | 10 |

Coffee fruits on open branches (with flying pollinators and ants), however, were significantly heavier than fruits on no-ant and bagged branches in only the high-shade site. These results could stem from positive influences of ants on coffee fruit weights or from an interaction between flying pollinators and ants. Because fruit weights were higher on open branches only in the high-shade site implies that some change in ants (diversity, activity, and/or species composition) may explain differences between high and low-shade sites. Especially because we did not include a treatment with ants, but without flying pollinators, we cannot rule out the likely possibility that some interaction between ants and flying pollinators may have resulted in increased coffee fruit weights – especially given that the number of flying pollinator visits was significantly higher in the high-shade site. Using our current data, we cannot determine which mechanisms are responsible for increased fruit weight, but suggest several potentially testable mechanisms either resulting from (1) indirect effects of ants via their interactions with flying pollinators and/or (2) direct effects of ants on pollination or fruit maturation.

### *Indirect effects of ants*

Given that ants are not often direct pollinators, it is more likely that coffee fruit weights increased as a result of an interaction between ants and flying pollinators. Most perplexing, perhaps, is determining how ants or ant-flying pollinator interactions may result in increased fruit weight without influencing fruit set. In a self-compatible plant such as arabica coffee, fruit set is often not reduced without pollinators (Free 1993). Pollen load, however, may affect size and numbers of seeds or fruits (Winsor et al. 1987; Quesada et al. 1993) or probability of seed and fruit abortion (Bawa and Webb 1984; Stephenson and Winsor 1986; Casper 1988; Lee 1988; Nakamura 1988; Niesenbaum and Casper 1994; Niesenbaum 1999) even in self-fertile species (Morandin et al. 2001). High pollen loads can also lead to faster pollen tube growth, earlier fertilization, and thus a longer maturation period (Niesenbaum 1999). Pollen deposition on coffee flowers is enhanced by pollinators (Ricketts 2004) and such increased pollen loads may influence fruit weights. Given the short time coffee flowers are open (up to 5 days) and relatively long maturation time of coffee fruits (up to 7 months), this explanation seems unlikely. Larger pollen loads may also provide plants with higher donor diversity and thus genetic diversity of pollen. Thus, one possibility in coffee, as in other plants, is that increased pollen diversity leads to more pollen competition, increased pollen vigor and subsequent increases in fruit weight and overall quality (Bjorkman 1995; Paschke et al. 2002). Ants may be aggressive towards flying pollinators (A. Klein, personal communication), and may increase relocation frequency of pollinators thereby increasing pollen transfer, pollen load, and number of pollen donors (Altshuler 1999). Thus, ant aggression may increase movement of flying pollinators increasing pollen diversity, and perhaps increasing fruit weights.

### *Direct effects of ants*

Ants may also increase coffee fruit weights either by directly depositing pollen on coffee stigmas or by influencing some aspect of fruit maturation. The coffee berry borer (*Hypothenemus hampei* Ferrari) attacks coffee fruits significantly reducing the weight of coffee beans (Damon 2000). We did not examine harvested fruits for berry borer attack, but ants prey on the berry borer (Velez et al. 2000, 2001). Thus, higher rates of berry borer attack on no-ant branches could account for lower fruit weights. Also, fruits on no-ant and bagged branches were nearly two times lighter than fruits on open branches in the high-shade site hinting to the pea-berry condition whereby only one of two coffee ovaries mature to a seed. Raw and Free (1977) and Klein et al. (2003b) reported pea-berry incidences of near 20% and 0.92%, respectively. We did not count number of seeds per fruit to verify the number of pea-berries, but based on previous numbers, this mechanism is an unlikely explanation for our results of lighter fruits in approximately 5/6 of fruits weighed.

### *Differences between management systems*

Average fruit weights were higher in the high-shade farm potentially resulting from many factors. Muschler (2001) found that under high-shade conditions, coffee fruits (and beans) were significantly heavier (and thus of better quality) than when grown in full sun. It is possible that smaller differences in shade, like those between our high-shade and low-shade sites, may influence fruit weight as well. Fruit weight effects may be due to pollination, but there are potentially many differences between high-shade and low-shade sites, including edaphic factors or nutrient availability, that may influence fruit weights. In fact, availability of some nutrients is higher in our high-shade site (K. Avilés-Vázquez, unpublished data). Fruit weights were significantly higher overall in the high-shade site, potentially pointing to limitation in nutrients necessary for fruit maturation in the low-shade site. These site differences may account for higher fruit weights in the high-shade farm, and may potentially have also affected increases in fruit weight specifically on the open branches compared with bagged or no-ant branches.

Biological differences in the pollinator community between sites, however, may also have played a role in increasing fruit weights. Ant species richness was slightly higher the high-shade site, but for data presented here was not significantly higher. Previous and more extensive ant sampling in the same sites, however, has revealed significantly higher ant richness in the high-shade site (Ibarra-Núñez et al. 1995; Perfecto and Vandermeer 2002; Armbrecht and Perfecto 2003). Ant activity was not significantly different between the two sites. Here, we observed nearly three times as many species in the high-shade site, and numbers of flying pollinator visits were twice as high in the high-shade site. Additionally, Ibarra-Núñez et al. (1995), working in the same sites, sampled coffee plants on a monthly basis for a 3 year period and collected twice as many ant individuals in the high-shade site (12,843) than in the low-shade site (6097). Furthermore, they collected 19 bee species in the high-shade site and 7 bee species in the low-shade site. Thus, there are potentially many differences in the pollinator communities (i.e. higher ant richness, presence of a particular ant species, increased visitation rates by flying pollinators) or particular interaction between ants and bees only in the high-shade site may account for fruit weight differences between the two sites. Changes in ant species composition may be especially important if ants function to directly increase pollen loads or cause flying pollinators to relocate more frequently, or prey on coffee berry borers.

### *Suggestions for future work*

Several tests could be carried out to begin to determine if ant-pollinator interactions or ants are responsible for increasing fruit weights via pollination or influencing aspects of maturation. To study if site differences (i.e. nutrient limitation or edaphic factors) or differences caused by pollination (i.e. pollen load, pollen diversity) are responsible for differences in fruit weights, hand pollination experiments could be conducted. If, for example, a lack of nutrients or water in low-shade sites is limiting fruit weights, pollen addition to flowers would not be expected to influence fruit weights in these sights, but would in high-shade sites where nutrients and water are not limiting. Furthermore, by using multiple pollen addition treatments (increase in self pollen, increase in outcross from only one plant, increase in outcross pollen from several plants), factors relating to pollen load and pollen diversity and their respective influences on fruit set and/or fruit weights could be elucidated. To determine if, ants may be responsible for increasing activity of flying pollinators, observations of visitation rates (in terms of both number and length of visits) could be compared for branches, or farm areas, with and without high ant activity. To distinguish if ants affect pollination or aspects of fruit maturation Tanglefoot® application could be applied only during flowering (pollination effect) and only after flowering until the harvest (maturation effect). In order to determine if ants affect fruit weights limiting berry borers, the number of attacked fruits, and individual weights of attacked berries could be quantified for branches with and without ants. Furthermore, to ensure that pea-berries do not account for differences, berries could either be checked for deformities or opened to check for presence of two seeds. Those differences potentially caused by differences in the pollinator community between the two sites are clearly more difficult to test. Nonetheless, by first clarifying some of the other potential mechanisms would likely shed light on the role of ants and flying pollinators under the two types of shade management system.

## Conclusion

In conclusion, we observed increased fruit weights in the high-shade site likely resulting from differences associated with management, including differences in the pollinator community. Higher fruit weights provide advantages to plants in terms of germination or seedling growth (Ngulube et al. 1997; Eriksson 1999) or in other aspects of plant reproductive biology (Tremayne and Richards 2000; Mukasa and Ogata 2001; Schippers et al. 2001). Furthermore, increased fruit weights confer obvious economic advantages to coffee farmers. Regardless of which mechanisms influence fruit weight on branches only with flying pollinators and ants in high-shade farms, this finding will have important implications for the maintenance of shaded coffee farms and biodiversity in general (Moguel and Toledo 1999).

## Acknowledgements

F. Hernández Gómez, L.A. Maldonado Alvarado, B.E. Chilel, F. Camposeco Silvestre, J. Cabrera Santos, G. López Bautista, J.A. García Ballinas, and A. González González provided field assistance. P. Bichier, A. Klein, C. Landry, I. Perfecto, B. Rathcke, M. Reiskind, S. Schueller, and J. Vandermeer provided thoughtful discussions and comments on the manuscript. G. Ibarra-Núñez and

El Colegio de la Frontera Sur in Tapachula provided logistical support. We thank the Peters and Edelmann families for permission to work on their farms. This study was partially funded by NSF grant #DEB-9981526, the Helen Olsen Brower Fellowship in Environmental Science of the University of Michigan, and an NSF Graduate Research Fellowship.

## References


Aizen M. and Feinsinger P. 1994. Forest fragmentation, pollination, and plant reproduction in a Chaco dry forest, Argentina. Ecology 75: 330–351.

Allen-Wardell G., Bernhardt P., Bitner R., Burquez A., Buchmann S., Cane J., Cox P.A., Dalton V., Feinsinger P., Ingram M., Inouye D., Jones C.E., Kennedy K., Kevan P., Koopowitz H., Medellin R., Medellin-Morales S., Nabhan G.P., Pavlik B., Tepedino V., Torchio P. and Walker S. 1998. The potential consequences of pollinator declines on the conservation of biodiversity and stability of food crop yields. Conserv. Biol. 12: 8–17.

Altshuler D.L. 1999. Novel interactions of non-pollinating ants with pollinators and fruit consumers in a tropical forest. Oecologia 119: 600–606.

Armbrecht I. and Perfecto I. 2003. Litter-dwelling ant species richness and predation potential within a forest fragment and neighboring coffee plantations of contrasting habitat quality in Mexico. Agric., Ecosyst. Environ. 97: 107–115.

Balvanera P., Daily G.C., Ehrlich P.R., Ricketts T.H., Bailey S.A., Kark S., Kremen C. and Pereira H. 2001. Conserving biodiversity and ecosystem services. Science 291: 2047–2047.

Bawa K. and Webb C. 1984. Flower, fruit and seed abortion in tropical trees: implications for the evolution of paternal and maternal reproductive patterns. Am. J. Bot. 71: 736–751.

Beattie A.J., Turnbull C., Knox R.B. and Williams E.G. 1984. Ant inhibition of pollen function – a possible reason why ant pollination is rare. Am. J. Bot. 71: 421–426.

Bjorkman T. 1995. The effect of pollen load and pollen grain competition on fertilization success and progeny performance in *Fagopyrum esculentum*. Euphytica 83: 47–52.

Calvo L. and Blake J. 1998. Bird diversity and abundance on two different shade coffee plantations in Guatemala. Bird Conserv. Int. 8: 297–308.

Casper B. 1988. Evidence for selective embryo abortion in *Crypantha flava*. Am. Nat. 132: 318–326.

Colwell R.K. and Coddington J. 1994. Estimating terrestrial biodiversity through extrapolation. In: Hawksworth D. (ed.), Bio-diversity: Measurement and Estimation. London.

Colwell R.K., Mao C.X. and Chang J. 2004. Interpolating, extrapolating, and comparing incidence-based species accumulation curves. Ecology 85: 2717–2727.

Cunningham S.A., FitzGibbon F. and Heard T.A. 2002. The future of pollinators for Australian agriculture. Aust. J. Agric. Res. 53: 893–900.

Damon A. 2000. A review of the biology and control of the coffee berry borer, *Hypothenemus hampei* (Coleoptera: Scolytidae). Bull. Entomol. Res. 90: 453–465.

Eriksson O. 1999. Seed size variation and its effect on germination and seedling performance in the clonal herb *Convallaria majalis*. Acta Oecol. Int. J. Ecol. 20: 61–66.

Free J.B. 1993. Insect Pollination of Crops. Academic Press, London.

Galen C. 1999. Flowers and enemies: predation by nectar-thieving ants in relation to variation in floral form of an alpine wildflower, *Polemonium viscosum*. Oikos 85: 426–434.

Galen C. and Cuba J. 2001. Down the tube: Pollinators, predators, and the evolution of flower shape in the alpine skypilot, *Polemonium viscosum*. Evolution 55: 1963–1971.

Garcia M.B., Antor R.J. and Espadaler X. 1995. Ant pollination of the palaeoendemic dioecious *Borderea pyrenaica* (Dioscoreaceae). Plant Systematics and Evolution 198: 17–27.

Ghazoul J. 2001. Can floral repellents pre-empt potential ant-plant conflicts? Ecol. Lett. 4: 295–299.

Gomez J.M. 2000. Effectiveness of ants as pollinators of *Lobularia maritima*: effects on main sequential fitness components of the host plant. Oecologia 122: 90–97.
Gomez J.M. and Zamora R. 1992. Pollination by ants – consequences of the quantitative effects on a mutualistic system. Oecologia 91: 410–418.
Gomez J.M., Zamora R., Hodar J.A. and García D. 1996. Experimental study of pollination by ants in Mediterranean high mountain and arid habitats. Oecologia 105: 236–242.
Gotelli N. and Colwell R.K. 2001. Quantifying biodiversity: procedures and pitfalls in the measurement and comparison of species richness. Ecol. Lett. 4: 379–391.
Guerrant E.O. and Fiedler P.L. 1981. Flower defenses against nectar-pilferage by ants. Biotropica 13: 25–33.
Ibarra-Núñez G., García J. and Moreno M. 1995. Diferencias entre un cafetal orgánico y uno convencional en cuanto a diversidad y abundancia de dos grupos de insectos. In: Ibarra-Núñez G., García J. and Moreno M. (eds), Primera conferencia internacional IFOAM sobre café orgánico. México, pp. 115–129.
Jiménez-Castano W. and Castillo-Zapata J. 1976. Observaciones sobre la polinización de *Coffea arabica* L. en la zona cafetera central de Colombia. Cenicafe 27: 51–66.
Kearns C.A., Inouye D.W. and Waser N.M. 1998. Endangered mutualisms: The conservation of plant-pollinator interactions. Ann. Rev. Ecol. Syst. 29: 83–112.
Klein A.M., Steffan-Dewenter I., Buchori D. and Tscharntke T. 2002. Effects of land-use intensity in tropical agroforestry systems on coffee flower-visiting and trap-nesting bees and wasps. Conserv. Biol. 16: 1003–1014.
Klein A.M., Steffan-Dewenter I. and Tscharntke T. 2003a. Bee pollination and fruit set of *Coffea arabica* and *C. canephora* (Rubiaceae). Am. J. Bot. 90: 153–157.
Klein A.M., Steffan-Dewenter I. and Tscharntke T. 2003b. Fruit set of highland coffee increases with the diversity of pollinating bees. Proc. Roy. Soc. Lond., B 270: 955–961.
Kremen C., Williams N.M. and Thorp R.W. 2002. Crop pollination from native bees at risk from agricultural intensification. Proc. Natl. Acad. Sci. USA 99: 16812–16816.
Lee T. 1988. Patterns of fruit and seed production. In: Doust J. and Doust L. (eds), Plant Reproductive Ecology: Patterns and Strategies. New York, New York, pp. 79–202.
Longino J.T., Coddington J. and Colwell R.K. 2002. The ant fauna of a tropical rain forest: estimating species richness three different ways. Ecology 83: 689–702.
Manrique A.J. and Thimann R.E. 2002. Coffee (*Coffea arabica*) pollination with Africanized honeybees in Venezuela. Interciencia 27: 414–416.
Mas A.H. and Dietsch T.V. 2003. An index of management4 intensity for coffee agroecosystems to evaluate butterfly species richness. Ecol. Appl. 13: 1491–1501.
Moguel P. and Toledo V.M. 1999. Biodiversity conservation in traditional coffee systems of Mexico. Conserv. Biol. 13: 11–21.
Morandin L.A., Laverty T.M. and Kevan P.G. 2001. Effect of bumble bee (Hymenoptera: Apidae) pollination intensity on the quality of greenhouse tomatoes. J. Econ. Entomol. 94: 172–179.
Mukasa Y. and Ogata N. 2001. Correlation of early growth with average germination period and true seed weight in sugar beet under direct sowing cultivation. Jpn. J. Crop Sci. 70: 510–514.
Muschler R.G. 2001. Shade improves coffee quality in a sub-optimal coffee-zone of Costa Rica. Agroforest. Syst. 51: 131–139.
Myers N. 1996. Environmental services of biodiversity. Proc. Natl. Acad. Sci. USA 93: 2764–2769.
Nakamura R. 1988. Seed abortion and seed size variation within fruits of *Phaseolus vulgaris*: pollen donor and resource variation effects. Am. J. Bot. 75: 1003–1010.
Ngulube M.R., Hall J.B. and Maghembe J.A. 1997. Fruit, seed and seedling variation in *Uapaca kirkiana* from natural populations in Malawi. For. Ecol. Manage. 98: 209–219.
Niesenbaum R. 1999. The effects of pollen load size and donor diversity on pollen performance, selective abortion, and progeny vigor in *Mirablils jalapa*. Am. J. Bot. 86: 261–268.
Niesenbaum R. and Casper B. 1994. Pollen tube numbers and selective fruit maturation in *Lindera benzoin*. Am. Nat. 144: 184–191.
Nogueira-Neto P., Carvalho A. and Antunes H. 1959. Efeito da exclusa ˜o dos insetos polinizadores na producç a˜ o do cafe' Bourbon. Bragantia 18: 441–468.

Norberg J. 1999. Linking Nature's services to ecosystems: some general ecological concepts. Ecol. Econ. 29: 183–202.

Oliveira P.S. 1997. The ecological function of extrafloral nectaries: Herbivore deterrence by visiting ants and reproductive output in *Caryocar brasiliense* (Caryocaraceae). Funct. Ecol. 11: 323–330.

Oliveira P.S., Rico-Gray V., Díaz-Castelazo C. and Castillo-Guevara C. 1999. Interaction between ants, extrafloral nectaries and insect herbivores in Neotropical coastal sand dunes: herbivore deterrence by visiting ants increases fruit set in *Opuntia stricta* (Cactaceae). Funct. Ecol. 13: 623–631.

Paschke M., Abs C. and Schmid B. 2002. Effects of population size and pollen diversity on reproductive success and offspring size in the narrow endemic *Cochlearia bavarica* (Brassicaceae). Am. J. Bot. 89: 1250–1259.

Perfecto I. and Snelling R. 1995. Biodiversity and the transformation of a tropical agroecosystem – ants in coffee plantations. Ecol. Appl. 5: 1084–1097.

Perfecto I. and Vandermeer J. 1996. Microclimatic changes and the indirect loss of ant diversity in a tropical agroecosystem. Oecologia 108: 577–582.

Perfecto I. and Vandermeer J. 2002. Quality of agroecological matrix in a tropical montane landscape: ants in coffee plantations in southern Mexico. Conserv. Biol. 16: 174–182.

Philpott S.M. 2004. The Diversity, Ecology, and Function of Arboreal Ants in Coffee Agroecosystems in Soconusco, Chiapas, Mexico. Department of Ecology and Evolutionary Biology, University of Michigan, Ann Arbor.

Philpott S.M., Perfecto I. and Vandermeer J. Effects of management intensity and season on arboreal ant diversity and abundance in coffee agroecosystems. Biodivers. Conserv.. In press.

Pimentel D., Wilson C., McCullum C., Huang R., Dwen P., Flack J., Tran Q., Saltman T. and Cliff B. 1997. Economic and environmental benefits of biodiversity. Bioscience 47: 747–757.

Quesada M., Winsor J.A. and Stephenson A.G. 1993. Effects of pollen competition on progeny performance in a heterozygous cucurbit. Am. Nat. 142: 694–706.

Ramsey M. 1995. Ant pollination of the perennial herb *Blandfordia grandiflora* (Liliaceae). Oikos 74: 265–272.

Rathcke B. and Jules E. 1993. Habitat fragmentation and plant–pollinator interactions. Curr. Sci. 65: 273–277.

Raw A. and Free J.B. 1977. Pollination of coffee (*Coffea arabica*) by honeybees. Trop. Agric. 54: 365–370.

Ricketts T.H. 2004. Tropical forest fragments enhance pollinator activity in nearby coffee crops. Conserv. Biol. 18: 1262–1271.

Ricketts T.H., Daily G.C., Ehrlich P.R. and Michener C.D. 2004. Economic value of tropical forest to coffee production. Proc. Natl. Acad. Sci. USA 101: 12579–12582.

Roubik D.W. 2002. Tropical agriculture – the value of bees to the coffee harvest. Nature 417: 708–708.

Schippers P., van Groenendael J.M., Vleeshouwers L.M. and Hunt R. 2001. Herbaceous plant strategies in disturbed habitats. Oikos 95: 198–210.

Sein F.Jr. 1959. [Do bees help coffee?] Hacienda 55: 36–50.

Sporleder M. and Rapp G. 1998. The effect of *Oecophylla longinoda* (Latr.) (Hym., Formicidae) on coconut palm productivity with respect to *Pseudotheraptus wayi* Brown (Hem., Coreidae) damage in Zanzibar. J. Appl. Entomol. – Zeitschrift Fur Angewandte Entomologie 122: 475–481.

Steffan-Dewenter I. and Tscharntke T. 1999. Effects of habitat isolation on pollinator communities and seed set. Oecologia 121: 432–440.

Stephenson A.G. and Winsor J.A. 1986. *Lotus corniculatus* regulates offspring quality through selective fruit abortion. Evolution 40: 453–458.

Tremayne M.A. and Richards A.J. 2000. Seed weight and seed number affect subsequent fitness in outcrossing and selfing *Primula* species. New Phytol. 148: 127–142.

Velez M., Bustillo A.E. and Posada F.J. 2000. Predacion sobre *Hypothenemus hampei*, (Ferrari) de las hormigas *Solenopsis* spp., *Pheidole* spp., y *Dorymyrmex* spp. durante el secado del café. In:

Vélez M., Bustillo A.E. and Posada F.J. (eds), Resúmenes XXVII Congreso. Sociedad Colombiana de Entomología, Medellín, Colombia, pp. 17.

Vélez M., Bustillo A.E. and Posada F.J. 2001. Hormigas de la zona central cafetera y perspectivas de su uso en el control de *Hypothenemus hampei* (Ferrari) (Coleoptera: Scolytidae). In: Vélez M., Bustillo A.E. and Posada F.J. (eds), Resúmenes XXVIII Congreso. Sociedad Colombiana de Entomología, Pereira, Colombia, pp. 51.

Wagner D. 2000. Pollen viability reduction as a potential cost of ant association for *Acacia constricta* (Fabaceae). Am. J. Bot. 87: 711–715.

Wagner D. and Kay A. 2002. Do extrafloral nectaries distract ants from visiting flowers? An experimental test of an overlooked hypothesis Evol. Ecol. Res. 4: 293–305.

Willmer P.G. and Stone G.N. 1997. How aggressive ant-guards assist seed-set in *Acacia* flowers. Nature 388: 165–167.

Winsor J.A., Davis L.E. and Stephenson A.G. 1987. The relationship between pollen load and fruit maturation and the effect of pollen load on offspring vigor in *Cucurbita pepo*. Am. Nat. 129: 643–656.

Yano S. 1994. Flower nectar of an autogamous perennial *Rorippa indica* as an indirect defense-mechanism against herbivorous insects. Res. Popul. Ecol. 36: 63–71.

Biodiversity and Conservation (2006) 15:489–502
DOI 10.1007/s10531-005-1132-6 

# Sampling to assess a re-established Appalachian forest in Ohio based on gelechioid moths (Lepidoptera: Gelechioidea)

SIBYL R. BUCHELI[1,*], DAVID J. HORN[2] and JOHN W. WENZEL[1]
*[1]Department of Entomology, The Ohio State University, 1315 Kinnear Road, Columbus, OH 43212-1157, USA; [2]Department of Entomology, The Ohio State University, 400 Aronoff Laboratories, 318 West 12th Avenue, Columbus, OH 43210-1242, USA; *Author for correspondence (e-mail: bucheli.1@osu.edu; phone: +1-614-292-9794; fax: +1-614-292-7774)*

Received 20 April 2004; accepted in revised form 18 July 2005

**Key words:** Appalachian forest, Biological surveys, Deforestation, Gelechioidea, Microlepidoptera, Ohio

**Abstract.** A list of microlepidoptera belonging to the superfamily Gelechioidea was produced from June trap samples from sites within the Wayne National Forest (Lawrence County), an Appalachian forest in Southern Ohio that was once a greatly disturbed area and has since re-established over a period of nearly 100 years. The composition and diversity of Lawrence county is compared to lists of gelechioid moths generated by other surveys in the eastern United States: the Great Smoky Mountains National Park, an Ohio study (Summerville and Crist 2003), and unpublished data from Connecticut (Wagner). From comparison with these studies, we address two questions: (1) How well do passive surveys of Gelechioidea compare to more labor intensive surveys? (2) How does the regenerated Wayne National Forest compare to other well documented areas with respect to gelechioid diversity? Our sample of diversity, though more narrow in time and area, compares favorably to more exhaustive sampling and demonstrates that it may be more efficient to focus on target groups in focal localities when time and resources are limited rather than conduct extensive sampling programs.

**Abbreviations:** ATBI – All Taxon Bioinventory; GSMNP – Great Smoky Mountains National Park; LAWCO – Lawrence County

## Introduction

Biological surveys have become increasingly popular. Discover Life in America's All Taxon Bio-Inventory (ATBI) and similar efforts are popular with both the scientific community and the public for a number of reasons. They aim to identify all species of organisms inhabiting specific areas, document the locations of organisms, record life histories and critical habitats for recorded organisms and make available a database for professionals and the public (Sharkey 2001). One universally understood problem with biological surveys is that it is difficult to know when, if ever, all potential diversity in an area has been sampled (Magurran 1988; Muona 1999; Sørenson et al. 2002; Scharff et al. 2003). Preston's (1948) classical example showed for moths at light traps that although species with intermediate numbers of individuals are most prevalent in communities, a few abundant species account for most of the records while many rare species account for few of the records. Thus, it is difficult to know when to stop sampling because many rare species may remain unmeasured. This problem can be severe, even under favorable circumstances, and an asymptotic approach to total diversity appears to be the best we can expect. Muona (1999) investigated whether sampling total beetle diversity was possible given various trapping methods including but not limited to car-net, handpicking, pitfall traps, and flight-intercept traps. He demonstrated that a variety of trapping methods was needed because the actual number of species in an area is dynamic and effectiveness of traps depends on the species (biology) of the beetle being trapped. He showed that even massive trapping schemes were not effective in recovering all species and in particular rare species that live in isolated and patchy habitats. Muona's (1999) results have the following implications: (a) to trap rare species, optimal search strategies should be devised; (b) trapping results cannot be used as proof of absence of species; and (c) because of large numbers of rare species, it is unlikely that trapping saturation curves ever reach the true maximum.

Sørenson et al. (2002) focused on the recovery of spider diversity using plotless and plot-based sampling methods. Their study showed that even intensive sampling was insufficient for recovery of the entire spider fauna in a given area and that, counter-intuitively, enlarging the sample area decreased, rather than increased, the number of rare species collected. In the case of plot-based sampling, Sørenson et al. (2002) recommended that long-term monitoring should focus on a single or few species and use standardized methods that are absolute and practical within standardized plots to provide a baseline for surveys.

Considering the above studies, one may conclude that due to the presence of many rare species in a given area at a given time, one can never anticipate recovering the total expected diversity using standard, cost efficient trapping efforts done by typical survey teams (Preston 1948; Muona 1999; Sørenson et al. 2002). Such a conclusion makes clear the importance of comparative survey data when we consider estimates of total biodiversity. But, comparing across surveys introduces complications of its own, such as unequal effort, unequal methods, and unequal landscape effects of community composition. To date, one of the few practical recommendations addressing these problems is to concentrate on target groups in focal localities rather than sampling broadly and extensively (Sørenson et al. 2002) when exhaustive, ATBI-scale efforts are not possible.

Lepidoptera have been demonstrated to be important indicators of habitat structure and community health for several diverse forest ecosystems, from Brazil (Brown and Freitas 2000) to Canada (Kerr et al. 2001), and Australia (Kitching et al. 2000). Eastern deciduous forests of North America support a high diversity of trees (Greller 1988), but today are fragmented due to destructive historical land use. Remaining or regenerated forests with a higher diversity of host-plants, older plant communities, or those with less disturbance should be expected to support a more diverse community of Lepidoptera (Usher and Keiller 1998; Summerville et al. 2003a, b; Summerville and Crist 2003). Following the recommendation of Sørenson et al. (2002), perhaps labor-efficient sampling focused on target groups can be used as a means to survey potential diversity, and also to provide standards of comparison with other surveys. Good candidates would be moths of the superfamily Gelechioidea, many of which come readily to light traps and are often closely associated with individual species of plants or narrow forest niches.

For practical, political, and management reasons, researchers measure diversity in local parks, regenerated natural areas, or regions undergoing succession. As such, these areas themselves represent samples of a greater, historical diversity that may not have been measured on site. Thus, initial estimates of total expected diversity must be derived from surveys of different areas. In this study, we survey gelechioid moths in the Wayne National Forest (Lawrence County, Ohio, USA), representing a large Appalachian area of forest regenerated in less than 100 years from landscape completely denuded for industrial charcoal. We compare results from our traps to three different kinds of more exhaustive inventories. The Great Smoky Mountain National Park ATBI (Wagner and Scholtens 2002) is a labor intensive effort representing a snapshot of Appalachian diversity for a 24 h period in June, as recorded by many expert lepidopterists. Summerville and Crist's (2003) study surveyed moths across different regions of Ohio throughout a summer season, and was based on passive traps only, but covered much more time than the GSMNP-ATBI. Finally, the most exhaustive inventory is that of state of Connecticut (DL Wagner, unpubl.). From comparison with these studies, we address two questions: (1) How well do passive surveys of Gelechioidea compare to more labor intensive surveys? (2) How does the regenerated Wayne National Forest compare to other well-documented areas with respect to gelechioid diversity?

## Materials and methods

### *Study sites*

Lawrence County, Ohio ("LAWCO") is the southernmost county of Ohio, located at the nexus of Ohio, Kentucky and West Virginia, and is Appalachian in its flora and fauna. Virgin Appalachian forests were logged extensively by early settlers to make charcoal to run furnaces to smelt iron ore. Deforestation was rapid and only the most resilient species of trees such as oak and hickory could withstand the repeated fires associated with the coal and iron industries. By 1920, virtually no primeval forest remained uncut in Ohio. In September, 1951, the Wayne National Forest was established, and today it forms the most heavily forested part of the state.

*Table 1.* Comparison of collection methods between LAWCO, Ohio, and the Great Smoky Mountains National Park Lepidopteran All Taxon Bio Inventory sponsored by Discover Life In America (GSMNP ATBI, data extrapolated from Wagner and Scholtens 2002). The LAWCO study represents a passive trapping scheme while the GSMNP ATBI represents a more labor-intensive trapping scheme. Both LAWCO and GSMNP are based on presence data, and frequency data were not collected.

| LAWCO study – Passive trapping scheme | GSMNP ATBI – Labor intensive trapping scheme |
|---|---|
| • June trap data for 1995, 1996, and 1997. | • 24 h trapping period starting June 9 2002 at 3:00 pm, 24 h identification session following. |
| • 6 trapping sites all within The Wayne National Forest. | • 30 trap sites placed throughout park. |
| • 1–2 collectors to retrieve traps. | • 30 lepidopterists, 24 volunteers and two llama teams. |
| • Plot-based sampling. | • Plot-less and plot-based sampling methods. |
| • Blacklight buckets. | • Blacklight buckets, blacklight sheets, mercury vapor traps, bait traps, pheromone traps, hand searching and picking, and net-collecting. |
| • 1 graduate student identifier. | • 1 professional identifier; 2 graduate student assistants. |
| • Identified over 3 months. | • Identified for the majority on site within 24 h and continued over a period of time. |

Other studies: Great Smoky Mountains National Park (GSMNP), on the border of Tennessee and North Carolina, is nearly 80% deciduous forest, some of it pristine, the remainder of the area being conifers and heath, with altitudes to about 2000 m. Thirty lepidopterists and 24 volunteers performed intense and directed sampling, both plot-based and plotless, including approximately 30 trap sites, for 24 h starting at 15:00 h on June 9, 2002. A detailed comparison of trapping methods with respect to Gelechioidea between the LAWCO study and the GSMNP Lepidopteran ATBI, is presented in Table 1. Summerville and Crist (2003) studied nine locations in Ohio (unglaciated and glaciated) from May to September, 1999 with passive blacklight traps. Sites were chosen to span forest biomes and prairies. Wagner's inventory (DL Wagner, unpubl.) for Connecticut includes samples of all life stages in most available habitats over 5 years, plus extensive museum specimens. This represents a cumulative inventory of the entire state.

### *LAWCO collection*

The diversity of nocturnal Gelechioidea was estimated by counting number of species present in the Wayne National Forest, LAWCO, Ohio, as part of an ongoing study monitoring selected insect populations (Sutherland and Hutchinson 2003). Six blacklight traps were placed in two locations for three years (1995, 1996, and 1997): Aid Township 38°36′ N, 82°31′ W (Sharps Creek/Bluegrass A, B, and C), and Decatur Township 38°43′ N, 82°41′ W (Young's Branch A, B, and C). To standardize comparison, only trapped Gelechioidea data from June for the LAWCO study are compared to the GSMNP Lepidopteran ATBI trap data.

### *Identification*

Gelechioidea were identified using genital and wing characters. Male and female abdomens were prepared for identification using a standard 10% potassium hydroxide solution. In some cases, structures were stained with Mercurochrome. Preparations were mounted in euparol on slides. Male structures were prepared using various techniques specifically aimed to preserve particular taxonomic features; females were left undissected and mounted whole. Moths were identified using the following: Forbes 1923; Hodges 1974, 1978, 1983, 1985, 1986, 1998; Covell 1984; Adamski and Brown 1989; Adamski and Hodges 1996; Landry 1998 as well as museum study and unpublished work (J.-F. Landry on *Coleophora*).

### *Cumulative totals for LAWCO Gelechioidea*

Actual and projected species accumulation curves of LAWCO Gelechioidea for June were generated. The actual accumulation was estimated by plotting total new species added per trap year and the potential diversity curve was produced by repeating the percent new species added until the asymptote. Frequency data were not collected.

## Results

### *Gelechioidea diversity of LAWCO*

Fifty-five species of Gelechioidea were collected at the LAWCO sites in June of 1995, 1996, and 1997. Ten species, or 19%, appear to be undescribed. For a summary of genera and species recovered per family, see Table 2 and Appendix 1. For LAWCO, most of the species diversity is within the families Coleophoridae and Gelechiidae. For a full species list, see Appendix 1. Of the identified and placed species collected, 25 Gelechioidea species were unique to LAWCO while two were unique to the GSMNP.

Aid and Decatur townships showed no difference in number of species recovered (Table 3). There was a disproportionate number of species recovered from one trapping location (Table 3).

*Table 2.* Comparison of Gelechioidea species collected during the LAWCO study and GSMNP Lepidoptera ATBI (data provided by Brian Scholtens). Number of genera and species recovered per study.

| | Gelechioidea LAWCO Study | Gelechioidea GSMNP ATBI |
|---|---|---|
| Number of genera per family, number of species per family | | |
| Species total | 52 | 76 |
| New county records | all | ca. 55 |
| Undescribed species | ca. 10 | ca. 34 |
| Amphisbatidae | 1,1 | 0, 0 |
| Coleophoridae | 6, 19 | 2, 2 |
| Cosmopterigidae | 2,2 | 2, 2 |
| Elachistidae | 2, 2 | 2, 2 |
| Gelechiidae | 11, 19 | 8, 12 |
| Oecophoridae | 2, 2 | 4, 4 |
| Unplaced Gelechioidea | 7 | 52 |

*Table 3.* Number of species for LAWCO study per township and site.

| Aid township | | Decatur township | |
|---|---|---|---|
| Total | 30 | Total | 22 |
| Percent total | 57% | Percent total | 42% |
| Sharps Creek/Bluegrass A | 5 | Young's branch A | 3 |
| Sharps Creek/Bluegrass B | 18 | Young's branch B | 9 |
| Sharps Creek/Bluegrass C | 7 | Young's branch C | 10 |

## Discussion

Gelechioid adults are generally short-lived and inconspicuous, but larvae are important components of most terrestrial ecosystems (Stehr 1987), are sometimes abundant, and are a major source of food for small predators and many parasites. Larvae have a great diversity of feeding habits: scavenging, gall-forming, leaf-mining, seed-mining, leaf-tying, leaf-rolling, stem-boring, flower-boring and case-making, mostly on gymnosperms and angiosperms (Powell et al. 1998). Larvae of each species tend to be specialists or oligophagous on only a few plant species, making it simple to predict what species of moths should be present in a given locality at a given time of the year if host plants are known (See also Lepš et al. 1998). As a group, Gelechioidea inhabit a great diversity of forest niches.

Ecological or conservation studies aspire to measure diversity, or perhaps underlying biological factors that produce it, but these measures are necessarily influenced by sampling protocols. Sampling protocols vary in intensity and method of catch, and all those reported here are superficial compared to thorough, long-term assessments. For example, DeVries et al. (1997) reports on 40 traps spread over 200 ha in Amazonian Ecuador, arranged as five replicate sampling sites in four habitat types, each site supporting an understory trap and a canopy trap, with traps baited for 7 days and then left without bait for 3 weeks (baited traps were cleared daily, unbaited traps were cleared weekly), run from August of 1992 to August of 1993. The entire study was essentially repeated (DeVries and Walla 2001) about 200 km away from August 1993 to August 1994. None of the sampling studies discussed here claim to represent this sort of effort.

For Gelechioidea, an inventory including all records spanning all seasons, several years, and many locations throughout Connecticut, (D. L. Wagner, personal communication) has recorded 205 species of Gelechioidea. Yet, no short-term study can achieve such a total, so the relevant comparison becomes the proportional merits of differing, imperfect sampling schemes. Our results are evaluated with reference to two different measures of what might be recovered in a sampling program in region of Eastern hardwood forest. First, we compare our results with those of intensive sampling from the GSMNP, estimated by a single 'snapshot' in June. We found similar species of Gelechioidea in both LAWCO, OH and particular sites in The GSMNP that have similar overall habitats and host-plant species. In the LAWCO study, we recovered 68% of the total Gelechioidea diversity of the GSMNP ATBI (Table 2). In the GSMNP, 21% of the diversity came from Gelechiidae and Oecophoridae. In LAWCO study, 73% of the total diversity came from Coleophoridae and Gelechiidae. Several families of Gelechioidea are greatly underrepresented in LAWCO, including Cosmopterigidae and Oecophoridae. Nearly half of the total diversity was recovered from each Ohio township (42% from Decatur and 57% from Aid, see Table 3); however, considerably more species were recovered from one site (34.6% from Sharps Creek, Bluegrass B in Aid Township).

Second, a 'total Ohio' estimate was produced through trapping methods more like those we used, but over a more extended period and across a greater area. Summerville and Crist (2003) recorded 93 species of Gelechioidea, 86% of the diversity coming from the families Gelechiidae and Oecophoridae. The LAWCO study recovered 56% of this diversity (see Table 4). Thus, our sample of diversity, though very much more narrow in time and area, produced an assessment about half of what Summerville and Crist produced in their much more exhaustive coverage.

### *Extrapolation to total species*

Assuming that our data represent the same progressive approximation of species shown by Muona (1999) and others (above), we can estimate that the actual diversity is higher than our measurements. Species accumulations show that for each year, roughly half as many additional species were added to the LAWCO total as the year before (32, 12 and 8, respectively). The projected diversity for LAWCO Gelechioidea for June reaches 100% in 4 years with a

*Table 4.* Comparison of the LAWCO study to the GSMNP ATBI (Wagner and Scholtens 2002), 'total Ohio' study (Summerville and Crist 2003) and Connecticut (D. L. Wagner, unpublished) for location, sampling methods, time of study and area of study; total and projected number of Gelechioidea and percent LAWCO diversity versus the other areas.

| Location, sampling method, time of study and area of study | Total number of Gelechioidea | Projected number of Gelechioidea | Percent LAWCO diversity versus other areas |
|---|---|---|---|
| LAWCO – Passive, June for 3 years in regenerated Appalachian forest. | 52 | 60 for the month of June | – |
| GSMNP – Exhaustive, June for 24 h in 'snap shop' Appalachian region (DLIA ATBI). | 76 | 100 for 24 h period | 68% |
| 'Total Ohio' – Passive, 1 year in all of Ohio (Summerville and Crist 2003). | 93 | — | 56% |
| Connecticut – Exhaustive, most habitats of Connecticut for 5 years (D. L. Wagner, unpubl.) | 205 | 300 total | 25% |

total of 60 species (Figure 1). Measured June diversity of LAWCO Gelechioidea is at 93% of the projected June diversity according to our simple estimates. We estimate that four more years of trapping would recover only four more species of Gelechioidea (Figure 1). Even with this modest correction, our results are surprisingly robust against the background of the enormously more extensive (and presumably diverse baseline) of the GSMNP. This

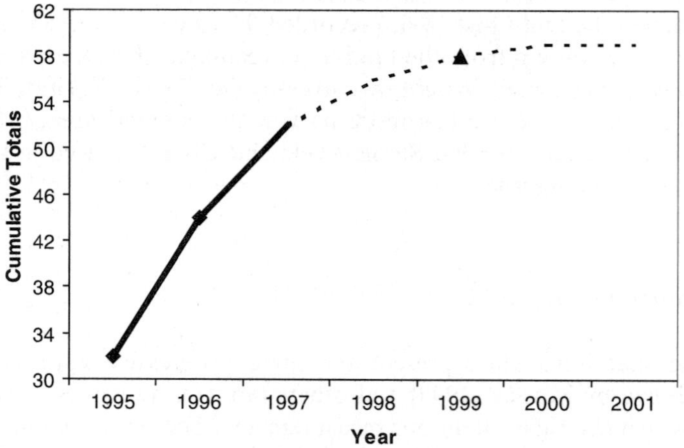


*Figure 1.* Actual and projected cumulative totals for LAWCO Gelechioidea recovered using passive, plot-based blacklight collection methods. Actual data shown with a diamond and solid line, projected data shown with a triangle and dashed line.

relationship represents exactly what Sørenson et al. (2002) predict: that it may be a more efficient use of time and resources to focus on target groups in focal localities rather than extensive sampling.

*Assessment of site quality*

Just as biologists present samples from sites of interest, regenerated habitats present samples from an historical total biodiversity. As a result, the exact problem of asymptotic approach to 'total' that we see in our site assessment protocols (above) must obtain also for the regeneration of historical biodiversity itself. Thus, the problem of assessment is two-fold because we must establish the species diversity on site (both recorded and projected) and we must estimate how this relates to a separate estimate of a regional, historical baseline, perhaps one for which there is only an approximate measure. We present here a provocative approach to this second approximation, based on our understanding of the first. Given that LAWCO itself represents only a sample of Appalachian diversity; we can assume that the regenerated forests can be plotted as points on as asymptotic approach to 100% of historical diversity. Assessment of what is 100% becomes a new challenge. Fortunately, we have several appropriate comparators.

If we take a generalized curve to represent an ideal, generated forest through time, we can imagine that LAWCO is moving along this curve toward a hypothetical value of 100% regeneration (Figure 2). If we take the exhaustive historical sampling of Connecticut as a measure of the total we could hope to find, then we see that LAWCO fares poorly at about one quarter regeneration of our index species (25%). It is important to note that Wagner was sampling many habitats at various times of the year while we are only interested in June samples. Of course, no one can actually trap total diversity in an assessment protocol (Muona 1999); so that it would be more appropriate to compare our trap data to other trap data more similar in habitat. Judging the LAWCO

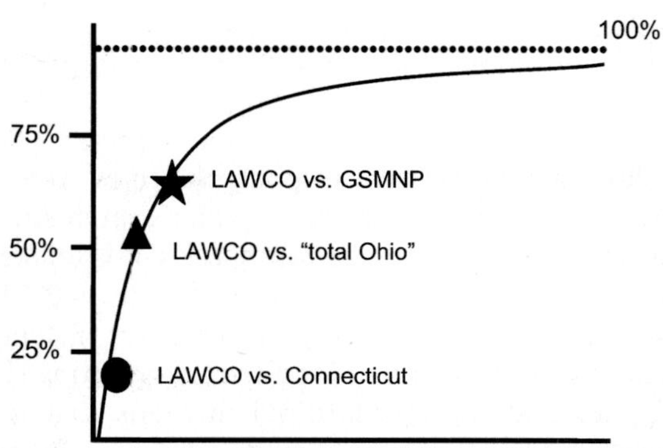


*Figure 2.* Species accumulations as a proportion of total in regenerated LAWCO, Ohio sites.

study against the 'total Ohio' survey (Summerville and Crist 2003), we see about half of expected measurable diversity present (56%). Yet, it still might be more appropriate to limit the comparison to moths caught in June in an Appalachian forest, given that this best describes the LAWCO data. Comparing with the GSMNP ATBI data (a 24-h snapshot of intense sampling), our data fare better than before, at 68% diversity. If this closet estimate of measurable diversity is used, then our low impact, low intensity samples appear to be very effective.

The areas of LAWCO that were severely deforested for the production of charcoal and allowed to re-establish under federal protection have regenerated just above 56% total Gelechioidea of the Summerville and Crist (2003) 'total Ohio' fauna in less than 100 years (Table 4). Little is known about the dispersal capabilities of most microlepidoptera except for pest species, which are usually introduced accidentally to an area and then spread from the point of introduction. Unlike many species of butterflies and larger moths, gelechioid moths are not considered robust animals. Individuals are short-lived, rather small and generally lack the capacity to fly great distances, as is the case with many microlepidoptera. Larvae are internal feeders that are unlikely or unable to leave the host-plant to travel. Potential for re-colonization is for Gelechioidea probably not as great as it is for more mobile Lepidoptera species (Holl 1996; Doak 2000; Gutierrez et al. 2001; Petit et al. 2001; Tscharntke et al. 2002; Wahlberg et al. 2002), demonstrated to re-colonize even areas that do not possess all necessary habitat requirements (Holl 1996). A possible scenario of re-establishment of Gelechioidea of LAWCO is that small patches of forest remained during a period when substantial industrial deforestation occurred (ETIB of MacArthur and Wilson 1967). Perhaps gelechioid moths persisted even in degraded forest patches and re-colonized regenerated forest. Potential for re-colonization is therefore dependent on distance between forest patches diversity and within each patch (MacArthur and Wilson 1967; Brotons et al. 2003). Small patches of forest have been shown to act as stepping-stones for species in colonization, or if incorporated into new forests, may serve as sources of species for immediate colonization (Usher and Keiller 1998).

## Conclusions

Muona (1999) showed that massive trapping schemes were not effective in recovering all species. To trap rare species, optimal search strategies should be devised but because of large numbers of rare species, it is unlikely that trapping saturation curves ever reach the true maximum. Sørenson et al. (2002) showed that even intensive sampling was insufficient to recover the entire spider fauna in a given area and recommended that long-term monitoring should focus on a single or few species and use standardized methods that are absolute and practical within standardized plots to provide a baseline for surveys. Here we demonstrate that the less intense sampling methods recommended by Muona

(1999) and Sørenson et al. (2002) are sufficient to recover the majority of Gelechioidea present in LAWCO during the month of June. Our passive survey of Gelechioidea compare favorably to more labor intensive surveys, recovering 68% of the total Gelechioidea diversity of the GSMNP ATB and 56% of the total Gelechioidea diversity of Summerville and Crist's (2003) Ohio study. With our simple projection, we demonstrate that it would take four more years of trapping to recover only four more species of Gelechioidea. Based on our findings, we conclude that less intensive sampling regimes are sufficient at recovering the majority, especially when the goals of the sampling project do not require sampling of rare species and that it may be a more efficient use of time and resources to focus on target groups in focal localities rather than extensive sampling. Our results also suggest that the regenerated forests of LAWCO, Ohio have reestablished roughly half the expected diversity of poorly dispersing species in about 100 years.

**Acknowledgements**

Many thanks to John Herbert and Foster Purrington for help with initial moth sorting, Brian Scholtens for data from the GSMNP ATBI, David Wagner for providing unpublished data for Lepidoptera of Connecticut, Jean-François Landry for providing unpublished data important for *Coleophora* identification, and Steven Passoa for assistance in identification of difficult taxa. Keith Summerville, Kurt Pickett, Hojun Song, Joe Raczowski, John Herbert, Foster Purrington, Chad Schone, Eric Dotseth and an anonymous reviewer provided helpful comments on this manuscript. This research was supported in part by funds provided by the USDA Forest Service, Northeastern Forest Experiment Station.

## Appendix

*Appendix 1.* List of species recovered from LAWCO for June.

| Family | Subfamily | Genus | Species |
|---|---|---|---|
| Amphisbatidae | Amphisbatinae | *Psilocorsis* | *reflexella* Clem. |
| Coleophoridae | Blastobasinae | *Blastobasis* | *glandulella* (Riley) |
| | | *Blastobasis* | 1 |
| | | *Blastobasis* | 2 |
| | | *Blastobasis* | 3 |
| | | *Holcocera* | 1 |
| | | *Holcocera* | 2 |
| | | *Holcocera* | 3 |
| | | *Hypatopa* | 1 |
| | | *Hypatopa* | 2 |
| | | *Pigritia* | 1 |
| | Coleophorinae | *Coleophora* | *tiliaefoliella* Clem. |
| | | *Coleophora* | *juglandella* McD. |

*Appendix 1.* Continued.

| Family | Subfamily | Genus | Species |
|---|---|---|---|
| | | *Coleophora* | *malivorella* Riley |
| | | *Coleophora* | *querciella* Clem. |
| | | *Coleophora* | *comptoniella* (McD) |
| | | *Coleophora* | 1 |
| | | *Coleophora* | 2 |
| | | *Coleophora* | 3 |
| Cosmopterigidae | Cosmopteriginae | *Cosmopteryx* | 1 |
| | | *Stagmatomorpha* | 1 |
| Elachistidae | Ethmiinae | *Ethmia* | 1 |
| | Stenomatinae | *Antaeotricha* | *schlaegeri* (Zell.) |
| Gelechiidae | | | 1 |
| | | | 2 |
| | | | 3 |
| | | | 4 |
| | | | 5 |
| | | | 6 |
| | | | 7 |
| | Dichomeridinae | 1 | |
| | | 2 | |
| | | *Dichomeris* | *georgiella* (Wlk.) |
| | | *Dichomeris* | *liguella* (Hub.) |
| | | *Dichomeris* | 1 |
| | Gelechiinae | 1 | |
| | | 2 | |
| | | 4 | |
| | | 5 | |
| | | 6 | |
| | | 7 | |
| | | *Arogalea* | 1 |
| | | *Chionodes* | *obscurella* (Cham.) |
| | | *Chionodes* | *formosella-group* |
| | | *Chionodes* | *formosella* (Murt.) |
| | | *Chionodes* | 1 |
| | | *Facista* | 1 |
| | | *Gelechia* | 1 |
| | | *Gelechia* | 2 |
| | | *Gelechia* | 3 |
| Oecophoridae | Oecophorinae | *Decantha* | 1 |
| | | *Martyringa* | *latipennis* (Wlsh.) |

## References


Adamski D. and Brown R.L. 1989. Morphology and systematics of North American Blastobasidae (Lepidoptera: Gelechioidea). Miss. Agric. Forest. Exp. Stat. Tech. Bull. (Mississippi Entomological Museum No. 1) 165: 1–70.

Adamski D. and Hodges R.W. 1996. An annotated list of North American Blastobasinae (Lepidopteran: Gelechioidea: Coleophoridae). Proc. Entomol. Soc. Wash. 98: 708–740.

Brotons L., Monkkonen M. and Martin J.L. 2003. Are fragments islands? Landscape context and density–area relationships in boreal forest birds. Am. Nat. 162: 343–357.

Brown K.S. and Freitas A.V.L. 2000. Atlantic forest butterflies: indicators for landscape conservation. Biotropica 32: 34–956.

Covell C.V. 1984. A Field Guide to the Moths of Eastern North America. Houghton Mifflin, Boston.

DeVries P.J., Murray D. and Lande R. 1997. Species diversity in vertical, horizontal, and temporal dimensions of a fruit-feeding butterfly community in an Ecuadorian rainforest. Biol. J. Linn. Soc. 62: 343–364.

De Vries P.J. and Walla T.R. 2001. Species diversity and community structure in neotropical fruit-feeding butterflies. Biol. J. Linn. Soc. 74: 1–15.

Doak P. 2000. Population consequences of restricted dispersal for an insect herbivore in a subdivided habitat. Ecology 81: 1828–1841.

Forbes W.T.M. 1923. Superfamily Gelechioidea. The Lepidoptera of New York and Neighboring States: Primitive Forms Microlepidoptera, Pyraloids Bombycoids. Cornell University, Agricultural Experiment Station, Ithaca.

Greller A.M. 1988. Deciduous forests. In: Barbou M.G. and Billings W.D. (eds), North American Terrestrial Vegetation, Cambridge University Press, New York, NY; Cambridge, UKpp. 287–316.

Gutierrez D., Leon-Cortes J.L., Menendez R., Wilson R.J., Cowley M.J.R. and Thomas C.D. 2001. Metapopulations of four lepidopteran herbivores on a single host plant, *Lotus corniculatus*. Ecology 82: 1371–1386.

Hodges R.W. 1974. Gelechioidea: Oecophoridae. In: Dominick R.B., Dominick T., Ferguson D.C., Franclemont J.G., Hodges R.W. and Munroe E.G. (eds), Moths of America North of Mexico. E. W. Classey and The Wedge Entomological Research Foundation, London.

Hodges R.W. 1978. Gelechioidea (in part): Cosmopterigidae. In: Dominick R.B., Dominick T., Ferguson D.C., Franclemont J.G., Hodges R.W. and Munroe E.G. (eds), Moths of America North of Mexico. E. W. Classey and The Wedge Entomological Research Foundation, London.

Hodges R.W., Dominick R.B., Dominick T., Ferguson D.C., Franclemont J.G., and Munroe E.G. (eds). 1983. Check List of the Lepidoptera of America North of Mexico Including Greenland. E. W. Classey and The Wedge Entomological Research Foundation, London.

Hodges R.W. 1985. A new species of *Dichomeris* from Costa-Rica (Lepidoptera, Gelechiidae). Proc. Entomol. Soc. Wash. 87: 456–459.

Hodges R.W. 1986. Gelechioidea: Gelechiidae (part) Dichomeridinae. In: Dominick R.B., Dominick T., Ferguson D.C., Franclemont J.G., Hodges R.W. and Munroe E.G. (eds), Moths of America North of Mexico. E. W. Classey and The Wedge Entomological Research Foundation, London.

Hodges R.W. 1998. The Gelechioidea. In: Kristensen N.P. (ed.), The Handbook of Zoology/ Handbuch der Zoologie. Walter de Gruyter, Berlin and New Yorkpp. 132–158.

Holl K.D. 1996. The effect of coal surface mine reclamation on diurnal lepidopteran conservation. J. Appl. Ecol. 33: 225–236.

Kerr J.T., Southwood T.R.E. and Cihlar J. 2001. Remotely sensed habitat diversity predicts butterfly species richness and community similarity in Canada. Proc. Entomol. Soc. Wash. 98: 11365–11370.

Kitching R.L., Orr A.G., Thalib L., Mitchell H., Hopkins M.S. and Graham A.W. 2000. Moth assemblages as indicators of environmental quality in remnants of upland Australian rain forest. J. Appl. Ecol. 37: 284–297.

Landry J.-F. 1998. Répartition géographique, plantes nourricières et notes taxonomiques sur 29 espèces de *Coleophora* (Lepidoptera: Coleophoridae) au Quebec. Fabreries 23: 25–104.

Lepš J., Spitzer K. and Jaroš J. 1998. Food plants, species composition and variability of the moth community in undisturbed forest. Oikos 81: 538–548.

MacArthur R.H. and Wilson E.O. 1967. The Theory of Island Biogeography. Princeton University Press, Princeton, NJ; Kent, England, U.K.

Magurran A.E. 1988. Ecological Diversity and its Measurement. Princeton University Press, Princeton, NJ; Kent, England, U.K.

Muona J. 1999. Trapping beetles in boreal coniferous forest – how many species do we miss? Fennia 77: 11–16.

Petit S., Moilanen A., Hanski I. and Baguette M. 2001. Metapopulation dynamics of the bog fritillary butterfly: movements between habitat patches. Oikos 92: 491–500.

Powell J., Mitter C. and Farrell B. 1998. Evolution of larval food preferences in Lepidoptera. In: Kristensen N.P. (ed.), The Handbook of Zoology/Handbuch der Zoologie. Walter de Gruyter, Berlin and New Yorkpp. 403–422.

Preston F.W. 1948. The commonness, and rarity, of species. Ecology 29: 254–283.

Scharff N., Coddington J.A., Griswold C.E., Hormiga G. and Bjorn de P. 2003. When to quit? Estimating spider species richness in a northern European deciduous forest. J. Arachnol. Soc. 31: 246–273.

Sharkey M.J. 2001. The All Taxon Biological Inventory of the Great Smoky Mountains National Park. Fl. Entomol. 84: 556–564.

Sørenson L.L., Coddington J.A. and Scharff N. 2002. Inventorying and estimating subcanopy spider diversity using semiquantitative sampling methods in an afromontane forest. Environ. Entomol. 31: 319–330.

Stehr F.W. 1987. Superfamily Gelechioidea. In: Stehr F.W. (ed.), Immature Insects. Kendall/Hunt, Dubuque, IApp. 379–399.

Summerville K.S., Boulware M.J., Veech J.A. and Crist T.O. 2003a. Spatial variation in species diversity and composition of forest Lepidoptera in eastern deciduous forests of North America. Conserv. Biol. 17: 1045–1057.

Summerville K.S. and Crist T.O. 2003. Determinants of lepidopteran community composition and species diversity in eastern deciduous forests: roles of season, eco-region and patch size. Oikos 100: 134–148.

Summerville K.S., Crist T.O., Kahn J.K. and Gerling J.C. 2003b. Community structure of arboreal caterpillars within and among tree species of the eastern deciduous forest. Ecol. Entomol. 28: 747–757.

Sutherland E.K. and Hutchinson T.F. (eds). 2003. Characteristics of mixed-oak forest ecosystems in southern Ohio prior to the reintroduction of fire. USDA Forest Service General Technical Report NE-299.

Tscharntke T., Steffan-Dewenter I., Kruess A. and Thies C. 2002. Characteristics of insect populations on habitat fragments: a mini review. Ecol. Res. 17: 229–239.

Usher M.B. and Keiller S.W.J. 1998. The macrolepidoptera of farm woodlands: determinants of diversity and community structure. Biodivers. Conserv. 7: 725–748.

Wagner D.M. and Scholtens B. 2002. Lepidoptera Blitz nets 860 and counting. ATBI Quart. 3: 6–8.

Wahlberg N., Klemetti T., Selonen V. and Hanski I. 2002. Metapopulation structure and movements in five species of checkerspot butterflies. Oecologia 130: 33–43.

Biodiversity and Conservation (2006) 15:503–524
DOI 10.1007/s10531-005-2353-4


# Diversity patterns of Bornean butterfly assemblages

DANIEL F. R. CLEARY[1,2,*] and MARTIN J. GENNER[3]
[1]*Institute for Biodiversity and Ecosystem Dynamics, University of Amsterdam, P.O. Box. 94766, 1090 GT Amsterdam, The Netherlands;* [2]*National Museum of Natural History, Department of Invertebrates, P.O. Box 9517, 2300 RA Leiden, The Netherlands;* [3]*Department of Biological Sciences, University of Hull, HU6 7RX, UK; *Author for correspondence (e-mail: cleary@science.uva.nl, cleary@naturalis.nnm.nl or dfrcleary@hotmail.com)*

Received 6 October 2004; accepted in revised form 4 August 2005

**Key words:** ENSO, Forest fires, Kalimantan, Logging, Species richness, Variance partitioning

**Abstract.** Borneo contains a diverse rainforest butterfly community, but its forests are under threat from logging and ENSO- (El Niño Southern Oscillation) induced fires. Contrasts in butterfly assemblage structure were examined in nine 450 ha landscapes in logged forest, primary unburned continuous and isolated forest, and forest affected by surface fires during the 1997/98 ENSO event. Temporally the effect of the 1997/98 ENSO event was followed in a single burned landscape from 1997 to 2004. In total, 517 species were present in 190 sampling sites. There was a five-fold difference in species richness among landscapes, with highest richness in continuous landscapes and lowest richness in burned landscapes. Richness was also higher in logged forest than proximate unlogged forest. Temporally, species richness dropped dramatically from 1997 to 1998, but afterwards increased remaining, however, substantially lower than pre-ENSO (1997) sampling. Sites in burned landscapes were distinct from other sites in terms of vegetation structure with the slash-and-burn area the most dissimilar to other landscapes. There was much less structure among unburned landscapes. The pattern of butterfly community composition was similar to that of vegetation structure with the community from the slash-and-burn area the most distinct. However, there was much less overlap among sites from different landscapes. Temporally, 1998 possessed the most distinct assemblage when compared to assemblages from other years. The community composition was, however, slowly returning to a pre-disturbance composition. Variance in community composition explained by environmental and spatial factors differed substantially among landscapes. The spatial fraction was the only explanatory component in recently burned landscapes and a proximate small unburned isolate, but explained no variation in logged landscapes. The environmental fraction explained substantial amounts of variation in logged landscapes and the slash-and-burn area. When all landscapes were pooled high proportions of variation in butterfly community composition were explained by both geographic distance between sites and environmental variables. In contrast when only unburned landscapes were considered, most variation was explained by the geographic distance among them. Despite differences among landscapes there was a general pattern of relatively sharp decline in similarity at short distances that levels out over greater distances, a result that agrees with previous studies on other tropical species assemblages.

## Introduction

Explaining spatial and temporal patterns of biological diversity is a fundamental goal in ecology (Williams et al. 2002). Tropical rainforests possess some of the most diverse communities on Earth, but factors influencing their spatial structure remain poorly understood. Spatial patterns of species composition (beta diversity) may be largely uniform, be spatially autocorrelated because of dispersal limitation, or be environmentally determined (Tuomisto et al. 2003). Over recent years there has been increasing interest in issues relating to tropical diversity and beta diversity of assemblages in particular (Condit et al. 2002; Potts et al. 2002; Longino et al. 2002; Tuomisto et al. 2003). Such studies have revealed the importance of environmental variables in structuring assemblages and have shown that sharp declines in community similarity over small spatial scales are typically followed by a lack of substantial change over larger spatial scales.

Large-scale studies of rainforest community spatial structure have focussed primarily on plant communities in pristine environments. There have been few replicated studies of disturbance and landscape-scale spatial variation in community structure. It is important to determine associations of environmental variables with patterns of community similarity to understand mechanisms that influence beta diversity (Spencer et al. 2002). This is also crucial for determining locations of nature reserves and for assessment of management strategies with respect to exploitation of forest products (Plotkin and Muller-Landau 2002; Tuomisto et al. 2003). An understanding of the main structuring agents of assemblages may help to predict effects on communities that result from human activity.

This study focussed on forest landscapes within Borneo, the largest of the Sundaland islands that are located to the west of Wallace's line. All of Sundaland including peninsular Malaysia and Southern Thailand, is categorised as a global hotspot of biodiversity, but its rainforests are severely threatened. On Borneo both logging and ENSO-induced fires have severely affected rainforests (Jepson et al. 2001; Putz et al. 2001; Siegert et al. 2001), and there is evidence that these factors have interacted compounding effects. For example, logging causes drought-resistant forests to become susceptible to fires by opening the canopy, increasing the fuel load and drying the forest interior (Nepstad et al. 1998). Little is known of long-term impacts of logging and fires on rainforest communities (Harrison 2000; Holmgren et al. 2001).

Butterflies are probably the best known invertebrate taxa with an estimated 20,000 species worldwide and a prominent place in conservation programmes and biodiversity assessments (Stork et al. 2003). They have also been identified as important indicator taxa for assessing biodiversity and monitoring ecosystem responses to environmental perturbations (Howard et al. 1998; Parmesan et al. 1999; Cleary 2004). Previous studies within Borneo have shown that butterfly diversity has been significantly affected by both logging and burning, and responses were dependent upon interspecific morphological, life-history and ecological differentiation (Cleary 2003; Cleary and Genner 2004). In this study, broader-scale patterns in butterfly community structure were examined. The aims were: (1) to compare species richness among differentially disturbed landscapes using sample-based rarefaction, Fisher's $\alpha$ and non-parametric richness estimators; (2) to compare the physical environment of each landscape and test whether butterfly species turnover (beta diversity) in different landscapes is consistently associated with the same environmental parameters; (3) to compare associations between butterfly beta diversity and environmental parameters at both landscape and regional scales.

## Material and methods

### *Study site and sampling*

Site descriptions and sampling methods are described in detail in Cleary (2003) and Cleary and Genner (2004). Three habitat classes were sampled; unburned continuous forest (C), unburned isolates surrounded by burned forest (I), and burned forest (B). All continuous forest landscapes were located in the province of Central Kalimantan, within the Kayu Mas concession, in the large unburned central core of Borneo that has not yet been affected by ENSO-induced fires. Unburned isolate landscapes were located in the province of East Kalimantan, and were not directly affected by the 1997/98 ENSO event, but were completely surrounded by forest that burned during this event. I1 and I3 were located in a 108,000 ha isolate located in part of the ITCI (International Timber Concessions Indonesia) and adjacent BFI (Balikpapan Forest Industries) concessions and including the Gunung Meratus Protected Forest Reserve. I2 was located in a 3500 ha unburned isolate that is all that remains of the Sungai Wain Nature Reserve. The burned landscapes were located in the province of East Kalimantan and surrounded the unburned forest isolates. B1 was located in the burned part of the Sungai Wain Nature Reserve, B2 in the Wanariset Samboja Research Forest and B3 in an area of frequently burned slash and burn agriculture along Km 30 of the Balikpapan to Samarinda highway. Both continuous forest and isolates contained primary unlogged forest (C1, I1 and I2) and logged forest (C2, C3 and I3). Burned forest included a landscape affected by one burn event (B1: burned 1997/98), a landscape affected by two burn events (B2: burned 1982/83 and 1997/98) and a landscape (B3) in an area dominated by slash-and-burn agriculture. Continuous forest was sampled in 1998, while isolates and burned forest were sampled in 2000. In addition landscape B2 was sampled from 1997 to 2004, a period coinciding with the most severe ENSO event recorded on Borneo (Harrison 2000; Siegert et al. 2001). Each landscape contained between 8 and 21 sampling sites (Table 1); in total there were 190 sites. Butterflies were sampled when encountered within *c.* 15 m on either side of a 300 m transect in each site. The transect was traversed repeatedly on foot from one end to the other at a steady pace, which was only broken to collect specimens. Within each site 200 butterflies were sampled, with the exception of B2 during 1997 and 2004 in which minimum sample sizes were 130 and 33 respectively.

*Table 1.* Summary of characteristics and sampling regimes for each landscape.

| Name | Landscape | Sites | Eff. (d) | *n* | Env | Habitat | Latitude | Longitude |
|---|---|---|---|---|---|---|---|---|
| Kayu Mas | C1 | 16 | 8 | 3200 | 14 | Primary continuous forest | 1° 17′ S | 112° 22′ E |
| Kayu Mas | C2 | 11 | 5 | 2200 | 10 | Logged continuous forest 1989–1990 | 1° 20′ S | 112° 20′ E |
| Kayu Mas | C3 | 16 | 6 | 3200 | 13 | Logged continuous forest 1993–1994 | 1° 16′ S | 112° 24′ E |
| Meratus | I1 | 16 | 7 | 3200 | 16 | Primary forest in 138000 ha isolate | 0° 58′ S | 116° 19′ E |
| Sungai Wain | I2 | 16 | 6 | 3200 | 16 | Primary forest in 3500 ha isolate | 1° 06′ S | 116° 49′ E |
| ITCI | I3 | 16 | 6 | 3200 | 16 | Logged forest in 138000 ha isolate | 0° 57′ S | 116° 21′ E |
| Sungai Wain | B1 | 16 | 6 | 3200 | 16 | Once-burned forest 1997–1998 | 1° 05′ S | 116° 48′ E |
| Wanariset | B2 | 21 | 5 | 4200 | 14 | Twice-burned forest 1982–1983 and 1997–1998 | 0° 59′ S | 116° 57′ E |
| Km 30 | B3 | 16 | 6 | 3200 | 16 | Slash-and-burn habitat cultivated since 1970s | 1° 03′ S | 116° 57′ E |
| Wanariset | 1997 | 8 | 6 | 1333 | – | Forest burned during 1982–1983 | 0° 59′ S | 116° 57′ E |
| Wanariset | 1998 | 9 | 5 | 1800 | – | Forest burned during 1982–1983 and 1997–1998 | 0° 59′ S | 116° 57′ E |
| Wanariset | 1999 | 9 | 5 | 1800 | – | Forest burned during 1982–1983 and 1997–1998 | 0° 59′ S | 116° 57′ E |
| Wanariset | 2004 | 11 | | 629 | – | Forest burned during 1982–1983 and 1997–1998 | 0° 59′ S | 116° 57′ E |

Sites refers to the number of sample sites in each landscape, Eff. (d) refers to the mean number of days spent sampling in a site, *n* refers to the number of individuals sampled over the whole landscape, Env refers to the number of sites in which environmental (vegetation and physiographical) variables were recorded and Habitat refers to the main habitat characteristics (i.e. whether landscapes were primary or logged and located in unburned continuous forest, unburned isolates or burned forest; the year of logging and burning is given). Note that C1, C2 and C3 were sampled in 1998 and the I and B landscapes in 2000.

Vegetation and physiographical characteristics were quantified within 200 $m^2$ ($10 \times 20$ m) subplots in each site (with the exception of B2 in which variables were only recorded in 14 of the 21 sites). In the continuous forest six subplots were established per site; elsewhere a single subplot was established per site. Physiographical variables included aspect (lower slope, middle slope and upper slope), relative elevation and presence of water. Aspect was visually estimated and the presence of water noted. Relative elevation was estimated with a Suunto clinometer (Suunto Oy, Vantaa, Finland). The following vegetation variables were measured in each site: density of trees (diameter at breast height 'dbh' $> 10$ cm); density of poles (stems taller than 130 cm; dbh 5 to 10 cm); sapling density (stems shorter than 130 cm; dbh $< 5$ cm); grass cover; herb cover and fern cover. Cover was visually estimated within a single 8 $m^2$ ($4 \times 2$ m) subplot nested within the larger 200 $m^2$ subplot in six classes: 0%; 1–10%; 11–30%; 31–70%; 71–90% and 91–100%.

*Analyses of species richness and assemblage structure*

Species richness estimates (rarefied species richness, Shannon–Weaver index, Fishers $\alpha$, incidence-based coverage estimator ICE and Chao2) were calculated using the programme Estimates (Colwell 2000). Dominance-diversity curves were also constructed for each landscape using species abundance data. These curves rank species in decreasing order of relative abundance (Whittaker 1965), and their shapes enable rapid visualisation of the extent to which species were distributed evenly or unevenly in abundance.

*Analyses of community composition*

Differences in vegetation structure among landscapes were assessed using multidimensional scaling (MDS) on a matrix of normalised Euclidean distances using $\log_{10}(x + 1)$ transformed data in PRIMER. Using this matrix, significant differences in vegetation structure among landscapes were tested using Analysis of Similarities (ANOSIM) in PRIMER 5 (Primer-E Ltd, Plymouth, UK). The matrix was then used for multidimensional scaling (MDS) ordination (using PRIMER). Multidimensional scaling has various advantages over other multivariate techniques for use in ecological studies; the results have been found to be robust under a wide range of conditions. MDS does not have stringent model assumptions, and any similarity measure can be used for ordination (Beck et al. 2002). MDS and ANOSIM analyses were repeated on butterfly community composition using $\log_{10}(x+1)$ transformed species abundance data and a similarity matrix constructed using the Bray–Curtis coefficient (Bray and Curtis 1957), a method often used for ordination of community species abundance data (Clarke and Gorley 2001; Legendre and Gallagher 2001; Ellingsen 2002). Ordinations were constructed using (1) data from the nine landscapes sampled in continuous forest, isolates and burned forest and (2) data sampled temporally in B2 from 1997 to 2004.

To examine whether community similarity was dependent upon environmental variables and geographic distance between sampling sites multiple matrix regression was used within the program PERMUTE! 3.4.9 (Casgrain 2001). Five predictor matrices were used: (1) $\log_{10}$ transformed distance between plots; (2) normalised Euclidean difference between sites in $\log_{10}(x+1)$ transformed sapling, pole and tree density; (3) Euclidean difference between sites in grass, fern and herb cover; (4) Euclidean difference between sites in physiography based on slope aspect and presence of water and (5) Euclidean difference in relative elevation between sites. Options for 999 permutations and Bonferroni-corrected $p$-to-enter value of 0.10 were selected. A quantitative variance partitioning technique described in Borcard et al. (1992) was used on results of multiple matrix regressions. This enabled the variation in community composition explained by geographic distance alone, environmental variables alone, and spatial and environmental variables combined to be quantified. Briefly, the procedure was as follows: first, the community similarity matrix was regressed against the total set of environmental and distance matrices to obtain the variance explained by all = $R_T$. Next, the community similarity matrix was regressed against the environmental matrices to obtain $R_E$, and the community similarity matrix was regressed against the distance matrix to obtain $R_S$. It was then possible to calculate the purely environmental fraction $R_{PE} = R_T - R_S$, the purely spatial fraction $R_{PS} = R_T - R_E$, the spatially structured environmental fraction $R_{SE} = R_E + R_S - R_T$, and the unexplained variation $R_{UN} = 1 - R_T$. Analyses were carried out for (1) each landscape separately, (2) for sites from all nine landscapes in continuous forest, isolates and burned forest and (3) for sites from continuous forest and isolates only (burned sites excluded).

## Results

### *Species richness and dominance diversity relationship*

In total 34,362 individuals of 517 species of butterflies were sampled in the 190 sites. Using the Chao2 nonparametric richness estimate, there was an almost five-fold difference in species richness among landscapes from the B3 landscape (74 species) to the C3 landscape (342 species). The only landscape in which a richness estimator (Chao2) clearly formed an asymptote was in B3 (Figure 1), indicating that further sampling would have resulted in more species in most landscapes. Species richness was highest in continuous forest, intermediate in isolates, and lowest in burned forest and the slash-and-burn area. All species richness estimators were significantly correlated (Pearson's product moment; $R = 0.962 - 0.999$; all comparisons $p < 0.001$). In the temporal comparison, species richness dropped dramatically from 1997 to 1998 (Table 2), but

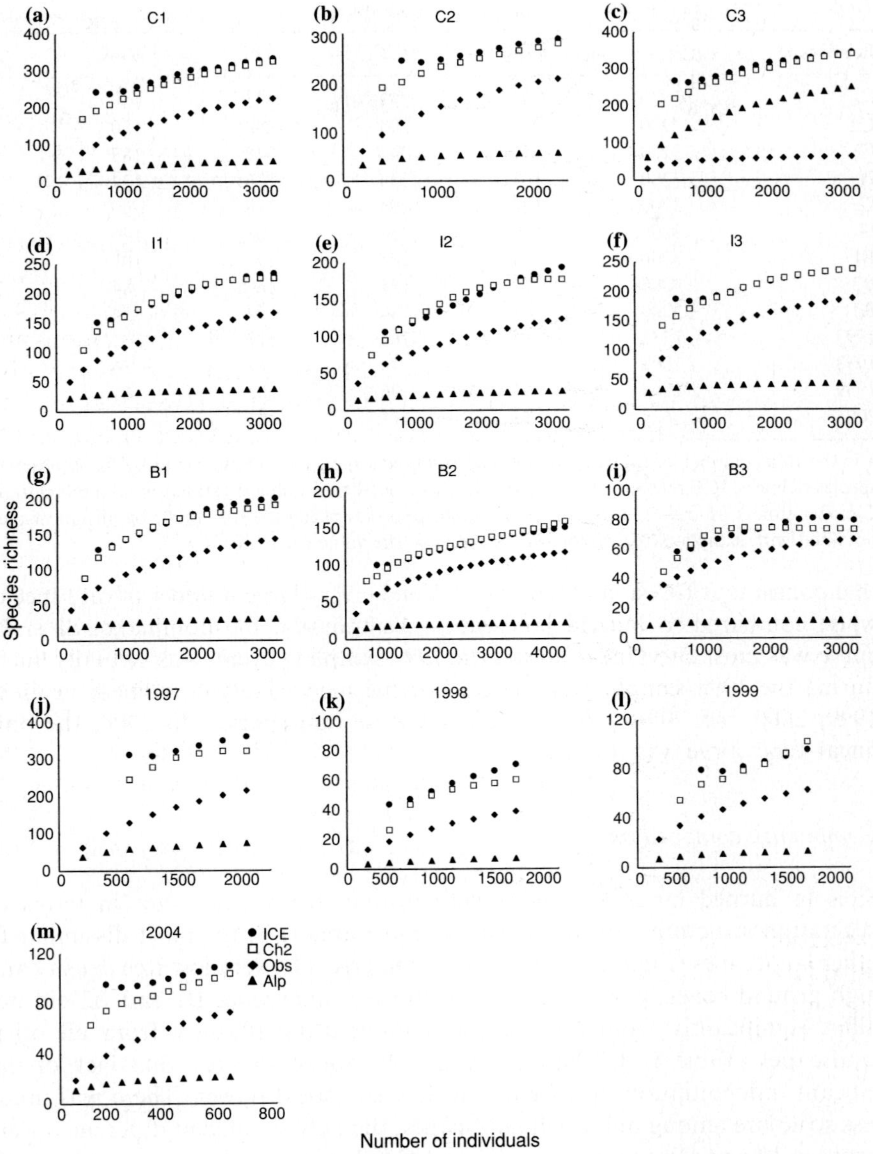


*Figure 1.* Species accumulation curves for all nine landscapes in 2000 and from 1997 to 2004 in B2. In each landscape, two nonparametric richness estimators [ICE (closed circles) and Chao2 (Ch2: open squares)], sample based rarefaction (obs: closed diamonds) and Fisher's $\alpha$ (Alp: closed triangles) are shown.

increased towards 2004. However, it was still substantially lower than pre-ENSO (1997) sampling. The dominance-diversity curves (Figure 1) became increasingly curvilinear from burned to continuous forest (Figure 1), indicating

*Table 2.* Summary of diversity measures for each landscape.

| Landscape | $n$ | Obs | ICE | Ch2 | Rare | $\alpha$ |
|---|---|---|---|---|---|---|
| C1 | 3200 | 225 | 333 | 325 | 159 | 46 |
| C2 | 2200 | 213 | 298 | 288 | 180 | 55 |
| C3 | 3200 | 251 | 346 | 342 | 183 | 56 |
| I1 | 3200 | 167 | 234 | 226 | 124 | 33 |
| I2 | 3200 | 124 | 194 | 178 | 90 | 21 |
| I3 | 3200 | 189 | 238 | 238 | 146 | 41 |
| B1 | 3200 | 144 | 203 | 193 | 107 | 27 |
| B2 | 4200 | 117 | 151 | 160 | 84 | 20 |
| B3 | 3200 | 67 | 80 | 74 | 54 | 11 |
| 1997 | 1333 | 216 | 362 | 323 | | 73 |
| 1998 | 1800 | 39 | 70 | 61 | 36 | 7 |
| 1999 | 1800 | 63 | 96 | 102 | 60 | 13 |
| 2004 | 629 | 73 | 111 | 104 | | 21 |

$n$ is the total number of individuals sampled per landscape (pooling all species). Obs is observed species richness. ICE refers to the value for the incidence-based richness estimator, Ch2 refers to the Chao2 richness estimator, Rare refers to rarefied species richness ($n$ = 1400 for all landscapes) obtained with sample-based rarefaction, $\alpha$ refers to the value for Fisher's $\alpha$.

that continuous forest landscapes contained a very large number of rare species when compared to burned landscapes. Temporally, the dominance-diversity curve was most curvilinear during the 1997 sampling event, was virtually linear during the 1998 sampling event and became increasingly curvilinear again in 1999, 2000 and 2004 indicating an influx of rare species. In 2000, the only linear-type curve was for B3.

### *Community composition*

Sites in burned landscapes were very distinct from other sites in terms of vegetation structure with the slash-and-burn area (B3) the most dissimilar to other landscapes (Figure 3a); B3 was characterised by very low tree density and high ground cover, particularly grass. Burned landscapes B1 and B2 did not differ significantly from C2, but both differed significantly from all other landscapes (Table 3). C2 had a low sample size of 10 sites, thus lack of significant differentiation may be due to low statistical power. There was much less structure among unburned landscapes; the only significant differences were between I2 and C1 and between C2 and C3 (Table 3). The pattern of butterfly community composition was similar to that of vegetation structure with the B3 community the most distinct. However, there was much less overlap among sites from different landscapes (Figure 3b). All pair-wise comparisons of landscapes differed significantly ($p < 0.001$). Temporally, all sampling periods differed significantly with 1998 possessing the most distinct assemblage when compared to assemblages from other years (Table 4).

Although dominant species from the unburned forest were present in the burned forest they tended to be rare. Dominants in burned forest included

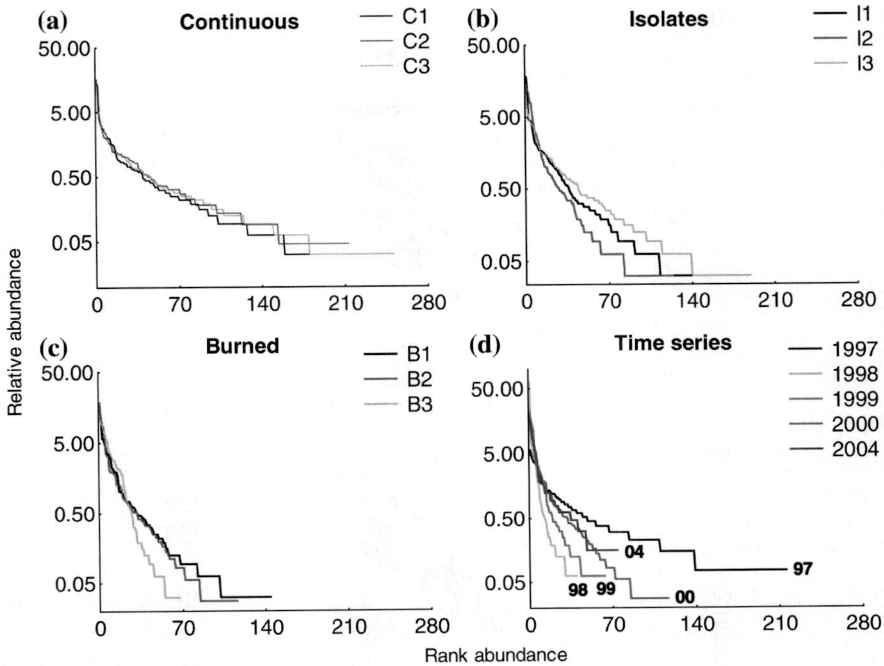


*Figure 2.* Dominance-diversity curves for each landscape in (a) continuous forest, (b) unburned isolates, (c) burned forest and (d) from 1997 to 2004 in B2 (Wanariset).

*Ypthima pandocus* and *Neptis hylas*, species that were invariably rare or absent in unburned forest. However, there were exceptions such as *Koruthaialos rubecula* that was relatively abundant in both unburned and burned forest. The slash-and-burn area housed the most unique faunal assemblage. Dominant species included the globally widespread *Zizina otis* and *Mycalesis perseus* that were unique to the habitat, and species such as *Taractrocera ardonia* and *Orsotriaena medus* that were rare in unburned forest.

In 2000, continuous, isolated and burned forest shared three dominant species, namely *Jamides pura*, *Koruthaialos rubecula* and *Drupadia theda* (Table 5). An additional dominant species, *Ragadia makuta*, was shared between isolates and continuous forest but was absent from burned forest. The remainder of species tended to be only dominant in one habitat class. For example, *Trogonoptera brookiana* and *Troides amphrysus* were dominant in continuous habitats, but were rare within forest isolates and absent from burned forest. Similarly, *Orsotriaena medus* was rare within continuous forest and absent in forest isolates. Temporally, the change in dominants in B2 was even more pronounced than spatially among continuous forest, isolates and burned forest (Table 6). From 1997 to 1998, only *Jamides celeno* remained dominant in B2 (Wanariset) and increased from 2.7 to 57.9% of total individuals (pooling all species). It then declined steadily in abundance to comprise

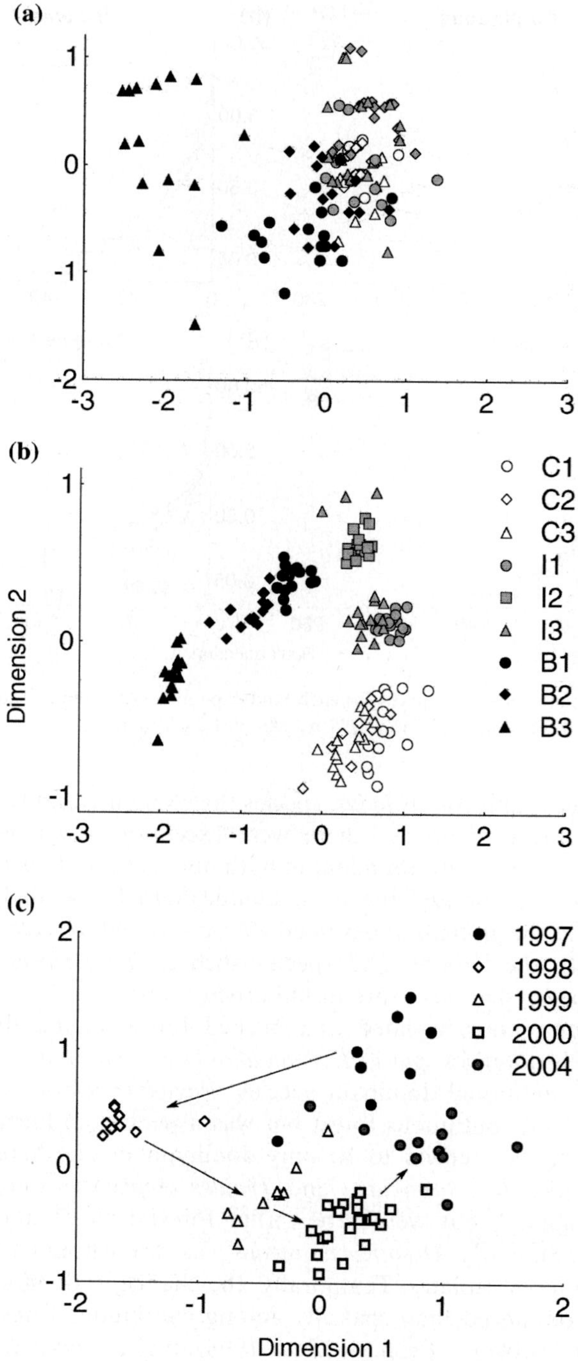


*Figure 3.* Ordination of (a) vegetation structure, (b) butterfly community composition in all nine landscapes in 2000 and (c) in B2 (Wanariset) from 1997 to 2004 using multidimensional scaling. Sites from different landscapes are indicated by symbols, closer symbols indicate more similar sites.

*Table 3.* Pairwise comparisons in vegetation structure and similarity of butterfly assemblages between landscapes.

| Pairwise comparisons | | Vegetation | | Butterflies | |
|---|---|---|---|---|---|
| $Group_1$ | $Group_2$ | $R_{ANOSIM}$ | $p$ | $R_{ANOSIM}$ | $p$ |
| C1 | C2 | 0.093 | 0.084 | 0.429 | < 0.001 |
| C1 | C3 | 0.412 | 0.002 | 0.478 | < 0.001 |
| C1 | I1 | 0.235 | < 0.001 | 0.796 | < 0.001 |
| C1 | I3 | 0.199 | 0.004 | 0.906 | < 0.001 |
| C1 | I2 | 0.484 | < 0.001 | 0.973 | < 0.001 |
| C1 | B1 | 0.590 | < 0.001 | 0.996 | < 0.001 |
| C1 | B2 | 0.511 | < 0.001 | 1.000 | < 0.001 |
| C1 | B3 | 0.976 | < 0.001 | 1.000 | < 0.001 |
| C2 | C3 | 0.067 | 0.135 | 0.311 | < 0.001 |
| C2 | I1 | −0.008 | 0.472 | 0.932 | < 0.001 |
| C2 | I3 | 0.009 | 0.369 | 0.890 | < 0.001 |
| C2 | I2 | 0.343 | < 0.001 | 0.960 | < 0.001 |
| C2 | B1 | 0.472 | 0.002 | 0.983 | < 0.001 |
| C2 | B2 | 0.306 | 0.003 | 0.999 | < 0.001 |
| C2 | B3 | 0.975 | < 0.001 | 1.000 | < 0.001 |
| C3 | I1 | 0.051 | 0.135 | 0.985 | < 0.001 |
| C3 | I3 | 0.167 | 0.010 | 0.961 | < 0.001 |
| C3 | I2 | 0.458 | < 0.001 | 1.000 | < 0.001 |
| C3 | B1 | 0.413 | < 0.001 | 1.000 | < 0.001 |
| C3 | B2 | 0.303 | < 0.001 | 1.000 | < 0.001 |
| C3 | B3 | 0.978 | < 0.001 | 1.000 | < 0.001 |
| I1 | I3 | −0.012 | 0.534 | 0.687 | < 0.001 |
| I1 | I2 | 0.178 | 0.004 | 0.998 | < 0.001 |
| I1 | B1 | 0.443 | < 0.001 | 1.000 | < 0.001 |
| I1 | B2 | 0.158 | 0.004 | 1.000 | < 0.001 |
| I1 | B3 | 0.974 | < 0.001 | 1.000 | < 0.001 |
| I3 | I2 | 0.028 | 0.211 | 0.943 | < 0.001 |
| I3 | B1 | 0.518 | < 0.001 | 0.987 | < 0.001 |
| I3 | B2 | 0.296 | < 0.001 | 0.999 | < 0.001 |
| I3 | B3 | 0.971 | < 0.001 | 1.000 | < 0.001 |
| I2 | B1 | 0.782 | < 0.001 | 0.993 | < 0.001 |
| I2 | B2 | 0.601 | < 0.001 | 1.000 | < 0.001 |
| I2 | B3 | 0.984 | < 0.001 | 1.000 | < 0.001 |
| B1 | B2 | 0.259 | < 0.001 | 0.909 | < 0.001 |
| B1 | B3 | 0.892 | < 0.001 | 1.000 | < 0.001 |
| B2 | B3 | 0.946 | < 0.001 | 0.953 | < 0.001 |

The $R_{ANOSIM}$ statistic values are an absolute measure of how separated the a priori defined groups (landscapes) are. A zero (0) indicates that there is no difference among groups (landscapes), while a one (1) indicates that all samples (here sites) within landscapes are more similar to one another than any samples from different groups (Clarke and Gorley 2001).

only 3.8% of all individuals in 2004. From 1998 to 1999 the dominant species also changed completely with the exception of *J. celeno*. From 1999 to 2000 and 2000 to 2004, six of the ten dominant species remained in the top ten, including *J. pura*, *K. rubecula* and *D. theda*.

*Table 4.* Pairwise comparisons in the similarity of butterfly assemblages between years in B2 (Wanariset).

| Pairwise comparisons | | $R_{ANOSIM}$ | $p$ |
|---|---|---|---|
| $Year_1$ | $Year_2$ | | |
| 1997 | 1998 | 0.987 | < 0.001 |
| 1997 | 1999 | 0.917 | < 0.001 |
| 1997 | 2000 | 0.939 | < 0.001 |
| 1997 | 2004 | 0.846 | < 0.001 |
| 1998 | 1999 | 0.976 | < 0.001 |
| 1998 | 2000 | 0.995 | < 0.001 |
| 1998 | 2004 | 1.000 | < 0.001 |
| 1999 | 2000 | 0.855 | < 0.001 |
| 1999 | 2004 | 0.945 | < 0.001 |
| 2000 | 2004 | 0.919 | < 0.001 |

All landscapes showed declining community similarity over increasing spatial scales within the landscape (Figure 4), and this relationship was significant in all cases except C2 and B2 in 1997, 1998 and 2004 although the lack of significance may be at least partially attributed to low power due to the limited

*Table 5.* The ten most dominant species in continuous forest, isolates and burned forest, including the number sampled and percentage of the total assemblage (between brackets).

| Species | Continuous | | Isolates | | Burned | |
|---|---|---|---|---|---|---|
| *Abisara geza* Fruhstorfer | 256 | (2.4) | | | | |
| *Allotinus leogoron* Fruhstorfer | | | 689 | (7.2) | | |
| *Allotinus unicolor* C. and R. Felder | | | 464 | (4.8) | | |
| *Arhopala epimuta* Moore | | | 456 | (4.8) | | |
| *Coelites eupythychioides* Felder | | | 434 | (4.5) | | |
| *Drupadia theda* Felder | 135 | (1.6) | 970 | (10.1) | 345 | (3.3) |
| *Eurema andersoni* Moore | 184 | (2.1) | | | | |
| *Eurema hecabe* L. | 152 | (1.8) | | | | |
| *Euthalia iapis* Godart | 216 | (2.5) | | | | |
| *Idea stolli* Moore | 200 | (2.3) | | | | |
| *Jamides celeno* Cramer | | | | | 262 | (2.5) |
| *Jamides pura* Moore | 1246 | (14.5) | 342 | (3.6) | 595 | (5.6) |
| *Koruthaialos rubecula* Plötz | 135 | (1.6) | 588 | (6.1) | 1259 | (11.9) |
| *Mycalesis anapita* Moore | | | 222 | (2.3) | | |
| *Neptis hylas* L. | | | | | 620 | (5.8) |
| *Orsotriaena medus* Wallengren | | | | | 421 | (4.0) |
| *Paralaxita orphna* Boisduval | | | 192 | (2.0) | | |
| *Ragadia makuta* Horsfield | 644 | (7.5) | 529 | (5.5) | | |
| *Spindasis lohita* Horsfield | | | | | 431 | (4.1) |
| *Taractrocera ardonia* Hewitson | | | | | 507 | (4.8) |
| *Trogonoptera brookiana* Wallace | 288 | (3.3) | | | | |
| *Troides amphrysus* Cramer | 347 | (4.0) | | | | |
| *Ypthima pandocus* Moore | | | | | 1049 | (9.9) |

*Table 6.* The ten most dominant species from 1997 to 2004 in B2, including the number sampled and percentage of the total assemblage (between brackets).

| Species | 1997 | | 1998 | | 1999 | | 2000 | | 2004 | |
|---|---|---|---|---|---|---|---|---|---|---|
| *Allotinus leogoron* Fruhstorfer | | | | | | | | | 38 | (6.0) |
| *Anosia melanippus* Cramer | | | | | | | | | 11 | (1.7) |
| *Appias albina* Boisduval | | | 150 | (8.3) | | | | | | |
| *Appias paulina* Cramer | | | 44 | (2.4) | | | | | | |
| *Arhopala atosia* Hewitson | 50 | (3.8) | | | | | | | | |
| *Arhopala epimuta* Moore | 64 | (4.8) | | | | | | | | |
| *Arhopala evansi* Corbet | 35 | (2.6) | | | | | | | | |
| *Arhopala hypomuta* Hewitson | 56 | (4.2) | | | | | | | | |
| *Arhopala norda* Evans | | | | | | | | | 109 | (17.3) |
| *Charaxes bernardus* Fabricius | | | | | 43 | (2.4) | | | | |
| *Drupadia theda* Felder | 48 | (3.6) | | | 40 | (2.2) | 249 | (5.9) | 33 | (5.2) |
| *Erites argentina* Butler | 26 | (2.0) | | | | | | | | |
| *Euploea algea* Godart | | | 104 | (5.8) | | | | | | |
| *Euploea crameri* Lucas | | | 17 | (0.9) | | | | | | |
| *Euploea modesta* Butler | | | 85 | (4.7) | | | | | | |
| *Euploea mulciber* Cramer | | | 79 | (4.4) | | | | | | |
| *Euploea sylvester* Fabricius | | | 14 | (0.8) | | | | | | |
| *Eurema sari* Horsfield | | | | | 275 | (15.3) | | | 12 | (1.9) |
| *Graphium antiphates* Cramer | | | 104 | (5.8) | | | | | | |
| *Hidari irava* Moore | | | | | 42 | (2.3) | | | | |
| *Ideopsis vulgaris* Butler | 43 | (3.2) | | | | | | | | |
| *Jamides celeno* Cramer | 36 | (2.7) | 1043 | (57.9) | 200 | (11.1) | 156 | (3.7) | 24 | (3.8) |
| *Jamides pura* Moore | 76 | (5.7) | | | 225 | (12.5) | 463 | (11.0) | 62 | (9.9) |
| *Koruthaialos rubecula* Plötz | 45 | (3.4) | | | 76 | (4.2) | 690 | (16.4) | 94 | (14.9) |
| *Neptis hylas* L. | | | | | 435 | (24.2) | 384 | (9.1) | | |
| *Orsotriaena medus* Wallengren | | | | | 66 | (3.7) | 97 | (2.3) | | |
| *Pachliopta aristolchiae* Fabricius | | | | | 58 | (3.2) | | | | |
| *Potanthus omaha* Fruhstorfer | | | | | | | 87 | (2.1) | | |
| *Saletara panda* Butler | | | 49 | (2.7) | | | | | | |
| *Spindasis lohita* Horsfield | | | | | | | 143 | (3.4) | | |
| *Taractrocera ardonia* Hewitson | | | | | | | 156 | (3.7) | | |
| *Ypthima fasciata* Hewitson | | | | | | | | | 12 | (1.9) |
| *Ypthima pandocus* Moore | | | | | | | 441 | (10.5) | 45 | (7.2) |

number of sites sampled in these landscapes. The spatial and environmental predictor matrices explained a significant amount of variation in all landscapes although the percentage explained differed substantially (range 6–70%; mean 24%) (Table 7). Variance partitioning revealed substantial differences among landscapes and type of disturbance. Purely spatial variation explained no variation in logged landscapes, but was the only explanatory component in burned landscapes (B1 and B2 in 2000) and the proximate small primary isolate (I2). The spatially-structured environmental component was most important in recently logged forest (C3) and the slash-and-burn landscape (B3). Finally the purely environmental component explained significant amounts of variation in all continuous forest landscapes, both landscapes in the large isolate (I1 and I3), and the slash-and-burn landscape B3.

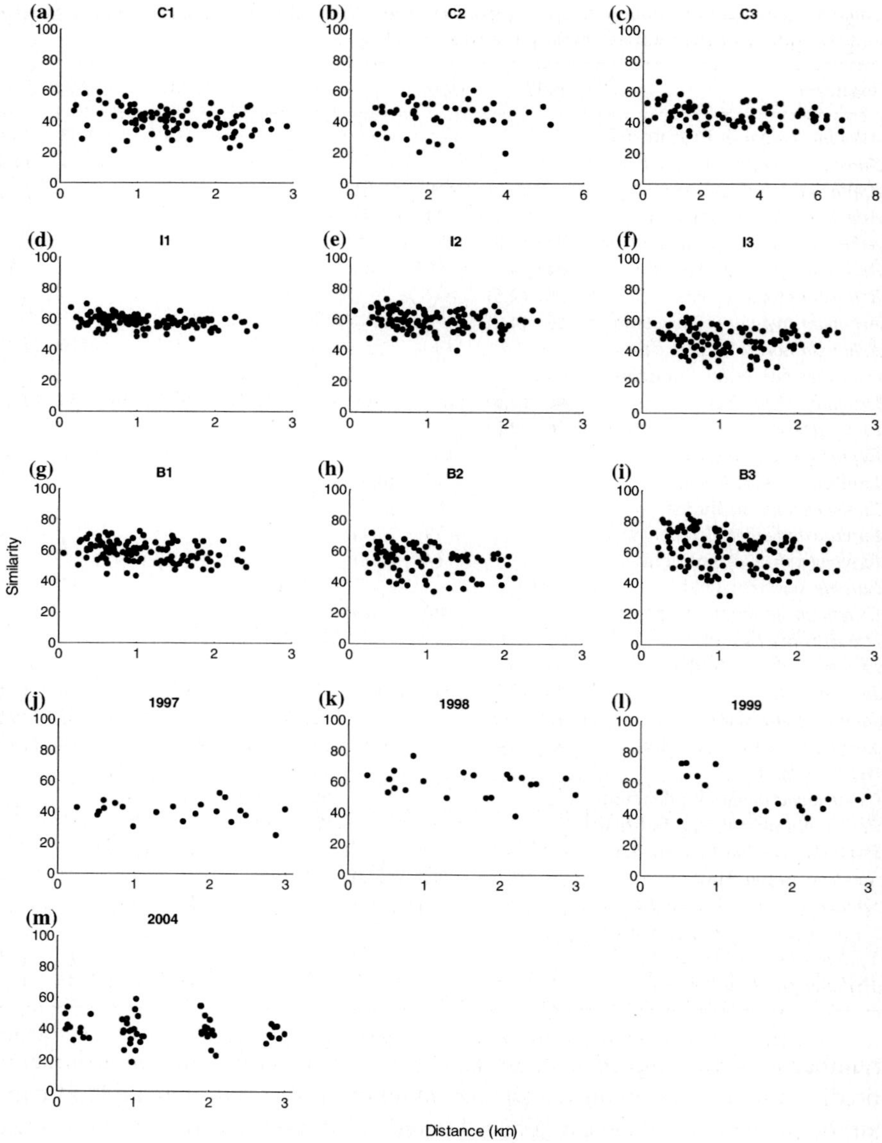


*Figure 4.* Variation in butterfly community similarity as a function of the geographic distance between sampling sites. Each point represents similarity between two pairs of sites based on the Bray–Curtis similarity index. Associations were significant for all landscapes except C2 and B2 in 1997, 1998 and 2004.

When all landscapes were assessed together over a scale of 520 km, 64.5% of the variation was explained by the full set of predictor variables of which 28.9% was related to the purely spatial component, 5.8% to the spatially structured environmental component and 29.9% to the purely environmental

*Table 7.* Summary of results using variance partitioning to assess the contribution of spatial and environmental processes in explaining variation in community similarity.

| Landscape | Habitat | Pure space | Spat. env. | Pure env. | Total |
|---|---|---|---|---|---|
| C1 | Primary continuous forest | 8.24 | 3.02 | 17.28 | 28.54 |
| C2 | Logged continuous forest (1989/90) | 0.00 | 0.00 | 38.72 | 38.72 |
| C3 | Logged continuous forest (1993/94) | 0.00 | 11.72 | 3.14 | 14.86 |
| I1 | Primary forest in 138,000 ha isolate | 5.72 | 3.35 | 5.28 | 14.35 |
| I2 | Primary forest in 3,500 ha isolate | 5.73 | 0.00 | 0.00 | 5.73 |
| I3 | Logged forest in 138,000 ha isolate | 0.00 | 3.84 | 19.34 | 23.18 |
| B1 | Once-burned forest (1997/98) | 7.94 | 0.00 | 0.00 | 7.94 |
| B2 | Twice-burned forest (1982/83 and 1997/98) | 11.55 | 0.00 | 0.00 | 11.55 |
| B3 | Slash-and-burn habitat cultivated since 1970s | 3.80 | 9.07 | 60.28 | 73.15 |
| Mean of landscapes | | 4.78 | 3.44 | 16.00 | 24.22 |
| Standard deviation of landscapes | | 4.17 | 4.30 | 21.03 | 21.14 |
| All landscapes pooled | | 28.87 | 5.80 | 29.87 | 64.54 |
| Unburned landscapes pooled | | 52.71 | 1.58 | 0.15 | 54.44 |

Pure space: pure spatial variation, Spat. env.: spatially structured environmental variation, Pure env.: pure environmental variation.

component (Figure 5a). When only unburned landscapes were considered 54.4% of the variation in community similarity was explained, of which 52.7% was related to the purely spatial component, 1.5% to the spatially structured environmental component and only 0.2% to the purely environmental component (Figure 5b). When burned forests are excluded, it would thus appear that purely spatial processes such as dispersal are of greater importance in structuring forest assemblages over large spatial scales than local habitat variables. However, at smaller spatial scales local variables achieve much greater importance within unburned forest. The pattern of a decline in similarity over small spatial scales that level out over larger scales was also prevalent when all landscapes were assessed together and when only the unburned landscapes were assessed (Figure 5).

## Discussion

### *Differential responses of butterfly species richness to logging and fires*

Disturbance can increase dominance and lower diversity by initiating succession with opportunistic species (Margalef 1968). Alternatively it may enhance diversity by providing additional ecological niches and preventing competitive exclusion of rarer species (Caswell 1976). In this study, primary forests were less species-rich than logged forests within the same habitat class categories

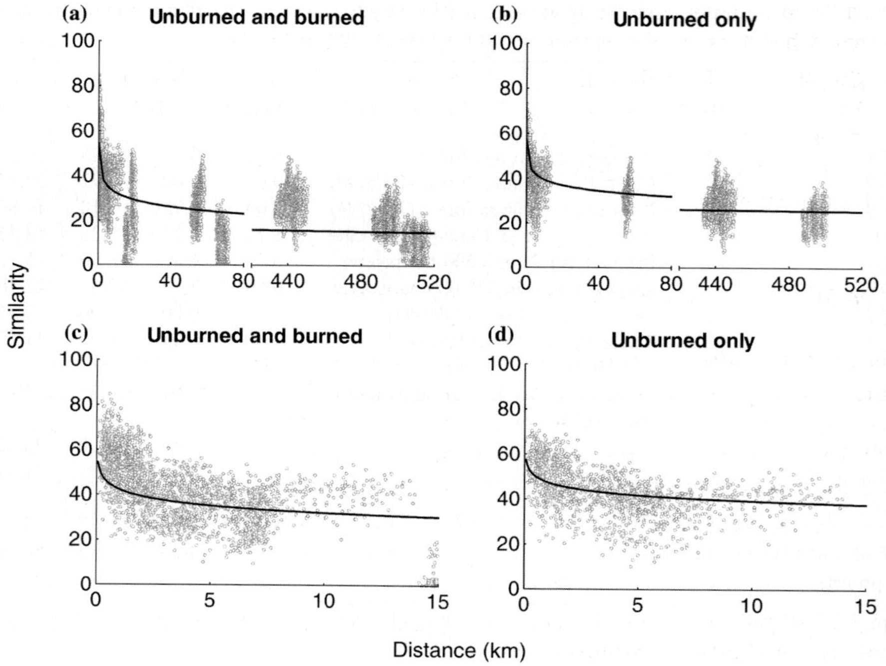


*Figure 5.* Variation in butterfly community similarity as a function of distance for: (a) all sites including sites from unburned and burned landscapes; (b) only sites from unburned landscapes; (c) unburned and burned sites, excluding all comparisons of sites great than 15 km apart; and (d) unburned sites only, excluding all comparisons of sites greater than 15 km apart. All fits are a logarithmic function obtained using Statistica 6.1. Note the decline in similarity at short spatial scales (< 2 km) and the levelling out over larger spatial scales.

(continuous forest or isolates), and dominance-diversity curves showed an increased prevalence of rare species in logged forest. Increases in diversity following selective logging have been reported in other studies on tropical butterflies (Spitzer et al. 1997; Willott et al. 2000). In the study area, logging has been highly selective, and may have opened up gaps in the canopy, leading to increases in habitat-heterogeneity. This may have provided new niches for rare opportunistic colonising species, and reduced abundance of previous common competitive dominants (Rozenzweig 1995).

In contrast to these results, there was a gradient from low diversity within burned landscapes through to higher diversity within continuous landscapes. Colonisers of these forests following fire have on average larger wingspans, indicating that they are likely to be better dispersers, they also tend to be predominantly specialised herb feeders or generalists, indicating burning has initiated a successional process (Cleary and Genner 2004). Thus, butterfly community responses to ENSO-induced fires and logging were fundamentally different. During the more catastrophic environmental changes associated with burning a successional process appears to have been initiated where a small number of widely dispersing opportunistic species colonised and dominated the burned habitats, while less dramatic changes associated with selective logging seems to have provided opportunities for otherwise rare species to establish within communities. Likewise, in Africa sites with lowest tree cover and greatest degree of disturbance had the lowest species richness and most dissimilar butterfly assemblages when compared to logged and uncleared forest sites (Stork et al. 2003).

*Spatial patterns of butterfly diversity over small (landscape) scales*

Nearly all landscapes showed significant declines in community similarity over increasing spatial scales, but variation explained by geographic distance and environmental variables differed among landscapes and disturbance classes. Three patterns were apparent: (1) within logged forest no spatial variation in butterfly community composition was explained by geographic distance alone, but there were significant components explained in unlogged forests; (2) environmental variables only failed to explain significant variation in butterfly community structure in burned forests and the small unburned isolate; and (3) the slash-and-burn landscape had the greatest component of variation in butterfly community composition explained by environmental variables compared to other landscapes.

The lack of association between butterfly community composition and geographic distance in logged forests may be related to the effects of logging on habitat structure. Openings in the canopy as a consequence of logging tend to increase the density of lianas, ferns and woody debris, but also the spatial distribution of gaps is a consequence of topographical conditions and stocking density (Cleary in press). It is thus likely that logged forests are more spatially heterogeneous in term of butterfly resource availability, perhaps more so than primary or burned forests. Hence, logged forests represent fundamentally different, but not necessarily resource poor, habitats compared with primary forest. The species that are able to thrive in these disturbed habitats may have life history characteristics rendering them more likely to search for favoured but patchy resources. Indeed, rainforest butterfly species that are more common within gaps tend to be better dispersers and have morphology designed for faster flight than those within the non-gap communities (Hill et al. 2001). Further study of dispersal, life history and ecological characteristics would determine if this is the case. Additional support for this hypothesis is in the high degree of butterfly community spatial variation explained by environmental variation within logged forest, implying that species within these communities tend to possess strong species-specific habitat preferences.

Environmental variables failed to explain significant variation in butterfly community structure in burned (B1 and B2) and isolated (I2) forests, but not in continuous forests. B1 and B2 were all low diversity habitats burned by ENSO-induced fires. I2 is a proximate forest that has a butterfly community also heavily affected, possibly because the burned structure of the surrounding landscape has acted as a barrier to dispersal for many species, preventing immigration from wider surroundings (Cleary and Genner 2004). The lack of environmentally-determined butterfly spatial structure may therefore be a temporal phenomenon; a consequence of colonisation by pioneering generalist and broadly dispersing large winged taxa (Cleary and Genner 2004).

The slash-and-burn landscape (B3) was the most distinct habitat in both vegetation and butterfly community composition. This area has been in cultivation since the late 1970s so the high degree of variance in the butterfly community may be due to resident fauna having sufficient time to adapt to the dominant environmental regime. Although largely grass-dominated, the area has considerable heterogeneity, with some areas dominated by herbs and swampy areas dominated by *Scleria* sp. (Cyperaceae). There were also sparse areas of tree growth, often along streams, creating a visually more abrupt environmental change than usually seen in forested areas.

*Dispersal between landscapes and succession*

Ricketts (2001) noted that although quality of the habitat-matrix affected butterfly movement between sites, habitat composition of site boundaries did not generally influence exit rates. As in Ricketts's (2001) study, butterflies such as the forest dependent *Arhopala* spp., that feed exclusively on trees as larvae, were frequently observed crossing boundaries between unburned and burned forest (from I2 to B1). However, most of these species were absent in the more geographically distant B2 landscape, and were very rare in B1. Before the fires (1997), 42 species (372 individuals) of *Arhopala* in B2 were recorded in over 1333 butterflies collected (Cleary and Mooers, 2004), but after the fires none were present in 1998 and 1999 from over 1800 butterflies collected in both years. Additionally, only 2 species (2 individuals) were present in over 3600 butterflies collected in 2000. In contrast, 119 individuals comprising five *Arhopala* species (*A. evansi*, *A. norda*, *A. phanda*, *A. pseudomuta* and *A. sintanga*) were collected in 2004 from a total of 629 individuals of all species, indicating that the B2 habitat has been slowly returning to a pre-disturbance composition.

*Spatial patterns of butterfly diversity over large (regional) scales*

There was a sharp decline in similarity at very small distances ($<2$ km) and little change in mean similarity over greater distances. This result is similar with those of recent studies of other tropical forest species assemblages (Condit et al. 2002; Potts et al. 2002; Tuomisto et al. 2003). This may be because contrasts in habitat structure become greater with increasing distance, and essential habitat and resources are less likely to be encountered as local habitat, topology and climate become more heterogeneous. Secondly there may be spatial and temporal intraspecific synchrony in life-history parameters that ultimately drive spatial contrasts in observed abundance. It has been found that butterfly population abundance tends to exhibit synchrony over local scales, this phenomenon has been linked to dispersal ability. Sedentary British butterfly species had significant synchrony up to 3–4 km, while mobile species had synchrony up to 8 km (Sutcliffe et al. 1996). However, the effect appears to reduce over larger regional spatial scales. Peltonen et al. (2002) found that spatial synchrony of population fluctuations of forest Lepidoptera were not directly associated with dispersal abilities over scales up to *c.* 1000 km. A practically non-dispersing species, the Gypsy moth had a "surprisingly similar" pattern of spatial synchrony to the strongly dispersing larch bud moth. Where synchrony is absent over large scales it may be linked to environmental heterogeneity, for example weather conditions may affect larval duration or availability of host plant resources.

When all landscapes were pooled high proportions of variation in butterfly community composition were explained by both geographic distance between sites and environmental variables. In contrast when only unburned landscapes were considered, most variation was explained by the geographic distance among sites. Since unburned landscapes were together less environmentally heterogeneous than all landscapes combined, the scale of environmental heterogeneity encountered may explain the effect. ENSO-induced burning is such a severe disturbance that it radically alters the physical environment and concomitantly butterfly assemblages present. It is possible that over regional scales more spatial environmental variation resulted in more variation in the focal butterfly assemblage.

### *Contrasts in variance explained by environmental variables over landscape and regional scales*

Although environmental heterogeneity was lower among the unburned landscapes over the regional scale than burned landscapes, it may still be expected that a high component of spatial variation in the butterfly assemblage would be linked to environmental variables given their apparent importance over smaller landscape scales. It is possible there is a threshold effect, the larger the spatial scale, the greater the contrasts in responses of butterfly populations, and thus communities, to the same environmental variables. Among landscape differences in response to environmental variables may also be a consequence of factors such as dispersal limitation, or historical stochastic processes (Caswell 1976; Hubbell 2001). Differences may also be related to unmeasured environmental variables, such as floristic similarity. In this study vegetation were not identified to species, and this may have varied spatially in response to differing soil characteristics. In northern Borneo, Potts et al. (2002) found that edaphic factors were more important in explaining large scale variation in floristic similarity than physiographic factors such as topography or geographic distance. However, Slik et al. (2003) found that sites in Indonesian Borneo, including those used in the present study, clustered together in a single distinct floristic zone. This was in contrast to northern Borneo where at least four distinct zones were present. This suggests a lack of pronounced heterogeneity in floristic composition of the study area, and hence may rule out among-landscape variation in butterfly associations with environmental variables being due to differences in plant composition.

## Acknowledgements

We thank all those who assisted with the study in Kalimantan and elsewhere. In particular we would like to thank H.J. Megens, D. Wielakker, A. Bijlmer, Y. Pitoy, M. Buntu, A. Pakala, I. Sharif, B. van Helvoort, T. de Kam, W. Smits, M. Oman, N. Boestani, K. Eichhorn, L. Eichhorn, I. Sharif, and L. du Pon. Rienk de Jong helped identifying specimens. The Indonesian Ministry of Forestry and Crop Estates, and Kayu Mas and ITCI forest concessions provided field facilities. The study was supported by grant 895.100.005 of the Netherlands Foundation for the Advancement of Tropical Research (N*W*O-WOTRO), within Priority Programme 'Biodiversity in Disturbed Ecosystems'.

## References

Beck J., Schulze C.H., Linsenmair K.E. and Fiedler K. 2002. From forest to farmland: diversity of geometrid moths along two habitat gradients on Borneo. J. Trop. Ecol. 18: 33–51.

Borcard D., Legendre P. and Drapeau P. 1992. Partialling out the spatial component of ecological variation. Ecology 73: 1045–1055.

Bray J.R. and Curtis J.T. 1957. An ordination of the upland forest communities of Southern Wisconsin. Ecol. Monogr. 27: 325–349.

Casgrain P. 2001. Permute! 3.4.9. User's Manual. Freely available at: http://www.fas.umontreal.ca/biol/casgrain/en/labo/permute/index.html.

Caswell H. 1976. Community structure: a neutral model analysis. Ecol. Monogr. 46: 327–354.

Clarke K.R. and Gorley R.N. 2001. Primer v5: user manual/tutorial. Primer-E Ltd, Plymouth.

Cleary D.F.R. 2003. An examination of scale of assessment, logging and ENSO-induced fires on butterfly diversity in Borneo. Oecologia 135: 313–321.

Cleary D.F.R. 2004. Assessing the use of butterflies as indicators of logging in Borneo at three taxonomic levels. J. Econ. Entomol. 97: 429–435.

Cleary D.F.R. and Genner M.J. 2004. Changes in rain forest butterfly diversity following major ENSO-induced fires in Borneo. Global Ecol. Biogeogr. 13: 129–140.

Cleary D.F.R. and Mooers A.Ø. 2004. Butterfly species richness and community composition in forests affected by ENSO-induced burning and habitat isolation in Borneo. J. Trop. Ecol. 20: 359–367.

Cleary D.F.R., Genner M.J., Boyle T.J.B., Setyawati T., Angraeti C.D. and Menken S.B.J. Dependence of bird species richness and composition upon local- and large-scale environmental factors in Borneo. Landscape Ecol. (in press).

Colwell R.K. 2000. EstimateS version 6.0b1: Statistical estimation of species richness and shared species from samples. Freeware published at http://viceroy.eeb.uconn.edu/EstimateS.

Condit R., Pitman N., Leigh E.G., Chave J., Terborgh J., Foster R.B., Nunez P., Aguilar S., Valencia R., Villa G., Muller-Landau H.C., Losos E. and Hubbell S.P. 2002. Beta-diversity in tropical forest trees. Science 295: 666–669.

Ellingsen K.E. 2002. Soft-sediment benthic biodiversity on the continental shelf in relation to environmental variability. Mar. Ecol. Prog. Series 232: 15–27.

Harrison R.D. 2000. Repercussions of El Niño, drought causes extinction and the breakdown of mutualism in Borneo. Proc. Roy. Soc. London B 267: 911–915.

Hill J.K., Hamer K.C., Tangah J. and Dawood M. 2001. Ecology of tropical butterflies in rainforest gaps. Oecologia 128: 294–302.

Holmgren M., Scheffer M., Ezcurra E., Gutierrez J.R. and Mohren G.M.J. 2001. El Niño effects on the dynamics of terrestrial ecosystems. Trends Ecol. Evol. 16: 89–94.

Howard P.C., Viskanic P., Davenport T.R.B., Kigenyi F.W., Baltzer M., Dickinson C.J., Lwanga J.S., Matthews R.A. and Balmford A. 1998. Complementarity and the use of indicator groups for reserve selection in Uganda. Nature 394: 472–475.

Hubbell S.P. 2001. The Unified Neutral Theory of Biodiversity and Biogeography. Monographs in Population Biology 32. Princeton University Press, pp.375

Jepson P., Jarvie J.K., MacKinnon K. and Monk K.A. 2001. The end of Indonesia's lowland forests? Science 292: 859.

Legendre P. and Gallagher E.D. 2001. Ecologically meaningful transformations for ordination of species data. Oecologia 129: 271–280.

Longino J.T., Coddington J. and Colwell R.K. 2002. The ant fauna of a tropical rain forest: estimating species richness three different ways. Ecology 83: 689–702.

Margalef R. 1968. Perspectives in Ecological Theory. University of Chicago Press, Chicago, pp. 111.

Nepstad D.C., Moreira A., Verissimo A., Lefebre P., Schlesinger P., Potter C., Nobre C., Setzer A., Krug T., Barros A.C., Alencar A. and Pereira J.R. 1998. Forest fire prediction and prevention in the Brazilian Amazon. Conserv. Biol. 12: 951–953.

Parmesan C., Ryrholm N., Stefanescu C., Hill J.K., Thomas C.D., Descimon H., Huntley B., Kaila L., Kullberg J., Tammaru T., Tennent W.J., Thomas J.A. and Warren M. 1999. Poleward shifts in geographical ranges of butterfly species associated with regional warming. Nature 399: 579–583.

Peltonen M., Liebold A.M., Bjørnstad O.N. and Williams D.W. 2002. Spatial synchrony in forest insect outbreaks: roles of regional stochasticity and dispersal. Ecology 83: 3120–3129.

Plotkin J.B. and Muller-Landau H.C. 2002. Sampling the species composition of a landscape. Ecology 83: 3344–3356.

Potts M.D., Ashton P.S., Kaufman L.S. and Plotkin J.B. 2002. Habitat patterns in tropical forests: a comparison of 105 plots in northwest Borneo. Ecology 83: 2782–2797.

Putz F.E., Blate G.M., Redford K.H., Fimbel R. and Robinson J. 2001. Tropical forest management and conservation of biodiversity: an overview. Conserv. Biol. 15: 7–20.

Ricketts T.H. 2001. The matrix matters: effective isolation in fragmented landscapes. Am. Nat. 158: 88–99.

Rozenzweig M.L. 1995. Species Diversity in Space and Time. Cambridge University Press, Cambridge, pp. 463.

Siegert F., Ruecker G., Hinrichs A. and Hoffmann A.A. 2001. Increased damage from fires in logged forests during drought caused by El Niño. Nature 414: 437–440.

Slik J.W.F., Poulsen A.D., Ashton P.S., Cannon C.H., Eichhorn K.A.O., Kartawinata K., Lanniari I., Nagamasu H., Nakagawa M., van Nieuwstadt M.G.L., Payne J., Purwaningsih, Saridan A., Sidiyasa K., Verburg R.W., Webb C.O. and Wilkie P. 2003. A floristic analysis of the lowland dipterocarp forests of Borneo. J. Biogeogr. 30: 1517–1531.

Spencer M., Schwartz S.S. and Blaustein L. 2002. Are there fine-scale patterns in community similarity among temporary freshwater pools? Global Ecol. Biogeogr. 11: 71–78.

Spitzer K., Jaros J., Havelka J. and Leps J. 1997. Effect of small-scale disturbance on butterfly communities of an Indochinese montane rainforest. Biol. Conserv. 80: 9–15.

Sutcliffe O.L., Thomas C.D. and Moss D. 1996. Spatial synchrony and asynchrony in butterfly population dynamics. J. Anim. Ecol. 65: 85–95.

Stork N.E., Srivastava D.S., Watt A.D. and Larsen T.B. 2003. Butterfly diversity and silvicultural practice in lowland rainforests of Cameroon. Biodiv. Conserv. 12: 387–410.

Tuomisto H., Ruokolainen K. and Yli-Halla M. 2003. Dispersal, Environment, and floristic variation of western amazonian forests. Science 299: 241–244.

Whittaker R.H. 1965. Dominance and diversity in land plant communities. Science 147: 250–260.

Williams S.E., Marsh H. and Winter J. 2002. Spatial scale, species diversity, and habitat structure: small mammals in Australian tropical rain forest. Ecology 83: 1317–1329.

Willott S.J., Lim D.C., Compton S.G. and Sutton S.L. 2000. Effects of selective logging on the butterflies of a Bornean rainforest. Conserv. Biol. 14: 1055–1065.

# TOPICS IN BIODIVERSITY AND CONSERVATION

1. David L. Hawksworth and Alan T. Bull (eds.): *Arthropod Diversity and Conservation*. 2006 ISBN 1-4020-5203-0